NONDESTRUCTIVE TESTING AND EVALUATION FOR MANUFACTURING AND CONSTRUCTION

NONDESTRUCTIVE TESTING AND EVALUATION FOR MANUFACTURING AND CONSTRUCTION

Edited by

Henrique L. M. dos Reis
University of Illinois, Urbana, Illinois

HEMISPHERE PUBLISHING CORPORATION
A member of the Taylor & Francis Group

New York Washington Philadelphia London

NONDESTRUCTIVE TESTING AND EVALUATION FOR MANUFACTURING AND CONSTRUCTION

1 2 3 4 5 6 7 8 9 0 BCBC 8 9 8 7 6 5 4 3 2 1 0 9

Cover design by Sharon DePass.
A CIP catalogue record for this book is available from the British Library.

Library of Congress Cataloging in Publication Data

Nondestructive testing and evaluation for manufacturing and construction / edited by Henrique L. M. dos Reis.
p. cm.
Includes papers presented at the Nondestructive Testing and Evaluation for Manufacturing and Construction Conference, held Aug. 9-12, 1988, at the University of Illinois at Urbana-Champaign.

1. Nondestructive testing—Congresses. I. Reis, Henrique L. M. dos. II. Nondestructive Testing and Evaluation for Manufacturing and Construction Conference (1988 : University of Illinois at Urbana-Champaign)
TA417.2.N674 1990
620.1'127—dc20 89-19777
ISBN 0-89116-926-1 CIP

CONTENTS

WOOD PRODUCTS

CONCRETE AND GEOTECHNICAL APPLICATIONS

PREFACE

Nondestructive testing and evaluation methods have undergone significant advances in recent years. These advances are manifested by the development of new methods and sophisticated mathematical models. The driving force behind these activities has been, and continues to be, the need to nondestructively evaluate and characterize newly engineered materials, as well as to accurately predict the fitness for service and the remaining useful life of a component or structural system. This rapid pace of development is largely due to the interdisciplinary nature of this field, (including physics, material sciences, applied mechanics, and computer sciences), and to the opportunities provided by the extensive advances in computer hardware and software.

Nondestructive testing and evaluation methods already have wide application in manufacturing, such as in the aerospace industry. Application of these methods is also beginning to develop in the construction industry because of the growing state of deterioration of our civil infrastructure. As a means of communicating recent advances and as a step toward stronger interaction among researchers in manufacturing and construction technologies, a three-day conference titled "Nondestructive Testing and Evaluation for Manufacturing and Construction," was held at the Illini Union Building on the campus of the University of Illinois at Urbana-Champaign. One hundred twenty-four people attended the conference. It brought together the leading experts and active researchers in areas which could impact future developments in manufacturing and construction. The formal presentations and personal interactions will serve to increase communication among researchers and foster development of new and innovative methods. This volume is a collection of some of the papers presented at the conference.

The editor is indebted to the many individuals who contributed to the planning of the conference, in particular to the Conference Secretary, Mrs. Edna Unfer, and to his secretary, Mrs. Carolyn Reed. Appreciation is expressed to A. Beattie, R. E. DeVor, W. C. Chew, W. A. Ellingson, C. S. Gardner, N. Gopalsami, N. N. Hsu, S. G.

Kapoor, H. W. Knoebel, W. J. McGonnagle, K. Ono, P. L. Ransom, R. A. Roberts, R. D. Streit, C. A. Tatro, A. Vary, D. E. Walker, J. W. Whittaker, and to all the authors, for reviewing manuscripts.

Additional thanks are due to the session chairmen, J. J. Antal, F. C. Beall, A. Beattie, J. C. Duke, D. M. Egle, A. Green, H. R. Hardy, Jr., J. C. Harris, E. Heiberg, W. J. McGonnagle, R. D. Streit, and A. Vary, for all their advice in organizing the conference and for keeping the conference timely. Also, a word of thanks is due to graduate students M. J. Chica, M. A. King, K. L. Koenig, and D. N. Slowinski for their help during the conference.

The financial support of R. F. Quattrone and F. W. Kearney of the US Army Construction Engineering Research Laboratory, Champaign, Illinois is gratefully acknowledged. The assistance of W. A. Dick of the College of Engineering at the University of Illinois at Urbana–Champaign is especially appreciated.

Finally a word of thanks to my colleague Warren J. McGonnagle. Without his help, encouragement, and enthusiasm, this conference would not have been a reality.

Henrique L. M. dos Reis

PARTIAL LIST OF ATTENDEES

Nondestructive Testing and Evaluation for Manufacturing and Construction Conference, August 9–12, 1988. Illini Union Building on the campus of the University of Illinois at Urbana–Champaign, Illinois.

(See photograph on next page)

BACK ROW (*Left to right*)

Joe Wong, Lynn Padilla, Robert Falk, Donna Herriott, Kenneth V. Cook, R. J. (Ron) Booth, Khaled M. Elbalawy, AMR Hassaballah, Ken-ichi Itakura, Ronald D. Streit, Robert E. Green, Sheryl Wallace, Michael J. Chica, Matthew A. King, Kurt L. Koenig, David N. Slowinski.

FOURTH ROW (*Left to right*)

Edna M. Unfer, Carolyn N. Reed, Evelyn H. Hirt, John G. Harris, Tim M. Whynot, Ziad Salameh, Kiho Choi, Tarun R. Naik, Henry Thornton, W. A. Ellingson, Alex Vary, Jasson Gryzagoridis, Clem Tatro, Keiichi Sato.

THIRD ROW (*Left to right*)

Frank C. Beall, Debbie Lawrence, Richard L. Lemaster, Greg Phillips, Thomas M. Heslin, E. James Chern, John C. Moulder, S. John Louisnathan, H. Reginald Hardy, Jr., Davis M. Egle, Gen-ichi Matsumoto.

SECOND ROW (*Left to right*)

Connie Presley, David W. Okey, Robert Ross, Luis Lorenzo, Ramesh Gupta, Erich J. Schwarz, Allen T. Green, Alan G. Beattie, L. Trace Nelson, Grant C. McClellan, Yasushi Ikeda.

FIRST ROW (*Left to right*)

Mike Ostroski, Marcia Patton-Mallory, Charles Hillary, Nicholas J. Carino, Larry D. Olson, Kent McDonald, Edvard Heiberg, John J. Antal, Richard L. Weaver, Henrique M. Reis.

Photograph taken August 10, 1988 at 9:00 a.m.

KEYNOTE

ADVANCED NDE TECHNIQUES FOR MANUFACTURING AND CONSTRUCTION APPLICATIONS

R. E. Green, Jr.
Center for Nondestructive Evaluation
The Johns Hopkins University
Baltimore, MD 21218

INTRODUCTION

The role of nondestructive evaluation has changed dramatically in the United States. Although historically nondestructive testing techniques have been used almost exclusively for detection of macroscopic defects (mostly cracks) in structures after they have been in service, it has become increasingly evident that it is both practical and cost effective to expand the role of nondestructive evaluation (NDE) to include all aspects of manufacturing and construction. Moreover, the present unfavorable status of a large segment of United States industry, coupled with the premature failure of structures, dramatically shows that our traditional approaches must be drastically modified if we are to be able to compete successfully in the global marketplace and to ensure that structural components are to perform safely throughout their expected life.

With the ever increasing requirements being placed on structural components, the necessity for increasing the educational level of nondestructive evaluation personnel and for transitioning new developments in basic research into practical applications has become of vital importance. This paper summarizes NDE techniques applied to examination of structures in the U.S. and cites specific examples of emerging applications for manufacturing and construction. Finally, a brief description will be given of the Center for Nondestructive Evaluation (CNDE), which was established at The Johns Hopkins University as an interdisciplinary center for research and instruction dedicated to the development of more sophisticated and accurate methods for the nondestructive evaluation of materials and structures and to the education of talented students who will enter the NDE field. Description of several CNDE research projects serve to illustrate how modern NDE techniques afford the possibility of improving both manufacturing and construction.

NDE TECHNIQUES FOR STRUCTURAL MATERIALS

As with most other applications of nondestructive evaluation, structural materials have been examined historically by one of the big five nondestructive testing techniques, namely: ET (eddy current), MT (magnetic testing), PT (penetrant testing), RT (radiographic testing), and UT (ultrasonic testing). However, in more recent times the number of techniques has greatly expanded to include those listed in Tables I and II. In addition, semi-nondestructive techniques have been developed for concrete testing as listed in Table III.

TABLE I. NDE Techniques for Manufacturing Process Control

Acoustic Emission
Acousto-Ultrasonics
Eddy Current
Dielectric Measurements
Electrical
Electron-Acoustic Imaging
Gamma Radiometry
Holographic Interferometry
Magnetic (Magnaflux)
Optical Interferometry
Penetrant Inspection
Real-Time Radiography
Thermal-Wave Imaging
Thermography, Vibrothermography
Ultrasonics
X-ray Diffraction

ACOUSTIC EMISSION MONITOR

Acoustic emission or stress wave emission is the phenomenon of transient elastic wave generation due to a rapid release of strain energy caused by a structural alteration. Generally these structural alterations are a result of either internally or externally applied mechanical or thermal stress. The importance of acoustic emission monitoring lies in the fact that proper detection and analysis of acoustic emission signals permits remote identification of source mechanisms and the associated structural alterations.

A unique acoustic emission monitoring system has been developed by Battelle Northwest for flaw detection in steel highway bridges. The system incorporates a digital memory microcircuit with source isolation to admit signals only from a pre-selected area of the bridge structure. The entire system consists of amplifiers, source location circuits, digital memories, and memory programming controls in a compact housing weighing only 5 pounds. A separate power supply provides power from either a standard 110 volt

TABLE II. NDE Techniques for Structural Materials

Acoustic Emission
Acoustical Birefringence
Acousto-Ultrasonics
Barkhausen Noise
Eddy Current
Dielectric Measurements
Electrical
Gamma Radiometry, Radiography, Back-Scattering
Holographic Interferometry
Impact Hammer Rebound
Impact-Echo
Magnetic (Magnaflux, Flux Leakage, NMR, ESR)
Microwave Absorption
Mossbauer Effect
Neutron Radiography, Tomography, Scattering
Optical Interferometry
Penetrant Inspection
Radar
Thermal-Wave Imaging
Thermography, Vibrothermography
Ultrasonics
X-ray Diffraction, Radiography, Tomography

TABLE III. Semi-Nondestructive and Destructive Tests for Concrete

(Property determined indicated in parentheses)

Pull-Out (Indirect Shear, Tensile Strength)
Pull-Off (Tensile Strength)
Break-Off (Flexural Strength)
Windsor Probe (Penetration Resistance)
Tescon Probe (Stress-Strain Relation)
Cores (Strength)
Maturity Method (Depends on Relationship Between Time and Temperature, Aging)
Permeability Testing (Chloride Ions, Gas Permeability)

A.C. source or from internal rechargable batteries. Two separate acoustic emission parameters are stored in solid state EPROM digital memories, which can be removed from the on-site recorder and taken to a data analysis center for read-out. The digital memories can be reused by erasing with ultraviolet light. It was suggested that such systems located strategically on bridge structures could monitor crack initiation and growth.

Similar systems could be economically placed on a variety of structures and the digital memory microcircuits collected and replaced with others in a regular scheduled fashion by relatively unskilled people and brought into a central facility for read-out, signal analysis, and erasing. A slightly more expensive installation would have a radio transmitter as an integral part of each monitor, with detected acoustic emission signals periodically broadcast to a recording and analysis instrumention van.

ULTRASONIC CRACK DETECTION

Ultrasonic techniques have been used more often in recent years than other nondestructive evaluation techniques for detection of cracks in metal structures. A major reason for this choice of inspection is the relatively portable nature of ultrasonic inspection equipment. However, even modern solid state devices usually require one hand to hold the electronic unit and the other hand to scan the transducer. This often makes it very difficult for the inspector to maintain a safe and comfortable position on the test structure.

An improved ultrasonic unit designed to eliminate this problem has been developed by Southwest Research Institute. This unit, called the Acoustic Crack Detector (ACD), is an ultrasonic system consisting of an electronic backpack and a hand-held transducer probe (Fig. 1). A unique feature of the transducer probe is that it can be operated in either one of two modes, by a switch on the probe. In the survey mode, two separate transducer channels operate alternately. One transmits longitudinal wave pulses of ultrasonic energy in a direction perpendicular to the inspection surface. When the ultrasonic energy reflected back to the transducer acting as a receiver is above a threshold level, illumination of a green light on the probe indicates that sufficient energy has been coupled into the inspection surface to permit reliable inspection.

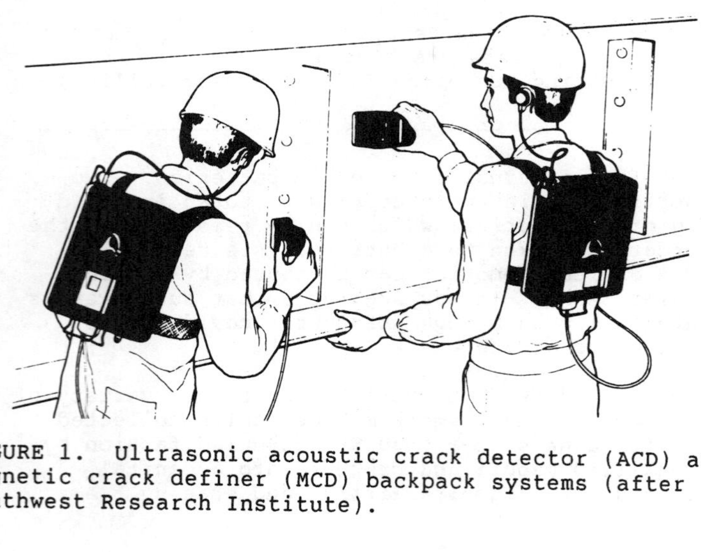

FIGURE 1. Ultrasonic acoustic crack detector (ACD) and magnetic crack definer (MCD) backpack systems (after Southwest Research Institute).

A more sophisticated ultrasonic inspection system has been developed by Sigma Research for automatic imaging of cracks in structures. This system consists of two major components linked by a communications cable, containing power and ultrasonic, audio, and video signals. One component carried by the inspector consists of a special hand-scanner unit, transducer, preamplifier, and a video display linked to the processor. The special portable hand-scanning unit can be mounted on the structure to be inspected and monitors the position of the ultrasonic transducer continuously as the operator hand scans the inspection area. The other major component, located at a data analysis station, consists of a computer system with a digitizer, graphics display, data analysis software and disk storage, and an ultrasonic mainframe system. The data acquisition and analysis component records the transducer position, amplitude of the ultrasonic signal, ultrasonic time-of-flight, and acoustic holographic coefficients. The software enables the system to determine the dimensions and depth of a flaw; extrapolate, fill in, and smooth out an image; and zoom in on a particular feature. Two operators, one at the inspection site and the other at the data analysis location, operate the system together. The scanner operator is provided with visual feedback via the television monitor and verbal feedback from the processor operator through a headset. The inspector uses an A-scan method to locate flaws, then the automated imaging system assesses the flaw and records a permanent image in computer storage. This record provides a baseline for succeeding inspections allowing growth of flaws to be measured and repair decisions to be easily made.

The feature of separation of relatively unskilled inspector and highly trained data analyzer, who are in continuous communication, can be optimally used in other nondestructive evaluation applications. Addition of a number of inspectors in communication with a single data analysis center could optimize inspection of large structures without the need for highly skilled inspectors.

OPTICAL GENERATION AND DETECTION OF ULTRASONIC WAVES

Although piezoelectric crystals have been predominantly used as ultrasonic transducer materials, a major problem associated with their use is the requirement that both the generating and receiving transducers be acoustically bonded to the test material with some sort of acoustical impedance matching coupling medium such as water, oil, or grease. In addition to modification of the ultrasonic signal and potential harm to the test structure by the coupling medium, the requirement of physical contact between transducer and test structure places limitations on ultrasonic testing in structural configurations which possess geometries with difficult to reach areas, as well as on testing materials at elevated temperatures or in hostile environments.

A number of techniques are presently available for non-contact generation and detection of ultrasound, including capacitive pick-ups and electromagnetic acoustic transducers (EMAT'S). However, since these techniques require miminum separation between probe and test structure and can only be used with electrically conducting materials, the only true practically useful non-contact technique is use of laser beam generation and detection. Moreover, laser beam probes afford the opportunity to make truly non-contact ultrasonic measurements in both electrically conducting and non-conducting materials, in materials at elevated temperatures, in corrosive and other hostile environments, in geometrically difficult to reach locations, in outer space, and do all of this at relatively large distances, i.e. meters, from the test structure surface.

An extensive effort is underway in the Johns Hopkins University Center for Nondestructive Evaluation to develop an optimum non-contact laser system for generation and detection of ultrasonic waves. Figure 2 shows a schematic of a system which used a pulsed laser to generate surface waves on metal plates whose surfaces had been microstructurally modified by scanning with electron beams or high-power lasers to either form an amorphous layer on the crystalline bulk, to produce different layers of transformed phases, or to induce case hardening. The surface displacements generated by the propagating waves were detected in a non-contact manner using a dual beam laser interferometer. Figure 3 shows a schematic of a proposed laser beam generation fiber optic system for non-contact ultrasonic scanning of large structures. Similar pulsed laser beam generating techniques have been coupled

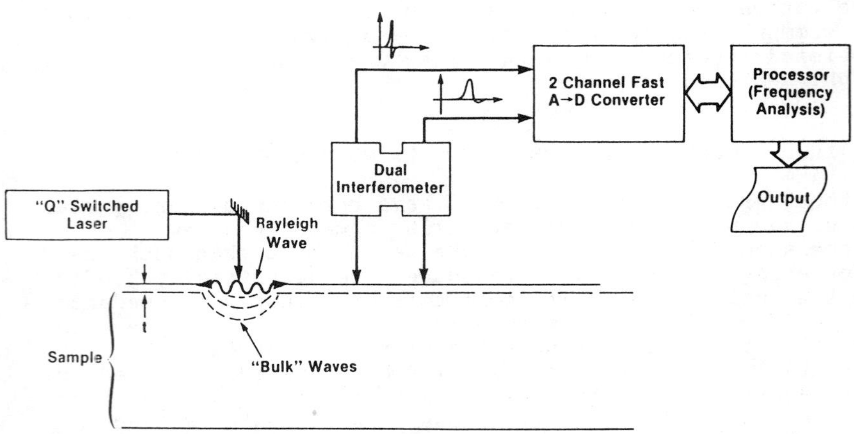

FIGURE 2. Schematic of laser beam generation and dual laser beam interferometric detection of ultrasonic surface waves (after Rosen, CNDE).

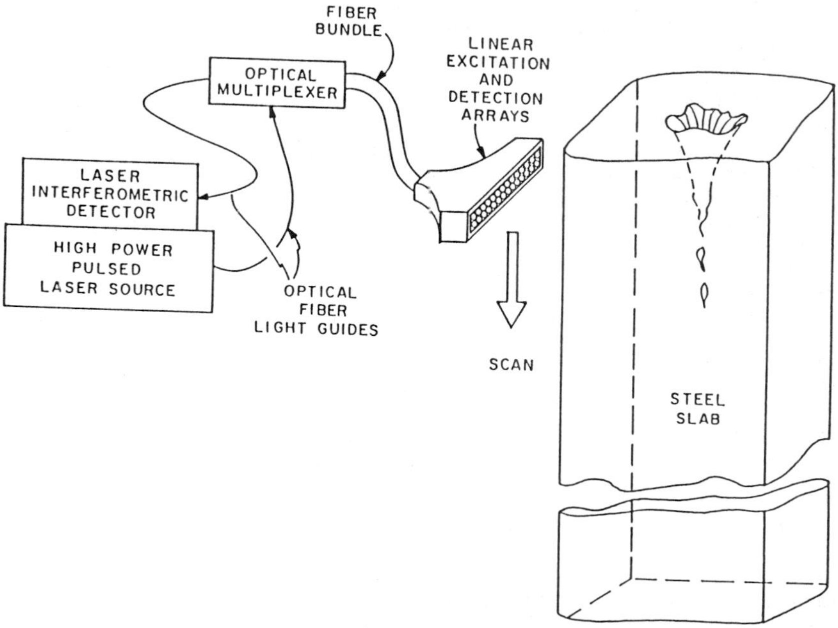

FIGURE 3. Schematic of laser beam fiber optic system for non-contact ultrasonic scanning of large steel slab prior to rolling operations (after Rosen and Green, CNDE).

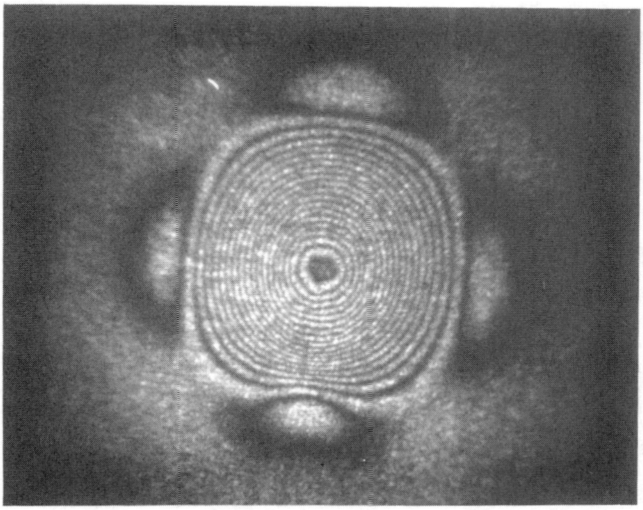

FIGURE 4. Heterodyne holographic interferometric full-field imaging of surface displacements due to ultrasonic waves propagating in graphite/epoxy composite plate (after Wagner, CNDE).

with holographic interferometric detection to obtain full-field images of surface displacements caused by ultrasonic waves propagating in both metals and polymer based composites. Figure 4 shows the results of a recently developed heterodyne holographic interferometric technique which permits full-field imaging of surface displacements due to ultrasonic wave propagation. The expanded beam from a pulsed laser was used to record a holographic image of a graphite/epoxy composite plate. Subsequently, a second narrow beam laser pulse was incident on the center (opposite side than shown in Fig. 4) of the plate and, after a selected time interval, an expanded beam from the pulsed laser was used to re-expose the holographic plate. The resulting interference pattern shows the surface displacements of the plate due to the ultrasonic wavefronts travelling outward from the source with the influence of the anisotropic character of the plate clearly evident. In addition, the fact that the plate has a defect, probably a delamination in the bottom quadrant is evidenced by the irregularity in the bottom lobe in comparison with the other three lobes. Laser beam interferometers have also been used for non-contact detection of acoustic emission signals.

MULTISENSOR GEOPHYSICAL SYSTEM

Unforeseen problems are frequently encountered in rock tunneling, since rock conditions cannot be predicted from surface geology, vertical borings, or excavations at tunnel portals. Ensco constructed a prototype highly mobile geophysical nondestructive system, which took pulsed acoustical, radar, and multi-spaced array resistivity measurements. The multisensor system was designed for pre-excavation subsurface inventigation for tunnels in rock by sensing from long, horizontal boreholes. The combination of sensors used in this system provided detailed information on rock characteristics close to a borehole and gross information to distances up to 100 feet away. In operation the sensors traversed along the borehole and data taken and stored on magnetic tape for subsequent analysis at a computational center.

Although this system was developed over 10 years ago, the concept of multisensor nondestructive instrumentation is an extremely important one that has not been optimally utilized in modern nondestructive evaluation systems. The importance of multisensor nondestructive evaluation probes derives from the fact that no single measurement system permits determination of all necessary characteristics of a structure sufficient to assure proper construction and in-service safety. With the vast decrease in physical size of electronic instrumentation, because of developments in microelectronics, the time is ripe for design and construction of multisensor nondestructive evaluation systems. Research directed toward development of such systems is currently in progress.

PENETRATING RADIATION TOMOGRAPHY

A number of investigators including Scientific Measurements have applied computer assisted gamma ray tomography to acquire three dimensional images of a variety of structures. Figure 5 shows an example of the use of computer assisted tomography to image aggregate and a corroded steel reinforcing bar in a concrete beam. Although computer assisted tomography has become a standard feature at all advanced medical institutions, it has only begun to make inroads into the nondestructive evaluation community. The expensiveness of the technique limits present usage, but future developments in detectors, computers, and reconstruction algorithms portends well for reduction of inspection cost and more use for structures in the future.

SPACE STATION HABITAT LEAK TEST SYSTEM

Increasingly, NDE techniques developed for nonstructural applications have come to be of direct use in testing and inspecting large structures. One such example is that of a nondestructive method which was initially developed by researchers at the Johns Hopkins University Center for Nondestructive Evaluation for rapid inspection of electronic microcircuit packages. The technique, which is based on the principles of optical holographic interferometry, provides a means to inspect for leaks in sealed packages which are difficult to inspect by standard tracer gas or bubble test methods. Often these packages are encased in a polymer coating in which tracer gases may be absorbed and adsorbed. Consequently, significant errors in leak rate measurements may result.

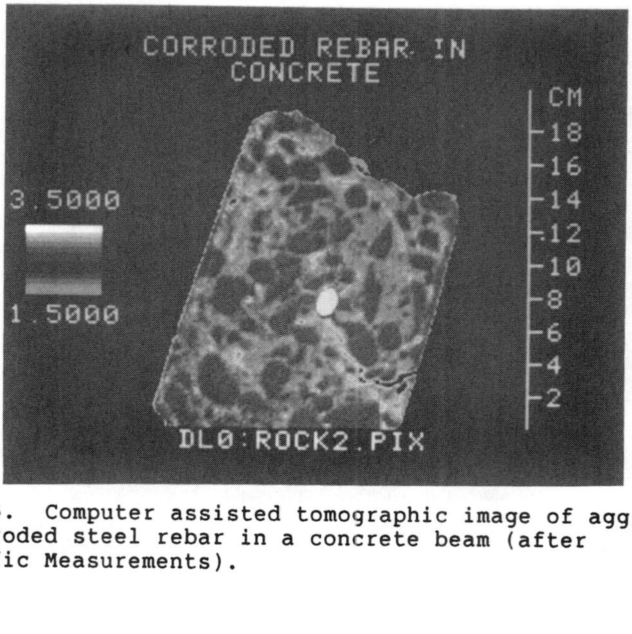

FIGURE 5. Computer assisted tomographic image of aggregate and corroded steel rebar in a concrete beam (after Scientific Measurements).

In lieu of tracer gases, optical leak rate measurement techniques take advantage of the fact that a sealed package placed in a test chamber deforms in response to an increase in chamber pressure. Over time, a package which leaks will permit gas to entire it and as the differential pressure across the package walls decreases, the package returns to its original shape. The ability of optical techniques to measure sub-micron displacements makes it possible to monitor the recovery of a leaking package with high sensitivity. Figure 6(left) is an optical photograph of a package, whose leak rate is to be tested, placed in a test chamber. Figure 6(right)) shows the holographic interferogram in which the displacement of the package surface is evidenced by the appearance of fringes superimposed on the image of the package as it is viewed through the glass faceplate of a pressurized test chamber. If the package does not leak the fringe pattern remains constant in time. If the package leaks, the rate at which the fringes disappear serves to measure the leak rate.

While the optical leak rate method was initially developed for microelectronics applications, efforts are now underway to apply the technology to large structures intended for use in manned space flight. For NASA's Space Station Habitat and Laboratory Modules, acceptable leak rate specifications must be met. In order to measure leak rates in such a large structure, the principle of the optical measurement method described above will be applied - only this time, in reverse. Rather than measuring the rate of recovery of a leaky package within a sealed test chamber, the recovery rate of a sealed package within a leaky chamber (the Space Station Module) will be measured.

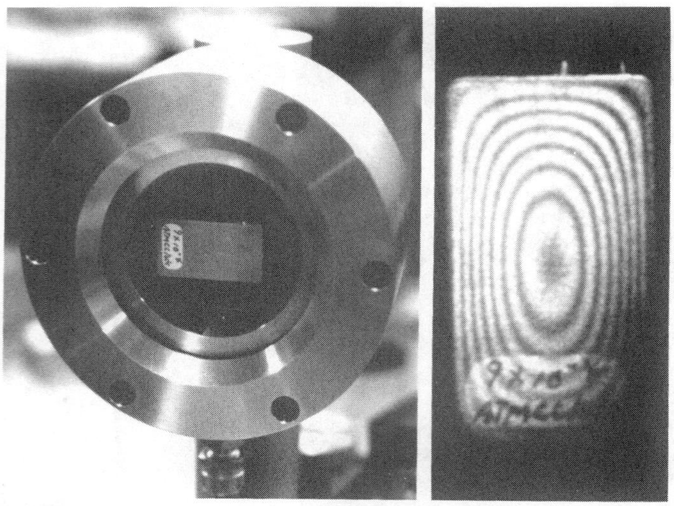

FIGURE 6. (left) Optical photograph of package in test chamber. (right) holographic interferogram of package undergoing leak test (after Wagner, CNDE).

MAGNETIC TECHNIQUES

A number of magnetic techniques have been used for detection of defects in ferromagnetic materials. Although the primary techique used is magnaflux, considerable work has been conducted with magnetic flux leakage and Barkhausen noise analysis. In addition, Southwest Research Institute, a major contributor of instrumentation and test methods in this field, has developed a magnetic crack definer (MCD) consisting of an electronic backback and a hand-held probe as shown in Fig. 1.

More recently magnetic resonance techniques, which are well established for medical applications, are being developed as nondestructive evaluation techniques for manufacturing and construction applications. Figure 7 shows a nuclear magnetic resonance system (NMR) developed by Southwest Research Institute for detection of moisture in concrete highway structures. Such systems are not only capable of measurement of the characteristics of moisture in concrete and composite structures, but can detect incomplete cure, elastic modulus variations, and environmental degradation and aging of polymer based composites. In addition, they may be used for process control of polymer based composites since NMR can validate starting materials, monitor resin impregnation, and monitor the cure cycle.

In biomedical and industrial NMR measurements and imaging the magnetic properties that are probed are due to the nuclear magnetic moments of protons in the body or material to be examined. Researchers in the CNDE have found, however, that there is a great deal of promise in investigating electron paramagnetic resonance (EPR) which is due to the magnetic moments associated with electron spins. An extremely large EPR signal is observed when certain polymeric materials are damaged. The fact that these EPR signals can be enhanced and controlled by adding metallic dopants to the polymers opens up the possibility to solve a problem that has been troubling polymer based composite materials for some time. When composite materials are damaged, for example by impact, the extent of damage or even its existence often cannot be accessed by examination of the outer surface. By doping each layer of a polymer based composite prior to lay-up with a different dopant, subsequent damage can be quantitatively determined as a function of depth into the composite by means of scanning EPR. Figure 8 shows an example of an EPR line scan across a polymer composite which had an organometallic dopant applied at four positions along the line as indicated in the upper left insert of the figure.

Research has also been in progess in the CNDE to develop a novel nondestructive method for remote non-contact detection and evaluation of corrosion in underground gas pipelines. Measurements on 100 ft. sections of buried pipe, under varied conditions of coatings and soil coverings, have been made using Electrochemical Impedance Spectroscopy (EIS) instrumentation with magnetometer sensing of pipe and earth

FIGURE 7. Nuclear magnetic resonance system for moisture detection in concrete highway structures (after Southwest Research Institute).

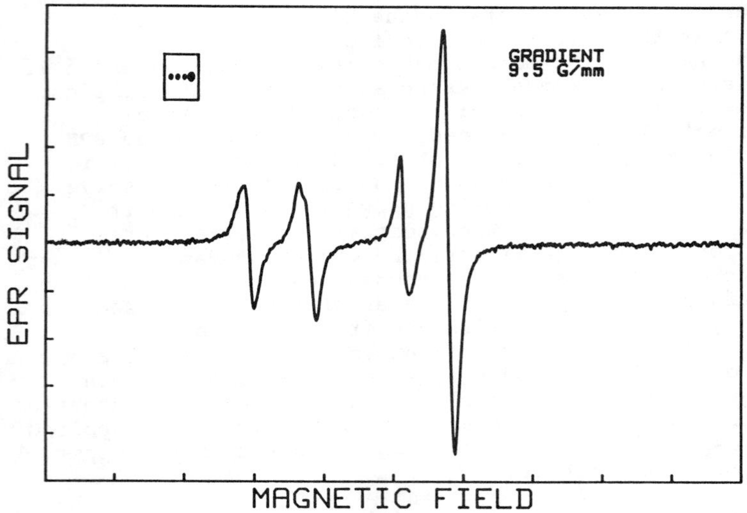

FIGURE 8. Electron paramagnetic resonance line scan across organometallic doped polymer based composite specimen (after Bryden and Poehler, CNDE).

current. Once optimized this methodology will allow the survey of underground pipelines by surface measurements, locate corroding regions in base pipelines, determine defective regions in coated pipelines including defect size and corrosion rate, and help assure the integrity and safe operation of the pipeline system.

THERMAL TECHNIQUES

Several nondestructive thermal measurement techniques are emerging into prominence which offer potential for manufacturing and construction applications. A particularly interesting technique vibrothermography, developed at Virginia Polytechnic Institute and State University, is a technique which uses a real-time heat imaging device such as an infrared thermographic television camera to observe the heat patterns generated at defects in solid materials due to mechanical vibration of the structure under test. Ideally the mechanical vibration system operates over a wide frequency range so that appropriate frequencies can be found to match the resonant frequencies of structural defects. This technique has been successfully applied to locate fatigue cracks and microstructural inhomogeneities in metals and a variety of structural defects and damage sites in complex composite laminates and structures. Figure 9 shows a vibrothermographic image of heat generated at a delamination in a composite helocopter blade.

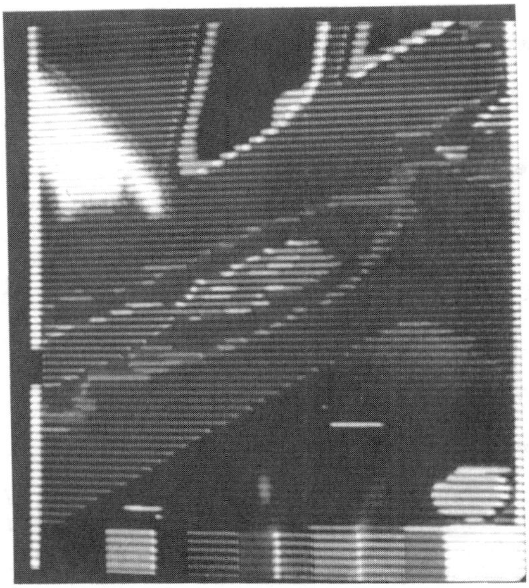

FIGURE 9. Vibrothermographic image of heat generated at a delamination in a composite helocopter blade (after Reifsnider).

THE JOHNS HOPKINS UNIVERSITY
CENTER FOR NONDESTRUCTIVE EVALUATION

The Johns Hopkins University Center for Nondestructive Evaluation (CNDE), established as an interdisciplinary center for research and instruction, exploits and integrates the resources and talent of the School of Engineering, the Applied Physics Laboratory, the School of Medicine, and the School of Arts and Sciences. Research in the CNDE has developed innovative NDE techniques which when employed are especially useful to control and improve product quality in manufacturing processes and to assure the in-service safety of devices and structures.

The CNDE is dedicated to research which will lead to more sophisticated and accurate methods for the non-destructive evaluation of materials and structures and to the education of talented students who will enter the NDE field. The Johns Hopkins University enjoys a reputation for developing innovative NDE techniques and is proud of the fact that more educators, engineers, and managers now working in the NDE field, have received graduate training at Johns Hopkins than at any other university in the world.

The CNDE provides both industrial and government collaborators with one "access port" to all NDE research and instruction at The Johns Hopkins University. The following institutions are sponsors of the CNDE:

Aluminum Company of America
Amdata/Combustion Engineering
Babcock & Wilcox Company
Baltimore Gas & Electric Company
Belvoir R.D.& E. Center
Bethlehem Steel Corporation
Department of Energy
E.I. DuPont de Nemours
Electric Power Research Institute
Fairchild Industries
General Dynamics Corporation
General Electric Company
Grumman Aircraft Systems
Hartford Steam Boiler Inspection & Insurance
IBM-Thomas J. Watson Research Center
Lockheed Missiles & Space Company
Martin Marietta Corporation
Physical Acoustics Corporation
United Technologies
Westinghouse Electric Company

Funding is based on an annual sponsor's membership fee, which is entirely dedicated to the CNDE program of research and education. Each member institution appoints an engineer or scientist to be the sponsor's liaison to the CNDE. The aggregate of the liaison representatives forms the committee which has oversight responsibility for the research and education goals of the CNDE. In addition to the sponsoring members, twenty-five government and industrial organizations have established formal research relationships with the CNDE. The strong collaborative support is evidence of the critical importance of NDE research and education for government science and technology programs and for industrial manufacturing and construction.

The research program of the CNDE has five major thrust areas focused on engineering problems, around which the program of research and education is built. These thrust areas are: Advanced Sensors, Process Control, Measurement Science for Microelectronics, Life-Cycle Management, and Artificial Intelligence. The goal of the program is to integrate advanced sensors, mathematical modeling, and computer based artificial intelligence for the development of expert systems to control manufacturing production and to assure safety of in-service structures.

CNDE RESEARCH PROJECTS PARTICULARLY APPLICABLE TO MANUFACTURING AND CONSTRUCTION

Much of the current NDE research performed by investigators at the CNDE is directed towards the inspection of materials used in manufacturing and construction rather than at completed structures. There are, however, a growing number of examples where innovative methods have been developed or applied specifically to structures. For example "smart structures" are being designed to permit continuous performance monitoring or portable bridge structures or space station trusses. These structures are designed to be inspectable and often incorporate permanently fixed NDE sensors. Polymer matrix composite materials used in these structures may have optical fibers embedded in them for detection of acoustic emission signals, strain monitoring, and fracture detection.

Other CNDE research activities with application to structures are as follows:

Acoustic Emission Proof-Load Testing of Composite Materials and Structures
Acoustic Emission Monitoring of Thermally Loaded Composite Materials
Noncontact Laser Generation of Narrow-Band Ultrasound
High Speed Heterodyne Holographic Interferometry
Laser Speckle Interferometry for Dynamic Fracture Studies
Ultrasonic Characterization of Composite Interfaces
Ultrasonic Measurement of Grain Size in Materials
Characterization of Composites with Laser-Generated Ultrasound
Ultrasonic Correlation for Characterization of Highly Attenuating Materials
Remote Corrosion Detection of Underground or Inaccessible Structures
Ultrasonic Residual Stress (Strain) Measurements in Structural Materials
Ultrasonic Characterization of the Nonlinear Elastic Properties of Graphite/Epoxy Composites

EXAFS Evaluation of the Effect of Alloying Elements on Corrosion Protective Films on Metals
Optical and Electrochemical Studies of the Corrosion Resistance of Films on Metal/Matrix Composites
Dynamic Fracture Toughness Measurements
Long Gage-Length Non-Contact Interferometric Strain/Displacement Gage
Application of Thermal Wave Imaging Techniques to the NDE of Engineering Structures
Thermal Wave Imaging Techniques for Coating Inspection

REFERENCES

The nondestructive evaluation techniques described in this paper listed immediately following were described in reports submitted to the Federal Highway Administration and provided to the author by Mr. Charles H. McGogney: Battelle Northwest acoustic emission monitor; Southwest Research Institute acoustic crack detector, magnetic crack definer, and nuclear magnetic resonance system; Sigma Research ultrasonic inspection system; Ensco multisensor geophysical system; Scientific Measurements gamma ray tomography system. The vibrothermographic image was provided by Professor Kenneth Reifsnider, Virginia Polytechnic Institute and State University. All other techniques referred to in detail were developed in The Johns Hopkins University Center for Nondestructive Evaluation. For additional information about these techniques, the author should be contacted.

COMPOSITES

NONDESTRUCTIVE EVALUATION OF COMPLEX GEOMETRY ADVANCED MATERIAL COMPONENTS

S. W. Bartlett and J. C. Duke, Jr.
Virginia Polytechnic Institute and State University
Blacksburg, VA 24061

INTRODUCTION

In recent years, the advances in the application of composite materials have been tremendous. The specific strength, weight, and stiffness advantages of composites have only begun to be fully utilized. As more and more advanced designs become reality, so do the problems of this new technology. Structural optimization and application tailoring have led to a group of composites that not only have complex geometries but complex fiber architectures.

Over the years, nondestructive evaluation has become an integral part of the fabrication and analysis stages in the development of composite structures. One technique that has shown great potential in this area is acousto-ultrasonics (AU). First defined at NASA-Lewis by Vary and co-workers [1], and later refined at Virginia Tech by Duke, et al. [2- 3] , the technique draws upon the knowledge of two classic NDT methods, acoustic emission and ultrasonics. Early work has shown a good correlation between the acousto-ultrasonic behavior and the strength / modulus of simple composite materials (uniform geometry, laminate-type architecture). Recent work has focused on examining the use of the AU value as an assessment of damage caused by impact and fatigue in graphite-epoxy laminates.

The new group of advanced engineering materials and structures has led to a renewed interest in the enhancement of the acousto-ultrasonic method. The manufacturing of complex geometries has created problems in evaluating the elaborate cure cycles and non-uniform fiber architectures generated that constitute these new structures. The databases of information compiled about the nature of metallic and simple structures are no longer of use, and the old techniques just are not suited to this complex technology. The need for new analysis methods is coupled with the need to enhance existing evaluation techniques. The future holds a bright promise of production oriented technology utilizing laser generation and detection of ultrasound, robotic manipulation of components and sensors, computerized expert systems to accumulate and analyze data, and real-time quality control analysis.

PROBLEM DEFINITION

Investigation of complex geometries is not just a problem for composite structures, but for metallic structures as well. One of the major problems, that ideally should be considered during the design of the structure, is the question of accessibility. Having access to an area of concern within the structure is critical, and often geometrical constraints tend to limit which type of inspection method can be implemented. Other complications exist if the surface curvature of the specimen is extreme. Trying to place a flat-face ultrasonic transducer on a curved surface not only reduces the effectiveness of the system, but often produces unrepeatable measurements. Curvature and non-uniform thickness in composite structures precipitate complex wave propagation paths for any form of ultrasonic evaluation, making theoretical models of performance difficult if not impossible.

While the geometry contributes to the complication of propagation paths, the major element is the fiber architecture. With the new understanding of these advanced materials has come an expanded design philosophy. No longer must the orientation be a combination of flat plates and fasteners, and most importantly the fiber geometry does not necessarily have to follow the geometrical shape. Indeed, this is the advantage of composite materials. Areas that carry little of the load can be made to suit the geometric dimensions required, while critical load carrying regions may encounter a complex load / stress condition. A variety of layer interactions can also occur in these advanced structures, where cloth and fiber laminates, and 3-D weaves interact to contribute to the character of the structure. The need to tailor these complex architectures to load considerations as well as geometric constraints further complicates the availability of theoretical or computational models of expected material behavior.

The complications of analyzing such complex structures often overlaps with the processing of complex geometries and fiber architectures. Elaborate fiber lay-up, using automated weaving and winding equipment, naturally adds some damage to the material. No longer is the cure cycle a straight forward calculation, indeed it will often require much iteration to develop the optimum processing method. With different thicknesses and fiber volume fractions in the material, the likelihood of processing complications increases; as does the need to assess their impact on the performance of the structure.

While cure condition and process monitoring and control are critical to the manufacture of complex composite structures, the primary application of acousto-ultrasonics is assessing the condition of the structure after production and during the service life. With the knowledge of how a component was fabricated, analysis and proof testing can show where it might no longer be within the limits of its design specifications. By monitoring the condition of the material within the structure, the strength and durability can be evaluated. This correlation becomes an accept/reject criteria or means of life estimation for the structure.

ACOUSTO-ULTRASONIC EVALUATION

By combining the passive nature of acoustic emission with the active elements of ultrasonics, a technique was developed that introduces a stress wave of known characteristics at one location that is detected at a different location within the structure. As originally defined, the stress wave factor was a product of several ultrasonic parameters: pulse repetition rate, time gate interval, and threshold count total. Further investigation into the mechanics of the technique has led to a different approach that is based on the energy content of the signal. By using a statistical computation of the energy in the signal , the amount of mechanical energy propagating through the structure can be infered. The energy approach takes the digitized time-voltage signal from the data acquisition system and converts it to an amplitude-frequency spectrum by means of a fast Fourier transform (FFT), [5] defined by the equation:

$$F(v) = (1/N) \sum_{\tau=\emptyset}^{N=1} f(\tau) \exp(-i2\pi v\tau/N)$$

The next step is to calculate the various statistical moments of the spectrum. Normally, the zeroth, first second, and fourth moments are calculated and recorded The general equation for the moments is given as [6] :

$$M_a = \sum_{i=1}^{N/2} [F_{i+1}(v) + F_i(v)][v^a \delta v/2]$$

Figures 1a and 1b illustrate the process flow of the techniques.

The zeroth moment of the signal is the area under the spectral density distribution, and is an indication of the ability of the material to transfer stress/strain energy. The first moment, when divided by the zeroth moment, is the centroid frequency of the spectrum. Studies have shown that different damage conditions in various materials affect these moments in a particular manner, such that they can be located and characterized. Under certain conditions the transfer function of the structure can yield information about the condition of a region, while eliminating some of the problems encountered with ultrasonic detection of stress waves. Multiple transducer arrays in conjunction with the transfer function concept have also been investigated as potential methods of damage assessment.

The AU measurement system in the Materials Response Laboratory consists of an PC based data acquisition board (GRC PC-DAS system) capable of sampling signals at up to 20 MHz, an ultrasonic pulser/receiver, an

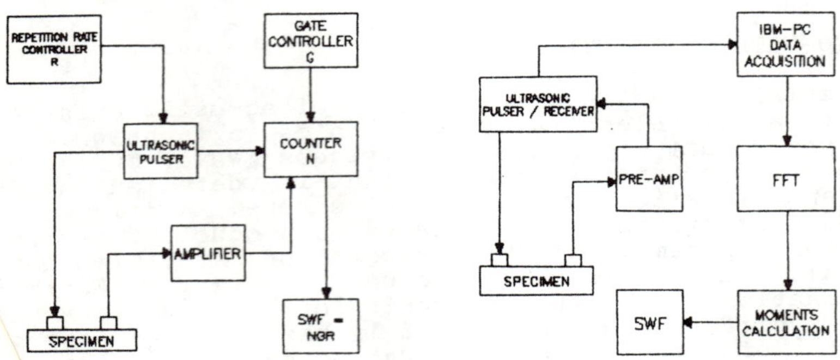

-e 1. Flow Diagrams of the a) original SWF procedure
b) VPI - AU procedure

-nic pre-amp, and a pair of 0.5" dia, 5.0 MHz -ic transducers. The software associated with the -lows the computer to act as a digital -ope, with the ability to store waveforms to -erform limited frequency analysis by means of -ier transform algorithm (FFT).

-o of signals is saved, they are input to a -am to perform the acousto-ultrasonic -s software is capable of performing FFTs, -ng of the signal in either the time or -ns, calculating the statistical moments, -e the transfer function of the system, -ve or batch mode. Once the time-voltage -d to an amplitude-frequency spectrum, -he spectrum are calculated. Figure 2 -nd frequency domain curves for a

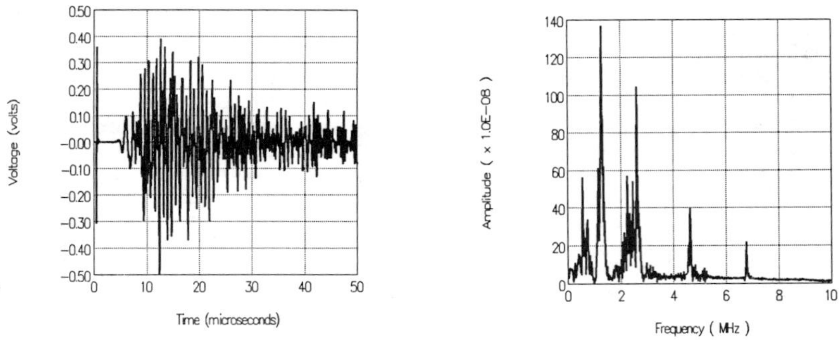

Figure 2. Signals from a composite structure a) time domain b) frequency domain

Since it is quite logical to optimize the sampling rate and transducer frequency range to the specimen in question, it would also be reasonable to optimize the input and output signals. By using computer-controlled waveform synthesizers, the input signal can be tailored to accentuate the characteristics of the structure. This can be particularly useful when performing transfer function analysis, and the input signal is of unknown or undesirable shape. In a similar manner the output signal may be digitally filtered to eliminate noisy regions that do not contribute to characterizing the nature of the structure. Certain imperfections change different parts of the signal, and as such it is useful to analyze portions of the signal independently. Thus the various modes of wave propagation(surface, bulk, and plate) may be evaluated together or separately.

APPLICATION CONSIDERATIONS

The simplicity of the technique and the lack of restrictions on implementation leads to a wide range of configurations. The sender-receiver pair can be any combination of 'transducers':

- piezoelectric
- dry couplant piezoelectric
- ultrasonic immersion
- capacitance
- laser interferometry
- pulsed laser
- impact
- EMAT

In many complex geometry/architecture configurations it is desirable to scan the object to create a map of the specimen, showing areas of property degradation. This presents the choice of moving either the object or the sensors. For laser-based systems the movement of the sensors is as simple as mounting a mirror on a multi-axis rotator, not too different from conventional FAX and copy reproduction systems. This is also the same type of approach used in analog and digital ultrasonic c-scan equipment. However, some c-scan systems put the sensors in a water-jet that is positionally controlled by robotic manipulator arms in a pulse-echo or pitch-catch mode, as illustrated in Figure 3. In other applications it may be simpler to move the specimen while the sensors remain stationary, similar to an assembly line procedure. Using lasers to excite and detect ultrasonic pulses is ideal for this high-speed environment, because there is no physical contact with the component. A fixed sensor inspection module utilizing these methods is shown in Figure 4.

The need for a flexible inspection system in essential for complex geometry and architecture structures. The non- uniform thickness and fiber arrangement may dictate the need to apply different guidelines and techniques to evaluate various regions.

ACCEPT/REJECT CRITERIA

A flexible inspection system has distinct advantages in not only where it can evaluate the structure but with what techniques. The complex geometry/architecture may also dictate that several regions must be inspected for various types of flaws. The types of flaws present would ideally be determined by a theoretical model or by computational methods, such as a finite element analysis. The significance of the flaws would be determined by a materials database, taking into consideration such details as size and type of flaw, geometrical constraints, and operating conditions. Due to the nature of the geometry/fiber architecture, and of composites in general, each location on the structure might be unique and a conformance criteria peculiar to it would be needed.

While it is obvious that several types of flaws at one location would raise concern, this concern might occur for the structure as a whole. A certain combination of factors at various locations occurring simultaneously could result in sub-standard performance. A flexible inspection system would allow the operator to use different techniques to complement and verify each other. In this manner, different types of imperfections would be evaluated by the best suited techniques, and then all of the data would be evaluated to determine the condition of the structure. While a materials database indicates what type of flaws might be expected and possibly where, mechanical testing during the prototyping phase would be required to determine their interaction and influence on the particular structure. Knowing this, a set of rules for an expert system can be defined and an accept/reject criteria developed. With the inspection procedure defined, equipment selection and the appropriate calibration standards would be determined.

By limiting the inspection process to areas of critical concern, as indicated from theoretical models, computer simulation, or prototype testing, the amount of data and the number of required inspections can be kept to a minimum. Since in many cases theoretical analysis is not possible, the accept/reject criteria must be established

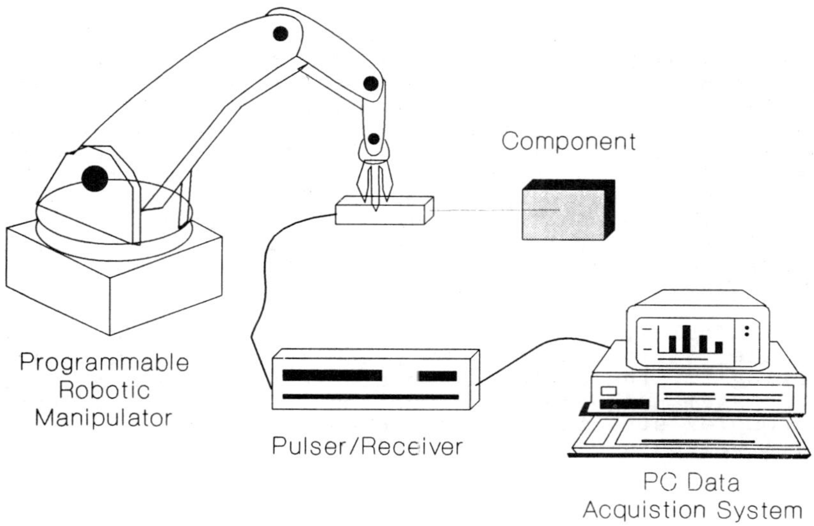

Figure 3. Fixed component with a movable sensor

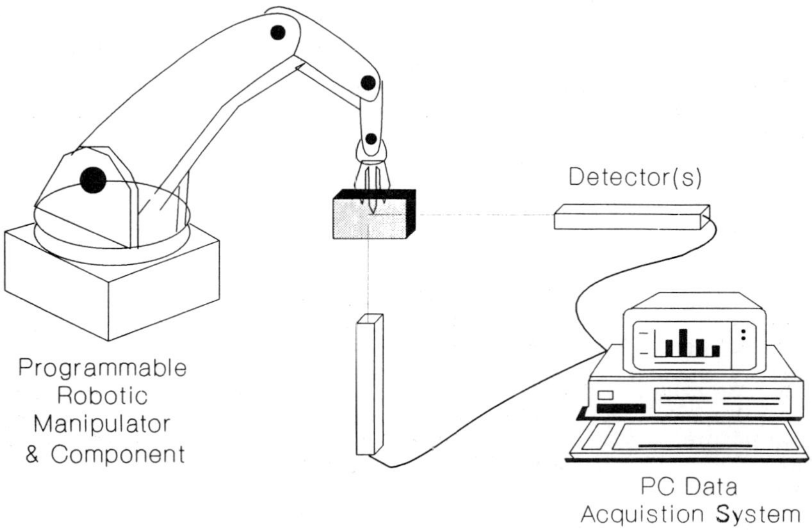

Figure 4. Fixed sensors with component in robotic manipulator

during the prototype phase. Using the baseline values developed during proof testing, further "fine tuning" would occur during the initial production run. Continuous monitoring during the production process would be used to expand the database of information and to track and verify the performance of the technique and the associated instrumentation.

SUMMARY

The complications encountered in trying to theoretically model complex geometry/fiber architectures necessitates the need for nondestructive evaluation of these structures. The acousto-ultrasonic method holds promise based on results obtained for simple geometries and laminate fiber architectures. Non-trivial engineering development is necessary to effectively overcome the challenges presented by these complex structures. It has been suggested that a computer based pulser-receiver data analysis system, in conjunction with robotic manipulators, would be needed to satisfy the demands of some advanced applications.

The flexibility of such an inspection procedure allows for the tailoring of the process to the structure by allowing for the optimization of sensors depending on the system requirements. Process tailoring is critical for complex structures since each new component will have a unique accept/reject criteria.

ACKNOWLEDGMENTS

This work has been supported in part through programs with Teledyne CAE, John Prati technical monitor, NASA/LeRC, Alex Vary technical monitor, and the Virginia Center for Innovative Technology Institute for Materials Science.

REFERENCES

1. A. Vary, K.J. Bowles, Ultrasonic Evaluation of the Strength of Unidirectional Graphite-Polyimide Composites, NASA TM X-73646, 1976

2. A.K. Govada, J.C. Duke Jr., E.G. Henneke II, and W.W. Stinchcomb, A Study of the Stress Wave Factor Technique for the Characterization of Composite Materials, NASA Contractor Report 174870, February 1985

3. Acousto-Ultrasonics: Theory and Application, ed. J.C. Duke, Jr., Plenum, New York, 1988

4. J. C. Duke, Jr. and M. T. Kiernan, "Impact Damage Development in Damaged Composite Materials," Proceedings of the 4th Japan-US Conference on Composite Materials, American Society of Composite Materials, Washington, July 1988.

5. R.N. Bracewell, The Fourier Transform and Its Applications , 2ed. rev. , McGraw-Hill, New York, 1986

6. J.S. Bendat, A.G. Piersol, Random Data - Analysis and Measurement Procedures, 2ed. rev. , Wiley Interscience, New York, 1986

NEW ACOUSTO - ULTRASONIC TECHNIQUES APPLIED TO AEROSPACE MATERIALS

H. E. Kautz
National Aeronautics and Space Administration
Lewis Research Center
Cleveland, OH 44135

ABSTRACT

The use of an NdYAG pulsed laser for generating ultrasonic waves for NDE in resin matrix composites was investigated. A study was conducted of the use of the 1.064 μm wavelength NdYAG pulsed laser with the neat, unreinforced resin as well as with graphite fiber/polymer composite specimens. In the case of neat resins it was found that, at normal incidence, about 25 percent of the laser pulse energy is reflected at the incident surface. An attenuation coefficient for the polyimide resin, PMR-15 was determined to be ~5.8 np/cm. This means, for example, that for a 1/16 in. panel thickness 24 percent of the incident energy is transmitted through the panel. Under these conditions much of the laser produced ultrasonic wave will originate throughout the bulk of the specimen rather than at the surface in contrast to the case with piezoelectric transducers. It was found in energy balance studies that graphite fiber/polymer specimens attenuate the laser beam more than do the neat resins. The increased absorption is in the graphite fibers.

The occurrence of laser induced surface damage was also studied. For the polymer neat resin, damage appears as pit formation over a small fraction of the pulse impact area and discoloration over a larger part of the area. A damage threshold was inferred from observed damage as a function of pulse energy. The 600 °F cured PMR-15 and PMR-11 exhibit about the same amount of damage for a given laser pulse energy. The damage threshold is between 0.06 and 0.07 J/cm^2. In the case of graphite fiber/polymer specimens damage studies showed that the fibers burn away to the extent of partially shielding the resin from pit formation and discoloration.

Acousto-ultrasonic signals were collected with a piezoelectric transducer over a wide range of laser pulse energies. It was found that signal strengths typical of those produced by piezoelectric sending transducers can be achieved at laser pulse energies well below the damage threshold.

INTRODUCTION

The use of lasers for stimulation and detection of ultrasonic waves in materials is of high interest for nondestructive evaluation (Refs. 1 to 5). Automation of laser beam scanning seems considerably more practical than direct coupling of piezoelectric transducers to the surface. Thus remote scanning of large areas becomes viable.

New techniques usually present limitation along with their advantages, however. One serious limitation of lasers is likely to be their potential for doing damage to surfaces. Development of laser ultrasonics ought to include examination of the potential damage to materials that are of aerospace interest.

Acousto-ultrasonics is a two transducer NDE technique for assessing mechanical strength in composites. It has an advantage of allowing one-side access to structures. The application of lasers to injection of the ultrasonic signal (laser-in) and also to the detection (laser-out) can lead to better mechanization of acousto-ultrasonic scanning of large specimens.

The present work focuses on laser-in ultrasonics. Laser pulse energy damage thresholds are examined and compared for energy ranges that produce useful acousto-ultrasonic signals. This is done for neat polymeric resin panels and fiber reinforced polymer matrix composite laminate samples.

MATERIALS, EXPERIMENTAL, AND PROCEDURE

A schematic diagram of the laser system used for these experiments is shown in Fig. 1. The laser is a solid state NdYAG (Neodymium Yttrium Garnet) pulsed laser with 1.064 μm wavelength and filtered to provide pulses of various energy. The digitizing oscilloscope was controlled to collect single sweep acousto-ultrasonic signals with a PDP/1134 computer and store them on disc.

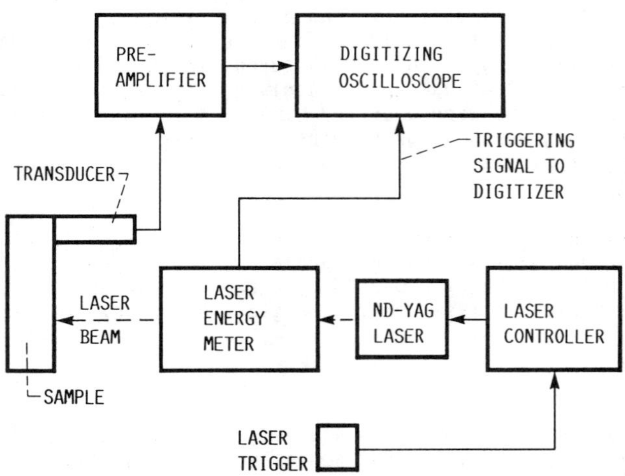

FIGURE 1. - SCHEMATIC OF LASER CALIBRATION AND MEASUREMENT SYSTEM.

Materials Studied

Table 1 lists the materials studied in these experiments. All were flat panels of various dimensions. The fiber-reinforced panels were composite laminates with ply orientations given in Table 1.

TABLE 1. Specimens studied

Material	Thickness, cm	Structure	Remarks, °F cure
PMR-15	0.145	Monolithic	600
PMR-15	.586	Monolithic	↓
PMR-11	.288	Monolithic	↓
Celion 6000/PMR-15	.533	28 ply 0/90	↓
Celion 6000/PMR-15	.550	30 ply uni-directional	↓
Celion 6000/PMR-11	.111	6 ply 0/45/90-center	↓

[a]PMR-15 is a polymide resin capable of service at 600 °F.
[b]PMR-11 is the same as PMR-15 except that it has lower molecular weight.
[c]Celion 6000 is a graphite fiber available from BASF.

Laser Damage Experiments

For the laser damage experiments positions on the surface of the specimen were repeatedly subjected to laser pulses of the same energy. Different positions received pulses at different energy levels but the number of pulses were also varied so that the total energy at all positions was the same. For these experiments no acousto-ultrasonic measurements were made.

Energy Balance Experiments

Energy transmitted through a specimen thickness was determined experimentally by placing the specimen between the laser and the energy meter in Fig. 1. Radiation energy meter readings were noted with the specimen present and absent.

Acousto-ultrasonic Experiments

For these experiments the specimen was placed at the output of the energy meter. A 1 MHz broad band transducer was coupled to the surface with bee's wax, melted into place. In some cases, for comparison, specimens were first excited by a 2.25 MHz broad band transducer before the laser-in experiment. In each case waveform data were collected and stored on disc for later examination of time domain waveform, frequency domain spectrum, and for calculation of stress-wave factor (Ref. 6). The transducer excited signals reported here were all produced with the same transducer, same pulser setting, and same geometry. The waveform digitizer for these experiments is shown in Fig. 1.

ENERGY BALANCE CALCULATION

Figure 2 shows a schematic for the energy balance for a laser beam impinging at normal incidence on the surface of a specimen of thickness x.

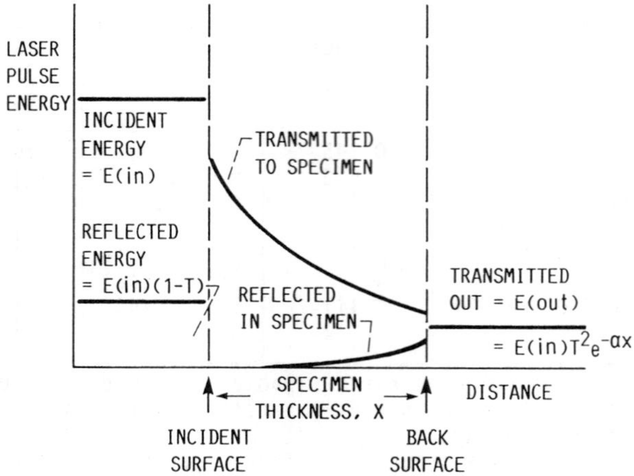

FIGURE 2. - ENERGY BALANCE FOR LASER PULSE TRANSMITTED THROUGH SPECIMEN.

Besides the specimen thickness, x, other measured quantities are the energy of the incident beam at the front surface, E(in), and the energy of the beam emerging from the back surface, E(out). These two energies are related as follows.

$$E(out) = E(in)T(in)AT(out) \qquad (1)$$

T(in) = transmission coefficient at the incident surface.
T(out) = transmission coefficient at back surface.
A = attenuation of beam through the thickness of the specimen.

It turns out that the transmission coefficient from a medium 1 to a medium 2, (such as air to a composite), is the same as from medium 2 back into medium 1, (Ref. 7). There is a difference in what happens to the phase of the wave but for our energy transmission coefficients at the surface:

$$T(in) = T(out) = T \qquad (2)$$

For the attenuation, A, within the specimen we assume Beer's law:

$$A = e^{(-\alpha x)} \qquad (3)$$

A is a factor that multiplies the intensity at zero thickness which is E(in)T of Eq. (1). "α" in Eq. (3) is the energy attenuation coefficient with distance.

Equations (1), (2), and (3) can be combined as shown below.

$$E(out) = E(in)T^2e^{(-\alpha x)} \qquad (4)$$

Equation (4) contains two unknowns T and α. If one applies Eq. (4) to data from two specimens that are identical except for thickness then Eq. (4) can be used as simultaneous equations to yield values of T and α. With this the energy balance can be calculated as below.

TABLE 2. Visible damage produced in PMR-15 neat resin material by Nd-YAG 1.064 μm laser radiation

[T = Nne, E = ne.]

N	n	e	E	T	Observed damage
Number of pulse trains	Number of pulses in train	Energy density of pulse, J/cm^2	Energy density of train, J/cm^2	Total energy density on specimen, J/cm^2	
50	5	0.42	2.10	105	Pitting Nodules Discolored
298	1	.35	.35	104	Pitting Nodules Discolored
100	5	.208	1.04	104	Small Pit Nodules Discolored
192	5	.11	.55	106	No pitting Nodules No discolor
1189	1	.087	.087	103	No pitting Nodules No discolor
282	5	.074	.37	104	No pitting Few Nodules No discolor
369	5	.056	.28	103	No pitting No nodules No discolor

Total laser pulse energy(measured) = E(in)
Energy reflected at incident surface = E(in)(1-T)
Energy absorbed in specimen = E(in)T-E(out)
Energy leaving back surface(measured) = E(out)

RESULTS

Laser Damage Experiments

Figures 3(a) to (b) show a sequence of laser radiated sites for a 0.145 cm thick PMR-15 panel at 50X magnification. Visible laser damage

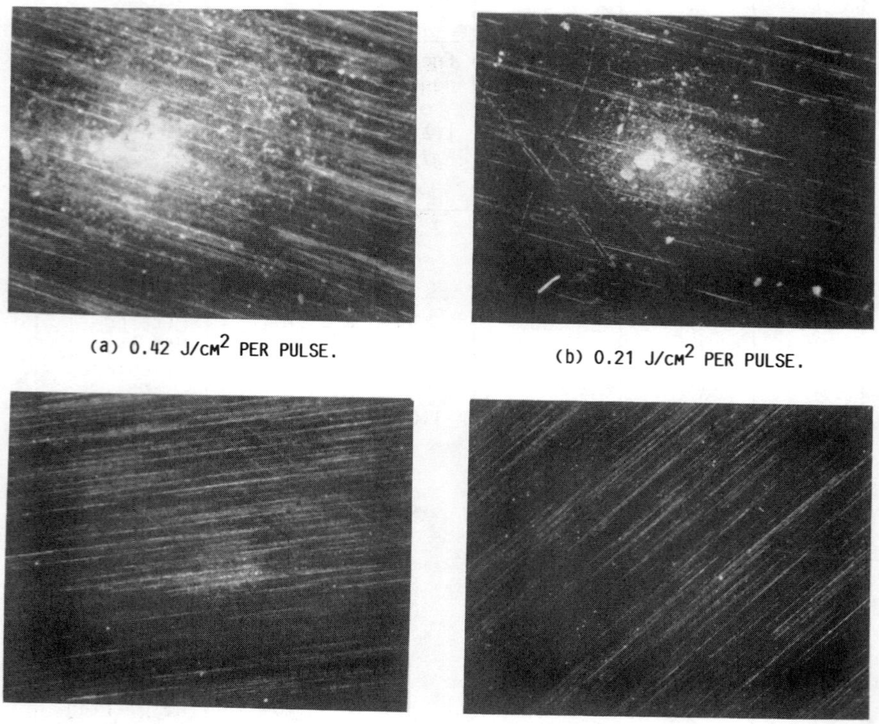

(a) 0.42 J/cm^2 PER PULSE.

(b) 0.21 J/cm^2 PER PULSE.

(c) 0.07 J/cm^2 PER PULSE.

(d) 0.06 J/cm^2 PER PULSE.

FIGURE 3. - LASER DAMAGE TO PMR-15 NEAT RESIN PRODUCED BY NdYAG PULSED LASER WITH CONTANT TOTAL DOSE.

was inferred from the presence of pitting, subsurface bubbles, and discoloring that was not present elsewhere on the panel. The same results are also described in Table 2. Each of the positions shown received repeated pulses at a given energy. The number of repeats at each position was chosen so that the damage at all positions could be compared at about the same total dose of 104 to 105 J/cm^2. In Fig. 4 a 0.35 J/cm^2 per pulse level is show at 200X for greater detail.

Results for the 0.288 cm thick PMR-11 neat resin panel were the same as for the PMR-15 presented in Figs. 3 and 4.

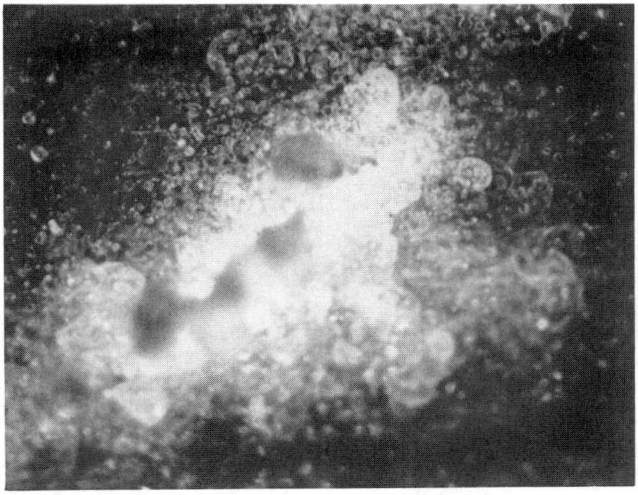

FIGURE 4. - VIEW OF DAMAGE FOR PMR-15 WITH 0.35 J/CM^2 PER PULSE.

Figures 5(a) to (d) show a sequence for the 28 ply composite panel of Celion 6000 graphite fibers and PMR-15 matrix for the same sequence of laser pulse energies as the Fig. 3 sequence for the neat resins alone.

Damage photos for these three panels should be compared with the same energy per pulse as follows:

Energy Balance Experiments

Among the specimen panels listed in Table 1 the two PMR-15 and the PMR-11 neat resin panels gave detectable levels of laser pulse energy exiting from the back surface. For these the ratio E(out)/E(in) was calculated. This was done for 6 to 8 pulse energy levels that spanned from below to above the observed damage threshold. The ratio did not vary significantly over this energy range. The averages for the panels with standard deviations is listed in Table 3.

In order to carry out an Eq. (4) calculation, the 0.145 cm thick and the 0.586 cm thick PMR-15 panels were assumed to have identical transmission coefficients and absorption coefficients. The result is:

Transmission coefficient T(air/PMR-15) = 0.750

absorption coefficient a(PMR-15) = 5.78/cm.

From this, the reflection coefficient R = 1-T = 0.25.

In Fig. 6 E(out)/E(in) is plotted against specimen thickness for the three neat resin panels. The line is constructed for the PMR-15 data used in the above calculation. The PMR-11 data point is also plotted for comparison.

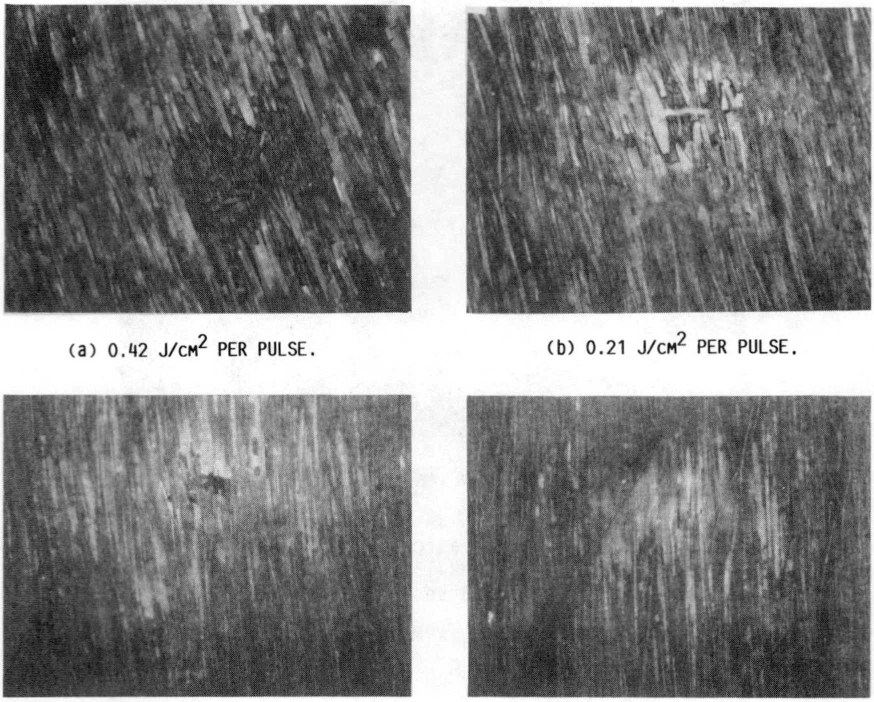

(a) 0.42 J/CM^2 PER PULSE.

(b) 0.21 J/CM^2 PER PULSE.

(c) 0.07 J/CM^2 PER PULSE.

(d) 0.06 J/CM^2 PER PULSE.

FIGURE 5. - LASER DAMAGE TO 28 PLY (0/90^0) COMPOSITE WITH CELIUM 6000 FIBERS AND PMR-15 MATRIX (CONSTANT TOTAL DOSE).

TABLE 3. Energy balance data for PMR neat resin panels

Material	Thickness, cm	Energy out/ energy in	Standard	Deviation, percent
PMR-15	0.145	0.243	0.008	3.3
PMR-15	.586	.0190	.0007	3.7
PMR-11	.288	.246	.003	1.2

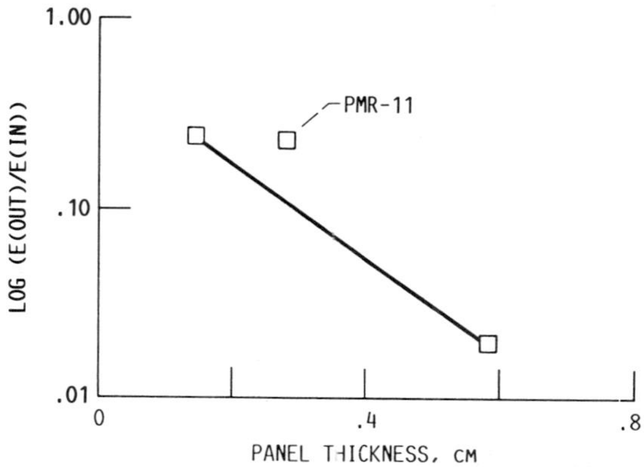

FIGURE 6. - LOG OF E(OUT)/E(IN) VERSUS PANEL THICKNESS FOR PMR 11 AND TWO PMR-15 NEAT RESIN SPECIMENS.

No measureable E(out) pulse energy values were found for the graphite/resin composite panels. This means that at the highest pulse energy E(out)/E(in) was <1 percent, the limit of sensitivity for the energy meter. This panel was free of visible voids or delaminations that could affect the passage of the laser pulse.

Acousto-Ultrasonic Experiments

Figure 7 shows acousto-ultrasonic signals collected on the 0.145 cm thick PMR-15 panel. Figure 7(a) is the time domain of a signal produced by a 2.25 MHz broad band piezotransducer coupled to the panel through an elastomer pad. Figure 7(b) is the frequency spectrum of 7(a). Figure 7(c) is the time domain of a signal produced by a 0.030 J/cm^2 laser pulse. Figure 7(d) is the frequency spectrum of 7(c). Both waveforms in Fig. 7 were collected by the same 1 MHz broad band piezotransducer fastened by melted wax to the panel at a point 1.5 in. from where the signal was introduced, either by transducer or laser pulse.

Figures 8 to 10 show acousto-ultrasonic signals collected from the graphite/polymer unidirectional 30 ply panel. As with the neat resin, all the signals were collected by a 1 MHz broad band piezotransducer fastened by wax to the panel at a point 3.8 cm from, and on the same side as, where the ultrasonic signal was introduced. Signals were collected with the point of introduction to 1 MHz receiver line parallel to the fiber direction and also with it perpendicular to the fibers.

Figures 8(a) and (b) show a signal produced with a 2.25 MHz transducer aligned with the receiver parallel to the fiber direction. Figures 8(c) and (d) are for the transducers aligned perpendicular to the fibers.

Figures 9(a) and (b) show a signal produced by the lowest laser pulse energy setting available. The 0.032 J/cm^2 target spot was aligned with the receiver parallel to the fiber direction. For Figs. 9(c) and (d) the lowest energy setting gave a 0.029 J/cm^2 pulse. The target spot was, in this case, aligned perpendicular to the fiber direction.

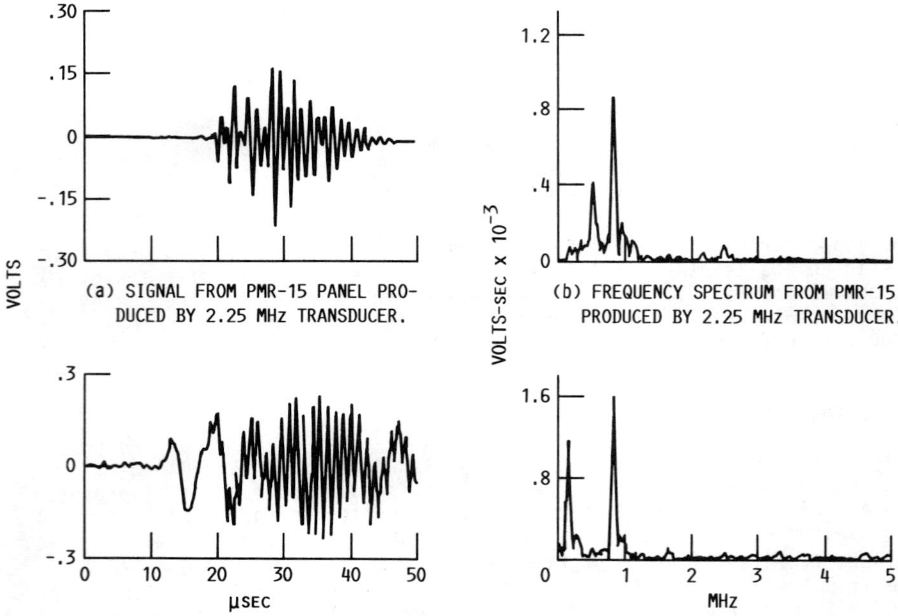

(a) SIGNAL FROM PMR-15 PANEL PRODUCED BY 2.25 MHz TRANSDUCER.

(b) FREQUENCY SPECTRUM FROM PMR-15 PRODUCED BY 2.25 MHz TRANSDUCER.

(c) SIGNAL FROM PMR-15 PANEL PRODUCED BY 0.03 J/CM^2 LASER PULSE.

(d) FREQUENCY SPECTRUM FROM PMR-15 PRODUCED BY 0.03 J/CM^2 LASER PULSE.

FIGURE 7. - ACOUSTO-ULTRASONIC SIGNALS COLLECTED FROM PMR-15 PANEL STIMULATED 2.25 MHz TRANSDUCER AND ALSO BY 0.033 J/CM^2 LASER PULSE.

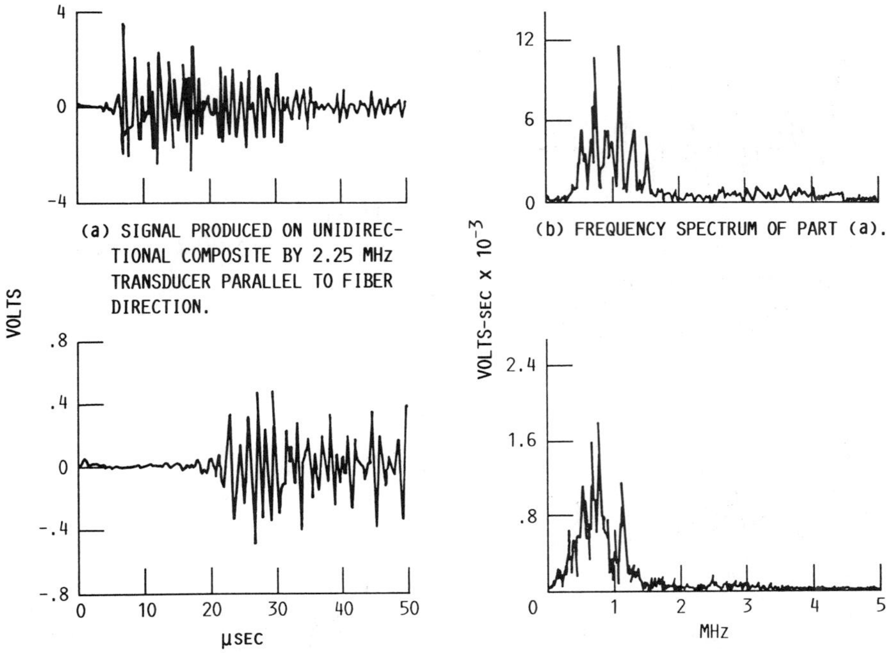

(a) SIGNAL PRODUCED ON UNIDIRECTIONAL COMPOSITE BY 2.25 MHz TRANSDUCER PARALLEL TO FIBER DIRECTION.

(b) FREQUENCY SPECTRUM OF PART (a).

(c) SIGNAL PRODUCED ON UNIDIRECTIONAL COMPOSITE BY 2.25 MHz TRANSDUCER PERPENDICULAR TO FIBER DIRECTION.

(d) FREQUENCY SPECTRUM OF PART (c).

FIGURE 8. - SIGNALS PRODUCED ON UNIDIRECTIONAL GRAPHITE/POLYMER COMPOSITE BY 2.25 MHz SENDING TRANSDUCER.

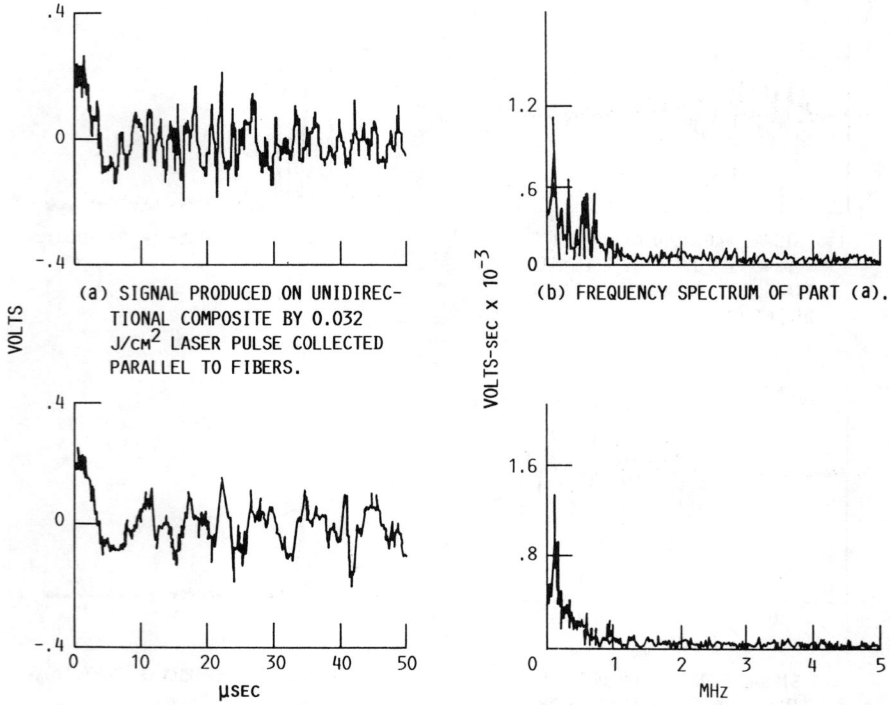

(a) SIGNAL PRODUCED ON UNIDIRECTIONAL COMPOSITE BY 0.032 J/CM^2 LASER PULSE COLLECTED PARALLEL TO FIBERS.

(b) FREQUENCY SPECTRUM OF PART (a).

(c) SIGNAL PRODUCED ON UNIDIRECTIONAL COMPOSITE BY 0.029 J/CM^2 LASER PULSE COLLECTED PERPENDICULAR TO FIBERS.

(d) FREQUENCY SPECTRUM OF PART (c).

FIGURE 9. - SIGNALS PRODUCED ON UNIDIRECTIONAL GRAPHITE/POLYMER COMPOSITE BY LASER LASER AT LOW PULSE ENERGY.

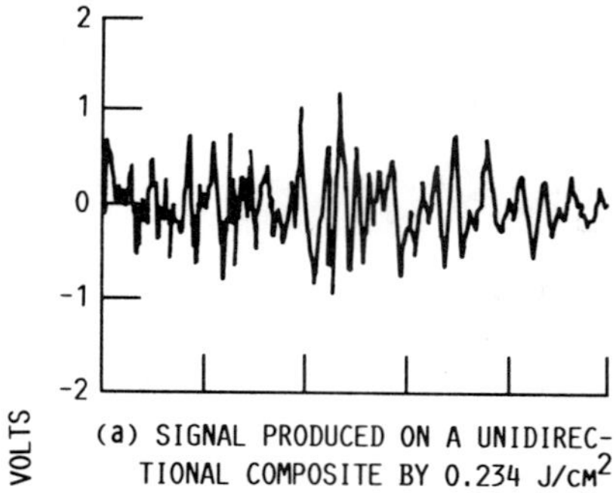

(a) SIGNAL PRODUCED ON A UNIDIRECTIONAL COMPOSITE BY 0.234 J/CM^2 LASER PULSE COLLECTED PARALLEL TO FIBERS.

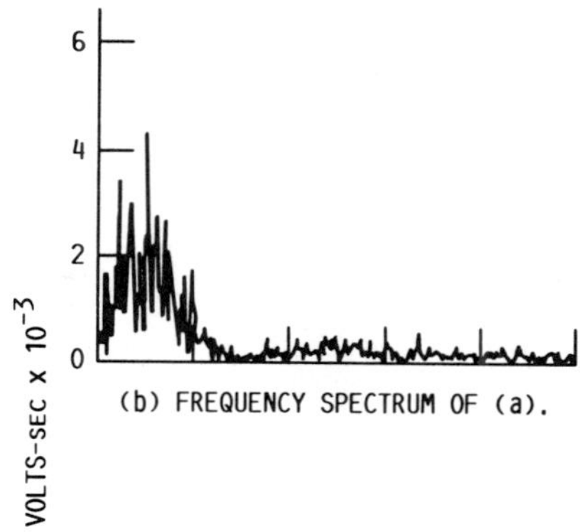

(b) FREQUENCY SPECTRUM OF (a).

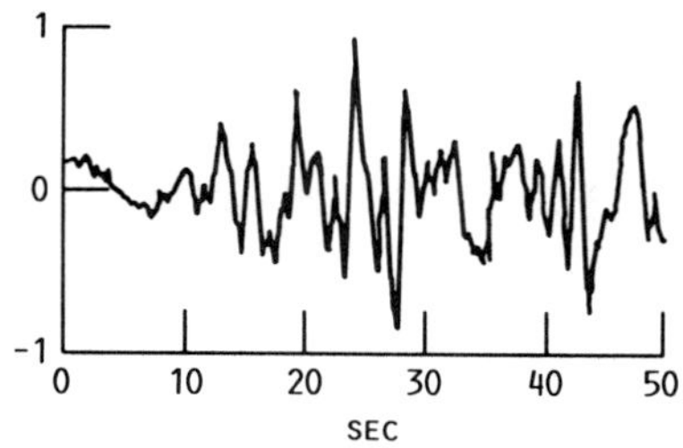

(c) SIGNAL PRODUCED ON A UNIDIRECTIONAL COMPOSITE BY 0.197 J/CM^2 LASER PULSE COLLECTED PERPENDICULAR TO FIBERS.

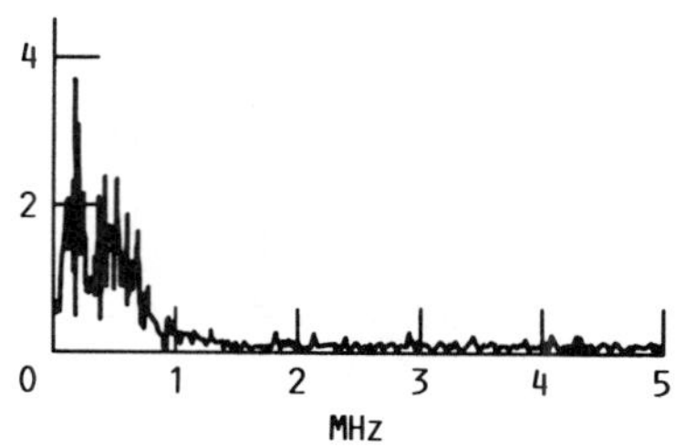

(d) FREQUENCY SPECTRUM OF (c).

FIGURE 10. - SIGNALS PRODUCED ON UNIDIRECTIONAL GRAPHITE/POLYMER COMPOSITE BY LASER AT HIGH PULSE ENERGY.

Figures 10(a) and (b) show a signal produced by the highest laser pulse energy setting available. The 0.234 J/cm^2 target spot was aligned with the receiver parallel to the fiber direction. For Figs. 10(c) and (d) the highest energy setting gave a 0.197 J/cm^2 pulse. The target spot was aligned perpendicular to the fiber direction.

Linear stress-wave factors were calculated from the waveforms collected in these experiments. The linear stress-wave factor, (SWF), is defined as:

$$SWF = \int_{t1}^{t2} V_{ab}(t)dt \qquad (5)$$

$V_{ab}(t)$ is the absolute value of the amplified receiver output voltage at time t. t1 and t2 are the beginning and end of the time record of the acousto-ultrasonic signal collected.

Figure 11 shows plots of linear SWF versus laser pulse energy for the energy settings available in these experiments. Figure 11(a) is for the signals produced on a PMR-15 neat resin panel. Figure 11(b) is for the

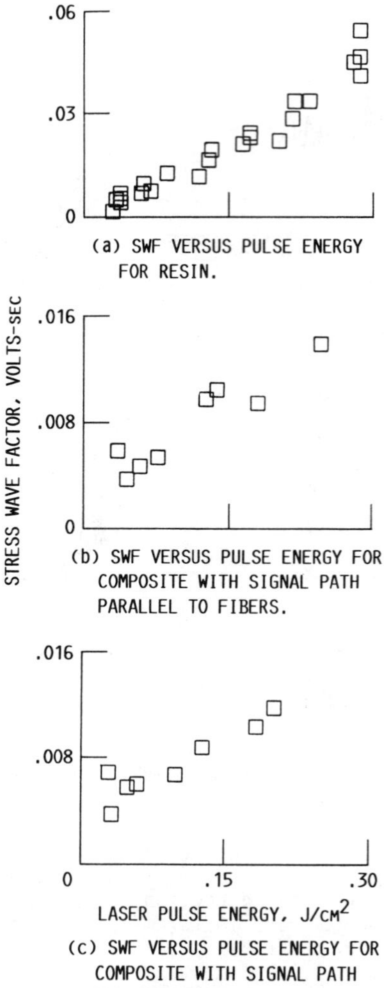

FIGURE 11. - STRESS WAVE FACTOR VERSUS LASER PULSE ENERGY.

unidirectional composite with signal path parallel to the fiber direction. Figure 11(c) is for the unidirectional composite with signal path perpendicular to the fiber direction.

All the acousto-ultrasonic waveform data presented is for geometries with the receiving transducer oriented on the same side of the panel as the ultrasonic energy source. Measurements were also made with the receiver on the opposite side from the source and offset 1.5 in. In the case of the neat resin panels, signals were recovered from the opposite side that were similar in strength as on the same side. In the case of the composites, however, the opposite side signal was greatly reduced in strength from that recovered from the same side.

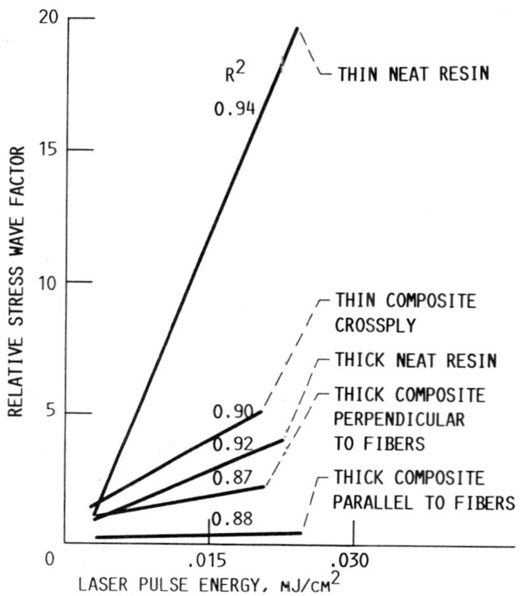

FIGURE 12. - LASER PRODUCED STRESS WAVE FACTOR DIVIDED BY TRANSDUCER PRODUCED STRESS WAVE FACTOR VERSUS LASER PULSE ENERGY.

DISCUSSION

Laser Damage Experiments

Earlier studies of laser damage to polymers indicate that two mechanisms are possible (Refs. 8 to 10). One is photothermal and the second is photochemical. The photochemical mechanism, photon breaking of molecular bonds, is not possible at the relatively low photon energy at the NdYAG 1.064 μm wavelength. This leaves the photothermal effect of heating the material at the site of laser beam incidence to account for our results.

Figures 3 and 4 shows laser damage to a neat resin panel. These photos show the pitting that occurs at the center of the laser spot as well as the presence of bubbles under the surface that imply some gas evolution. (The bubbles are particularly evident in the 200X view of Fig. 4.) Being black and white, the photos do not show the yellowish discoloration of the damage area. This discoloration, like the other manifestations, lessens with decreasing energy per pulse. The discoloration probably indicates the presence of charring of the polymer. Gas evolution and charring are two signs that decomposition is taking place along with vaporization. It should be pointed out that these damage sites are much smaller than the laser beam diameter. The beam diameter is actually about twice the width of these photos.

As Table 2 shows, in some cases trains of five laser pulses were used where as in other cases they were single pulses. Single pulses were separated in time by at least 2 sec. Adjacent pulses in a train were separated about 10 μsec. There is no difference in the total damage caused by pulses 10 μsec apart from those separated by 2 sec. This implies that the heating of a pulse has effectively dissipated in 10 μsec. If this were not so one would expect that experiments employing pulse trains to cause the material to reach higher temperatures and thus greater damage.

Decreasing individual pulse energy level leads to decreasing damage even though the accumulated energy incident on all the spots is the same. This means that the damage per increment of energy decreases with pulse level. This implies that a damage threshold pulse energy exists. The photos indicate that this threshold is probably between 0.06 and 0.07 J/cm^2. The 0.06 J/cm^2 photos show no sign of damage.

Figure 5 shows that damage in the graphite/polymer composite is much greater than in the neat resins. The primary effect in the composite is the vaporization or burning of graphite fibers. The area affected is much larger than for the resins with the same laser pulse energy. The fibers apparently absorb the laser energy much more than the resin and rise to much higher temperatures. In addition to vaporizing or burning graphite at the center the heat vaporizes the surface layer of resin over a very large area. On the other hand, resin below the surface seems to be shielded. No bubbles or yellow discoloration was observed with the composite. The composite panel in Fig. 5 is a 0/90 cross ply. The greatest damage shown, in Fig. 5(a), does not appear to penetrate to the second ply. This makes the depth <0.02 cm. The width of the fiber burn area is about 0.08 cm. Thus the depth to width ratio for the crater is <1 to 4. This is in contrast to monolithic materials where deep craters are usually found (Ref. 9).

Energy Balance Experiments

The transmission and absorption coefficients found for PMR-15 are estimates. Although the energy ratio E(out)/E(in) reproduces within 3 to 4 percent at one point on one specimen it may vary more than that from place to place. There are visible variations in color and roughness which are probably comparable to variations in the infrared reflecting properties of the materials. The line drawn in Fig. 6 is based on the data for PMR-15. The PMR-11 data point does not fall on this line. It is not obvious whether this represents a difference in infrared properties between the two polymer systems or simply a display of the scatter among three data point.

Of greater significance is the fact that graphite/polymer composites in the same thickness range transmitted pulse energy levels that were below detectability. This is in agreement with the laser damage experiments where there were indications of very strong absorption by graphite fibers very near the surface. This would lead to an absorption coefficient much larger than calculated for the neat resins and therefore a much smaller value of E(out). The rate in which infrared is absorbed is apparently so different between graphite fibers and polymer resin that the exponential decay model may not be appropriate in the composite case.

Note in Eq. (4), that if the attenuation coefficient, α, is large E(out) will be insignificant. Also, most of the laser energy absorbed in the specimen will be near the incident surface. Any ultrasonic waves produced by the conversion of laser energy to mechanical energy will appear near that surface simulating the propagation path conditions that result from a piezotransducer coupled at that spot on the surface. On the other hand, a low value of α will favor a large E(out). It will also mean absorption of a significant amount of energy deeper into the bulk of the specimen. Under this condition ultrasonic waves may appear from the bulk of the specimen rather that solely near the surface. The distribution of wave propagation paths in this case may be quite different from the piezotransducer situation.

Acousto-Ultrasonic Experiments

Figures 7 to 10 show typical acousto-ultrasonic signals collected on the specimens of these experiments. There are three cautions associated with comparing transducer produced and laser produced signals. The first is that the triggering of the waveform digitizer is different between the two systems. Thus there is some offset between their time domain positions. The laser produced signals seem to trigger the system 5 to 10 μsec later than do the transducer signals. The second caution is that absolute signal amplitudes ought not be compared. The amplitude in both cases depends upon the amount of ultrasonic energy introduced. But in the laser case one does not know the conversion efficiency between electromagnetic and ultrasonic energy. Certainly the transducer signal amplitudes can be varied at will by adjusting the pulser settings and the laser pulse can also be adjusted with some range. The third caution is associated with the means of mounting the specimens. For laser measurements they were clamped in one spot in a vertical orientation. For sending transducer they were held horizontally against a corrugated backing sheet by transducer pressure as described in Ref. 6. Boundary conditions for the plate wave component of the signal may have been different for the two orientations.

With the above in mind, we continue by observing that laser produced signals exhibit lower frequency spectra. In the case of the neat resin there is a distinct bimodal spectrum. The high frequency component is much like the spectrum produced by transducer. The laser signal has a different low frequency in the spectrum.

Figures 8 to 10 each show contrasts between wave paths parallel to and perpendicular to the graphite fiber orientation in the unidirectional composite. The transducer produced signals show the early arriving higher frequency component in the parallel signal and absent in the perpendicular measurement. This is in agreement earlier studies (Ref. 6).

The laser produced signals are much less distinct. However, the parallel orientation signals do have higher frequency components that are, again, absent from the perpendicular to fibers signal.

The neat resins exhibit significant through transmission acousto-ultrasonic signals and the composites do not. This is in agreement with the damage and energy balance results. Certainly with the neat resin absorbing laser energy deep in the bulk it will have bulk waves that can be observed on the back side. On the other hand, the composites absorbed laser energy very near the front surface. This apparently favors surface waves that do not appear in the back.

Figure 11 shows that the linear stress wave factor calculated with Eq. (5) for the various specimens is generally linear with laser pulse energy. The linear stress wave factor is not, however, an energy function. Substituting the quantity $V(t)^2$ for $V_{ab}(t)$ in Eq. (5) gives a quantity that is proportional to the energy of the acousto-ultrasonic signal. The fact that acousto-ultrasonic energy is not linear with laser pulse energy may be related to the fact that damage is being produced at the higher laser energies. The acousto-ultrasonic energy grows faster than expected with high laser energy. In order to explain this by damage one must assume that energy is released to the ultrasonic pulse from the exothermic decomposition of the organic materials of the specimens.

One way to compare the efficiency of producing laser acousto-ultrasonic waves between different materials is to plot a relative stress wave factor. This is the stress wave factor from the laser pulse divided by the stress wave factor from the 2.25 MHz transducer. This is done in Fig. 12 where, as with Fig. 11, the horizontal axis is laser pulse energy. To avoid confusion only the linear regression lines are shown. (R^2 values are included to indicate the degree to which the actual data agrees with those lines.) It can be seen that thin specimens have higher relative energy transfer efficiencies than thick ones.

Figure 12 also shows that neat resins are more efficient than composites. This probably means the polymers convert laser energy to ultrasonic better than graphite fibers.

The thin neat resin has much higher energy transfer efficiency (from laser to ultrasonic) than the thick specimen. This thin panel is the one that transmitted 24 percent of the laser energy out the back surface. The key to its high efficiency is probably that it absorbs laser energy to a large degree throughout its bulk from front to back. Thus it distributes ultrasonic waves throughout the bulk whereas the other specimens produce ultrasonic waves only near the incident surface where the laser pulse is absorbed.

The lowest efficiency is for the thick composite with wave path parallel to fiber direction. It is so low because it is relative to a large transducer produced stress wave factor. This can be inferred by comparing Figs. 8(a) to (c). The parallel to fiber waveform is much larger than the perpendicular one. On the other hand, comparing Figs. 11(b) to (c), we see that the advantage of the parallel orientation is lost with the laser produced signals. This is another indication that the resin converts laser energy to ultrasonic energy more efficiently than do the fibers.

CONCLUSIONS

The 1.064 μm wavelength pulsed NdYAG laser produces acousto-ultrasonic signals in neat polymer resin and in graphite/polymer composite specimens that are similar to those obtained when employing a piezoelectric sending transducer.

The efficiency of inserting ultrasonic energy into a resin matrix relative to insertion into graphite fibers is much higher for laser produced waves than for the piezoelectric transducer case.

The NdYAG pulsed laser can be adopted to the acousto-ultrasonic NDE of graphite/polymer composites provided that pulse energies are employed that are at or below the energy levels studied in the present work. Higher energies can lead to damage at the laser beam impingement site.

A threshold energy value per pulse exists for the onset of laser damage to a composite or neat resin. For pulse energies greater than the threshold damage accumulates with succeeding pulses. For pulse energies less than the threshold no damage accumulates for repeated pulses. The value of the threshold energy is lower for graphite/polymer composites than for neat resins.

ACKNOWLEDGEMENTS

We wish to thank K.J. Bowles and P.J. Cavano of Materials Division, NASA Lewis Research Center for the neat resin and composite specimens used in this study.

REFERENCES

1. South, H.M., and Palmer, C.H., Analysis of Two-Beam Interferometry for Bulk Wave Measurements, Appl. Opt., vol. 22, pp. 2854-2859, 1983.

2. Palmer, C.H., Circumventing Laser Relaxation Oscillations in Interferometers, Appl. Opt., vol. 23, pp. 3510-3512, 1984.

3. Drake, A.D., and Leiner, D.C., Fiber-Optic Interferometer for Remote Subangstrom Vibration Measurement, Rev. Sci. Instrum., vol. 55, pp. 162-165, 1984.

4. Bourkoff, E., and Palmer, C.H., Low-Energy Optical Generation and Detection of Acoustic Pulses in Metals and Nonmetals, Appl. Phys. Lett., vol. 46, pp. 143-145, 1985.

5. Burger, C.P., Dudderar, T.D., and Gilbert, J.A., Thermal Acoustic Optic NDE for Non-Contact Inspection in Remote Locations, Presented at the Acousto-Ultrasonics: Theory and Application Conference, Blacksburg, VA, July 12-15, 1987.

6. Kautz, H.E., Ray Propagation Path Analysis of Acousto-Ultrasonic Signals in Composites, Presented at the Acousto-Ultrasonics: Theory and Application Conference, Blacksburg, VA, July 12-15, 1987.

7. Von Hipple, A.R., Dielectrics and Waves, John Wiley and Sons, New York, 1954.

8. Garrison, B.J., and Srinivasan, R., Laser Ablation of Organic Polymers: Microscopic Models for Photochemical and Thermal Processes, J. Appl. Phys., vol. 57, pp. 2909-2914, 1985.

9. Braren, B., and Srinivasan, R., Optical and Photochemical Factors Which Influence Etching of Polymers by Ablative Photodecomposition, J. Vac. Sci. Technol. B, vol. 3, pp. 913-917, 1985.

10. Gorodetsky, G., Kazyaka, T.G., Melcher, R.L., and Srinivasan, R., Calorimetric and Acoustic Study of Ultraviolet Laser Ablation of Polymers, Appl. Phys. Lett., vol. 45, pp. 828-830, 1985.

EVOLUTION OF THE WAVE ENVELOPE IN COMPOSITE LAMINATES IN ACOUSTO - ULTRASONIC TECHNIQUE

S. M. Moon
Washington University
St Louis, MO 63130

H. T. Hahn
Pennsylvania State University
University Park, PA 16802

K. L. Jerina
Washington University
St Louis, MO 63130

ABSTRACT

In the acousto-ultrasonic technique, a wave packet is injected onto the surface at one location and the displacement normal to the same surface is measured at another location. The characteristics of the wave packet propagating through the material are not fully understood. The main objective of the present investigation is to study the propagation of a wave packet in cross-ply laminates.

An S-2 glass/epoxy cross-ply laminate was used in this investigation. A wave packet with fast rise and decay was injected into the laminate. Wave velocities were measured in different directions to the fiber of the contacting ply orientations. To characterize the wave envelope, signal parameters such as peak amplitude, ringdown counts, RMS amplitude and energy content were measured at different locations. A Fourier analysis of the received signal was also carried out.

The investigation revealed that the wave velocities in the cross-ply laminate could be predicted by Lamb wave equations based on the average laminate properties. The theory could not predict the effect of stacking sequence. Variation of velocity with external layer fiber orientations was predicted. It was found that the peak amplitude of the signal may not reveal all the information concerning the damage. To obtain detailed information, energy should be monitored. In the acousto- ultrasonic technique, reflections from the boundaries of the laminate were present for a tone burst signal. This led to higher ringdown counts in the received signal. The reflections were studied with time and Fourier analyses. The received signal was gated to eliminate the reflections and to obtain realistic information.

INTRODUCTION

The analysis of the response of composite laminates to elastic stress waves is complex because of inherent viscoelasticity, anisotropy, and heterogeneity. The particle displacement cannot be indentified as longitudinal or transverse except when the direction of propagation corresponds to a material symmetry axis. The heterogeneity results in higher dispersion when the wave length is comparable to the characteristics dimension of the internal structure of the composite [1-3].

For longitudinal and shear wave propagation [4-7] there exist both dispersion and attenuation in composite laminates. Further, dispersion and attenuation increase with frequency. The main contribution to this phenomenon is the viscoelastic nature of the composite material. In dispersive and attenuating media the wave will continuously change its shape and frequency content during propagation [5]. Pulse distortion in longitudinal wave propagation decreases its amplitude and the frequency of the dominant component [8-9]. Lower frequency will result in longer wave length and lower detectability of a flaw.

In the acousto-ultrasonic technique a wave is injected onto the surface at one location with the help of a longitudinal transducer. The injected wave strikes the free boundary of the plate normally and at an angle. The longitudinally polarized stress wave incident at an angle gives rise to a longitudinal and shear wave on reflection from the boundary. The interaction between the longitudinal and shear waves in relatively thin isotropic and composite plates produces Lamb waves [10-19]. Lamb wave propagation in an aluminium plate revealed that the velocity of the leading edge corresponded to that of the first longitudinal (symmetric) mode. The elastic energy in the plate can be divided into an infinite number of modes. The mode corresponding to the leading edge carries most of the energy [13]. Specimen resonances are also possible, resulting in better response of the longitudinal transducer being found around the resonant frequencies of the specimen [14].

Microscopically composite laminates may be considered as homogeneous orthotropic material. This assumption is valid only if the wavelength is longer than the thickness of the laminate. The low frequency approximation results in almost nondispersive symmetric and dispersve antisymmetric waves. The computed velocities correlate well with experimental results [15, 17-18]. For velocity measurement, only the leading edge of the signal is considered. However, for assessment of damage, it is necessary to characterize the complete signal.

The main objective of the present investigation is to understand through analysis and experiment in an S-2 glass/epoxy cross-ply laminate the evolution of the tone burst signal to be used in acousto-ultrasonic testing.

EXPERIMENTAL PROCEDURES

An S-2 glass/epoxy cross-ply laminate filament-wound at the Lawrence Livermore National Laboratory was used in this investigation. The properties of this material are listed in Table 1. The experimental setup is shown in Figure 1. The laminate was supported on a cellular rubber pad. The emitting transducer was mounted at the center of the laminate with silicone grease under contact pressure of 0.072 MPa. The receiving transducer was positioned at distances of 45, 70 and 95 mm from the emitter, forming different angles with fiber orientation in the outer ply of the plate, Figure 1. Contact pressure of 0.16 MPa was maintained between the receiving transducer and the laminate.

The AES-1L acoustic emission simulator was used to drive the emitting transducer AET FC-500. The excitation amplitude was set at 1.78 volts (85 dB). A wave envelope with a fast rise and decay rate of 999 dB/ms was generated with the simulator, Figure 2. The carrier frequency was adjusted to 430 kHz for maximum reception. The delay time between the input and received signals was measured with a digital wave analyzer Data 6000 to compute the velocities. The velocity was read six times in each external ply fiber orientation. The experiment was repeated with different distances between the transducers. Parameters such as peak amplitude, root mean square (RMS) amplitude, energy content, and Fast Fourier Transform (FFT), were measured for external ply orientations 0, 22.5, 45, 67.5 and 90 degrees and at locations 45, 70 and 95 mm from the emitting transducer.

RESULTS AND DISCUSSION

Wave Propagation in Cross-Ply Laminate

Perturbations created when the transducer is bonded to the upper surface of the laminate travel as plate/Lamb waves [11,15-19]. The velocity of a such a wave in the low frequency range is approximated by [3]:

$$V_s = [(C_{11} - C_{13}^2/C_{33})/p]^{1/2} \; [1-(C_{13}wh)^2/(24yC_{33}^2)] \qquad (1)$$

where

V_s = Symmetric mode plate/Lamb wave velocity

C_{ij} = Stiffness tensor

p = Density

h = Thickness of the laminate

$y = (C_{11} - C_{13}^2/C_{33})/p$

w = Circular frequency

Effective elastic constants were used to calculate the velocities in different external ply orientations. These in plane and out-of-plane moduli were determined on the assumption that the out-of-plane stresses and in-plane strains remain constant through the laminate thickness. The details of this formulation and evaluation can be found in Reference [19].

Figure 3 shows the variation of wave velocity with frequency in the S-2 glass/epoxy cross-ply laminate. The change in velocity in cross-ply laminate is 14% as compared to 12% in the unidirectional laminate for the frequency range of 0 to 1.5 MHz, indicating that the cross-ply laminate is more dispersive. The measured values for the symmetric mode correlate well with predicted values for the cross-ply laminate. The change in the distance between the transducers from 65 mm to 100 mm kept the velocities unaltered, indicating that the type of wave did not change with distance. Velocities were found equal on the lower and upper surfaces of the specimen. The predicted and measured values indicate that the dispersion in the low frequency range was small.

The wave velocities in different directions are shown in Figure 4. As the angle increased from the external ply fiber direction the velocity decreased. The measured velocities correlate well with those computed. A minimum value of 3.34 mm/usec was obtained at 45 degrees. At this fiber orientation all the plies in the cross-ply laminate are oriented at + 45 degrees. At 0 and 90 degrees the predicted velocities are equal. However, a velocity of 3.80 mm/usec was measured at 0 degree compared to 3.70 mm/usec at 90 degrees. The difference between these two wave velocities indicates that external ply fiber orientation in contact with the transducers influences the wave velocity. A 0-degree outer ply allows the wave to travel faster, acting as a wave guide. The 90-degree outer ply constrains the wave propagation. These experimental values illustrate the effect of stacking sequence. However, difference in measured values is not significant, and the theoretically computed values can be used in both situations.

Evolution of Wave Envelope

Figure 5 shows the signal envelopes obtained at distances 45, 70, and 95 mm from the emitter in the external ply directions. These envelopes are different from the input signal, Figure 2. The modification of the wave envelope continues as it propagates. The peak amplitude of the wave decreased from 5.07 to 4.07 volts as the distance between the transducers increased from 45 mm t0 70 mm. The peak amplitude alone cannot describe the wave envelope which carries the imprints of the state of the material. The RMS value represents all amplitudes in a wave envelope and thus can be considered an appropriate measure. The measured RMS value decreased from 2.035 to 1.105 volts over a distance of 25 mm. The amount by which the RMS value decreased was less than the peak amplitude. Another important parameter that represents the complete wave envelope is energy. It is defined as:

$$E = \int_0^T V^2 dt \qquad (2)$$

where
V = Amplitude in mV

T = Duration of signal

This measure of the wave envelope takes into consideration not only the amplitudes in the wave envelope but also the duration of the signal. The energy measured in the received signal for the duration of 102 µsec showed a rapid decrease in magnitude as the wave propagated. An attenuation based on the energy can be defined as :

$$r = \frac{\ln (E_1/E_2)}{2(X_2-X_1)} \quad (3)$$

where
r = Attenuation coefficient
E_1 = Energy at distance X_1
E_2 = Energy at distance X_2

In the derivation of expression (3) it was assumed that amplitude V varies as $V_0 \exp(-rx)\cos wt$. Similarly an attenuation coefficient on the basis of the peak amplitude of the signal is given by:

$$r = \frac{\ln (V_1/V_2)}{X_2-X_1} \quad (4)$$

where
r = Attenuation coefficient
V_1 = Peak amplitude at distance X_1
V_2 = Peak amplitude at distance X_2

Attenuation coefficients based on energy measurements were lower than those based on peak amplitude except at 45 degrees, Figure 6. The attenuation coefficient increased with an increase in fiber orientation and was maximum for 45 degrees fiber orientation. It is known that the attenuation coefficient for a plate wave in the low frequency region is dominated by the damping present in the laminate [17,20]. Maximum damping at 45-degree fiber orientation for glass reinforced plastics was also observed by Schultz and Tsai [21].

The variation of peak amplitude, RMS, and energy with the fiber orientation was almost parabolic. It can be represented as:

$$Y = a+b\theta+c\theta^2 \quad (5)$$

where
Y = Peak amplitude/ RMS amplitude /Energy in wave envelope
θ = External ply fiber orientation

a, b, c = Constants

The values of a, b, and c were evaluated with regression analysis for peak amplitude, RMS amplitude and energy. From Figures 7-9, it can be inferred that a parabolic fit to experimental data can predict the behavior of these signal parameters for any other orientations.

The wave envelopes received at location 70 mm from the emitting transducer for external ply fiber orientations 0, 22.5, 45, 67.5 and 90 degrees are shown in Figure 10. The wave envelope for the 0 degree orientation had 5.07 volt peak amplitude whereas the 90 degree orientation had only 3.90 volts, Figure 10. In addition, RMS amplitude and energy also showed a marked difference in their magnitudes for 0 and 90 degrees, Figure 8-9. The difference in waveform parameters and shapes of the wave envelope clearly indicates the influence of external layer fiber orientation on wave propagation.

The wave envelopes for 22.5 and 67.5 degrees were almost the same, Figure 10. There was a small difference in peak amplitudes. The RMS amplitude and energy were equal in magnitude, Figure 8-9. The peak amplitude, RMS amplitude and energy for the 45-degree were lower than any other orientations indicating that the maximum damping occurs at 45 degrees, Figure 8-9.

Identification of Reflections with Time Domain Analysis

A close examination of the received waves in Figures 5 and 10 revealed that after the first peak, which corresponds to the peak in the input wave, there is a second peak of smaller amplitude. This second peak corresponds to the reflection of the wave from the boundary of the laminate. As the location of the transducer from the center of the plate was changed from 45 mm to 95 mm, the first and second peaks merged. The observed delay time between the two peaks and the one predicted based on the distance between the transducer and the boundary of the laminate are in close agreement, Table 3.

One consequence of the reflection is to increase the ringdowns, Figure 11. For 22.5-and 67.5-degree oreintations higher ringdown counts were registered at location 70 mm than at 45 mm. The wave envelope at 70 mm location contained a reflected wave of substantial magnitude as compared to 45mm. However, lower ringdown counts were registered at 95 mm due to spatial attenuation of the wave. Almost equal ringdown counts were obtained at 45 mm and 70 mm for 0-and 90-degree orientations. At 45 degrees, higher ringdown counts were noted at 45 mm than at 70 mm. The higher damping at 45 degrees attenuates the wave sufficiently, resulting in a smaller reflected signal. Therefore, the ringdown counts were not affected. These reflections may lead to difficulties in characterizing damage in the material when ringdown counts are used in acousto-ultrasonic testing. Another method is to identify the reflections of the signal through a spectrum analysis.

Identification of Reflections with Spectrum Analysis

Simpson [22] has explained the method to obtain the delay time between the signal and the reflection of the signal with spectrum analysis. This approach is valid for longitudinal wave propagation. The velocity of the plate wave is not very much different from the longitudinal wave in the low frequency region. Hence, spectral analysis can be applied to indentify the reflections. The signal and the reflected signal are separated in time by $2t_0$, and are represented by $f(t-t_0)$ and $f(t+t_0)$. A Fourier spectrum of such signals will be

$$|F[f(t+t_0)+f(t-t_0)| = \exp(-jwt_0)F(w)+\exp(jwt_0)F(w) \quad (6)$$
$$= |\ 2\cos wt_0||F(iw)|$$

This spectrum has the shape of any one of the signals but its magnitude is changed by factor $2\cos wt_0$. The maximum will occur when $wt_0 = n\pi$ or

$f_n = n/2t_0 = n/\Delta t$

where

f_n = Frequency at maximum

t = Delay time between pulses

n = 0, 1, 2

This condition is valid for the zero slope of the spectrum. The separation between the maxima is therefore:

$$\Delta f = 1/2t_0 = 1/\Delta t \quad (7)$$

This equation was used to determine the delay time between two signals.

The input signal transmitted with the broadband transducer contains no reflection and its Fourier spectrum has only one dominant component, Figure 2. The spectrum of the received signal was for the S-2 Glass/epoxy cross-ply laminate, Figure 12. The spectrum obtained contains frequency components below 500 kHz. This was done to take advantage of the sensitivity of the transducer.

The spectral analysis of the signal showed that the dominant frequency component was near the input signal frequency. Components at other frequencies were due to reflections of the signal from the boundary. The multiple peaks in the spectrum correspond to the reflection from different boundaries of the laminate. The magnitude of the reflected signal depends on the distance from the boundary and the attenuation present in the material. From the spectrum, f corresponding to the first and second maximum amplitude was

obtained. Equation (7) was used to determine the delay time. Knowing the delay time and the velocity in the various materials, the path length traveled by the signal was evaluated. The location of the source of reflection can be determined from the path length. The sources of reflection were the boundaries of the plates, Table 4. The experimentally obtained distances correlated well with the actual distances of the boundaries from the transducer location.
The spectrum analysis method is more useful when the reflected signal does not show predominant peaks in the time domain. A signal overlapped in the time domain can be separated. The equation for f is valid when the slope of the envelope of spectrum (F′(w)) is zero. When the slope of the spectrum is not zero there is variation in f. That introduces a slight error in the source location of reflection. For the Fourier spectrum obtained, the maxima were not well separated, Figure 12. The minimum between the peaks is not zero. This behavior is attributed to the attenuation present in the laminate.

The problem of reflection from the boundaries of the laminate in acousto-ultrasonic testing can be avoided by using a large laminate. However, this may not be possible in all situations, especially in the laboratory. The received signal was displayed on the signal analyzer. The reflections were identified in the time and frequency domain. Then signal analysis was performed on a portion of the received signal. Figures 13-14 show the received signal without reflection and its spectrum obtained with a Data 6000 spectrum analyzer cursor. The spectrum contains only one component corresponding to input signal.

CONCLUSIONS

The plate wave equation used to predict the velocity in various fiber directions can be used to predict the velocity in the laminate with different ply orientations. The equation cannot take into account the fiber orientation influence of the external layer.

The peak amplitude of the wave envelope may not reveal all the necessary information concerning the presence of damage. Energy in the wave envelope can give detailed information on the state of the material. RMS amplitude and energy showed a parabolic variation with the external ply fiber orientation studied.

In the acousto-ultrasonic technique, reflections from the boundaries of the laminate were present for the tone burst signal. These reflections can be identified with time domain and spectral analysis. Reflected signals can be eliminated by gating. The signal containing reflections will lead to difficulties in characterizing damage in the material when ringdown counts are used in acousto-ultrasonic testing.

ACKNOWLEDGMENT

This paper is based on the work supported by the Office of Naval Research through Grant No. N00014-87-K-0143 with Dr. Y. Rajpakse as Program Director.

REFERENCES

1. C. T. Sun, J. D. Achenbach and G. Herman, " Time Harmonic Waves in a Stratified Medium Propagating in the Direction of Layering," Journal of Applied Mechanics, Vol. 35, 1968, pp. 408-411.

2. J. D. Achenbach, "Waves and Vibration in Composites," Mechanics of Composite Materials, Vol. 2, ed. by G. P. Sendeckyj, Academic Press, New York, 1974.

3. F. C. Moon, "Wave Propagation and Impact in Composite Materials," Mechanics of Composite Materials, Vol. 7, ed. by C. C. Chamis, New York, 1974.

4. C. Sve, "Time-Harmonic Wave Travelling Obliquely in Periodically Laminated Medium," Journal of Applied Mechanics, Vol. 38, 1971, pp. 477-482.

5. W. Kohn, "Propagation of Low-Frequency Elastic Disturbances in Composite Materials," Journal of Applied Mechanics, Vol. 41, 1974, pp. 97-100.

6. J. C. Peck and G. A. Gurtman, "Dispersive Pulse Propagation Parallel to the Interfaces of a Laminated Composite," Journal of Applied Mechanics, Vol. 36, 1969, pp. 479-484.

7. J. S. Whitter and J. C. Peck, "Experiments on Dispersive Pulse Propagation in a Laminated Composites and Comparision with Theory," Journal of Applied Mechanics, Vol. 36, 1969, pp. 485-490.

8. S. Serabian, "Influence of Attenuation Upon the Frequency Content of Stress Wave Packet," Journal of Acoustical Society of America, Vol. 42, 1967, pp. 1052-1059.

9. S. Serabian, "Implication of the Attenuation-Produced Pulse Distortion Upon the Ultrasonic Method of Nondestructive Testing," Material Evaluation, Sept. 1968, pp. 173-179.

10. H. Lamb, "On the Waves on Elastic Plates," Proceeding of Society of America, Vol. 93, 1917. pp. 114-128.

11. I. A. Viktrov, "Rayleigh and Lamb Wave," Plenum Press, New York, 1967, pp. 67-117.

12. D. C. Worlton, "Experimental Confirmation of Lamb Waves at Mega Cycle Frequencies," Journal of Applied Physics, Vol. 32, No. 6, 1962, pp. 967-997.

13. K. A. Fowler and Papadakis, "Observation and Analysis of Simulated Ultrasonic Acoustic Emission Waves In Plates and Complex Structures," Acoustic Emission, ASTM, STP 505, 1972, pp. 222-237.

14. D. M. Egle and A. E. Brown, "Consideration for Detection of Acoustic Emission Waves in Thin Plates," Journal of the Acoustical Society of America, Vol. 57, No. 3, 1975, pp. 591-597.

15. C. C. Habeger, R. W. Mann and G. A. Baum, "Ultrasonic Plate Waves in Paper," Ultrasonics, March 1979, pp. 57-62.

16. J. H. Hemann and Baaklini, "The Effect of Stress on Ultrasonic Pulses in Fiber Reinforced Composites," NASA CR 3724, August 1983.

17. S. M. Moon, H. T. Hahn and K. L. Jerina, "Acousto-Ultrasonic Wave Propagation in Composite Laminates," First International Symposium on Acousto-Ultrasonics: Theory and Application, July 1987.

18. B. Tang and E. G. Henneke, II, "Low Frequency Flexural Wave Propagation in Laminated Composite Plates," First International Symposium on Acousto-Ultrasonics: Theory and Application, July 1987.

19. S. M. Moon, "Damage Assessment By Acousto-Ultrasonic Technique in Composites," D.Sc. Dissertation, Washington University, St Louis, 1987.

20. A. B. Schultz and S. W. Tsai, "Dynamic Moduli and Damping Ratios in Fiber-Reinforced Composites," Journal of Composite Materials, Vol. 2. No. 3, 1968, pp. 368-379.

21. T. R. Tauchert, "Internal Damping in Fiber-Reinforced Composite Material," Journal of Composite Materials, Vol. 7, 1973, pp. 516-520.

22. W. A. Simpson, "Time-Frequency-Domain Formulation of Ultrasonic Frequency Analysis," Journal of the Acoustical Society of America, Vol. 56, 1974, pp 1776-1781.

TABLE 1. Material Properties

Material	E_1 GPa	E_2 GPa	E_3 GPa	v_{12}	v_{23}	Density gms/cm^3
S-Glass/Ep	58.80	17.50	17.5	0.27	0.38	2.023

TABLE 2. Acoustic Emission Preset Parameters

Preamplifier Gain	60	dB
Amplifier Gain	6	dB

Threshold Voltage .... 1.0 volts (fixed)
Event Duration clock .... 125 ns
Rise Time Clock .... 125 ns
Acceptance Criteria:
-Ringdown Count .... 1<n<4095
-Peak Amplitude .... 40<I<64 dB
-Event Duration .... 1<T<480 µS
-Rise Time .... 1<RT<480 µS
-Slope .... 0<S<65520 dB/uS

TABLE 3. Determination of Source of Reflection with Time Domain Analysis

Material S-2 Gl/Ep Outer Ply Orientation	Velocity (mm/µsec)	Distance From Boundary (mm)	Predicted Delay Between Peaks (µsec)	Observed Delay Between Peaks (µsec)
θ = 0	3.792	77.0	40.61	40.00
	3.792	51.5	27.16	31.25
θ = 22.5	3.539	80.0	45.21	46.28
	3.539	52.0	29.38	27.43
θ = 45	3.348	83.5	49.88	40.00
	3.348	67.0	40.02	30.85
θ = 67.5	3.591	78.0	43.68	37.57
	3.591	55.0	30.80	31.42
θ = 90	3.703	75.0	40.50	37.14
	3.703	50.0	27.00	21.71

TABLE 4. Evaluation of the Source of Reflection with Spectrum Analysis

Material S-2 Gl/Ep Outer Ply	Frequency Separation Between Peaks (kHz)	Delay Time µsec	Experimentally Obtained Distance (mm)	Actual Distance (mm)
θ = 0	29.29	34.13	64.72	51.5
θ = 22.5	39.06	25.60	45.30	52.0
θ = 45.0	19.53	51.20	85.70	67.0

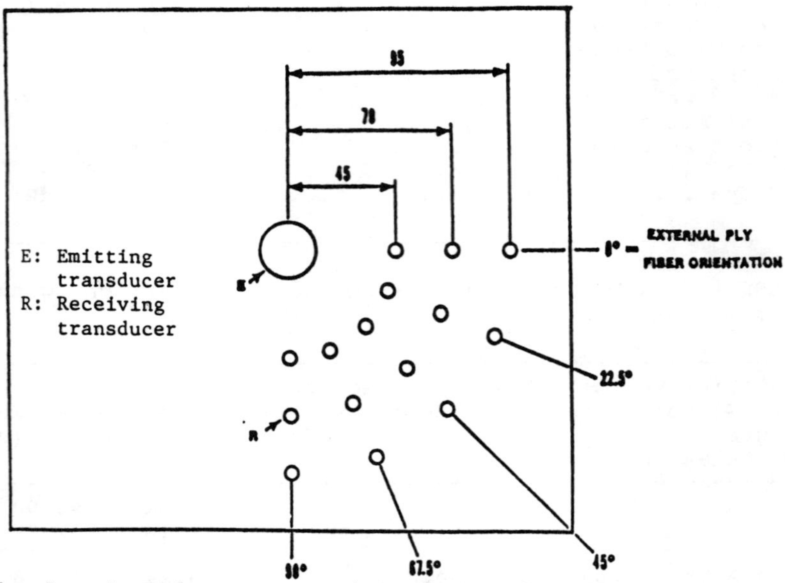

FIGURE 1. Experimental setup to study evolution of wave envelope in S-2 glass/epoxy cross-ply laminate

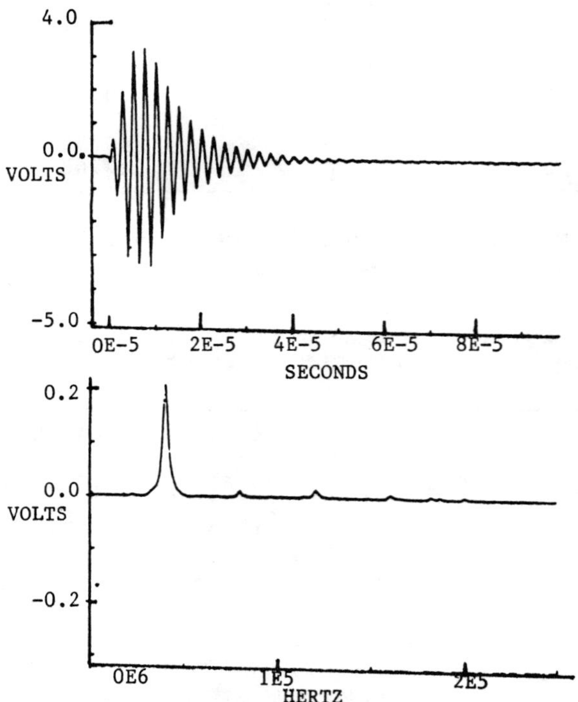

FIGURE 2. Input signal and its Fourier spectrum

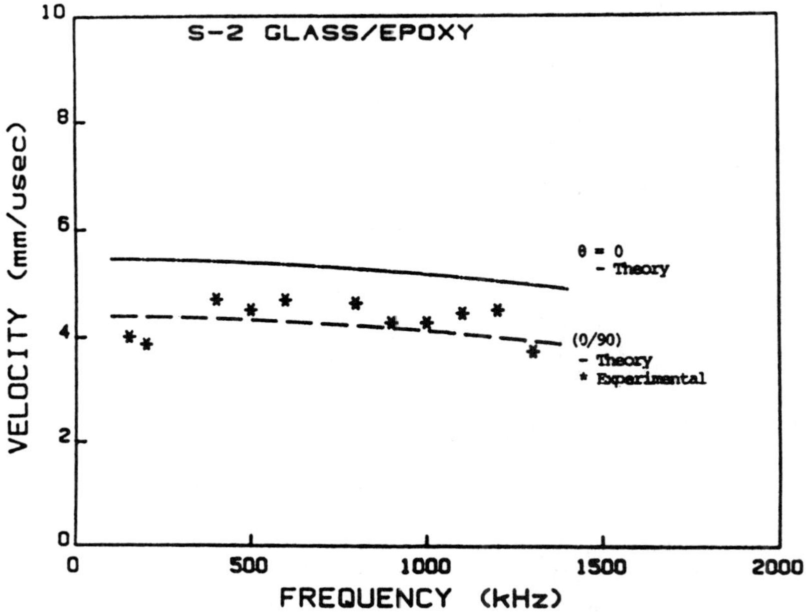

FIGURE 3. Dispersion of symmetric plate wave velocity with frequency for S-2 glass/epoxy

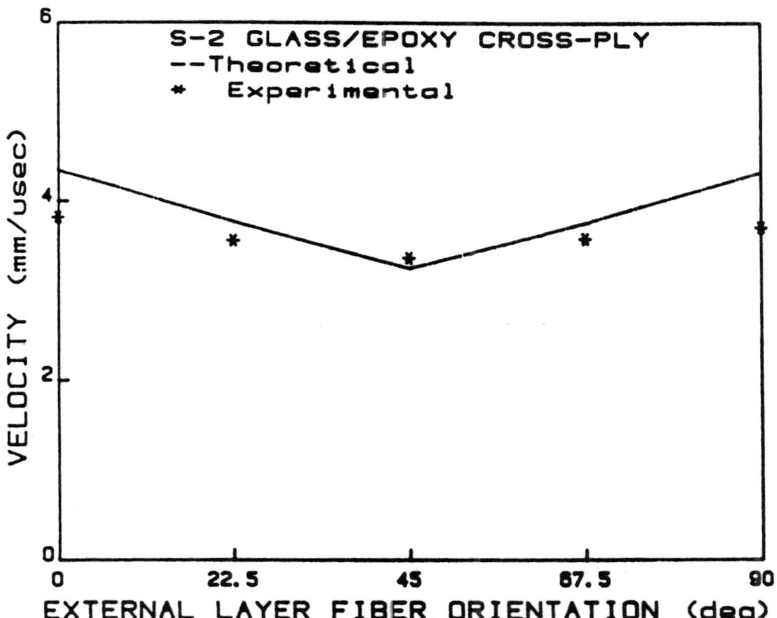

FIGURE 4. Velocity as a function of external layer fiber orientation for S-2 glass/epoxy cross-ply laminate.

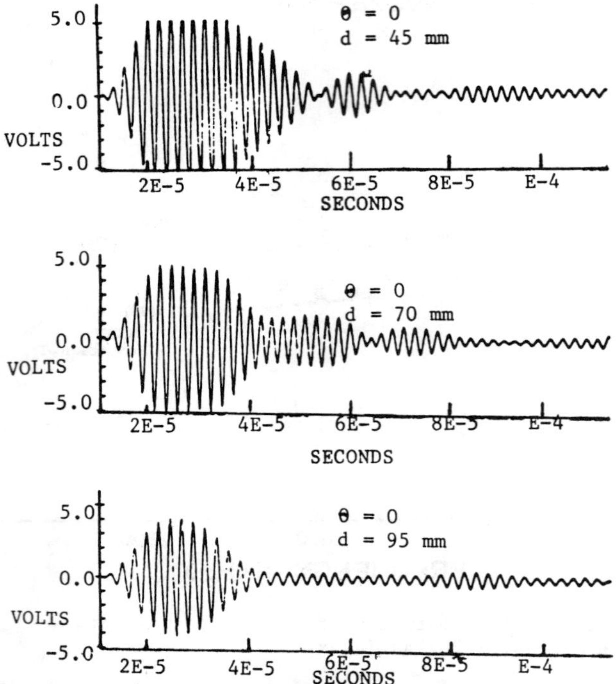

FIGURE 5. Received wave envelope at 0 external layer fiber orientation at distances 45,70 and 95mm. for S-2 glass/epoxy cross-ply laminate

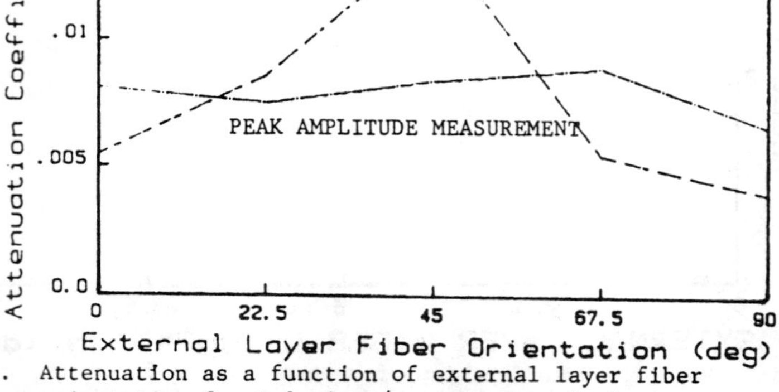

FIGURE 6. Attenuation as a function of external layer fiber orientation for S-2 glass/epoxy cross-ply laminate.

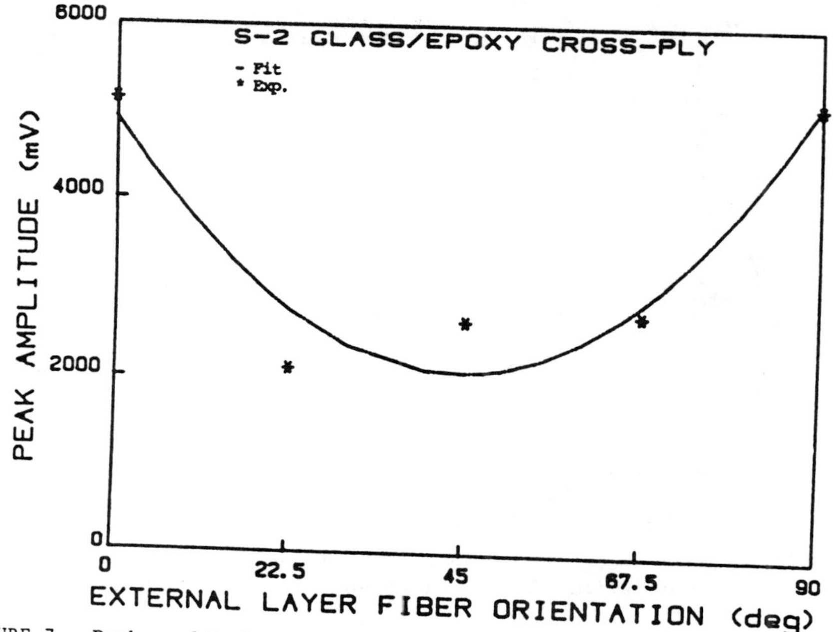

FIGURE 7. Peak amplitude as a function of external layer fiber orientation for S-2 glass/epoxy cross-ply laminate.

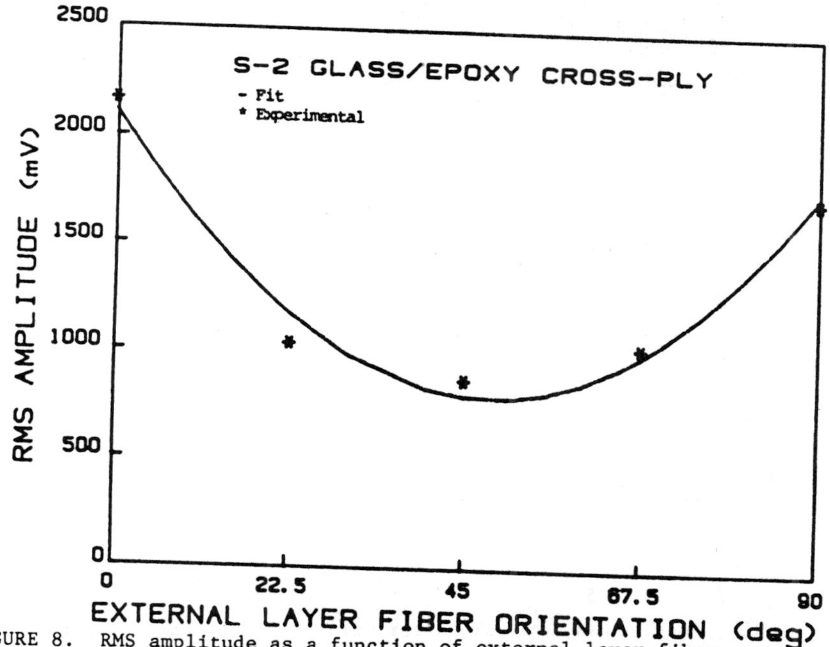

FIGURE 8. RMS amplitude as a function of external layer fiber orientation for S-2 glass/epoxy cross-ply laminate.

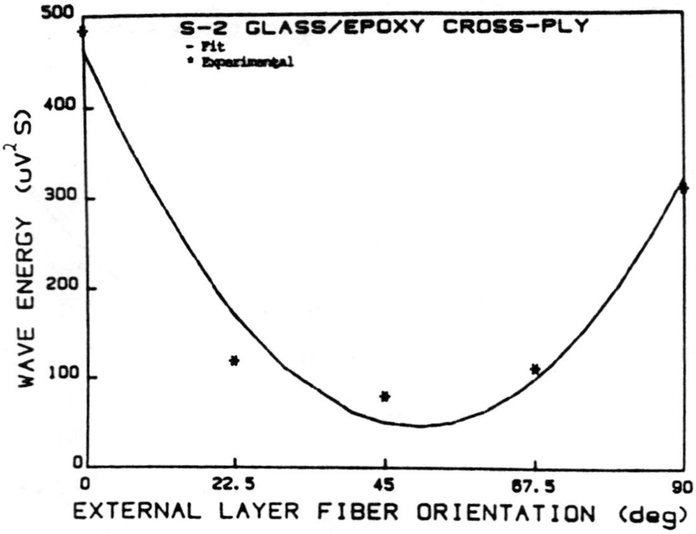

FIGURE 9. Wave energy as a function of external layer fiber orientation for S-2 glass/epoxy cross-ply laminate.

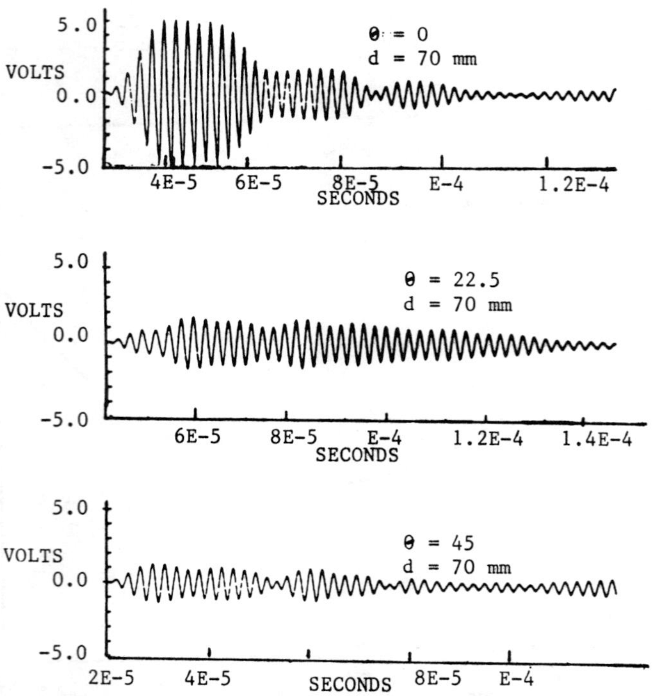

FIGURE 10. Received wave envelope for different external layer fiber orientation for S-2 glass/epoxy cross-ply laminate.

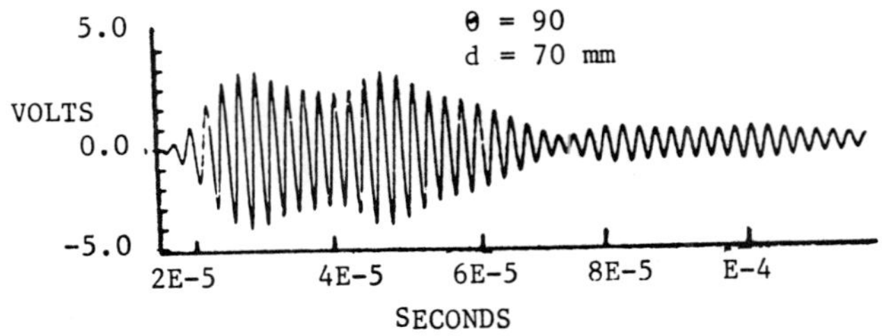

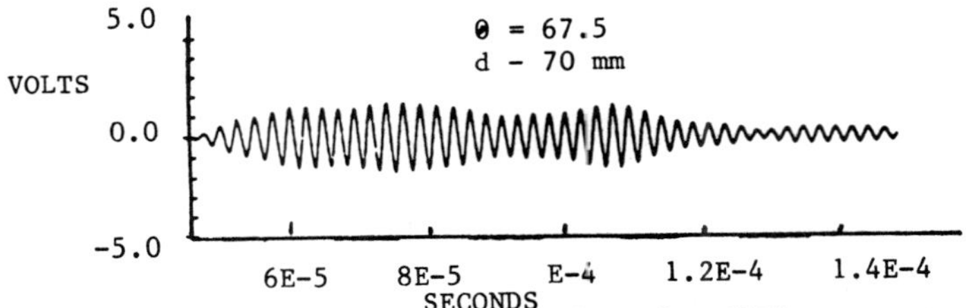

FIGURE 10. (Cont.) Received wave envelope for different external layer fiber orientation for S-2 glass/epoxy cross-ply laminate.

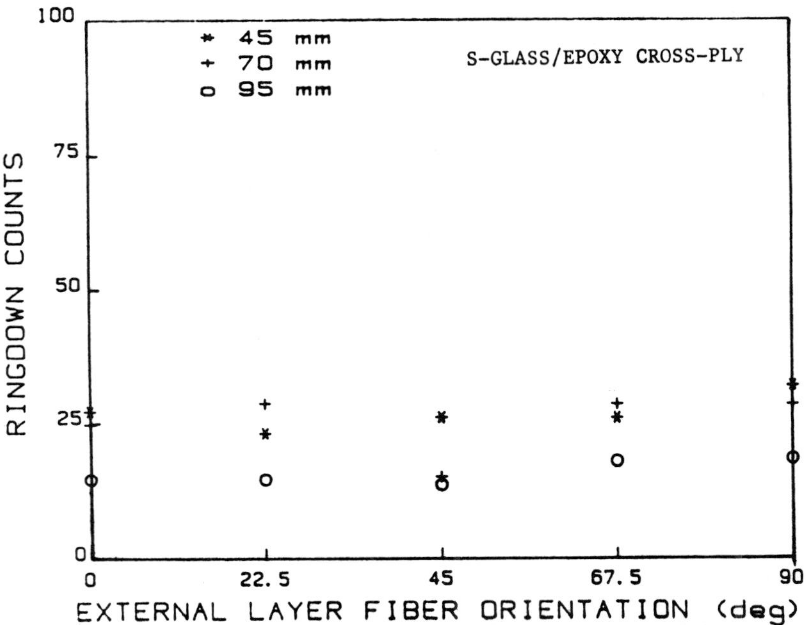

FIGURE 11. Ringdown counts as a function of external layer fiber orientation for S-2 glass/epoxy cross-ply laminate.

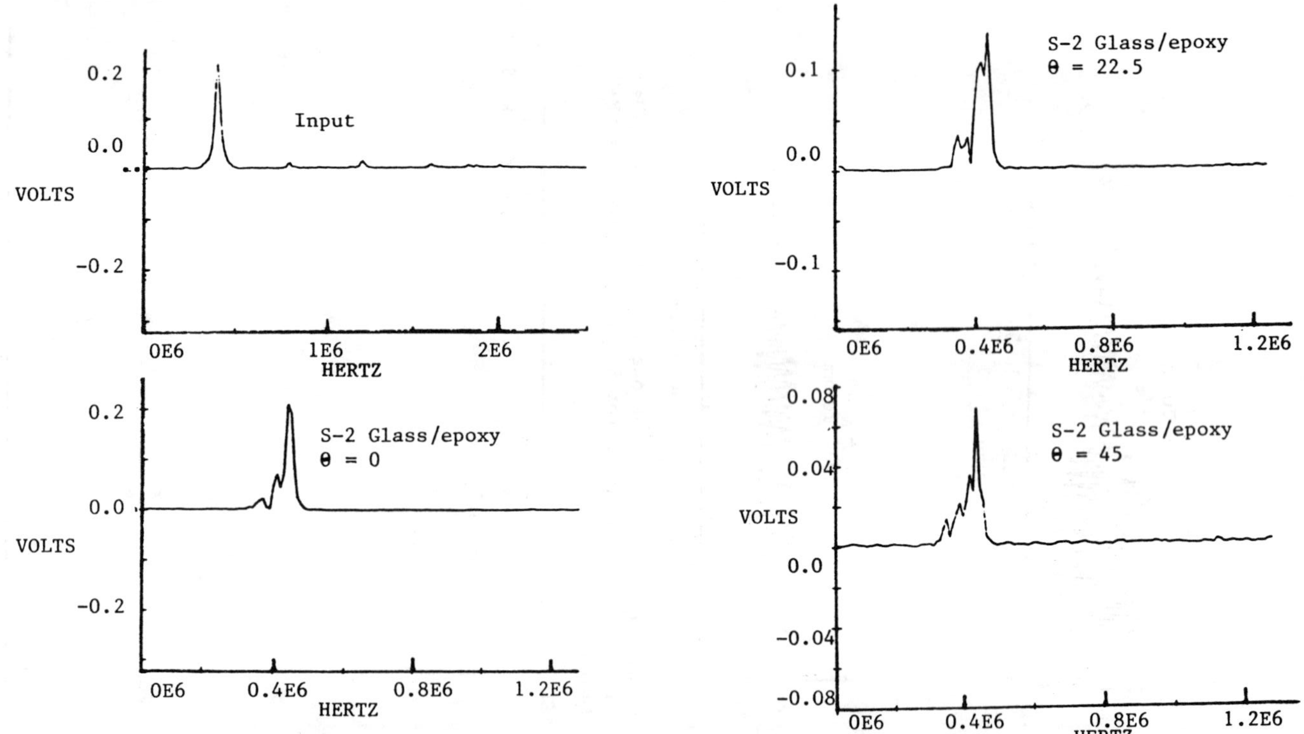

FIGURE 12. Fourier spectrum of received wave for S-2 glass/epoxy cross-ply laminate.

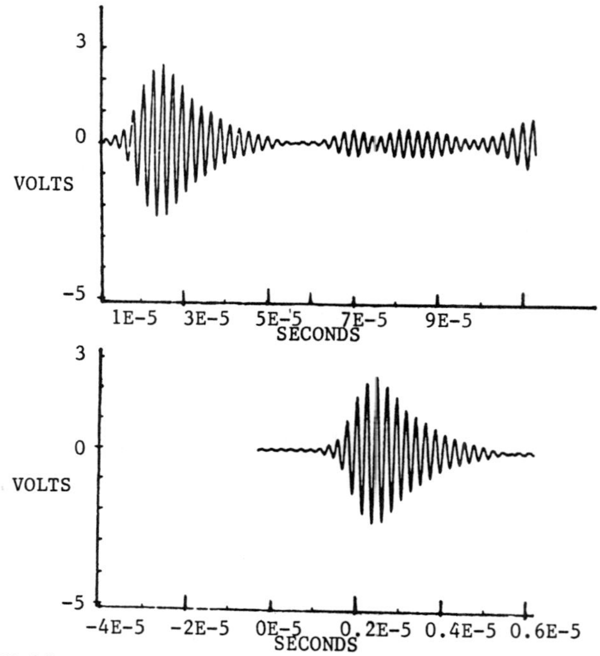

FIGURE 13. Typical received wave with its reflection and without reflection.

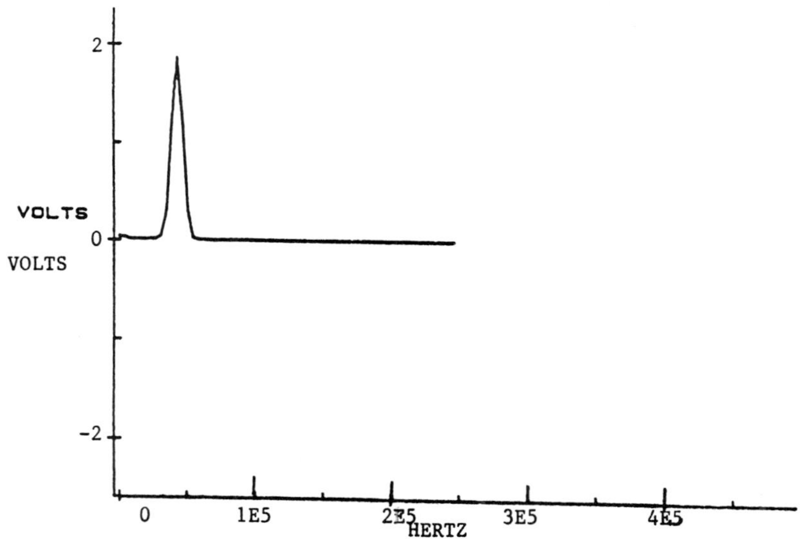

FIGURE 14. Fourier spectrum of received signal without reflection.

STATIC AND FATIGUE FRACTURE MONITORING IN UNIDIRECTIONAL COMPOSITES BY ACOUSTIC EMISSION

L. Lorenzo
Michigan Molecular Institute
Midland, MI 48640

H. T. Hahn
The Pennsylvania State University
University Park, PA 16802

ABSTRACT

Acoustic emission (AE) was used to monitor the failure sequence in model unidirectional composites under static and tension-tension fatigue loading. Specimens consisted of glass and graphite layers of bundles encapsulated in ductile and brittle epoxy resins. Stable matrix microcracking normal to fibers was detected at early stages of deformations and under fatigue loading. Microcracks appeared before fibers started to fail and did not induce fiber breaks. Fatigue failure of both glass and graphite fibers was observed. Acoustic emissions from fiber failures were distinct. Despite the matrix influence on the overall fracture behavior, no clear differentiation between fiber and matrix failure mechanisms was observed. However, when matrix and fiber failures occurred simultaneously peak amplitudes were affected. In fatigue acoustic activity increased between 90 and 98% of the fatigue life giving a warning of impending failure, this being more marked in ductile-resin specimens. Amplitude distribution descriptors changed monotonically during cycling due to the accumulation of events with a wider range of amplitudes.

INTRODUCTION

Unidirectional composites are the basic building block and main load carrying member in advanced composites. The incorporation of unidirectional laminates in structures can lead to significant improvement in performance and cost when the material is tailored for given loading conditions. However, the application of unidirectional composites requires a good understanding of the individual failure mechanisms activated during exposure to the service environment and of the sequence of failure mode accumulation. Moreover, it is necessary to be able to monitor the fracture mode activation in composites as damage accumulates. Work has been done addressing the subject of using real-time AE monitoring to assess structural degradation in metals and composite laminates [1-3]. However, the field is still far from developed.

Unlike in metals, the fracture process in composites involves many interdependent failure modes that lead to final collapse. Currently, more fundamental studies are being conducted addressing the failure process of simple model composites [4-7]. The main thrust has been to define the failure process recognizing basic individual failure modes and their relationship to constituent materials and to establish the effect that the different fracture mechanisms have on acoustic emission signatures.

The present study addressed the use of AE to monitor the failure sequence in model unidirectional composites under static and tension-tension fatigue loadings. Layers of glass and graphite fiber bundles were used in ductile and brittle epoxy resins to look at the difference in acoustic response induced by material behavior. The fracture process was monitored in-situ by optical microscopy. Acoustic activity was recorded during static and fatigue tests. Event accumulation and peak amplitude distributions were considered as the main AE parameters for correlation with the fracture process.

EXPERIMENTAL PROCEDURE

Materials and Specimens

Ductile Epon 815/Versamid 140 (60/40 mixture by weight) and brittle Epon 828/Epon Z (80/20 mixture by weight) epoxy resins were combined with OCF-433AE-113 E-glass and T300 graphite fibers. The 828/Z formulation is stiffer and has a tensile strength twice as high as the 815/V140. Properties of neat epoxies are given in Reference [8].

The glass bundle contains 204 fibers of 13 μm nominal diameter and has a finish compatible with polyester resins. The graphite bundle has 3000 filaments of 7 μm nominal diameter. The glass bundle has a rectangular cross section approximately 1 mm wide and 0.2 mm thick while the graphite yarn is approximately cylindrical with a diameter of 1.2 mm. Specimens with a layer of glass or graphite bundles embedded in the resin were fabricated. Either 15 glass bundles (E-Gl,15B) or 12 graphite bundles (T300,12B) were arranged in a single layer at the center of the specimen cross section. The layer of bundles was completely encapsulated in resin. The epoxy was cured for 24 hours at room temperature (24°C) and then postcured for 12 hours. Postcuring temperature was 65°C and 100°C for 815/V140 and 828/Z, respectively. Strips containing a layer of bundles were milled down to a dogbone shape. The specimen had a gage length 30 mm long and 12 mm wide with a grip-to-gage width ratio of 3 to prevent failure at the grips.

Test Methods

Static tests were run on an Instron machine while fatigue tests were performed on a 2000-lb (8900 N) MTS hydraulic testing machine. Static tests were run at a crosshead speed of 5 mm/min (initial loading rate = 20 MPa/s). Load-controlled tension-tension fatigue tests were run at a constant average loading rate of 100 MPa/s. The applied load had a sinusoidal waveform. The frequency was adjusted in order to maintain the same average loading rate at different maximum fatigue stresses. All fatigue tests were run at a stress ratio R = 0.1. During testing, ambient temperature was kept constant at 24°C.

Acoustic emissions were monitored by an AET Model 5000 operated in Sensor Mode. The transducers were standard omnidirectional AET AC375L with resonant frequency of 375 kHz and sensitivity better than -70 dB referred to 1V/μbar. A highly viscous silicone grease was used as the coupling agent. The sensor was secured with a specially made clamp that provided a uniform pressure. The detected signals were filtered in a band-pass filter FI-25X with flat frequency response between 250 kHz and 500 kHz and preamplified in an AET Model 160B preamplifier with a total gain of 60 dB and flat frequency response between 1 kHz and 2 MHz. Final amplification was attained in an AET 208 Signal Processing Unit. Table 1 shows the preset parameters used during AE monitoring.

TABLE 1. AE Preset Parameters

Parameter	Static	Fatigue
Preamplifier Gain (dB)	60	60
Amplifier Gain (dB)	30	20
Threshold Voltage (V,Fixed)	1.0	1.0
Event Duration Clock (ns)	125	125
- Longest Event (μs)	480	480
- Dead Time (μs)	32	32
Rise Time Clock (ns)	125	125
Acceptance Criteria:		
- Ringdown Counts	4<N<4095	4<N<4095
- Peak Amplitude (dB)	20<Ip<64	25<Ip<64
- Event Duration (μs)	2<T<480	2<T<480
- Rise Time (μs)	$2<T_r<480$	$2<T_r<480$

RESULTS AND DISCUSSION

Mechanical Behavior

Figure 1 shows stress-strain relations for representative layer composites. Table 2 summarizes failure mechanisms and fracture sequences as a function of the applied strain level. Failure modes are comparable with those reported previously for single layer specimens which had fiber bundles spaced approximately one bundle diameter apart [8]. In the present case, the larger number of fibers in the layers with no spacing resulted in higher stiffness and strength. However, failure strains were fairly independent of bundle spacing.

Both glass and graphite specimens showed matrix microcracks normal to the fibers. Resin microcracks between fibers appeared at strain levels of 1.2% and 0.8% in glass and graphite specimens, respectively, regardless of the resin used. The difference in strain level for microcracking in glass and graphite specimens is mainly the result of a larger tensile residual stress in the matrix between fibers because of the larger mismatch in longitudinal modulus between the resin and the graphite fibers.

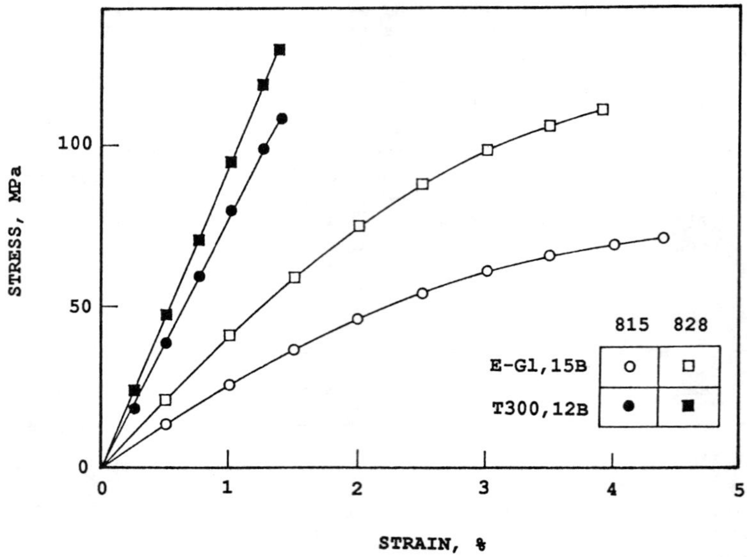

FIGURE 1. Static Stress-Strain Relations

TABLE 2. Static Failure Processes of Single-Layer Composites

Fiber	Strain Level	Failure Mode
E-Glass	below 1.2%	no fiber breaks or matrix damage
	1.2% to 3.6%	matrix microcracks among fibers normal to applied load isolated fiber breaks followed by interfacial failure
	3.6% to fracture	transverse accumulation of fiber breaks leading to layer failure at one location followed by extensive interfacial failure and matrix cracking initiation of unstable matrix crack with layer as fracture source bundle pullout no plastic flow of matrix
T300	below 0.8%	no fiber breaks or matrix damage
	0.8% to 1.2%	matrix microcracks along fibers scattered fiber breaks with no interfacial damage
	1.2% to fracture	transverse accumulation of fiber breaks leading to layer failure at one location initiation of unstable matrix crack with layer as fracture source no plastic deformation of matrix no bundle pullout

Microcracks were normal to the applied load and nucleated in regions of densely packed fibers and did not grow fracturing or bridging surrounding fibers. In graphite specimens, they were preferentially at the fiber/matrix interface. The size of the cracks as observed under a stereo microscope was smaller in graphite specimens mainly due to the smaller diameter of the graphite fibers. The maximum microcrack dimension is of the order of a fiber diameter and, since the crack opening displacement is constrained by the fibers, there is no driving force to propagate microcracks [9]. Moreover, since crack size depends on crack opening displacement, the lower the strain capability of the specimen, the smaller the size of the microcrack. Thus, smaller cracks were expected in graphite specimens, as observed in the present study.

Initial fiber breaks were isolated at low and moderate strain levels. At higher strains, fiber breaks started to accumulate at one cross section and propagated across the layer fracturing the glass and graphite specimens. The failure of the fibers triggered unstable matrix crack propagation in both ductile and brittle resin specimens.

Interfacial failure was extensive in the glass specimens of the ductile resin. Interfacial yielding and matrix cracking propagated along the broken fibers. In the brittle resin, however, interfacial damage was not as extensive although some debonding was detected. In polished cross sections of 828/E-Gl,15B specimens, longitudinal splitting of the bundles was clearly identified. Yet, no interfacial damage occurred in graphite specimens no matter whether ductile or brittle resin was used.

Figure 2 shows maximum cyclic strain-life relations for layer specimens, where straight line fits are also included. Three regions are indicated in the low, medium and high cycle regimes to correlate with the observed fracture processes and AE outputs.

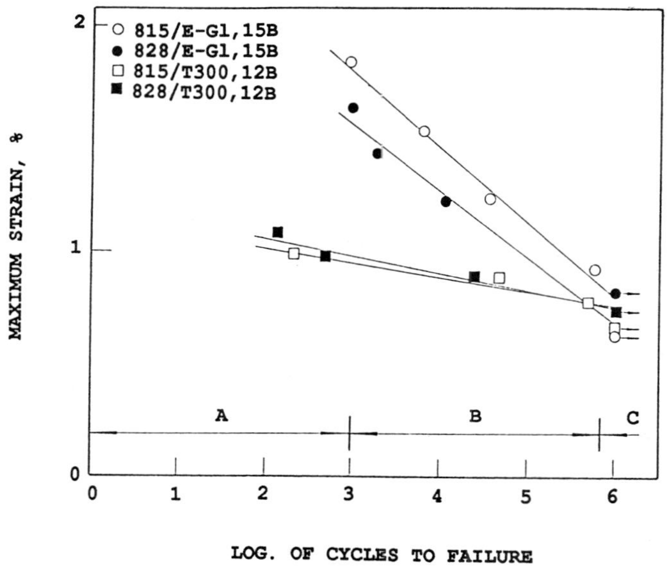

FIGURE 2. Maximum Cyclic Strain-Life Relations

Fracture of both glass and graphite specimens was the result of fiber breaks accumulating in a transverse plane along the layer. Inspection of failed cross sections showed a region of unstable crack growth in the matrix extending away from the failed layer. Creep strains were monitored during cycling of each specimen. Typically, maximum permanent strains were low enough, i.e., below 0.8% and 0.2% for glass and graphite composites, respectively, not to allow static overload of the fibers.

Fiber breaks were detected along the glass and graphite bundles in ductile- and brittle-resin specimens. However, fiber breaks were not evenly distributed along the bundle. For a given reinforcement, ductile-resin specimens presented more fiber breaks than brittle-resin specimens cycled at the same initial strain level.

In 815/E-Gl,15B specimens tested at 1.528 and 1.833% maximum strains, isolated fiber breaks appeared during the first load cycle. Upon further cycling, more fibers failed randomly along the bundle and fiber breaks started to accumulate in zones. At 0.917 and 1.222% maximum strains, however, no transverse accumulation was observed except at the fracture section. Failure of glass fibers was followed by extensive interfacial yielding extending from the fiber breaks that grew along the fibers during cycling.

The fracture sequences in 828/E-Gl,15B specimens were similar to those in 815/E-Gl,15B specimens.

Brittle-resin specimens containing glass fibers cycled at medium and high stresses showed interfacial damage in the form of debonding extending from the fiber breaks. This was more extensive at higher stresses. However, little growth of debonding occurred during cycling.

Transverse matrix microcracks were observed between glass fibers in ductile and brittle resin specimens at all strain levels. After the first load cycle, the extent of microcracking was proportional to the maximum strain level. However, in fatigue microcracking was more extensive in the high-cycle region than in the low-cycle region. A few microcracks were detected to turn and start to grow by Mode II in the matrix along the fiber.

Graphite specimens also presented fiber breaks. As it was the case in glass specimens, graphite bundles in the ductile resin showed more fiber failures than bundles in brittle resin specimens cycled at the same strain level. At high strain levels, transverse accumulation of breaks occurred rapidly and the specimen fractured in a few loading cycles. Brittle resin specimens showed only isolated fiber breaks without large zones of accumulation at medium strain.

No interfacial failure was detected in graphite specimens. Rather, at larger zones of fiber failures, matrix cracks originated mainly due to high shear stresses. At points where only a few fiber breaks were close together, no matrix cracking was observed.

The fracture behaviors described previously are depicted in Figures 3 and 4 for glass and graphite composites, respectively.

In Region A with applied strains larger than 1.5% and 0.9% for glass and graphite specimens, respectively, fatigue failure resembles static failure. Fiber breaks and matrix cracks appear during the first loading cycle. Thereafter, fiber failures accumulate very rapidly leading to specimen fracture.

In Region B the failure process is progressive. A few fiber breaks and matrix cracks appear initially. Upon further cycling damage accumulates slowly: more microcracks, isolated fiber failures, and zones of transverse accumulation of fiber breaks across the bundles appear. Also, interfacial failure grows during cycling from the zones of fiber break accumulation. The amount of damage suffered by the fibers depends on the applied strain: the higher the maximum deformation, the more matrix cracks and fiber breaks. Brittle-resin specimens show fewer isolated fiber breaks and fewer zones of fiber failure accumulation. When in the brittle resin, the glass bundles show less interfacial degradation and the graphite bundles show no matrix cracking. Thus, the stress concentration in brittle-resin specimens is more effective in breaking neighboring fibers.

Region C corresponds to run-outs. Damage progression during cycling is very slow; only a few scattered fiber failures and matrix cracks appear.

In all three regions, matrix microcracks appear very early and accumulate throughout fatigue life. The same is true in static tests: matrix cracks usually appear much earlier before fiber breaks start to

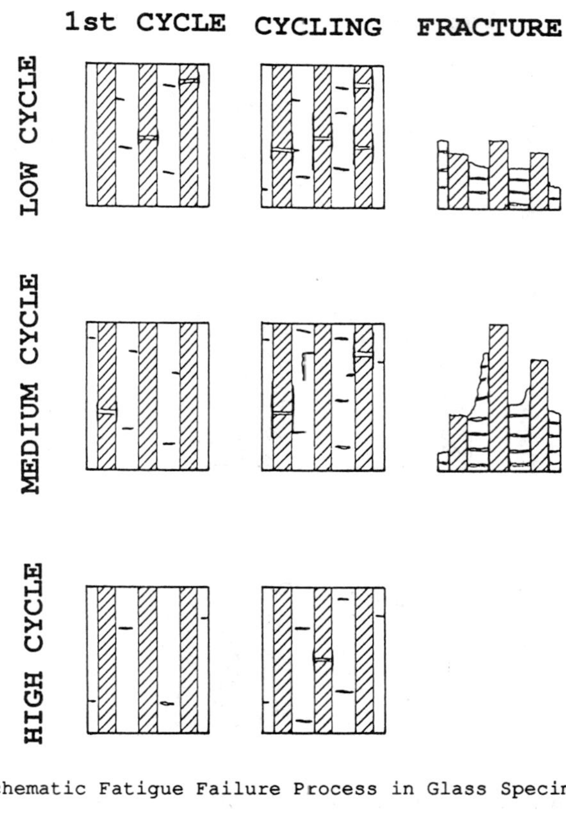

FIGURE 3. Schematic Fatigue Failure Process in Glass Specimens

accumulate. The actual interaction between matrix cracks and fibers is rather complex and not fully known yet.

After nucleation, matrix cracks grow only in the resin-rich zones between the fibers and remain dormant overstressing neighboring fibers. Neighboring fibers do not fail immediately for the following reasons. First, fibers can support high local overstress over a short length. For a fiber having a probability of failure represented by a two-parameter Weibull distribution [10], the ratio of fiber stresses for two different gage lengths L_1 and L_2 at the same probability of failure is

$$\frac{\sigma_2}{\sigma_1} = \left(\frac{L_1}{L_2} \right)^{\frac{1}{\alpha}} \tag{1}$$

Using a shape parameter α equal to 10, it is seen that the fiber can support one and a half times as much load over a length L_2 one hundredth of the length L_1. Therefore, a fiber will fail only when the local fiber stress exceeds the local fiber strength. Second, crack tip blunting may occur during cycling, especially in ductile resin specimens, redistributing the stress over a larger area. When several fiber breaks accumulate in a region and are linked by interfacial failure, the stress concentration on the remaining unbroken fibers increases substantially. Eventually, the overloading leads to specimen failure.

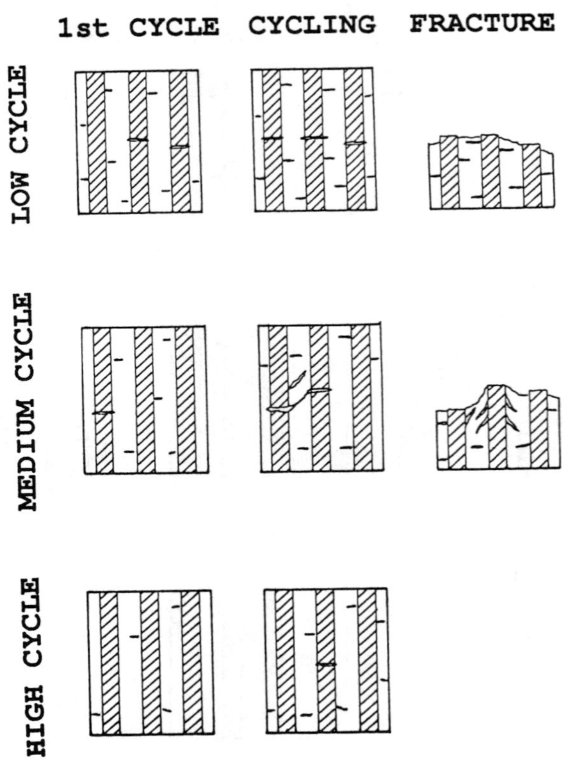

FIGURE 4. Fatigue Failure Process in Graphite Specimens

Acoustic Emission Behavior

Figures 5 and 6 show cumulative number of events as a function of crosshead displacement for glass and graphite specimens, respectively. The acoustic activity begins earlier in ductile-resin specimens regardless of the reinforcement. Sporadic emissions accompanied initial accumulation of subcritical damage. However, as deformation proceeds, emission rates increase up to a maximum value at fracture. This behavior agrees with previous results from specimens containing single fibers and single bundles embedded in similar matrices [4].

For a given displacement, a lower number of events was registered in brittle-resin specimens. Since failure strains were fairly close, less acoustically detectable damage accumulated in brittle-resin specimens. The more numerous emissions of ductile-resin specimens indicate that bundles in this specimens can accommodate a larger number of fiber breaks without triggering fracture probably due to a better relief of stress concentration by the matrix. In a less tough and more notch sensitive matrix such as the 828/Z, a few fiber breaks can easily lead to fracture. This behavior is exactly opposite to that observed when a single fiber or a single fiber bundle was used [4]. In that case, more events were detected at low strains due to the contribution of the more acoustically active brittle resin.

When glass fibers were used in the brittle resin, peak amplitudes were lower (Figure 7). However, peak amplitudes remained independent of the matrix type in graphite specimens (Figure 8). Average peak amplitudes were 41 and 37 dB for 815/ and 828/E-Gl,15B specimens while both 815/and 828/T300,12B specimens showed an average of 35 dB. Two reasons may have contributed to this effect. The 828/E-Gl,15B specimens underwent less deformation than the 815/E-Gl,15B specimen prior to fracture producing fewer events. Also, the simultaneous occurrence of fiber failures and debonding in the 828/E-Gl,15B specimen may give events with lower peak amplitude than those from fiber breaks only. Matrix yielding at the interface as observed in 815/E-Gl,15B specimens is likely to produce stress waves of only very low amplitudes that can hardly be detected as reported for plastic flow of pure epoxy specimens [11]. In graphite specimens, failure strains were quite close and no interfacial damage was observed which resulted in similar histograms. It is worth noting that lost events in 815/T300,12B specimen are not included in the histogram.

Figures 9 and 10 show cumulative events-fatigue cycle relations for layer specimens tested at different maximum cyclic strains. In each case, specimens are classified according to the regions in Figure 2. The total number of events during static tests was usually larger than in fatigue. The only exception was a 815/T300,12B specimen cycled at 0.877% maximum strain where there were twice as many cumulative events in fatigue as in static tension. Note that in this case events accumulated more rapidly than in the other specimens in Region B.

For all material combinations, higher event rates were associated with higher cyclic strain. The more extensive and rapid accumulation of damage in specimens in Region A resulted in more emissions regardless of material combination.

In 815/E-Gl,15B specimens the event rate increased rapidly close to final fracture. The sudden increase in emission rate immediately before fracture is more pronounced in the ductile resin than in the

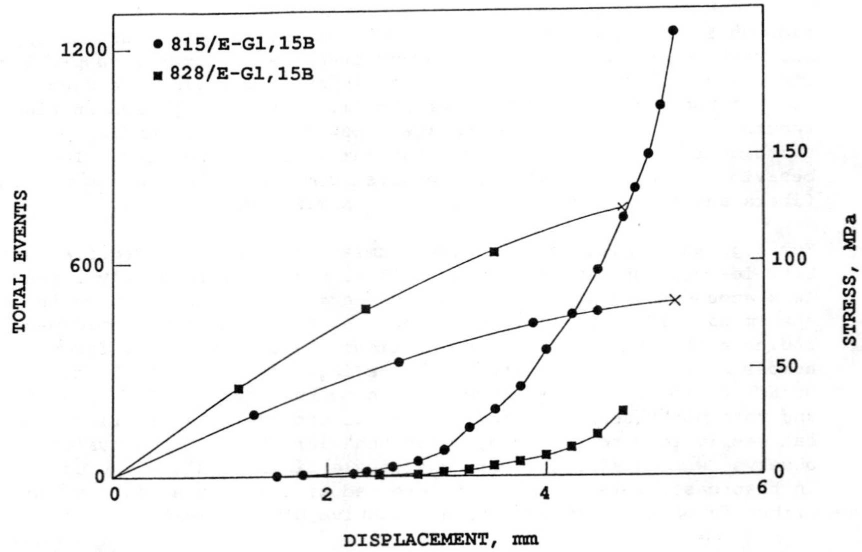

FIGURE 5. Cumulative Events vs. Displacement in Glass Specimens

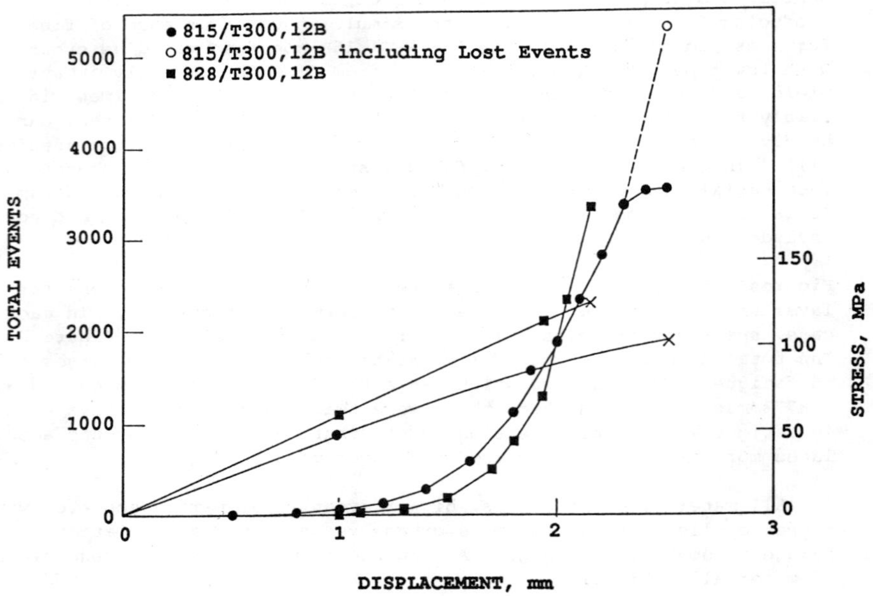

FIGURE 6. Cumulative Events vs. Displacement in Graphite Specimens. Events in 815/T300,12B specimen were lost due to a high emission rate close to fracture.

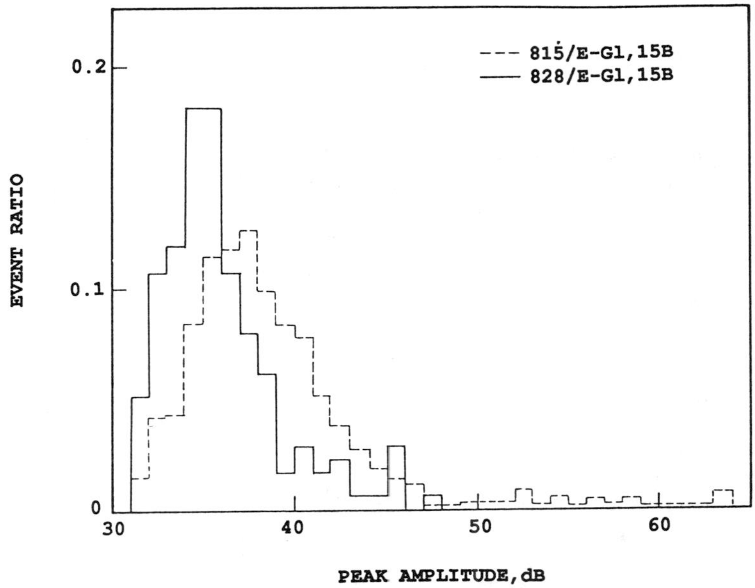

FIGURE 7. Event Histograms in Glass Specimens

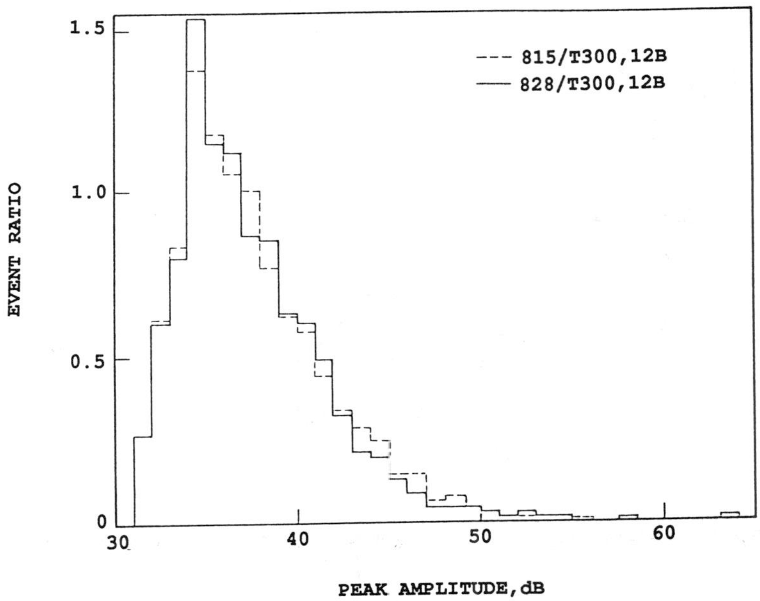

FIGURE 8. Event Histograms in Glass Specimens

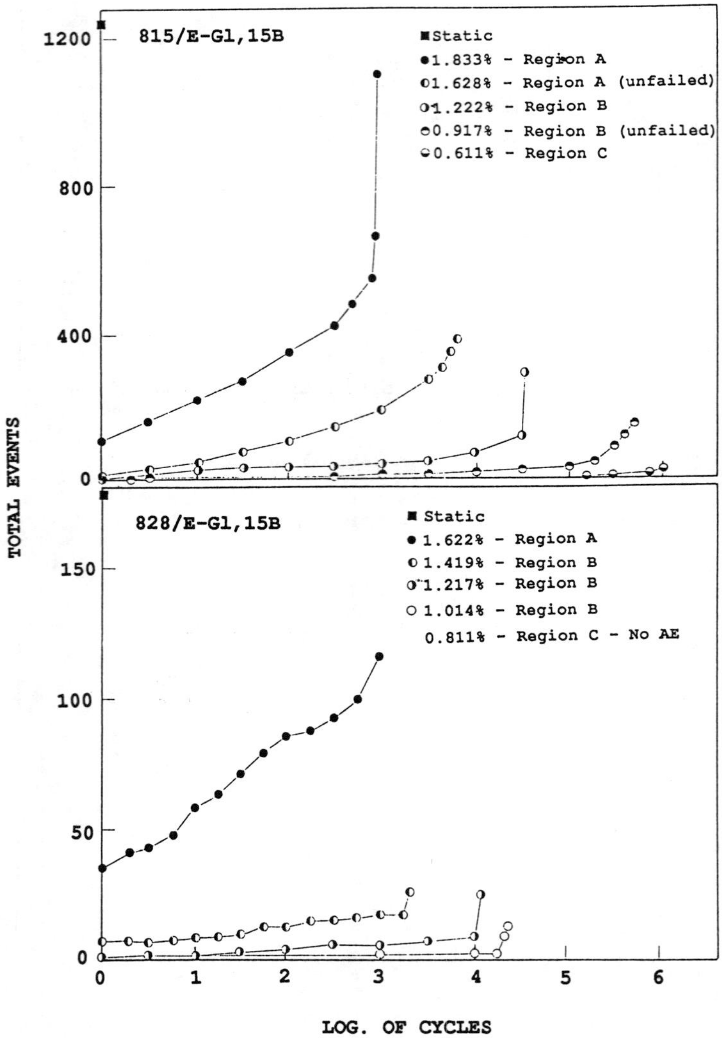

FIGURE 9. Cumulative Events vs. Number of Cycles in Glass Specimens

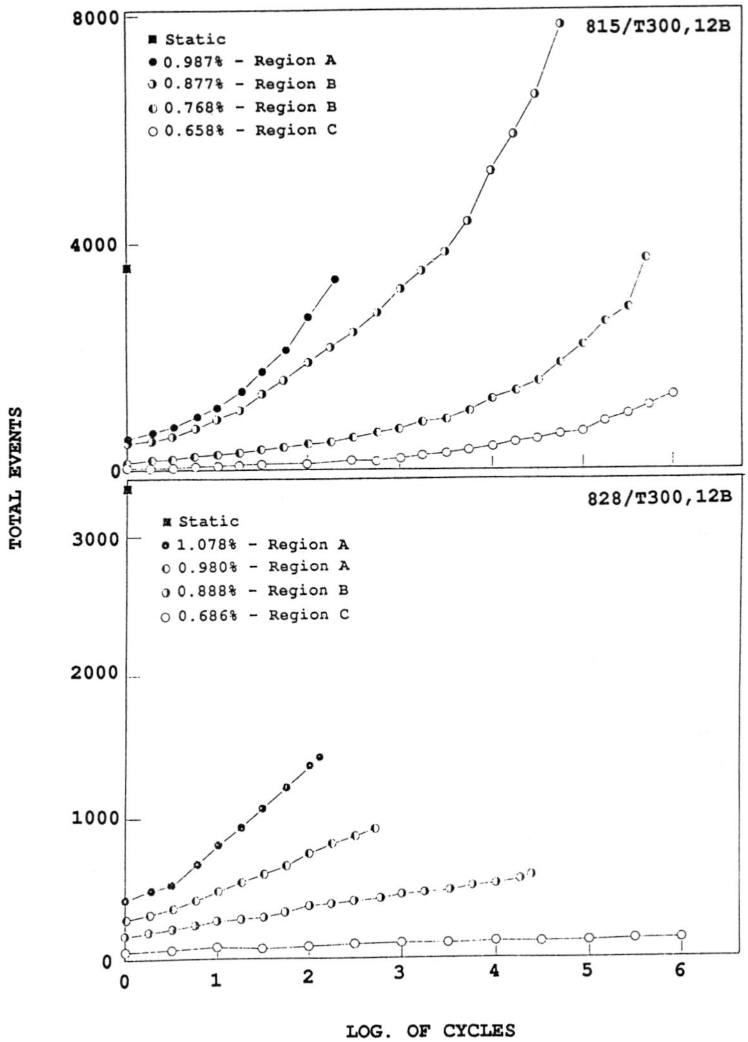

FIGURE 10. Cumulative Events vs. Number of Cycles in Graphite Specimens

brittle resin. In 828/E-Gl,15B specimens the event rate also increased but much more moderately. The stronger AE activity in ductile-resin specimens is due to the larger zones of transverse accumulation of fiber breaks and the growth of matrix damage from zones of fiber break accumulation. The ductile resin creeps more and is tougher than the brittle resin, and hence, allows more damage accumulation before final fracture.

Of the two types of fibers tested the glass fiber is more amenable to the detection of imminent failure. Specimens containing graphite fibers showed no sudden increase in emission rate close to fracture regardless of the resin type. In the extreme case of 828/T300,12B specimens the cumulative event plots are almost linear up to final fracture with no warning of incipient failure. However, the increase in AE activity close to fracture was always very sudden and occurred in the final 2 to 10% of the fatigue life.

The initial damage detected after the first loading cycle correlates properly with AE outputs. Figure 11 shows the number of events recorded during the first loading cycle as a function of the maximum cyclic strain. The number of detected events in glass specimens increases markedly as Region A is approached. The stronger initial acoustic activity in Region A results from the large number of

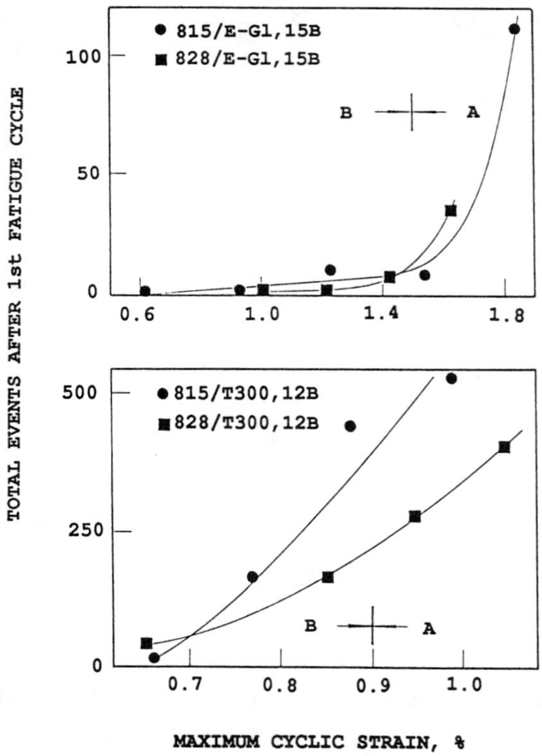

FIGURE 11. AE Events during First Loading Cycle vs. Maximum Cyclic Strain

isolated fiber breaks along the bundles. Note that the initial activity is almost the same in both ductile and brittle-resin specimens. In graphite specimens there is no such sharp increase at Region A. Instead, the increase occurred at maximum cyclic strains in Region B. Moreover, the number of events during the first cycle was larger than in glass specimens. Maximum cyclic strain closer to the static failure strain of graphite fibers and the larger number of fibers in the graphite layer are responsible for this behavior. Having an average failure strain of only 1.34%, the graphite fiber begins to fracture at a strain level as low as 0.60%. For the glass fiber which has a much higher failure strain, however, fiber fracture does not start until the applied strain reaches 1.50%.

Amplitude distributions were monitored at regular intervals during the fatigue life of the specimens. They were observed to change in a consistent pattern. After the first loading cycle event amplitude ranged from 30 to 40 dB (0 dB = 5.54 mV after preamplification). However, the actual sequence of accumulation varied from specimen to specimen within each material group. Higher-amplitude events became more prevalent as final fracture approached. At fracture some events had an amplitude as high as 63 dB.

In the 815/E-Gl,15B run-out specimen, amplitude changed more or less randomly. No events were detected during testing of the 828/E-Gl,15B run-out at a maximum strain of 0.811%. Run-out graphite specimens showed some accumulation of events. However, no emissions were detected between 300000 and 1000000 cycles.

A two-parameter Weibull distribution of the form

$$F(V) = 1 - \exp\left[-\left(\frac{V}{V_o}\right)^a\right] \tag{2}$$

was used to fit peak amplitude distributions. Here, a, V and V_o are the shape parameter, event amplitude and scale parameter, respectively. The shape parameter a determines the shape of the distribution. If $a > 1$, the distribution is unimodal. Peak amplitude distributions from unidirectional composites are typically unimodal [6]. When $a \leq 1$, there is no mode in the distribution. On the other hand, the scale parameter V_o relates to the absolute values of the detected amplitudes. Statistically, V_o specifies the $100[1 - \exp(-1)]$th distribution percentile of V. The shape and scale parameters are related to the distribution mode V_M by

$$V_M = V_o\left(1 - \frac{1}{a}\right)^{\frac{1}{a}} \tag{3}$$

Shape and scale parameters were estimated using the method of maximum likelihood [12].

The shape and scale parameters for specimens in Region A are shown in Figure 12. Figure 13 shows similar results for specimens in Region B. Both parameters have been normalized with respect to their values at fracture.

The change of amplitude distribution parameters with cycles for the graphite specimens is different from that for the glass specimens.

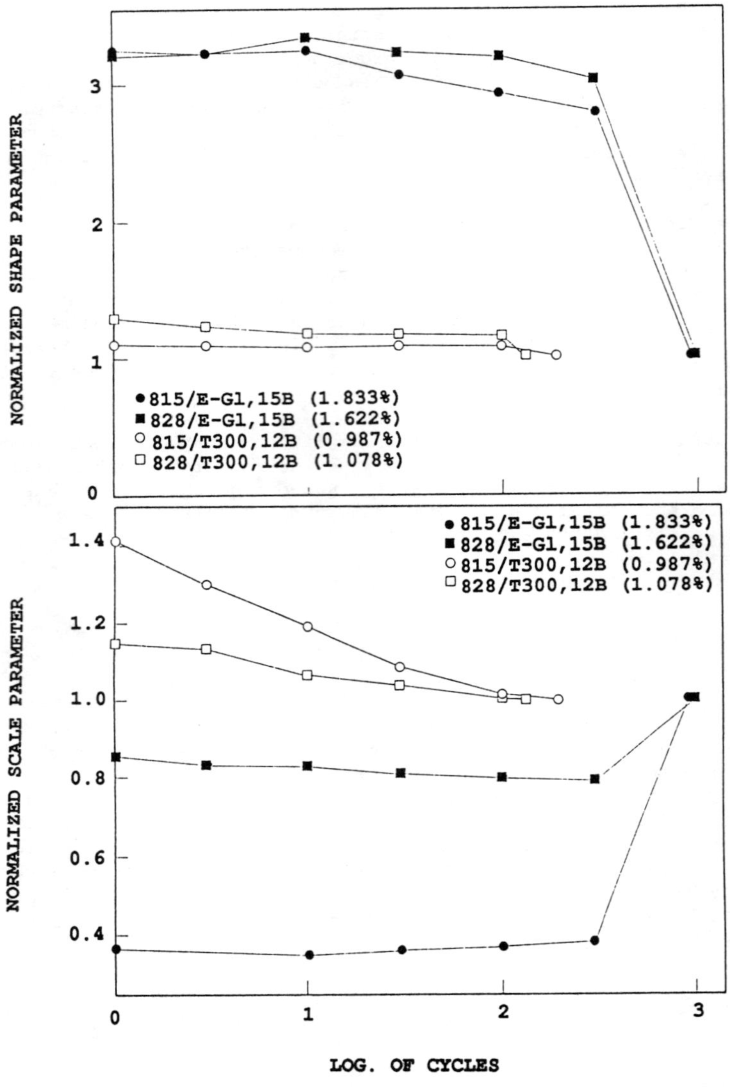

FIGURE 12. Shape and Scale Parameter Estimators vs. Number of Cycles in Specimens in Region A

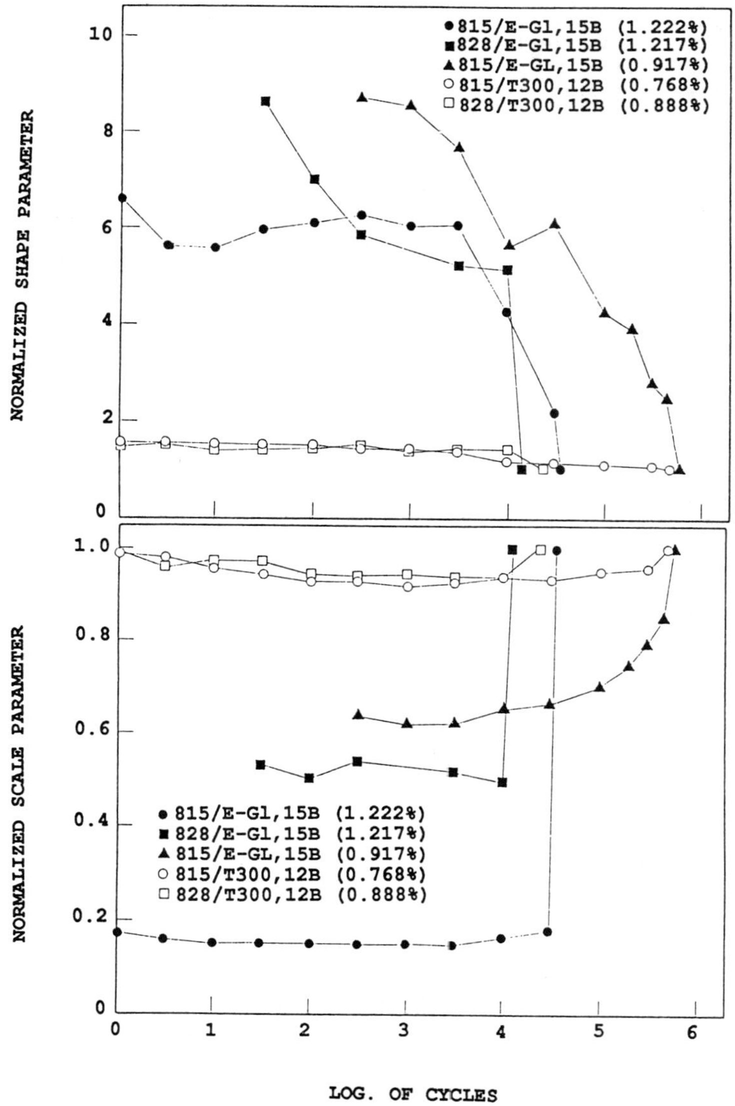

FIGURE 13. Shape and Scale Parameter Estimators vs. Number of Cycles in Specimens in Region B

When the cyclic strain levels are in Region A, both graphite and glass specimens show only a slight decrease in shape parameters until immediately before final fracture. However, final fracture of the glass specimens is preceded by a sudden change in the shape parameter while the graphite specimens show hardly any changes. The scale parameters also show a sudden increase near final fracture for the glass specimens, but they decrease monotonically for the graphite specimens. Thus, while acoustic emissions in the graphite specimens uniformly become weaker with fatigue cycles, they remain fairly constant in strength and scatter in the glass specimens until final fracture.

In Region B despite cyclic strains are lower, the changes in the distribution parameters are essentially of the same type as in Region A. However, the magnitudes of the changes are much higher for the glass specimens. Again, the parameters for the graphite specimens remain fairly constant throughout fatigue. Whether in Region A or B, no correlation could be found between the values of the amplitude distribution parameters and the cyclic strain levels.

A decrease in shape parameter indicates production of emissions over a wider range of amplitudes. Therefore, it may be concluded that with increasing cycles, a wider variety of failure mechanisms become operative.

Glass and graphite run-out specimens had distribution parameters which fluctuated with fatigue cycles. Since damage accumulation is very slow in run-out specimens, it is likely that different failure mechanisms become active one at a time, thereby leading to a random change of the amplitude distribution. In 815/E-Gl,15B specimens, scale parameters were always less than 1, which indicates the random nature of the detected amplitudes and the fact that the distribution departs from the unimodal behavior.

CONCLUSIONS

Damage progression in unidirectional composites was monitored under static and fatigue loadings by acoustic emission. Emissions from fiber failures were distinctively identified. Matrix microcracking could not be clearly distinguished. On the other hand, interfacial related events did not appear differentiated but rather influenced the peak amplitude of the emissions generated by fiber breaks when they were energetic enough. This is usually due to the almost simultaneous occurrence of fiber breaks and associated interfacial damage. In fatigue, acoustic activity increased between 90 and 98 percent of the fatigue life of the specimen, the increase being more differentiated in ductile-resin specimens. Also, ductile-resin specimens showed more accumulation of events as more isolated and clustered fiber breaks accumulated.

Estimators of a two-parameter Weibull distribution used to fit peak amplitude distributions were observed to change smoothly with fatigue life. Shape parameter estimators generally decreased monotonically at high and medium cyclic strains due to the appearance of events with a wider range of amplitudes near fracture. At medium strains the change was more pronounced since initial events had typically lower amplitudes. Moreover, the change appeared more markedly in glass specimens since cyclic strains were much lower than the fiber failure strain.

REFERENCES

1. Williams, R.S., and Reifsnider, K.L., Investigation of Acoustic Emission During Fatigue Loading of Composite Specimens, *Journal of Composite Materials*, vol. 8, pp. 340-355, 1974.

2. Hamstad, M.A., and Chiao, T.T., Acoustic Emission from Stress Rupture and Fatigue of an Organic Fiber Composite, *Composite Reliability*, STP 580, American Society for Testing and Materials, pp. 191-201, 1975.

3. Harris, B., Fatigue and Accumulation of Damage in Reinforced Plastics, *Composites*, vol. 8, pp. 214-220, 1977.

4. Lorenzo, L. and Hahn, H.T., Acoustic Emission Study of Fibers Embedded in Epoxy Resin, *Proc. First Int. Symposium on Acoustic Emission*, SPI, 1983.

5. Hamstad, M.A. and Moore, R.L., Acoustic Emission from Single and Multiple Kevlar™ 49 Filament Breaks, *Journal of Composite Materials*, vol. 20, pp. 46-66, 1986.

6. Lorenzo, L. and Hahn, H.T., On the Applicability of Amplitude Distribution Analysis to the Fracture Process of Composite Materials, *Journal of Acoustic Emission*, vol. 5, pp. 15-24, 1986.

7. Lorenzo, L. and Hahn, H.T., Evaluation of Fatigue Failure in Unidirectional Composites by Acoustic Emission Technique, *Proc. 6th Int. Symposium on Offshore Mechanics and Arctic Engineering*, ASME, 1987.

8. Lorenzo, L. and Hahn, H.T., Fatigue Failure Mechanisms in Unidirectional Composites, in *Composite Materials: Fatigue and Fracture, ASTM STP 907*, ed. H.T. Hahn, American Society for Testing and Materials, Philadelphia, pp. 210-232, 1986.

9. Hahn, H.T., A Synergistic Effect in Fatigue of Composites, 1983 Winter Annual Meeting, ASME (unpublished).

10. Rosen, B.W., Mechanics of Composite Strengthening, *Fiber Composite Materials*, American Society for Metals, pp. 37-75, 1964.

11. Hollman, K, Acoustic Emission Behavior of Epoxies under Tensile Loading, Dept. of Mech. Engineering, Washington University, M.Sc. Thesis, December, 1983.

12. Mann, N.R., Schafer, R.E. and Singpurwalla, N.Z., Methods for Statistical Analysis of Reliability and Life Data, John Wiley and Sons, New York, 1974.

ACOUSTIC EMISSION TESTING OF COMPOSITE POWER TRANSMISSION COMPONENTS

D. W. Okey
Sundstrand Corporation
Rockford, Il 61125

J. M. Rodgers
Hartford Steam Boiler Inspection Technologies
Sacramento, CA 95815

ABSTRACT

Power transmission shafts, also known as drive shafts or torque tubes, offer an ideal application for composite materials. Strength and stiffness can be tailored to the application, and weight savings of 50 percent over aluminum are possible. For several years the difficulty in application has been the termination of the shaft into appropriate end fittings. The use of typical metallic fittings reduces the weight savings gained by the application of composites. Minimal weight savings will not justify the extra expense incurred in bringing composite parts into present aerospace systems (acceptance tests, reliability assurance, and procedures that assure vendor and customer of the performance of the component under all required conditions, especially fatigue and in a hot/wet environment). These problems are solved using polygonal, molded-in end fittings for composite fiber/resin structures. Acoustic emission was used in conjunction with destructive testing to determine the end fitting design best suited for power transmission. A molded-in hexagonal end fitting provides a better structure and lower stress concentration than either square or triangular fittings.

1. INTRODUCTION

Quality assurance and nondestructive testing are two hurdles that composite materials must pass before their acceptance in flight hardware by the aerospace industry. Procedures and standards will vary at least slightly for each part. Because of the nature of acoustic emission tests, each part should be loaded in the manner in which it will be used in service. For composite torque tubes, the application of the load is particularly simple, and straightforward. The test articles are placed in torsion. In this way, the load is applied, acoustic emission data are recorded, and the tube may be tested nondestructively or destructively depending on the load applied.

Metallic torque tubes typically have riveted and/or splined end fittings to transmit power or torsion from one structure to another. In composites, however, where the goal is frequently to reduce weight, such end fittings can easily weigh more than the composite portion of the structure. The weight advantage gained by using composites may be lost by use of metal fittings. A logical solution is to achieve the transmission by molding an insert into the composite. The insert will have a polygonal exterior for contact with the inside of the composite structure. The interior of the insert may be splined or polygonal as necessary for power transmission and weight savings.

The difficulty in the use of molded-in end fittings is that stress concentration can occur at the interface between the composite and the insert. It is at this interface that the torsion (shear) is transmitted from the shaft to the end fitting. This paper describes the use of acoustic emission in differentiating the reactions of a series of constructions at this critical interface.

2. EXPERIMENTAL PROCEDURES

A series of composite tubes was fabricated with a number of different terminations. End fittings were fabricated with tri-lobal (triangular), quadri-lobal (square), and hexagonal-shaped ends. In actual manufacture, delaminations or disbonds may occur. Therefore, another series of tubes was produced with mold release between plies, to simulate delamination.

The tubes were manufactured with Fiberite HyE-1234A graphite/epoxy composite material, with seven plies of ± 15 degrees along the length of the tube, and two plies of hoop (90 degrees) at each end. The first series was tested with no bond between tube and end fitting. The second series was tested with mold release applied to simulate a failed bond. Properly bonded samples were also made, as were several samples with Mylar "flaws" inserted into the plies to act as delaminations. These flaws function as voids, with no chemical or structural bonding between the plies so affected. The tubes were 78 inches long, with a 1.625 inch diameter (Figure 1).

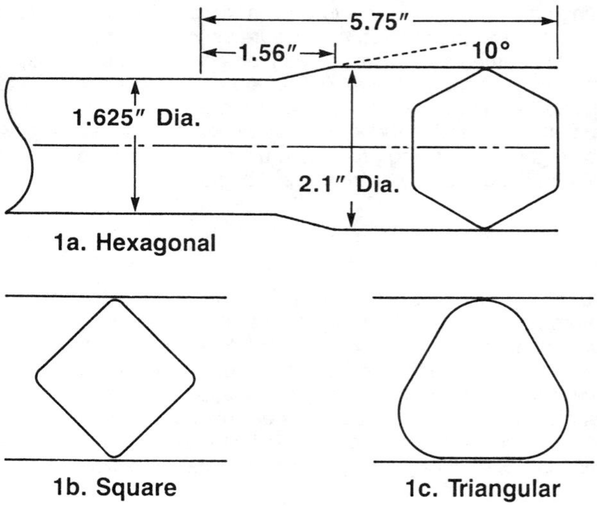

FIGURE 1. Torque tube end fittings

In all of the samples, the presence of flaws and disbonds was expected to increase the acoustic emission and lower the load at which failure would occur. This is because flaws, disbonds and stress concentrations are expected to increase the effective loads on a given area of the structure, leading to earlier failure.

3. EXPERIMENTAL RESULTS

An ECHO GT 16C microprocessor based acoustic emission system was used for the tests. Sensors with resonant sensitivity at 110 kHz were used, with one mounted on each end of the tube, designated Channel 1 and Channel 2. Low threshold events were those above 45 dB, high threshold events were those above 70 dB. Low threshold events are associated with matrix crazing and cracking, and fiber debonding. High threshold events are associated with fiber breakage and higher energy delamination or macro-cracking phenomena. The specimens were loaded gradually with 2-minute holds, at 100 and 200 lb, before being loaded to failure. All testing was accomplished by applying the given load through a 7 inch moment arm.

TABLE 1. Unflawed, Unbonded End Fittings

End Fitting	Cumulative Low Threshold Events 100-lb Load	Cumulative Low Threshold Events 200-lb Load	Cumulative High Threshold Events 100-lb Load	Cumulative High Threshold Events 200-lb Load	Failure Load (lb)
Hexagonal	222	967	20	156	470 (average of 2)
Square	538	2,864	63	415	397 (average of 3)
Triangular	596	4,852	121	1040	340 (average of 3)

Figures 2a and 2b show cumulative emissions for three shafts with unbonded triangular end fittings, for high and low threshold events, respectively. Figures 3a and 3b present the data for square, unbonded end fittings. Figures 4a and 4b present the data from the hexagonal, unbonded end fittings. The trend in the data is clear. Triangular end fittings produce more emissions and have lower failure loads than the square fittings, and square end fittings produce more emissions and have lower failure loads than the hexagonal. Table 1 summarizes the combined data from Channels 1 and 2, for the unflawed, unbounded end fittings.

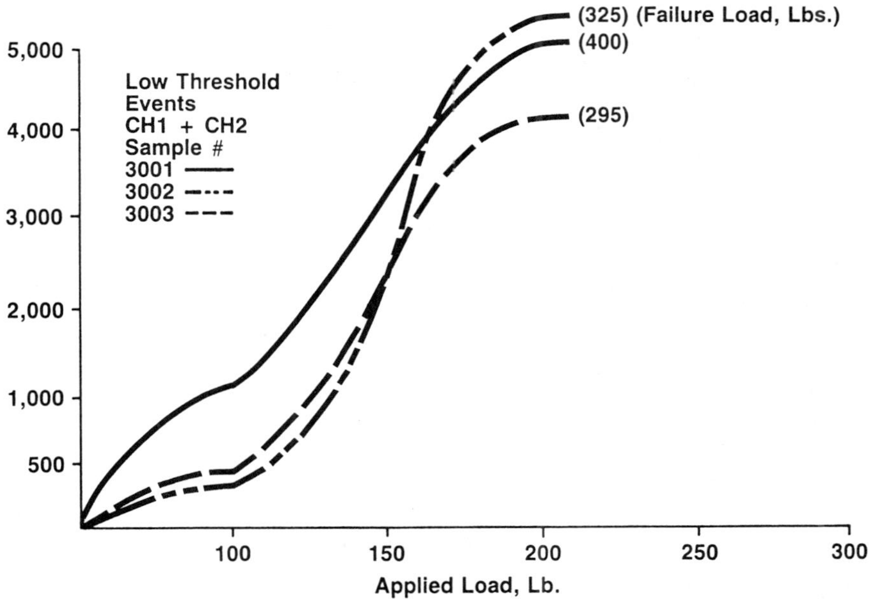

FIGURE 2a. Unbonded triangular end fittings

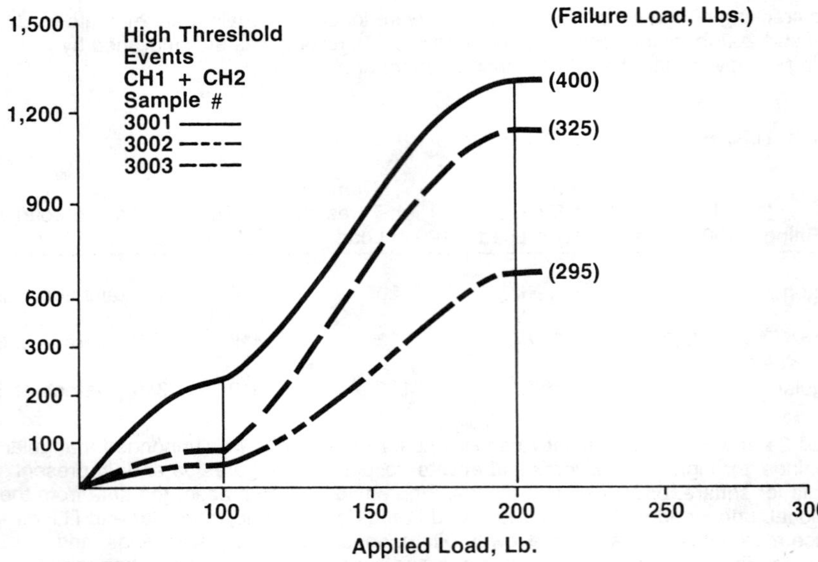

FIGURE 2b. Unbonded triangular end fittings

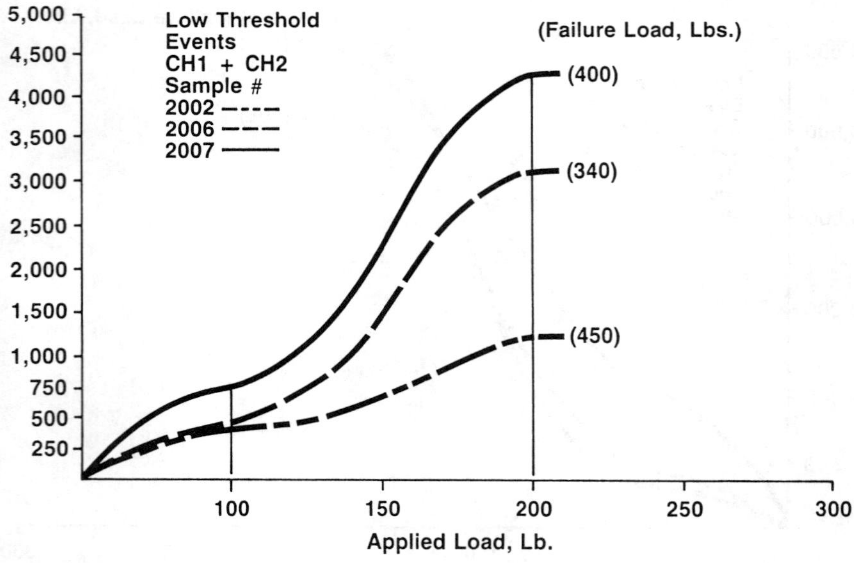

FIGURE 3a. Unbonded square end fittings

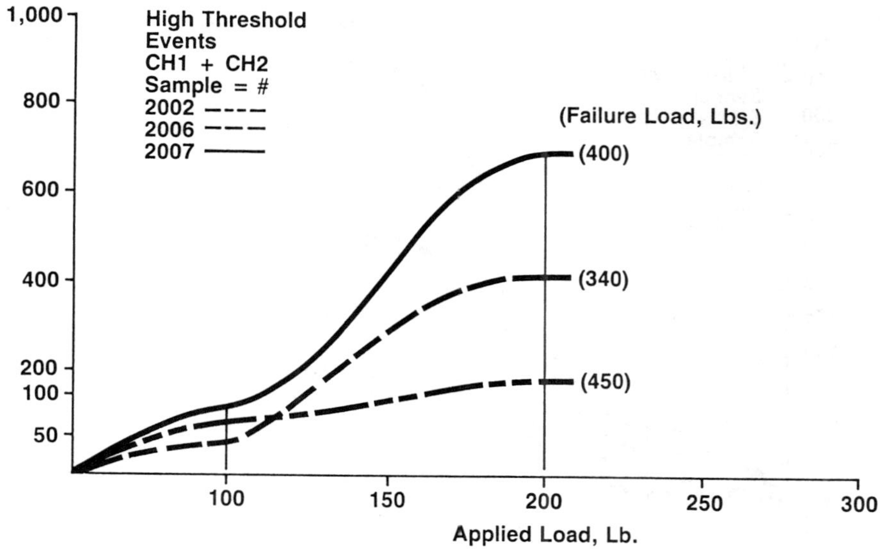

FIGURE 3b. Unbonded square end fittings

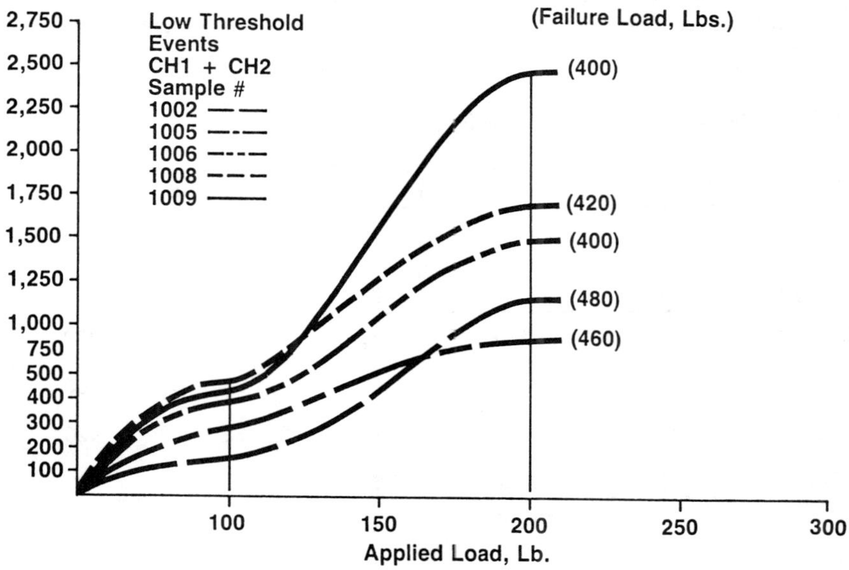

FIGURE 4a. Unbonded hexagonal end fittings

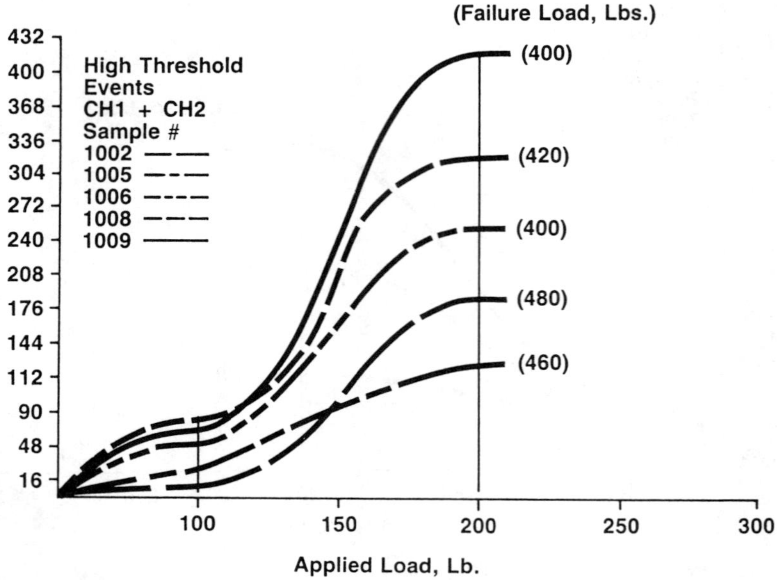

FIGURE 4b. Unbonded hexagonal end fittings

Another series of tubes was tested, in which the end fittings had a 0.5 inch mylar disc inserted between plies to act as a disbond. Table 2 presents the data for unbonded hexagonal end fittings.

TABLE 2. Unbonded, Hexagonal End Fittings

Sample No	Low Threshold Events		High Threshold Events		Failure Load
	100-lb Load	200-lb Load	100-lb Load	200-lb Load	(lb)
1002 (unflawed)	164	1139	11	186	480
1005 (unflawed)	280	794	29	125	460
Average	222	967	20	156	470
1006 (flaw)	396	1460	52	251	400
1008 (flaw)	458	1699	76	321	420
1007 (flaw)	403	2420	69	414	400
Average	419	1860	66	329	407

The presence of a .5" mylar disc makes a significant difference, even in this limited data, on all tubes studied. The area occupied by the disk is not available to help carry the torsional load, putting the remainder of the structure under greater effective stress. High threshold events, low threshold events, and failure load all reflect the presence of the flaw.

Finally, one tube was examined in detail for the generation of acoustic emission events during a prolonged hold at 200 lb load. Table 3 summarizes the continuing emissions resulting from prolonged application of a load representing 50% of the ultimate strength. (This tube failed at 400 lbs.) From the data, it is clear that damage is occurring, (implied by production of acoustic emissions) in the tube by testing it at 50% of its ultimate strength. The data suggest that failure will occur near Sensor 1, in looking both at total and high-threshold events.

TABLE 3. Unflawed, Unbonded, Square End Fittings

Minutes During 200-lb Load Hold	Cumulative Low Threshold Events		Cumulative High Threshold Events	
	Ch 1	Ch 2	Ch 1	Ch 2
0	1760	1347	334	226
2	2411	1844	410	279
4	2570	1963	425	290
6	2751	2086	436	300

4. DISCUSSION

The data presented in the figures and tables above are consistent. Greater load carrying capacity and fewer emissions are observed from hexagonal end fittings than for square ones and for square ones compared to triangular ones. Peak stresses are reduced by distributing them over a larger number of interferences (corners) on the end fittings. The presence of flaws or disbonds has the effect of decreasing the load carrying capacity of end fittings, and a clear correlation exists between failure strength, existence of flaws, and type of end fitting.

The other point of interest is the correlation of flaws and type of end fittings with acoustic emission (AE) data. Figures 2, 3, and 4 show clear trends of increasing AE count with applied load. The tubes with higher failure loads (greater torsional strength) produced fewer emissions. Table 1 and 2 make this correlation clearer. The trends are clear at 100 lb of applied load, and also at 200 lb. A general average of failure for these tubes is 400 lb. Thus, it follows that a clear trend is available for AE data at 25 percent of ultimate load. This is consistent with the spirit of nondestructive testing, and consistent with generally accepted concepts of AE practice. In metals, fatigue strengths are typically 30 percent of ultimate strength. In composites, a higher figure may be applicable. However, limiting the load to 25-30 percent of ultimate strength with low levels of acoustic emission, will generally prevent reducing the fatigue life of the sample.

5. CONCLUSIONS

1. Torque tubes made with hexagonal end fittings are stronger than those with square end fittings, which in turn are stronger than those made with triangular end fittings.

2. Acoustic emission clearly distinguishes flawed specimens from properly manufactured specimens.

3. Correlation of AE data with failure loads is clear at low loading, about 25 percent of ultimate, which is consistent with nondestructive testing goals.

4. Correlation of AE data with loads is best with the hexagonal fittings, consistent with lower stress concentration and better distributed loads.

Acoustic emission was a useful tool in these tests for correlation with the presence or absence of manufacturing flaws, stress distribution, and ultimate strength. This technique will be further investigated as a quality assurance tool on similar structures.

6. ACKNOWLEDGMENTS

This work was supported by Sundstrand Corporation IR&D funding.

QUANTITATIVE CHARACTERIZATION OF IMPACT DAMAGE IN COMPOSITE MATERIALS - A COMPARISON BETWEEN COMPUTERIZED VIBROTHERMOGRAPHY AND X-RAY TOMOGRAPHY

P. Potet, F. Lesbre, and C. Bathias
French DOD STCAN MSN 8
Boulevard Victor
Paris, France

INTRODUCTION

Composite materials are ideally suited for light weight structures, as for aircrafts, because of their high strength-to-stiffness ratios. High levels of reliability must be ensured in this type of structure, and therefore nondestructive inspection techniques (NDI) play a major role in the quality control of such materials. Infrared (IR) thermography has been intensively used by nondestructive evaluation engineers. Some investigations were directed toward the location of sub-surface flaws by applying an external heat source to the structure and observing the resulting thermal images at the surface[1]. A new technique, vibrothermography was developed and proposed for application to composites[2]. It utilizes high frequency low-amplitude mechanical vibrations to induce localized heating in a material. The frequency of excitation is varied in such a way that local flawed regions are set into local resonance, and serve as generators of heat.

The purpose of this work is to compare some experimental results obtained by vibrothermography with those obtained by X-ray tomography in the characterization of impact damage in composite materials. The X-ray tomography data will serve as reference to evaluate the possibilities offered by vibrothermography.

PRINCIPLES OF COMPUTERIZED VIBROTHERMOGRAPHY

During vibrothermographic experiments, a strong source of energy dissipation is created by damaged regions in the specimen. The resulting bidimensional thermal field, for a thin plate, observed by the infrared camera is given by the following equation[3] :

$$t\partial/\partial x_i(k_{ij}\,\partial x_j) - \alpha\theta + Q = \rho c\,\partial\theta/\partial\tau \qquad \text{for } i,j=1,2 \qquad (1)$$

where

- t is the tickness of the specimen,
- x_i and x_j spatial coordinates,
- k_{ij} thermal conductivity tensor,
- θ temperature field,
- α convective and radiative heat loss coefficient,
- Q heat source field per unit area,
- ρ density,
- c specific heat and
- τ time

During vibrothermographic tests a steady-state field of temperature is obtained (at least for the time scale of the camera) and is given by :

$$t\partial/\partial x_i(k_{ij}\,\partial\theta/\partial x_j) - \alpha\theta = -Q \qquad \text{for } i,j=1,2 \qquad (2)$$

This equation describes the different heat transfer mechanisms involved during a vibrothermographic test. The interaction of internal flaw surfaces during vibration is represented by the term Q, directly related to the damage state in the specimen.

As infrared cameras allow measurement only of the temperature field, the resulting thermal picture is not a real representation of the energy dissipated during vibration. In order to characterize quantitatively the amount of damage from such measurements, it is necessary to solve the heat transfer equation (Eq.2). This will give an estimated map of the dissipated energy at the surface of the specimen, obviously closely related to the real damage state of the specimen[4].

A schematic of the experimental set up is shown in figure 1. A standard IR camera, AGA 780, was used in this study. The video signal coming from the camera is digitized by an 8 bits analog-digital converter at a 500 kHz sample rate and then sent at the same rate into a screen buffer. The data are then transferred into a personal computer by DMA (Direct Memory Acces) channel.

A specific computer software was developed for the digital processing of thermograms. It features many basic functions such as spatial filtering, temporal averaging, image substraction and some other specific vibrothermographic functions, such as the resolution of eq.2. In this case, the main difficulty is to calculate the double spatial derivative in Eq.2, without amplifying the noise introduced by measurement. A finite difference method has been employed but it gives poor results, therefore a specific algorithm, based upon the divergence theorem has been developed. Let us consider a small surface S made of a grid of n by n pixels. Equation 2 can be integrated over the surface S :

$$t\int_S \partial/\partial x_i(k_{ij}\,\partial\theta/\partial x_j)\,ds - \alpha\int_S \theta\,ds - \int_S Q\,ds \quad \text{for } i,j=1,2 \tag{3}$$

By the divergence theorem, the first term of eq. 3 is equal to :

$$\int_S \partial/\partial x_i(k_{ij}\,\partial\theta/\partial x_j)\,ds = \int_{\partial s} k_{ij}\,\partial\theta/\partial x_j n_i d(\partial s) \tag{4}$$

where ∂s represents the boundary of S, n_i are the components of the unit normal to the surface boundary s . The double space derivative problem now reduces to a simple space derivative (made by finite difference in the computer program), and an integration over the surface of the grid.

This algorithm was successfully tested with a synthetic closed-form temperature field, and was found to give good results in vibrothermographic experiments.

The specimen was attached to a piezoelectric shaker and then vibrated. The frequency of excitation was chosen by sweeping it between 10 and 20 kHz until the best thermal contrast around damaged regions was obtained. The specimen was then vibrated at a fixed frequency. When a steady-state field of temperature was obtained, the resulting thermal patterns were recorded.

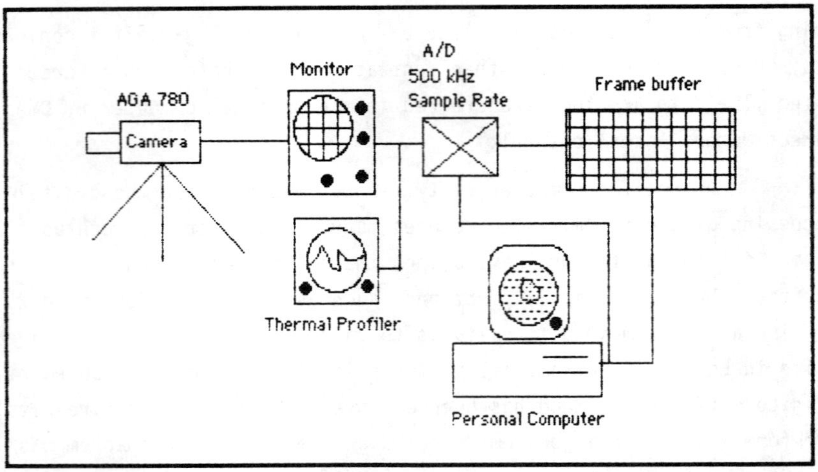

Figure 1 - Schematic of the experimental set-up used for computerized vibrothermographic experiments.

DAMAGE CHARACTERIZATION BY X-RAY TOMOGRAPHY

Computerized tomography (CT) is mainly used for medical applications. It utilizes X-ray source and detector associated to an automated displacement system. By measuring the absorption of X-rays in several directions throughout the specimen, it is possible to compute at any point the value of the absorption coefficient, and therefore the density of the specimen.

Because of the non-homogeneous and anisotropic nature of composite materials, the type of damage occuring can take many forms, such as delaminations, matrix cracking or fiber breaking.Furthermore, in most cases, damages are located over the entire thickness of the structure and not equally distributed over it, depending upon the relative orientation and position of each elementary ply. The determination of damage state severity can be made by the knowledge of its nature and position in the structure. CT is particularly well adapted to observe this three-dimensional state of damage ; therefore density measurements are used as a quantitative evaluation of damage[5].

The scanner used in this study is a ND 8000 C.G.R., and was designed for medical studies. It is calibrated with Hounsfield density scale (H), which gives - 1000 H for air, 0 H for water, 1000 H for dry bone ; it sensitivity is about 10^{-3} H. The minimum slice thickness that can be obtained with this scanner is 0.5 mm, which is an important limitation for composite materials since it does not make it possible to study each elementary ply (0.14 mm thick). Therefore only 5 slices were made in the plane of the specimen, and 11 in the cross-section, as shown in figure 2.

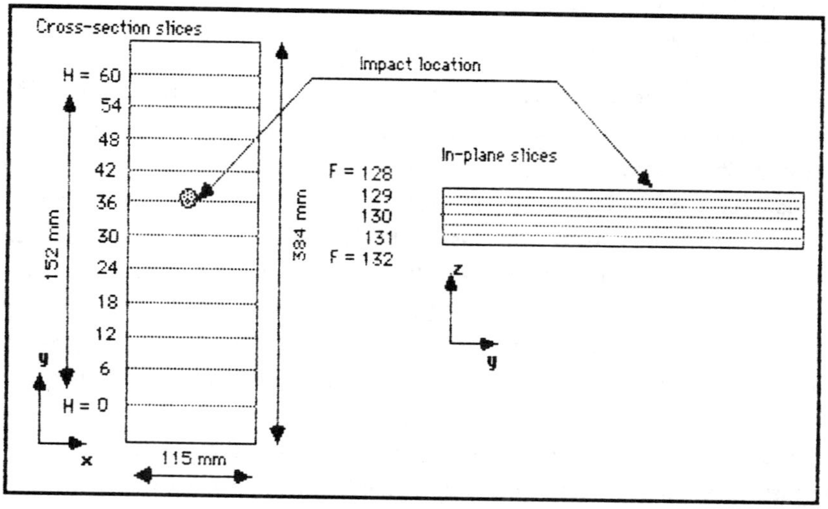

Figure 2 - Drawing of slice locations and references used for CT experiments.
Each slice is represented a dotted line and numbered.

RESULTS AND DISCUSSION.

The material used in this study is a carbon-epoxy composite, $(-45/0/45/0-45/90/45/0)_{2S}$. A specimen was impacted by dropping an aluminium sphere, while it was under compression load. A 1 mm diameter hole was drilled at the impact location and an X-ray opaque penetrant injected through the hole. Therefore, regions of low density (damaged regions) are represented in the tomographic picture by high Hounsfield coefficients.

CT EXPERIMENTS

Figure 3 represents in-plane slices taken at 3 different levels : F=128, F=129, F=130 (from top to bottom). An ovoid-shaped damaged zone, (delaminated area) oriented in the compression load direction, is clearly located at the center of each picture, Its diameter is about 15 mm for F=128 and 30 mm for F=130. Its density is 500 H.

A cracked area of lower density (660 H), in the 0 plies, is also shown, the maximum crack length is 16 mm. Figure 4 represents in-plane slices for F=130, F=131, F=132. The damage patterns are nearly the same than fig. 3, however for F=132 the damaged aera is clearly reduced. A cracked area extending around the delaminated aera is also located by cross-section slices at H=54.

VIBROTHERMOGRAPHIC EXPERIMENTS

The bidimensional heat transfer model given by eq.2 results from a thin plate approximation ; its validity decrease when increasing specimen thickness. Therefore, in order to obtain a good representation of the specimen damage state, it is necessary to observe the thermal patterns developped at both sides of the specimen. The frequency of excitation was 21 kHz. The two curves at top and left of fig. 5 are cross-sections of the temperature field for an abcissa and ordinate given by the black window at the center of the picture. This vibrothermogram was made by averaging 100 thermograms taken from a steady-state temperature field, and by filtering the resulting average by a low-pass filter, in order to cut down the high frequency noise. The damage area located by the hot spot at the center of the picture. A heat source picture was computed from fig. 5 temperature field, and the result is shown in fig.6. Edges of heat source are clearly delineated, and shape of the damage area can be compared with fig. 3. Figure 7 is a vibrothermogram of the opposite impacted side, taken at the same frequency of excitation as fig. 5. Again, the impacted area is clearly located. Figure 8 is a heat source map computed from fig. 7 temperature data. It indicates the direction of damage, in the 45° ply, in this case.

DISCUSSION

In this study a comparison between CT and vibrothermography was made. It first shows the high potential offered by vibrothermography for the NDE of composite materials. From CT data, three different damaged areas were detected : (i) a strongly cracked area, in the 0° plies, (ii) a delaminated area, with a variable diameter over the thickness, and (iii) a lightly cracked region extending over the delaminated area.

From vibrothermographic tests, the location of damaged area can be made accuratly, in real time, by observing the thermal patterns developed during testing. However, due to the diffusion of heat, the observation of temperature field permits only the detection of damage, but do not give any information upon its size and shape. The heat source field, computed from temperature data, as shown in fig. 6 and 8, gives a map of the dissipated energy during testing. In the case of fig. 6, shape of damaged area corresponds to the delaminated region indicated by CT pictures. The strongly cracked area can also be detected by the presence of two "hot spots" (high energy areas) in the 0° ply direction.

The observation of fig. 8, (opposite impacted side), indicates that damage direction is 45°, ie the surface ply direction. A visual inspection of the specimen surfaces confirmed that cracks have developed at the opposite impacted side, in the surface ply. These cracks were not revealed by CT experiments in the in-plane slice (see fig. 4, F=132) ; this is due the minimum slice thickness obtained by this scanner (0,5mm) which correspond to about 4 plies.

The differences in thermal patterns, and therefore in heat source maps, observed at both sides can be attributed to the non-negligible thickness of the specimen. During testing, the heat generated in the volume, is transfered to the surface by conduction and thus strongly attenuated. Therefore, thermal pictures observed by vibrothermography result principally from damages in the first plies of specimen. The lightly cracked area, extending around the delaminated region, was not detected by vibrothermography.

CONCLUSIONS

The purpose of this work was to make a comparison between CT and computerized vibrothermography for the detection of impact damage in composite materials. For matter of comparison, CT was taken as a reference method.

A specific computer program was developed in order to compute heat source maps from the experimental temperature fields. On the impacted side of the specimen, damages indicated by the heat source map clearly correspond to those detected by CT. On the opposite impacted side, damages were located in the surface ply, which was confirmed by visual examination. These cracks were not detected by CT.

Due to the attenuation of heat through the thickness, vibrothermography permits only the detection of damage in the first plies of the specimen, wich is an important limitation of this technique.

Acknowledgment.

This research was funded by the Service Technique des Constructions et Armes Navales, Paris, France. The authors wishes to thank T.K. O'Brien of Langlais NASA Research Center for the supply of the specimen. We also acknowledge M. Gagnasso of CNRS Marseille for his help in CT experiments.

References.

1. Williams J.H., Mansouri S.H., and Samson S.L., "One dimensional analysis of thermal NDT", *British Journal of NDT*, Vol 22, May 1980, pp 113-118.

2. Reifsnider K.L., Henneke E.G., and Stinchcomb WW., "The mechanics of vibrothermography.", *The mechanics of nondestructive testing*, W.W. Stinchcomb Ed, Plenum Press, New York, 1980, pp 249, 276.

3. Nayroles B., Bouc R., Caumon H., Chezeaux J.C. and Giacometti E., "Téléthermographie infrarouge et mécanique des structures.", *Int. J. Engng. Sci.*, Vol 19, 1981, pp 929-947.

4. Potet P, Jeannin P. and Bathias C., "The use of digital image processing in vibrothermographic detection of impact damage in composite materials.", To be published in *Materials Evaluation*, April 1987.

5. Vancon M., Odorico J. and Bathias C., "A fractographic study of delamination in carbon epoxy laminates.", *Comptes rendus des 4em journées nationales sur les composites*, Paris, 11-13 septembre 1984, G. Verchery Ed., Edition Pluralis.

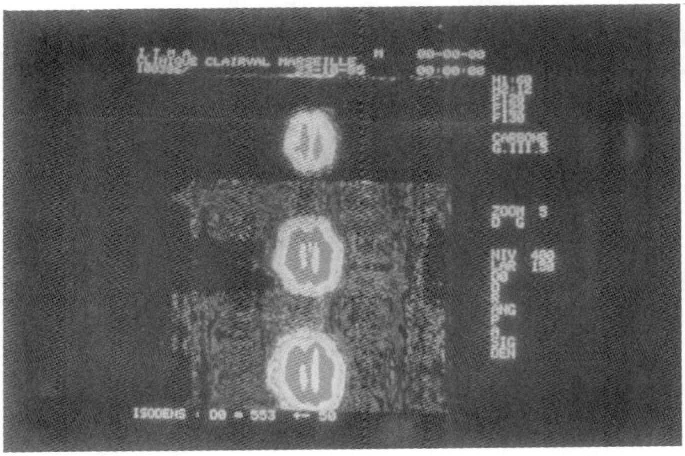

Figure 3 - In-plane tomographic slices taken at three levels : F=128, F=129, F=130. The red zone is of 553+/-50H.

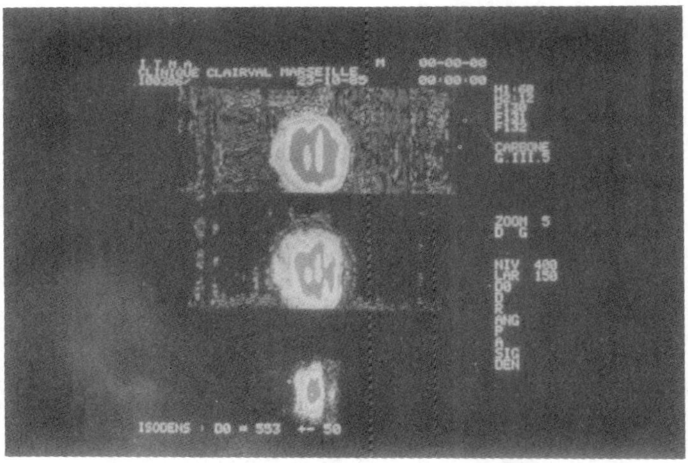

Figure 4 - In-plane tomographic slices taken at three levels : F=130, F=131 and F=132. The red zone is of 553+/-50H.

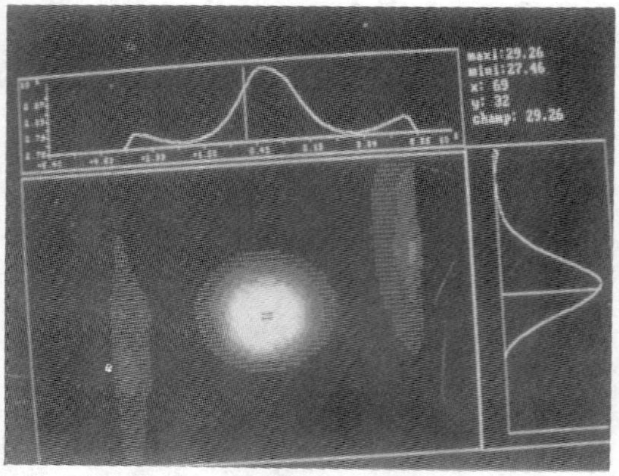

Figure 5 - Vibrothermogram of the impacted side
of the specimen taken at 21 kHz.
The damaged area is detected by the hot spot
at the center of the picture.
The curves at top and left are cross-sections
of the temperature field for an abcissa and ordinate given
by the black window at the center of the picture
(at impact location).

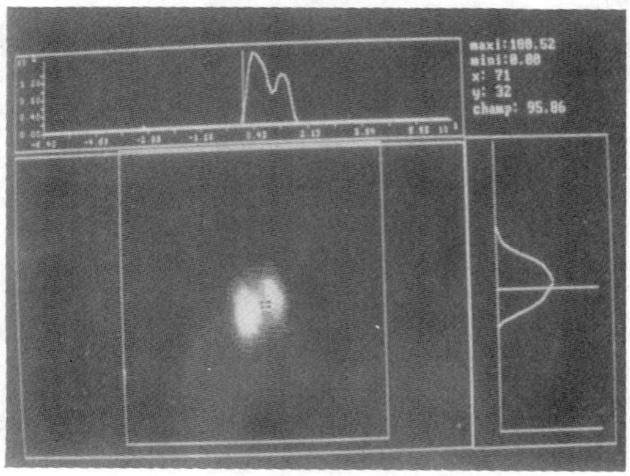

Figure 6 - Heat source map computed by using fig. 5
data and equation 2.
Edges and shape of heat source are clearly delineated
and can be compared with fig. 3 and fig. 4.

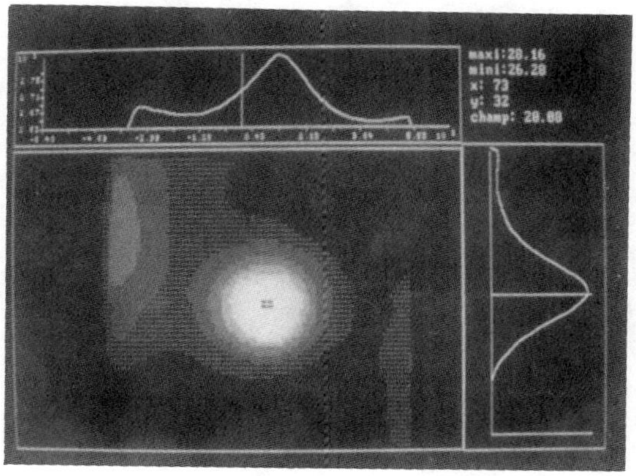

Figure 7 - Vibrothermogram of the opposite impacted side taken at 21 kHz.

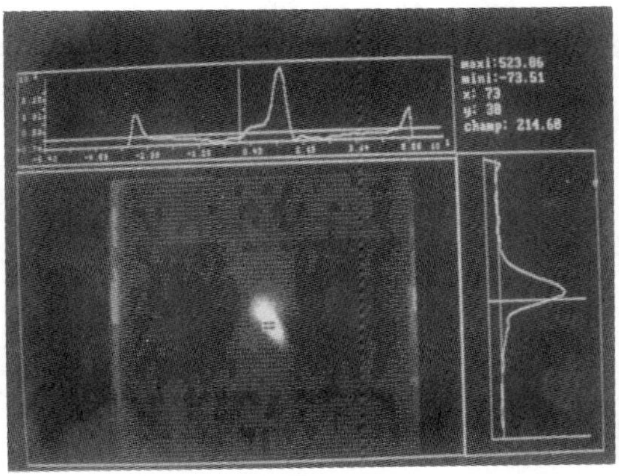

Figure 8 - Heat source map computed by using fig. 7 data. The direction of damage is found ie in the 45° surface ply.

ACOUSTIC EMISSION PHENOMENA IN POLYMERIC MEDIA

C. P. Hsiao, R. A. Kline, and D. M. Egle
University of Oklahoma
Norman, OK 73019

ABSTRACT

Fundamental mechanisms of the generation of acoustic emission (AE) signals in composite materials were investigated in this research. In particular, efforts were focused on examination of the differences (epicentral response) in the AE signals resulting from matrix fracture and fracture of fibers embedded in the matrix. A specially-designed specimen was used to isolate the first arrivals of the longitudinal and the shear waves from the subsequent reflections. Four different specimen compositions were utilized to accomplish the goal. One consisted solely of pure matrix and the other three contained a single fiber with diameter of 0.254, 0.635, or 0.762 mm embedded in the matrix. The techniques developed by Kline and Ali were employed to obtain theoretical predictions in order to make comparisons with the experimental data. Good agreement was found by using a synthetic forcing function to represent the behavior of the source.

INTRODUCTION

In studying the acoustic emission (AE) problem, much work [1-11] has been done on AE signal generation, source discrimination, etc. for elastic media. But relatively little research attention [12-15] has been paid to materials exhibiting dissipation. The complexity of the problem is the principal reason for the slow progress in this area. The introduction of dissipative terms in the constitutive equation presents a tremendous task in mathematics. The nonlinearity of the equations precludes the possibility of obtaining closed-form solutions. Furthermore, the introduction of complex valued and frequency dependent material moduli enhances the computational difficulty in deriving the solutions.

With the extensive use of composite materials in engineering applications in recent years, the study of polymeric media has become increasingly important. Solutions which omit the dissipative effects may not be sufficient to explain the phenomena observed in real materials. The objective of this research is to investigate the differences between the AE signals (epicentral response) from matrix fracture and the fracture of a fiber embedded in the matrix.

As a start toward a solution of the source discrimination problem in composite materials, a solution for the buried source problem with dissipation effects is needed. However, attempts at derivation of a solution using previous approaches [15] proved fruitless due to the existence of a singularity at the origin. Instead, numerical techniques as proposed by Kline and Ali [14] were employed to incorporate attenuation and dispersion effects into the elastic solution to predict the epicentral response for comparison with experimental results. The experiment was performed on a specially-designed Yobell specimen [11] to isolate the arrivals of the first longitudinal (P) wave and the shear (S) wave from the subsequent reflections in the specimen. Two different specimen configurations, one with pure matrix and another with a single glass fiber embedded in the matrix, were used to investigate the differences between the AE from matrix fracture and fiber fracture.

THEORY

The approach used in this research follows that of Shibata [10] and Kline and Ali [14]. This technique enables the incorporation of attenuation as well as dispersion effects into the elastic solution to obtain an approximation of the viscoelastic solution. Shibata has shown that using the infinite space Green's function for the half-space problem by constructing an image force to maintain the stress free condition at the surface introduces a relatively small error into the problem and simplifies the calculation considerably. The infinite space Green's function is given by

$$G_{33}^{\infty}(h,\underline{t}) = \frac{1}{4\pi\rho}\left(\frac{\delta(\underline{t} - h/C_L)}{C_L^2 h} + \frac{2t}{h^3}\left[H(\underline{t} - \frac{h}{C_L}) - H(\underline{t} - \frac{h}{C_S})\right]\right) \tag{1}$$

where C_L : longitudinal wave speed

C_S : shear wave speed

h : source depth

ρ : density

$\delta(t)$: delta function

$H(t)$: Heaviside step function

$\underline{t} = t - t'$

This expression can be separated into two parts corresponding to the longitudinal and shear contributions.

$$G_{33}^{\infty}(h,\underline{t}) = G_L^{\infty}(h,\underline{t}) + G_S^{\infty}(h,\underline{t})$$

$$= \frac{1}{4\pi\rho}\left(\frac{\delta(\underline{t} - h/C_L)}{C_L^2 h} + \frac{2\underline{t}}{h^3} H(\underline{t} - \frac{h}{C_L})\right) - \frac{\underline{t}}{2\pi\rho h^3} H(\underline{t} - \frac{h}{C_S}) \tag{2}$$

The longitudinal and shear contributions to the vertical surface displacement then can be obtained by the convolution of the Green's function with the appropriate forcing function. It was found that using a more elaborate functional representation between 0 and τ instead of a simple ramp as the forcing function (Fig. 1), better agreement between the theoretical and the experimental results could be obtained [4].

$$
\begin{aligned}
f(t') &= 0 && t' < 0 \\
f(t') &= F_o \left\{ \left[1 - \left(1 - \frac{t'}{\tau} \right)^m \right]^\omega \right\} && 0 < t' < \tau \qquad (3) \\
f(t') &= F_o && t' > \tau
\end{aligned}
$$

where F_o : steady state amplitude

$t' = t - \underline{t}$

τ : rise time

Then the displacement is given by

$$
\begin{aligned}
U_z(h,t) &= U_z^L(h,t) + U_z^S(h,t) \\
&= \int_0^t G_L^\infty(h,\underline{t})\, f(t-\underline{t})\, d\underline{t} + \int_0^t G_S^\infty(h,\underline{t})\, f(t-\underline{t})\, d\underline{t} \qquad (4)
\end{aligned}
$$

where $U_z^L(h,t)$: longitudinal displacement

$U_z^S(h,t)$: shear displacement

$U_z(h,t)$: total displacement

To incorporate the dissipation effects into the solution, the Fourier transform was applied to the longitudinal and shear contributions separately to yield

$$\bar{U}_z^L(h,\omega) = \int_{-\infty}^{\infty} G_z^L(h,t)\, e^{-i\omega t}\, dt = A^L(h,\omega)\, e^{i\phi^L(h,\omega)} \qquad (5)$$

$$\bar{U}_z^S(h,\omega) = \int_{-\infty}^{\infty} G_z^S(h,t)\, e^{-i\omega t}\, dt = A^S(h,\omega)\, e^{i\phi^S(h,\omega)} \qquad (6)$$

where $\bar{U}_z^L(h,\omega)$: Fourier transform of longitudinal displacement

$\bar{U}_z^S(h,\omega)$: Fourier transform of shear displacement

$A^L(h,\omega)$: amplitude spectrum of longitudinal wave

$A^S(h,\omega)$: amplitude spectrum of shear wave

$\phi^L(h,\omega)$: phase spectrum of longitudinal wave

$\phi^S(h,\omega)$: phase spectrum of shear wave

Then the attenuation and dispersion are introduced by modification of the amplitude spectra for attenuation and the phase spectra for dispersion.

$$
\begin{aligned}
\bar{A}^L(h,\omega) &= A^L(h,\omega)\, e^{-\alpha_L(\omega)h} \\
\bar{A}^S(h,\omega) &= A^S(h,\omega)\, e^{-\alpha_S(\omega)h}
\end{aligned}
\qquad (7)
$$

$$\bar{\phi}^L(h,\omega) = \phi^L(h,\omega) + \omega\tau_L(\omega)$$
$$\bar{\phi}^S(h,\omega) = \phi^S(h,\omega) + \omega\tau_S(\omega) \qquad (8)$$

where α_L : longitudinal attenuation factor

α_S : shear attenuation factor

τ_L : longitudinal transient time difference factor

τ_S : shear transient time difference factor

It was found by Kline and Ali [14] that dispersion has a relatively small effect on the waveform obtained by this technique. After modifying both the amplitude and the phase spectra, the dispersed and attenuated displacement can be obtained by taking the inverse Fourier transform.

$$U_z^L(h,t) = \frac{1}{2\pi}\int_{-\infty}^{\infty} \bar{A}^L(h,\omega)\, e^{i\bar{\phi}^L(h,\omega)}\, e^{i\omega t}\, d\omega$$
$$U_z^S(h,t) = \frac{1}{2\pi}\int_{-\infty}^{\infty} \bar{A}^S(h,\omega)\, e^{i\bar{\phi}^S(h,\omega)}\, e^{i\omega t}\, d\omega \qquad (9)$$

EXPERIMENTAL PROCEDURES

A specially-designed specimen (Yobell specimen)[11] was chosen and its configuration is presented in Fig. 2. By using this specimen, it is possible to isolate the first arrivals of the longitudinal (P) and the shear (S) waves generated by the fracture of the specimen from the multiple reflections created by waves travelling within the specimen itself.

Two different sets of constituents were used to form the matrix. One is D.E.R. 383 epoxy resin (a Bisphenol A-type resin) and Epon V-50 curing agent (Amidoamine). The second set consisted of alpha resin (Polyester resin) and DDM-9 catalyst. The low viscosity associated with these combinations provided a method of specimen fabrication with low void content. Fibers (Specialty Glass, Inc. SP467 and SP297) in three diameters (0.762, 0.635, and 0.254 mm) were used.

The mixed resin was outgassed in a vacuum chamber to remove most of the air bubbles entrapped in the resin during the mixing process. The outgassed resin was then poured into a silicon-based mold to create the Yobell specimen. If a fiber was required in the specimen, the fiber was fixed into position before the resin was poured into the mold. After casting, the mold was placed in a vacuum chamber for initial cure under vacuum. It was then removed from the chamber to continue the curing process at room temperature for 24 hours.

Two different specimen configurations were used. One consisted of pure matrix (neat resin) for examination of epicentral response resulting from matrix fracture. Another consisted of a fiber embedded in the matrix (fiber diameter ranged from 0.254 mm to 0.762 mm). Since the fiber usually fails prior to the matrix, it is possible to separate the fiber fracture signal from the matrix signal (for configuration 2). Therefore, comparisons between the signals generated by the fiber fracture and the matrix fracture were possible in the post-test analysis.

The specimens were placed in a specially designed fixture (Fig. 3) on the MTS machine and failed in tension using dead-weight loading. The complete experimental apparatus is shown in Fig. 4. The signal generated by either matrix fracture or fiber fracture was captured by an NBS point-contact transducer placed on the top of the specimen. To circumvent the problem of the impedance mismatch between the transducer and digitizer, the signal was first sent through a unit gain operational amplifier (a voltage follower). The digitized signal was stored in the minicomputer and displayed on both the graphics terminal and the oscilloscope simultaneously. The results were transferred to the IBM main frame for comparison with the theoretical predictions.

RESULTS AND DISCUSSIONS

Both the longitudinal and shear contributions as well as the combined reponse are plotted in Fig. 5. The material properties of the matrix were found to lie between that of polymethyl methacrylate (PMMA) and polyethylene (PE). Table 1 shows the velocities and attenuation factors for both the longitudinal and the shear waves at 5 MHz for PMMA, PE, and the specimen. The PMMA and PE data were based on the data reported by previous researchers [15-17]. Figure 6 shows the difference between the epicentral responses calculated from the elastic solution and from the modified solution with attenuation effects.

The epicentral response was obtained via convolution of the delta function solution and the appropriate forcing function (Eq. 3). By adjustment of two parameters, $m = 2.5$ and $\omega = 3.2$, in the Weibull distribution function, excellent agreement was achieved. Comparisons between the theoretical prediction and the experimental result are given in Fig. 7 for pure matrix signal, in Fig. 8 for a 0.762 mm fiber signal, and in Fig. 9 for a 0.254 mm fiber signal. In addition, the three recovered forcing functions are plotted together in Fig. 10 for comparison. Different rise times and amplitudes are associated with the matrix and fiber signals. The experimental results also show that the amplitude of the signal is determined by the amount of strain energy released when fracture occurred. In Fig. 11, signals were taken from the failures of pure matrix specimens

TABLE 1. Material Properties at 5 MHz

	PMMA	PE	Specimen
C_L (cm/sec)	2.69×10^5	2.43×10^5	2.52×10^5
C_S (cm/sec)	1.34×10^5	9.50×10^4	1.11×10^5
α_L (1/cm)	0.4066	0.9476	0.8376
α_S (1/cm)	1.2458	7.2713	2.0344

with different cross-sectional areas. A linear relationship exists between the amplitude and the diameter of the fracture surface (Fig. 12). Since the signal amplitude squared is proportional to the area of the fracture surface (based on the strain energy considerations alone), amplitude is linearly proportional to the diameter.

However, stress concentrations were also introduced by reduction of the specimen cross section (Fig. 2). According to published data for geometrical stress concentration [18], Fig. 13 presents a plot of stress concentration factor (K_t) vs. diameter for the Yobell specimens. A linear relation was found to exist between K_t and the diameter. Using these data to compensate for the effect of stress concentration, the failure stress data for the Yobell specimens can be brought into good agreement. The failure stresses are plotted against the diameter in Fig. 14 while Fig. 15 shows the failure stress vs. diameter after taking the stress concentration effect into account.

Because of the low fiber volume fraction (V_f = 0.5% to 2%) and the relative ductilities of the materials involved, adding the fiber to the matrix was not found to increase the load carrying capability of the specimen. Instead, the specimen became weaker with a single reinforcing fiber embedded in the matrix. The failure stresses of the pure matrix, with a 0.635 mm fiber, and with a 0.254 mm fiber are listed in Table 2 for two different combinations of resin and curing agent. It can be seen that the failure stresses of the specimens with fiber are lower than that of the pure matrix specimen in both cases (Table 2).

For very low fiber volume fractions (V_f), ductility considerations are the dominant factor in determining the load carrying capability of the composite. According to the theory of composite materials [19] (Fig. 16), in the case of matrix being more ductile than fiber, composite materials can gain strength once V_f has exceeded a critical value. To explore this conjecture further, several uniaxial tension specimens were made of both compositions to determine the stress-strain curve, the failure stress, and the failure strain for the matrix materials used. It was found that the properties of the epoxy resin are highly dependent on the displacement rate of the tests. Table 3 shows both the failure stresses and the failure strains for the polyester resin and the epoxy resin. The corresponding stress-strain curves are presented in Fig. 17(a) - 17(c).

Viscoelastic effects were also found with epoxy resin specimens. Specimens tested at higher displacement rates appeared to be more brittle than those tested at lower rates. The latter showed a considerable amount of plastic deformation before fracture.

TABLE 2. Failure Stress and Fiber Volume Fraction

	D.E.R. 383 & V-50			Alpha resin & DDM-9		
	Pure matrix	0.254mm fiber	0.635mm fiber	Pure matrix	0.254mm fiber	0.635mm fiber
Failure stress (x 10^7 N/m^2)	4.023	1.162	0.865	3.957	1.310	1.749
V_f	0.0	~0.0028	~0.018	0.0	~0.0026	~0.016

TABLE 3. Material Properties

	Alpha Resin & DDM-9	D.E.R. 383 & V-50	
Failure stress (x 10^7 N/m^2)	3.887	1.936	1.565
Failure strain (%)	4.3	5.91	10.71
Displacement rate (cm/sec)	8.3853×10^{-3}	1.6445×10^{-2}	8.3871×10^{-3}

The failure data for the epoxy resin obtained from the uniaxial tension specimen is not consistent with that obtained from the Yobell specimens (after compensation for K_t). To further investigate this problem, Yobell specimens with different curing time were tested. It was found that the properties of the epoxy resin are also time dependent. The results are given in Table 4.

Because the properties of the epoxy resin are functions of curing time, the results obtained from the uniaxial tension specimen tests (which were cured for only one week) cannot be used to directly verify the results obtained from the aged Yobell specimens. Therefore, the results from the polyester resin (a more stable system) were used to make the verification.

Another ambiquity comes from the fact that the material properties of the glass fibers are unavailable from the manufacturer. The actual V_{min}, $V_{crit.}$, and σ_{Cmax} for different V_f (refer to Fig. 16) cannot be calculated. However, using the properties of glass from the handbook [20], an attempt was made to evaluate V_{min}, $V_{crit.}$, and σ_{Cmax} for different V_f for the polyester resin-glass fiber system (Table 5).

It was found that for the 0.254 mm fiber specimen ($V_f = 0.0026$), the fiber volume fraction is indeed lower than V_{min}. This leads to a failure stress lower than that of the pure matrix specimen. However, the failure stress is lower than the calculated value by a factor of three. Similar problems are found in the 0.635 mm fiber specimen with $V_f = 0.016$. Again, the experimental result is lower than the calculated result by a factor of 3. Note that once the fiber fails, it may create a void within the matrix, introducing a stress concentration (estimated stress concentration factor $\sim$ 2-3). This stress concentration effect along with the misalignment and imperfections of the glass fibers may contribute to the observed discrepancy. Further research is needed to address the problem.

TABLE 4. Time Dependent Properties of D.E.R. 383 & V-50

Curing time	One Week	Two Months	One Year
Failure stress (x 10^7 N/m^2)	1.936	3.074	4.023

TABLE 5. Properties of the Polyester Resin-Glass Fiber System

	V_{min}	$V_{crit.}$	σ_{Cmax} at $V_f = 0.0026$	σ_{Cmax} at $V_f = 0.016$
Calculated	0.00728	0.00742	3.876×10^7 N/m^2	5.668×10^7 N/m^2
Experiment			1.310×10^7 N/m^2	1.749×10^7 N/m^2

CONCLUSIONS

In this research program, AE generation in composite materials has been studied using Yobell specimens to isolate the source mechanisms. Results were compared with theoretical prediction for the epicentral response of a viscoelastic medium to a buried source within the sample. Numerical techniques were employed to incorporate attenuation and dispersion effects into the solution. The source-time history was mathematically modeled with two parameters which were adjusted to best fit the experimental data. In this way, estimated source-time histories (after compensation for attenuation and dispersion) could be obtained.

For pure matrix fracture, signal amplitude was found to be linearly proportional to the area of the fracture surface. Due to the reduced energy release associated with fiber fracture with our single fiber specimen geometry, smaller signal amplitude was observed for the fiber reinforced specimens. No significant differences in rise time between two source mechanisms were found.

FUTURE WORK

Further study is required to fully address the problem of low fiber volume fraction.

ACKNOWLEDGMENTS

This work was sponsored by the American Society for Nondestructive Testing (ASNT) Fellowship program. Also, thanks are due to Shell Chemical Co. and Dow Chemical Co. for supplying some of the resin and curing agent, to Specialty Glass, Inc. for providing the glass fiber.

REFERENCES

1. Breckenridge, F., Tschiegg, C., and Greenspan, M., "Acoustic Emission: Some Applications of Lamb's Problem," *J. Acoust. Soc. Am.*, vol. 57, no. 3, pp. 626-631, 1975.

2. Pekeris, C.L. and Lifson, H., "Motion of the Surface of a Uniform Elastic Half-Space Produced by a Buried Pulse," *J. Acoust. Soc. Am.*, vol. 29, no. 11, pp. 1233-1238, 1957.

3. Pinney, E., "Surface Motion due to a Point Source in a Semi-Infinite Elastic Medium," *Bull. Seism. Soc. Am.*, vol. 44, pp. 571-596, 1954.

4. Kim, K.Y. and Sachse, W., "Acoustic Emissions from Penny-Shaped Cracks in Glass. II. Moment Tensor and Source-Time Function," *J. Appl. Phys.*, vol. 59, pp. 2711-2715, 1986.

5. Mencher, A.G., "Epicentral Displacement Caused by Elastic Waves in an Infinite Slab," *J. Appl. Phys.*, vol. 24, no. 9, pp. 1240-1246, 1953.

6. Knopoff, L., "Surface Motions of a Thick Plate," *J. Appl. Phys.*, vol. 29, no. 4, pp. 661-670, 1958.

7. Sinclair, J.E., "Epicentre Solutions for Point Multipole Sources in an Elastic Half-Space," *J. Phys. D: Appl. Phys.*, vol. 12, pp. 1309-1315, 1979.

8. Hsu, N., Simmons, J., and Hardy, S., "An Approach to Acoustic Emission Signal Analysis-Theory and Experiment," *Material Evaluation*, vol. 35, pp. 100-106, 1977.

9. Pao, Y.H., Gajewski, R., and Ceranoglu, A., "Acoustic Emission and Transient Waves in an Elastic Plate," *J. Acoust. Soc. Am.*, vol. 65, pp. 96-105, 1979.

10. Shibata, M., "A Theoretical Evaluation of Acoustic Emission Signals-the Rise-Time Effect of Dynamic Forces," *Material Evaluation*, vol. 42, pp. 107-115, 1984.

11. Wadley, H.N.G., Scruby, C.B., and Shrimpton, G., "Quantitative Acoustic Emission Source Characterization During Low Temperature Cleavage and Intergranular Fracture," *Acta Metallurgica*, vol. 29, pp. 399-414, 1981.

12. Knopoff, L., "The Seismic Pulse in Materials Possessing Solid Friction II. Lamb's Problem," *Bull. Seism. Soc. Am.*, vol. 49, no. 4, pp. 403-413, 1959.

13. Newlands, M., "Lamb's Problem with Internal Dissipation, I.," *J. Acoust. Soc. Am.*, vol. 26, pp. 434-448, 1954.

14. Kline, R.A. and Ali, S.S., "Methods of Calculating Attenuation and Dispersion Effects on Acoustic Emission Signals," *J. Acoust. Emission*, vol. 4, no. 4, pp. 107-114, 1985.

15. Hsiao, C.P., "The Propagation of Acoustic Emission Signals in Polymeric Media," Ph.D. Dissertation, University of Oklahoma, 1988.

16. Hartmann, B. and Jarzynski, J., "Ultrasonic Hysteresis Absorption in Polymers," *J. Appl. Phys.*, vol. 43, no. 11, pp. 4303-4312, 1972.

17. Sutherland, H.J. and Lingle, R., "An Acoustic Characterization of Polymethyl Methacrylate and Three Epoxy Formulations," *J. Appl. Phys.*, vol. 43, no. 10, pp. 4022-4026, 1972.

18. Peterson, R.E., *Stress Concentration Design Factors*, John Wiley & Sons, Inc., New York, 1953.

19. Jones, R.M., *Mechanics of Composite Materials*, pp. 128-130, McGraw-Hill, New York, 1975.

20. Baumeister, T., Avallone, E.A., and Baumeister, T. III, *Marks' Standard Handbook for Mechanical Engineers*, 8th Ed., pp. 6-115 - 6-158, McGraw-Hill, New York, 1979.

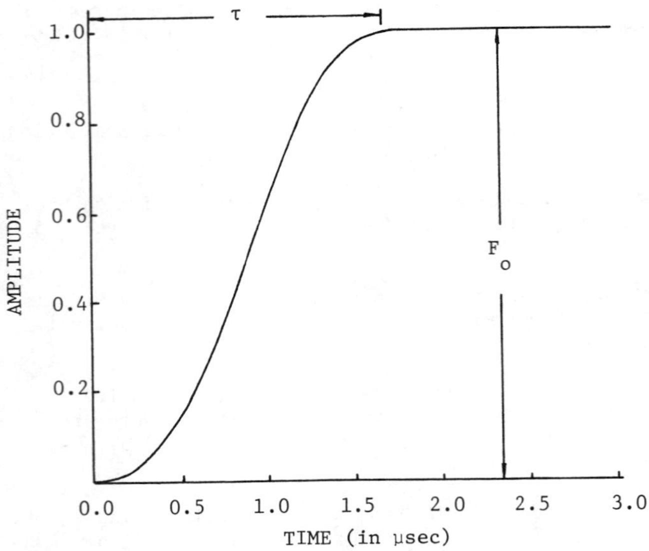

FIGURE 1. Forcing function.

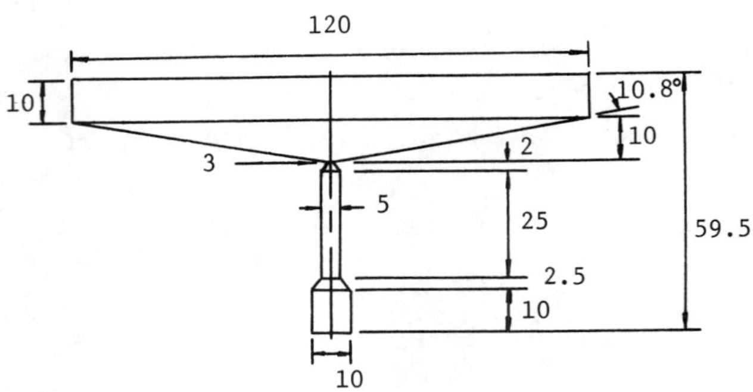

FIGURE 2. Specimen configuration (in mm).

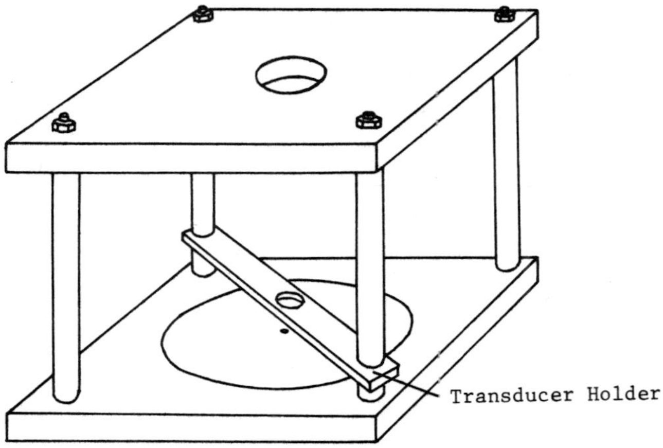

FIGURE 3. Specimen grip.

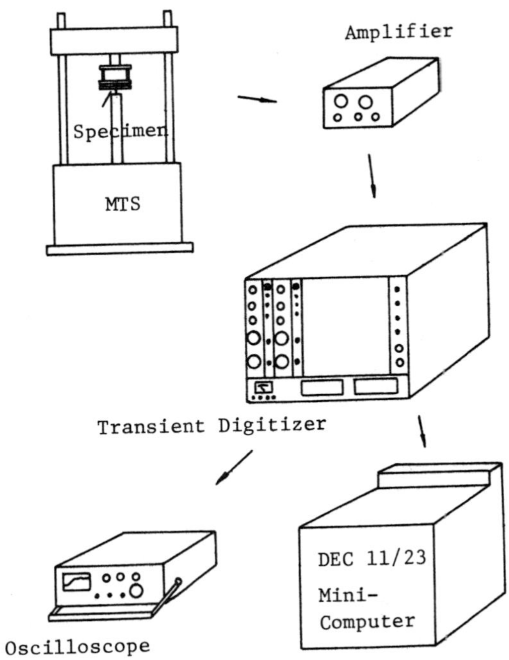

FIGURE 4. Experimental apparatus.

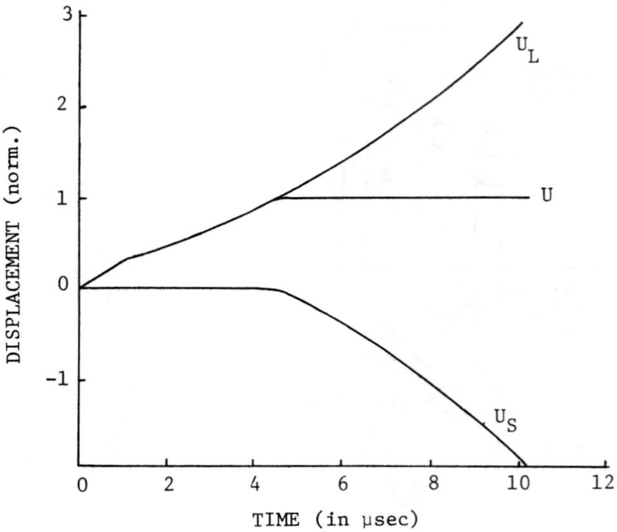

FIGURE 5. Longitudinal U_L, shear U_S, and combined displacement U as a function of time.

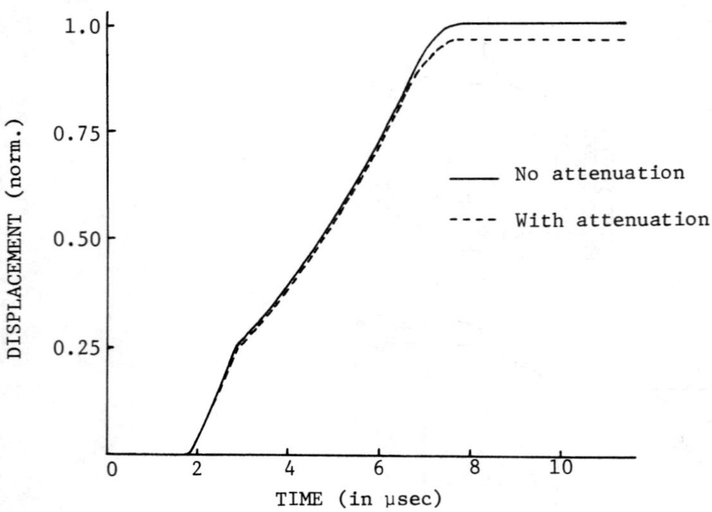

FIGURE 6. Elastic and modified solutions.

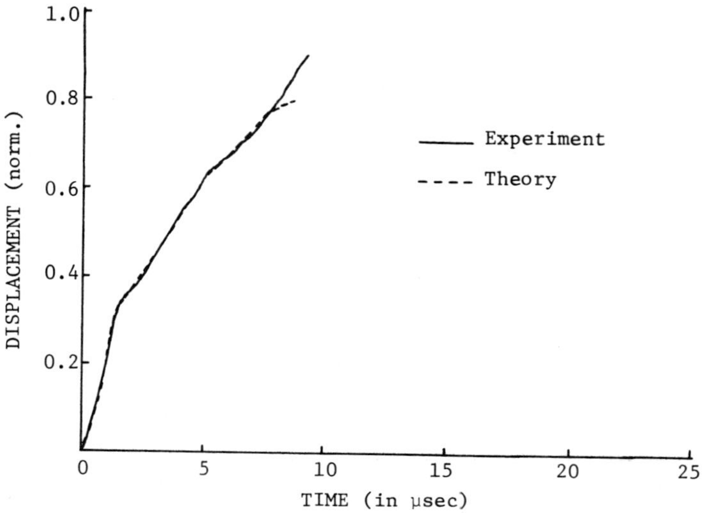

FIGURE 7. Matrix signal.

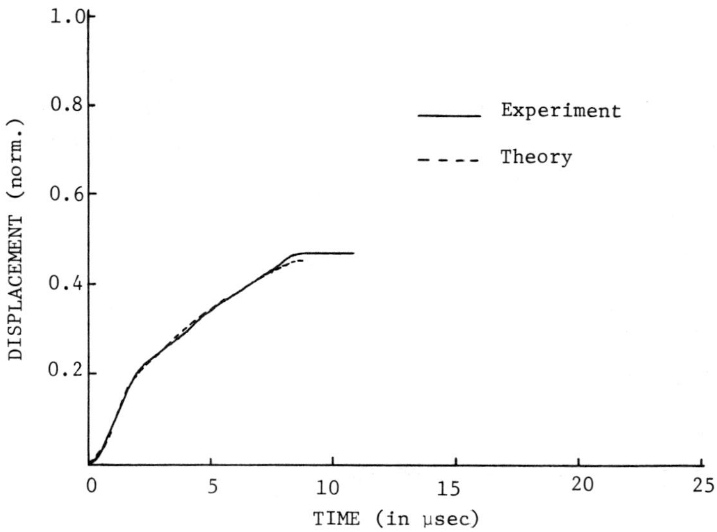

FIGURE 8. 0.762 mm fiber signal.

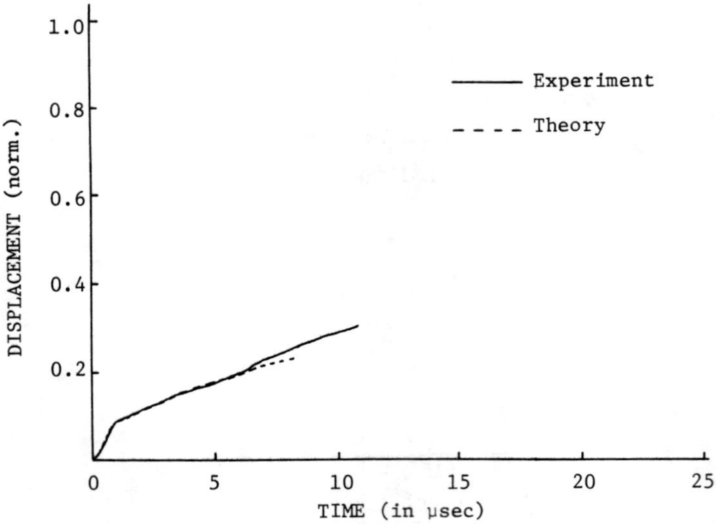

FIGURE 9. 0.254 mm fiber signal.

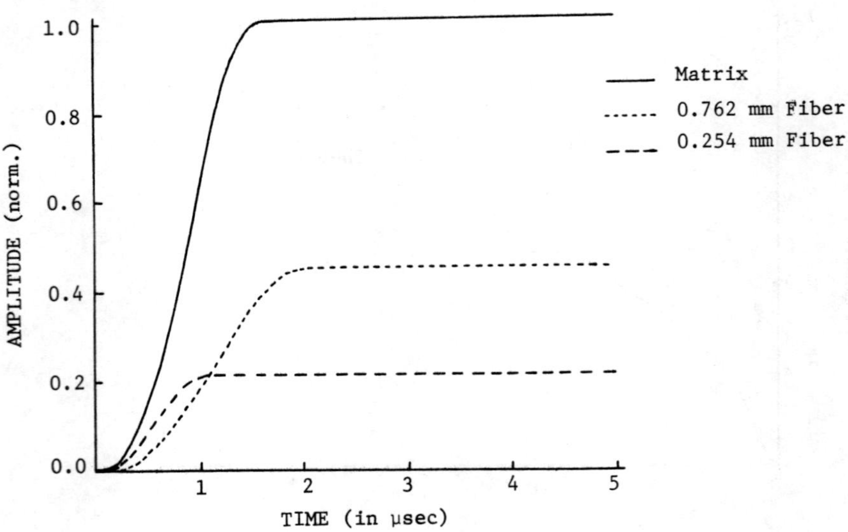

FIGURE 10. Forcing function.

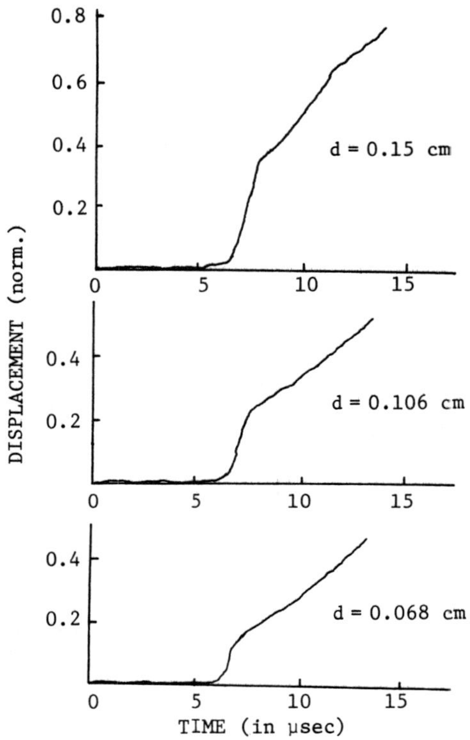

FIGURE 11. Signals from various sizes of the fracture surface.

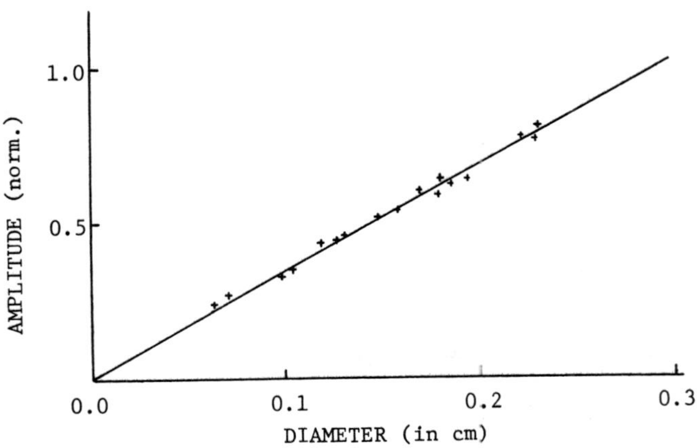

FIGURE 12. Amplitude vs. diameter of the fracture surface (matrix).

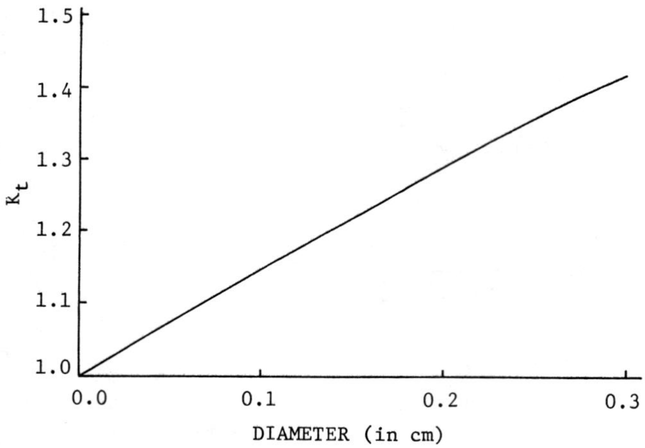

FIGURE 13. Stress concentration factor K_t vs. diameter.

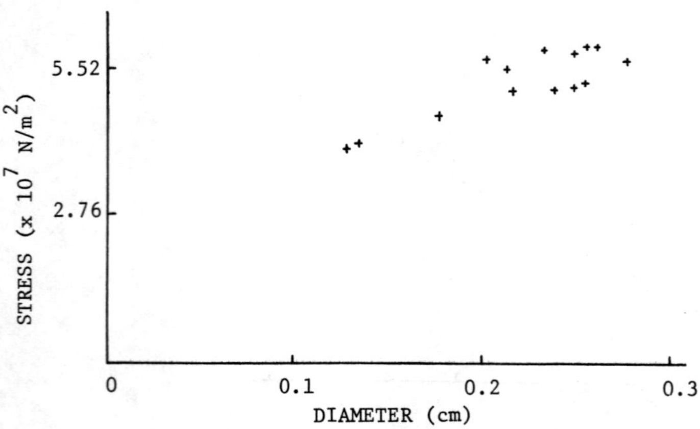

FIGURE 14. Failure stress vs. diameter.

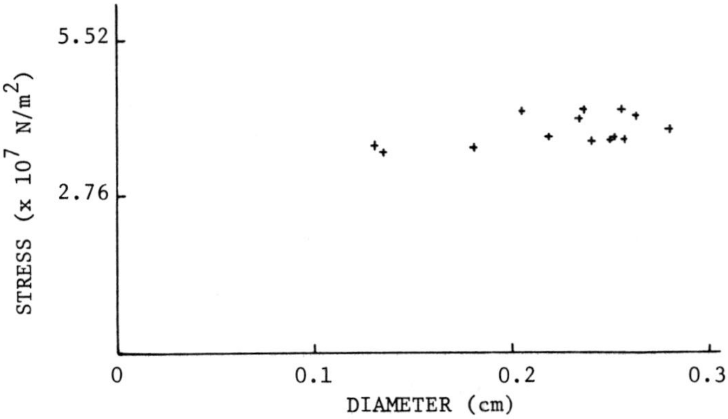

FIGURE 15. Failure stress vs. diameter (compensated for stress concentration).

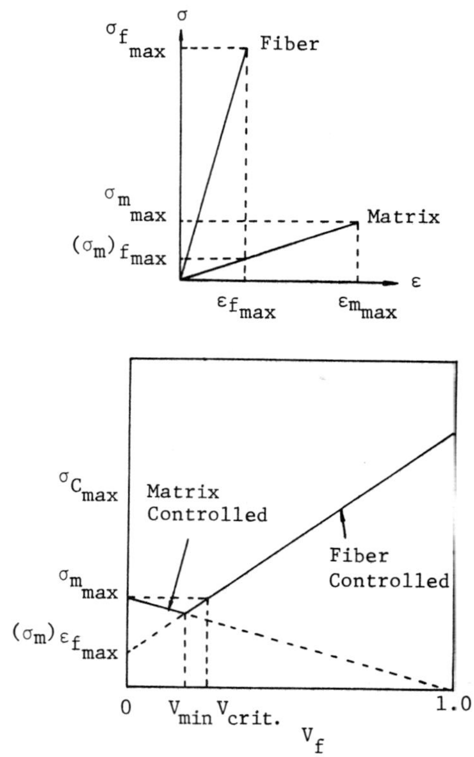

FIGURE 16. The effect of fiber volume fraction (V_f) on composite strength (after Jones).

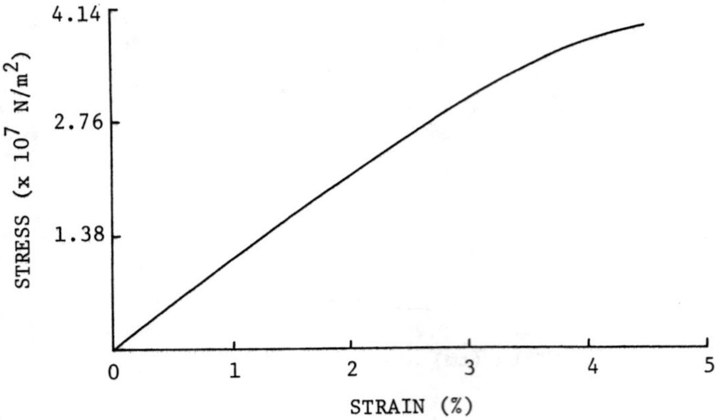

FIGURE 17(a). Stress vs. strain (alpha resin & DDM-9) (displacement rate = 8.3853 x 10^{-3} cm/sec).

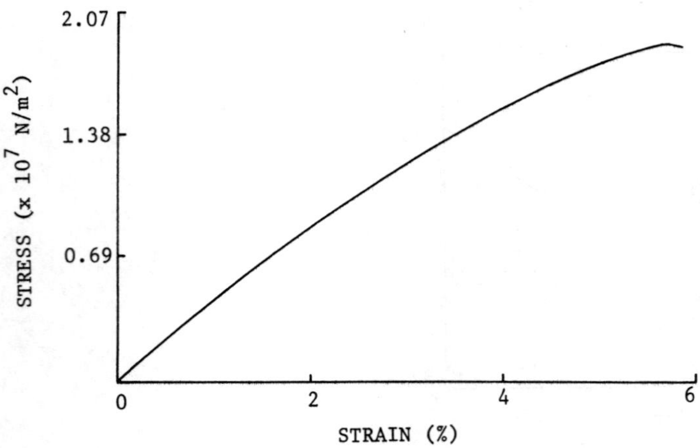

FIGURE 17(b). Stress vs. strain (D.E.R. 383 & V-50) (displacement rate = 1.6445 x 10^{-2} cm/sec).

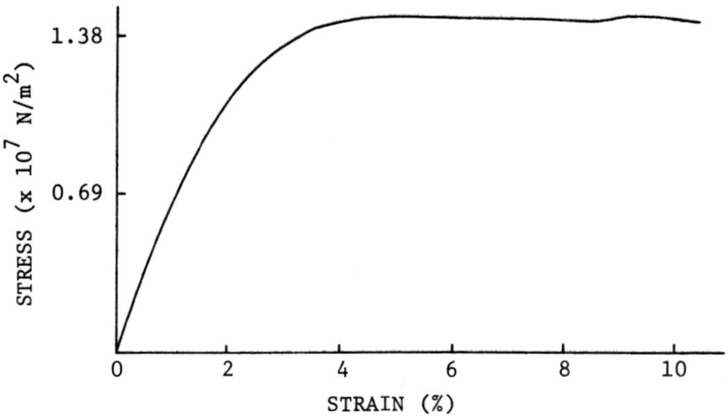

FIGURE 17(c). Stress vs. strain (D.E.R. 383 & V-50) (displacement rate = 8.387 x 10^{-3} cm/sec).

NONDESTRUCTIVE TESTING AND EVALUATION FOR WOOD PRODUCTS

F. C. Beall
University of California Forest Products Laboratory
Richmond, CA 94804

INTRODUCTION

Production and Consumption of Wood-based Materials for Construction

For the purpose of this paper, the term wood-based materials is restricted to lumber and panels. A number of other products, including poles, piling, and railroad ties, have not been included but have specific efforts underway in NDE. Table 1 shows the hardwood and softwood production and consumption by species in the U.S. For the most part, softwoods dominate structural applications while hardwoods are used for appearance products. There are some exceptions such as where low density hardwoods are graded for structural use. One of the largest growth areas for softwood lumber is in treated material (preservative and fire retardant), which in 1986 exceeded 5 billion board feet (BF), 80% of which was southern pine, representing about one-third of all southern pine lumber produced. Treated wood is the fastest growing segment in the industry. The panel market continues to be dominated by plywood, which accounts for about two-thirds of the total panel production. However, the most rapid growth area in panels is the flake-type structural panels, which mimic plywood in many properties, but have a much larger range of possible combinations of size, thickness, density, and strength.

The manufacturing sequences for different wood-based materials can be outlined as follows:

<u>Lumber</u>

stem → log → green lumber → (treatment) → dry lumber → systems

<u>Plywood</u>

stem → ply block → green veneer → dry veneer → plywood

TABLE 1. Wood-based Products Production and Consumption in the U.S.

Lumber (Production and Consumption in Billion Board Feet in 1986) (Anon 1987)	
Softwood Production	
Southern Pine	11.6
Douglas-fir	9.0
Western Pine	10.5
Redwood	1.9
Other	1.7
Total Softwood Production	34.7
Total Softwood Consumption	46.8
Hardwood Production	
Southern	2.7
Appalachian	1.4
Other	2.9
Total Hardwood Production	7.0
Total Hardwood Consumption	7.0
Total Softwood and Hardwood Lumber Production	41.7
Total Softwood and Hardwood Lumber Consumption	53.8
Panels (Estimated Production in Billion Sq Ft in 1987 at Given Thickness Basis) (Anon 1988)	
Structural (3/8 in. basis)	
Plywood	23.3
Waferboard, Oriented Strandboard	4.0
Total Structural	27.3
Non-structural	
Particleboard (3/4 in. basis)	3.8
Hardboard (1/8 in. basis)	5.5
Medium Density Fiberboard (3/4 in. basis)	0.9
Total Non-structural	10.2
Total Panels	37.5

Structural Panels

roundwood → green flakes → dry flakes → panels

Non-structural Panels

green sawdust, chips → dry sawdust, chips → panels

Table 2 gives a list of some key material properties for various products and raw materials in manufacturing. The sequence of processing from raw material to finished product is somewhat different from other basic materials in that, for different products, there is either a reduction or an increase in size. This sequence introduces special NDE requirements for defect location or shape detection, which affect subsequent value, yield, or productivity within the constraints of desired lengths of logs for downstream processing.

TABLE 2. Candidate Wood-Base Materials Properties and Applications for NDE.

Materials Property	Application Examples
Defects	
Surface	Veneer, Lumber, Composites Panels
Internal	Logs, Plyblocks, Installed Structural Members
Basic Properties	
Grain Orientation	Lumber
Density	Composites, Lumber
Moisture Content	All
Strength/Stiffness	All
Permeability/Diffusion	Lumber
Surface Roughness	Veneer
Geometry	
Profile, Shape	Logs, Plyblocks, Lumber
Thickness	Lumber, Veneer, all Panels
Bonding Quality	
Particulates	Panels
Laminates	Laminated Beams, Plywood, Laminated Veneer Lumber
Fasteners	Trusses, Beams, Joists, etc.

INTERNAL DEFECT SCANNING

In order to extract the greatest value from logs, it is currently necessary for an operator to visually scan the surface and ends for defects together with using some technique to establish the cutting pattern for both the particular geometry and the market. The latter functions have been automated to some degree, where geometric scanning is combined with lookup tables to provide the operator with alternatives. If there are no hidden defects, the operator can push the limits of value of a particular log. Since internal defects substantially confound the process, there has been an on-going effort to introduce some internal scanning process that can be done on a real-time basis plus processing this data into an operator-assisted output. Most research approaches have taken advantage of CT (computer tomography) scan equipment available in the medical field. Recent improvements in data processing have made the CT technique possible at least conceptually. The limitation appears to be in sensing with sufficient resolution. The initial research on scanning was done with radiological sources, but the scan time requirements were unacceptable, even with the best theoretical analysis. Several other sensing techniques have been considered--NMR and microwave. NMR is receiving the most concentration, presumably because of the availability of medical scanners, but also because of the resolution and differentiation of types of defects or materials. As with radiological scanning, the limit appears to be unacceptably long scan times. Microwave tomography has been proposed as a candidate and could substantially lessen the scan time, although it is unclear if the resolution is adequate. At this point, internal defect scanning has not been demonstrated to be technically or economically feasible, especially if it is to be used on a real-time basis.

GRADING TECHNOLOGIES

For dimension lumber intended for use as individual pieces (primarily in framing) the major process for value determination is visual grading. The grade is based on association standards as to (primarily) geometry, defects, and moisture content. For the most part, the grades assigned are an attempt to weigh the effect of these factors on the reduction of utility of the entire piece. In some of the more critical applications, some of the material is machine graded for assignment into "stress classes." If machine grading is considered the first generation of commercially-accepted stress grading technologies, then the second generation is developing in three major technological areas; optical/laser scanning, microwave transmission, and ultrasonic, none of which involve application of stress. To some degree, all of these newer technologies depend on relating local grain orientation to local or overall strength. Each method still requires human visual override for defects not sufficiently quantifiable by the method.

Machine Stress Rating (MSR)

MSR is used for lumber intended for critical applications such as in trusses and glulam beams. The volume processed by MSR is only about 1% of the total production of softwood volume in the U.S. and is concentrated primarily in the West. In most machines, the material is fed longitudinally and flatwise, and loaded elastically to obtain a dynamic modulus of elasticity (MOE), either at constant deflection or constant force. Modulus of rupture (MOR) values are assigned based on lookup tables of correlations of MOR to MOE. Additionally, a visual override is provided for certain defects such as size and position of knots.

There are limitations caused by the machine test configuration (primarily flatwise flexure) and the portion of the piece that is stressed (excluding the free ends). While the overall correlations of MOR/MOE may be high, the stress class boundaries can have substantial overlap.

Laser Defect Scanning

A laser scanning system has been developed to provide real time information on location and size of a number of defects on surfaces of softwood lumber (Soest 1984). The technique uses the "tracheid effect" (tracheids are the dominant fibers in softwoods), which is named from the anisotropic scattering and transmission of light in the tracheids. By using the tracheid effect signal together with color detection and retroreflected laser light, the results can be related to major types of defects. The scanner will operate at 6 m/s with a resolution of 0.75 mm in the scan direction and 3.1 mm in the feed direction. An alternative to the above approach is the use of a vision system to differentiate defects in lumber (Conners, McMillin, and Ng 1985). Although black and white scanning had been shown to provide reasonable gray-scale discrimination, color (blue, green, red) showed about a 20% improvement in defect identification. In a controlled study, workers identified defects (position and type) at a 58% accuracy. The computer vision system ranged in identification accuracy from 58 to 76%, depending on the number and combination of black/white and colors processed. Since defects represent only 5 to 13% of the board area in the "common" grades, a challenge in scanning and data processing is the discrimination within clear wood.

Microwave Technologies

Microwave transmission is a sensitive means of obtaining average local grain orientation and has been used in commercial machines such as the Finnograder (Parker 1984). As with the MSR technique, grain orientation correlates primarily with stiffness, requiring a similar process of data base generation. For the Finnograder, additional signal information is obtained from a nuclear gauge to compensate for the effect of density and moisture content on microwave attenuation. There is a proposed system that can provide moisture content, density, and grain orientation by simultaneously obtaining attenuation, phase change, and depolarization of a linearly-polarized microwave source (King, James, and Yen 1985).

Sonic and Ultrasonic Analysis

Sonic analysis was one of the earliest technologies explored in attempting to interrelate MOE to wave speed. Ross and Pellerin (1988) have shown conflicting results on wood-base composites panels. Preliminary work on similar types of specimens but having a wide range of bonding quality produced correlation coefficients (r) of approximately 0.9 for wave attenuation and 0.6 for wave speed vs. elastic and strength properties. When these parameters were analyzed for commercially-made panels, the correlation coefficients were approximately reversed. It is not clear if such conventional parameters are sufficiently sensitive to the influence of local strength-reducing defects. More recently, ultrasonic analysis has moved into a subsequent stage of multiparameter analysis of the transmitted waveform in both the time and frequency domain. The additional parameters analyzed should reduce the variability in correlation with MOE and increase the sensitivity to grade-reducing defects. A companion technology within the family of ultrasonics, acoustic emission, has potential as a supplementary means of sensing flaw development or growth while the material is under stress, such as in flexural or tensile proof loading. Laboratory research has been used to identify the effect of timber breaks (failures from axial compression occurring naturally in standing trees or from felling and/or transportation) that have little effect on MSR stiffness values, but contribute to strength reduction.

PROCESS AND MATERIALS VARIABLES

Moisture Content

In nearly every wood products operation, one of the major material properties that must be sensed is moisture content (MC). MC information is used for process monitoring and control, and product standards. Additionally, moisture content of the material in service is an excellent indicator of potential performance problems from biological deterioration, shape distortion, and strength reduction. The development of on-line sensors for MC measurement continues to be one of the major areas of NDE development and need.

Lumber. The measurement of moisture content after drying lumber serves several functions: (1) meeting grading association standards for grade, (2) sorting for end users or redrying, and (3) monitoring dry kiln performance. Dry kiln performance monitoring is the most complex of the three functions because of the difficulty of relating moisture content of an individual piece of lumber to its specific location in the dry kiln(s). Also, the accuracy of moisture content assignment to a piece of lumber is affected by the area sampled, the sampling technique, and the moisture content distribution (and other wood-related parameters). If the maximum moisture content were the only consideration, the simplest solution would be to dry all material below that value. However, the value of lumber is considerably affected by warp produced during drying such that about 1 to 2% value loss occurs from each point of moisture content loss below the standard. Therefore, the optimal moisture content distribution for material from the dry kiln would require the smallest possible standard deviation with the mean value just below the moisture content standard. The practical result is that an acceptable operating strategy is to allow about 5% "wets" (with the concomitant loss in value) in order to keep the mean as close as possible to the prescribed standard. Controlling the degrade from warp by reducing over-drying is a major opportunity in lumber manufacturing. In order to meet the optimal distribution of moisture content, the kilns must be as uniform as possible in drying conditions, and the drying schedule (temperature and relative humidity vs. drying time or moisture content) must be controlled as well as the endpoint detection to reach the proper mean. However, most dry kilns are operated in an open loop manner since the needed sensors for distributional and endpoint control are not generally considered as state of the art, although several techniques have shown promise. Consequently, the NDE burden on the dry lumber moisture meter is even greater.

Veneer. Softwood plywood is one of the oldest reconstituted wood products having several key attributes: convenient-size panels of structural quality, relatively smooth surfaces, and excellent in-plane dimensional stability (plywood shrinks and swells volumetrically about the same as solid wood, producing nearly all of the dimensional change in thickness). Most of the performance problems of plywood are related to poor adhesive bonding, which in turn is often caused, directly or indirectly, by the moisture content of the veneer. In many respects, the optimal drying strategy for veneer is similar to that for lumber -- a narrow moisture content distribution and a "suitable" mean value. Ideally, the mean should be high enough to promote good adhesion, reduce shrinkage (controlling volume loss), and permit greater throughput (very strongly related to the final moisture content value). However, moisture content must be low enough to prevent the generation of entrapped steam at the end of the pressing cycle (causing "blows," which are delaminations from release of entrapped steam). However, the major problem in typical veneer dryers is non-uniformity in moisture content, making the only practical operating strategy one in which veneer is substantially over-dried to minimize the distribution. The over-drying, in addition to the negatives mentioned, causes physical loss of material because of brittle fracture during conveyance, surface inactivation (presumably from loss in hydroxyl bonding sites), and suboptimal adhesive formulations. The dry veneer moisture meter therefore is a major NDE tool to use for moisture content distributional analysis and control, endpoint control, and to identify out-of-specification veneer that could result in future delamination problems. A number of dry veneer moisture meters have been used in the industry, spanning the frequency range of 60 Hz to the microwave region, with most current meters operating near line frequency. In a Catch-22 scenario, the line-frequency meters are used because most other types are less accurate at low moisture contents, whereas the reverse logic applies for high moisture content. The consequences of over-dried veneer leading to over-dried plywood have not been generally reported, although the green peel thickness must be increased to compensate for lower product thickness, and thicker material not only takes longer to dry, but also causes a poorer moisture content distribution. Other forms of veneer drying, direct contact (in presses) and microwave introduce somewhat different, but related, problems of moisture content endpoint and distribution.

Particles. Particle drying is generally accomplished in direct-fired systems in which the hot gases are used both in drying and (to varying degrees, depending on dryer design) conveyance. The outfeed moisture content is generally measured with IR (infrared) meters using surface reflectance and narrow wavelength filters. The IR meters are used as supervisory gauges, controlling outfeed moisture content by adjusting the inlet temperature of the gases. In many cases, the surface and core materials in the resulting product are different in geometry, size, and moisture content as required for the process or product. Unlike solid materials (lumber and veneer), both the process and conditions can be designed and operated to obtain close to optimal conditions for downstream operations (needed for the particular adhesive and pressing conditions desired). For some selected products, panel moisture content may be increased or sorts made according to end use requirements.

Density

Measurement of density has not been identified as a necessary variable for processing or utilization of lumber or veneer. One commercial use of density data is in the Finnograder in which density is required to improve the prediction equation for strength. A proposed use of density (or mass) is in presorting green lumber into drying classes for more uniform product moisture content. For the most part, other uses for density are primarily to establish one of the growth characteristics for specific sources of lumber and timber, and the techniques are largely destructive. In contrast, density is one of the more important controllable product characteristics for particulate-based panels. Both average density and density gradients are important processing variables. Average density is controlled by both the wood species used and the pressure in pressing. As an operating strategy, the panels should be fabricated to meet the required physical properties, but at the lowest density in order to reduce raw material consumption and final panel mass. Density gradients are controlled primarily by the pressing schedules, which normally cause denser surfaces than cores. The density gradient is one of the most important characteristics of the final panel in terms of surface properties and strength. There exists no present on-line means of measuring density gradient and this remains a substantial NDE opportunity. Such on-line knowledge of density gradients could provide useful feedback since the gradient can be affected by the press schedule.

Treatability and Treatment

Treated solid wood and plywood represents the fastest growth segment within the wood products market. However, there are no NDE methods to evaluate treatability of material or the quality/quantity of treatment. The development of a means of presorting material by degree of treatability would be a major advancement. An experimental method has been reported (Humphrey and Morrell 1987) in which the penetration of water into dry Douglas-fir has been measured as a means of sorting. In general, some technique must be developed before the Western species can be fully accepted as candidates for treatment. Even more pervasive is the problem of determining the effectiveness of treatment. All current techniques require destructive assays and are essentially laboratory methods. Measurement of the distribution of treatment chemicals, which in many cases is more important than the average values, is beyond state of the art.

Surface Defects

Apart from natural defects in wood-base materials, there are a number of other types of surface characteristics or manufacturing defects as future candidates for NDE. One type could be categorized as surface roughness, which applies to both lumber and veneer. For lumber, roughness applies to the resulting surface after sawing and/or planing. Veneer roughness is caused largely by the texture of the wood being peeled. Panels, particularly non-structural, have product requirements which preclude visual defects, including those that could telegraph through overlays in subsequent processing. In all of these areas, optical techniques have been studied, but the research efforts are primarily exploratory. It appears that there are several laser reflectance candidate techniques, depending on the size of the surface disturbance.

Adhesive Bonding

A number of wood products involve reconstitution of raw materials using adhesives, including all of the panels and certain parallel-laminated and/or end-jointed strength members. For most of these materials, stress testing (such as MSR) represents the current state of the art and is only in limited practice. Acoustic emission and acousto-ultrasonics have been found to provide bond strength correlations for destructive and non-destructive evaluation of non-structural panels, and a commercial system is under development for NDE testing of particleboard (Beall 1987). Acousto-ultrasonics shows promise for both laboratory and on-line sensing of bond quality for veneer and lumber assemblies.

FIELD NDE

The most widely-used tool for field NDE of wood-based materials is the hand-held moisture meter. Of these meters, the predominant one used is the DC conductance type, capable of measurement from about 6% (corresponding to about 30% RH at ambient temperature) to approaching the fiber saturation point (FSP, about 27% for solid wood), which is the theoretical condition of equilibrium at 100% RH. Wood-base materials in use generally fall in the 6 to 12% moisture content range. Since decay can only occur at moisture content levels at FSP or greater, the conductance meter is an excellent, albeit tedious, means of surveying for potential decay in materials in the field. Surface contact meters are also used to a more limited extent and with considerable risk since most react primarily to surface conditions of the material. Assuming that no free water exists on initial installation (not always the case with dimension lumber), such a condition can occur only through means other than the normal vapor sorption process. The most common sources of free water are condensation during vapor transmission through walls and roofs, and leakage of liquid water from internal or external sources. Unfortunately, the most common cause of decay in commercial/industrial wooden materials is improper design and installation. Structural wooden members (glulam, trusses) in service have excellent safety records, due in a large part to very conservative standard design practices. When decay is suspected, it can usually be inferred from residual water stains from water incursion. There are exceptions when the intrusion of water has occurred at some hidden point and has concentrated in the center of the wooden member. Evaluation techniques for decay of in-service members or systems are (1) visual, (2) mechanical penetration, (3) coring for subsequent testing, (4) moisture content measurement, and (5) X-ray. In general, the detection of fungal degradation is much simpler than the corrective measures, primarily because the extent of damage is difficult to determine. For example, until degradation from decay reaches about 5 to 10% of the initial oven-dry mass of wood, field detection (or even laboratory detection) is unlikely. However, at 10% mass loss, wood may lose about 50% of its toughness, meaning that subsequent normal dynamic loading could cause unanticipated failure. For this reason, structural engineers who specialize in corrective measures tend to be very conservative. The lack of any substantive field tools for NDE of structural members continues to limit the breadth and intensity of inspection. One developing area of promise uses ultrasonic evaluation, either AE or AU. A second area of concern is the adhesive bond strength of laminated materials (glulam beams, laminated veneer lumber, plywood, etc.). The only current alternative is sampling of material and laboratory testing of the glueline shear strength. A technique is needed for survey inspections of these products, with destructive testing only for verification of suspect areas. Some moisture sensing elements (including embedded types) have been used for remote periodic measurement of the environmental conditions to imply the state of the material.

REFERENCES

Abbott, A.R., and Elcock, G., Pole Testing in the European Context, Proc. Sixth Nondest. Testing of Wood Symp., Pullman, WA, pp. 277-302, 1987.

Anon, North American Lumber Output Hits New High, Forest Ind. 114(7), pp. 12-14, 1987.

Anon, Booming Panel Industry Shatters Output Records, Forest Ind. 115(4), pp. 18-19, 1988.

Bach, L., Machine Stress-Rating Panel Products, Proc. Fifth Nondest. Testing of Wood Symp., Pullman, WA, pp. 75-97, 1985.

Beall, F.C., Future Applications of Acoustic Emission and Acousto-ultrasonics, Proc. Sixth Nondest. Testing of Wood Symp., Pullman, WA, pp. 369-375, 1987.

Conners, R.W., McMillin, C.W., and Ng, C.N., The Utility of Color Information and Identification of Defects in Surfaced Hardwood Lumber, Proc. First International Conference on Scanning Technology in Sawmilling, San Francisco, Paper 18, pp. 1-33, 1985.

Conners, R.W., McMillin, C.W., and Vasquex-Espinosa, R., A Prototype Software System for Locating and Identifying Surface Defects in Wood, Proc. Scanning Technology for the Eighties, Vancouver, pp. 107-114, 1984.

Galligan, W.L., and Barrett, J.D., Nondestructive Testing for Structural Purposes, 42nd Annual Meeting, Forest Prod. Res. Soc. Exec. Sum. pp. 49-50, 1988.

Humphrey, P.E., and Morrell, J.J., Distinguishing Between Refractory and Permeable Wood at Mill Production Speeds: A Possible Approach, Proc. Sixth Nondest. Testing of Wood Symp., Pullman, WA, pp. 269-276, 1987.

King, R.J., James, W.L., and Yen, Y.H., A Microwave Method for Measuring Moisture Content, Density, and Grain Angle of Wood, Proc. First International Conference on Scanning Technology in Sawmilling, San Francisco, Paper 16, pp. 1-9, 1985.

McDonald, K.A., Cramer, S.M., and Bendtsen, B.A., Research Progress in Modeling Tensile Strength of Lumber from Localized Slope of Grain, Proc. Sixth Nondest. Testing of Wood Symp., Pullman, WA, pp. 113-123, 1987.

Parker, R.S., Electrical Scanning in Wood Application to Moisture Content Gauging, Proc. Scanning Technology for the Eighties, Vancouver, pp. 48-62, 1984.

Ross, R.J., and Pellerin, R.F., NDE of Wood-Based Composites with Longitudinal Stress Waves, For. Prod. Jour. 38(5), pp. 39-45, 1988.

Samson, M., and Hailey, J.R.T., Scanning and Imaging Techniques for Assessing Decay and Wood Quality in Logs and Standing Trees, Proc. Sixth Nondest. Testing of Wood Symp., Pullman, WA, pp. 83-93, 1987.

Samson, M., and Hailey, J.R.T., International Accomplishments in Nondestructive Testing and Evaluation, 42nd Annual Meeting, Forest Prod. Res. Soc. Exec. Sum. pp 46-47, 1988.

Soest, J.F., Optical Scanning Technique for Defect Location, Proc. Scanning Technology for the Eighties, Vancouver, pp. 120-123, 1984.

NONDESTRUCTIVE TESTING OF WOOD PRODUCTS AND STRUCTURES: STATE-OF-THE-ART AND RESEARCH NEEDS

R. H. Falk, M. Patton-Mallory, and K. A. McDonald
USDA Forest Service, Forest Products Laboratory
Madison, WI 53705

INTRODUCTION

Though much of the research on nondestructive testing and evaluation has focused on manufactured materials such as metals, polymers, ceramics, and concrete, many applications have been found for the use of nondestructive testing techniques in the inspection and evaluation of wood products and structures. In the case of existing wood structures, these techniques are used for in-place material property determination and detection of decay. In the case of new wood products, nondestructive methods are often used for process control, quality assurance measurements, and grade classification (during and after processing).

Nondestructive techniques have also found application in the evaluation of existing wood buildings and built-up components and in the periodic inspection of other essential facilities and structures such as wood utility poles, cooling towers, dock piling and piers, scaffold planks, and mine timbers.

This paper overviews nondestructive testing research as it is applied to wood products and structures and discusses areas of research needed to more fully utilize this technology. Specifically, the discussion includes acoustic techniques such as sonically and ultrasonically induced stress waves, acoustic emissions, and acousto-ultrasonics. Other methods, including optical scanning, electrical resistance, and those relying on the ability of wood to transmit waves in the form of x rays or microwaves, are discussed. Methods incorporating mechanical vibration characteristics, such as transverse vibrations, are also included.

BACKGROUND

Wood has been used for thousands of years as a construction material for dwellings, ships, and bridges. While availability and ease of workability have historically made wood a logical choice for construction, more recent uses are inspired by its low cost and high strength to weight ratio compared with metals and cementous materials.

Because the material properties of wood are not tightly controlled by its natural manufacturing process, these properties exhibit inherent variability. The presence of natural defects such as knots, splits, and checks, and the wide diversity of species available for various end uses, add to this variability.

Wood is a material with orthotropic material properties. The structural properties of a wood member depend not only on its orientation when cut from a log, but also on the distribution, size, and shape of the strength-reducing characteristics within the piece and the cellular makeup of the clear wood, which varies between species. The environmental conditions in which wood is used are very important because moisture content of the wood dramatically affects its mechanical properties and its susceptibility to degradation by decay. As a result of these factors, users of wood have long recognized the need to grade or assess quality of wood building materials before they are used in construction.

Nondestructive evaluation (NDE) of wood has been performed for generations. Visual inspection for strength-influencing characteristics has been traditionally used to grade wood into end-use categories. This assessment of wood quality is based only on the size and distribution of anomalies that can be seen on the surface of the lumber. Using this technique, strength-reducing characteristics and species are empirically related to member tensile, bending, or compression performance.

Machine stress rating (MSR) has evolved as a means of measuring and evaluating mechanical properties (strength and stiffness) of lumber (Hoyle 1968, Logan 1978). This nondestructive technique is based on the fact that a beam in bending will deflect in proportion to its average modulus of elasticity (MOE) when subjected to a known load. Lumber bending strength and other properties are then empirically related to MOE by species class. Visual grading and MSR have received considerable research attention and are well established in the solid wood products industry.

In addition to solid wood products, many flake, particle, fiber, and veneer products have been developed that are a combination of wood and adhesive. Following the development of these products there has been a growing need for NDE methods for grading and quality assurance. These methods usually involve identifying internal voids undetectable by visual means, such as "blows" in particleboard. These methods are also used for sorting veneers in the manufacture of parallel laminated veneer lumber.

The concept of "sound" wood has been used for generations to describe another basic form of NDE--aural assessment. Often used to assess the condition of wood utility poles, this method relies on the ability of an inspector to recognize sound patterns, produced by hammer blows, that are indicative of the presence of decay. Besides being rather subjective, this method is limited by the rather limited frequency range perceptible by the human ear, the upper limit of which is about 20 kHz. In response to this limitation, research has incorporated the use of electronic sensors.

Electronic sensors containing piezoelectric crystals are sensitive to much higher frequencies of sound vibration, and mechanical accelerometers are available for detecting lower frequency sound vibrations. These sensors replace the human ear and serve as the basis of most acoustic techniques used for NDE of new wood products for quality assurance and for the in-place evaluation of wood structures.

NONDESTRUCTIVE EVALUATION METHODS BASED UPON ACOUSTIC MEASUREMENTS

Differences in terminology exist in the literature on acoustic methods applied to the nondestructive testing of wood. Though referred to by several names, such as sonic stress waves, ultrasonic stress waves, ultrasonics, and ultrasonic pulse methods, these methods rely on the theory of acoustic wave propagation and usually differ only in the mode and frequency of excitation. Typically, these methods incorporate two sensors to determine wave speed, the first at one location on the member used to start a timer and the second at another location on the member used to stop the timer. Some devices include signal processing hardware for time and frequency domain analysis.

The basic theory of acoustic wave propagation in a solid material incorporates the fact that energy storage (wave speed) and energy dissipation (damping or attenuation of the wave) characteristics are controlled by the same mechanisms that determine the mechanical properties of the material (Jayne 1959).

Sound Velocity Techniques

The determination of wood material properties using acoustic wave propagation has attracted considerable research effort. Bertholf (1965) described the use of elementary stress wave theory for the prediction of dynamic strain in wood. Galligan and Courteau (1965) applied this theory for the measurement of MOE in structural lumber by inducing stress waves and measuring wave passage time. This work lead to the use of sonic stress wave velocity as a predictor of MOE, modulus of rupture (MOR, ultimate bending load capacity), and other material parameters (Kaiserlik and Pellerin 1977, Pellerin 1965, Ross and Pellerin 1988, Vogt 1985).

A high correlation (correlation coefficient of about 0.90) has been observed between stress wave velocity predictions of MOE and bending MOE from bending tests. It is more difficult to predict MOR using stress wave velocity because the presence (and location) of defects (such as knots) and the slope of grain have a significant effect on MOR but much less of an effect on longitudinal stress wave velocity over the length of a typical piece of structural lumber. Because the defects in wood affect the straightness of the grain (fiber direction) in the wood, any method that is sensitive to the slope of grain has high potential for assessing the strength of wood. Gerhards (1980) investigated the effects of crossgrain on stress wave velocity and found that wave speed varied by as much as 1% per degree of grain angle up to about 30°. This study concentrated on local measurements with respect to grain deviations.

Using stress wave velocity measurements, Bucur and Archer (1984) determined the elastic compliance matrix for various species of wood. Shear moduli, Young's moduli, and Poisson's ratios for the longitudinal, radial, and tangential directions of the wood correspond to various coefficients of this matrix.

The relative change in moisture content of wood as it dries is reported to be well correlated with the speed of ultrasonic pulses (James et al. 1982). This fact has potential as a means to monitor and affect control of the drying process of wood.

Stress wave velocity measurements have been used to predict material properties in a variety of wood products including structural composite lumber, particleboard, roof and floor sheathing, floor underlayment, and medium density fiberboard (Ross and Pellerin 1988). Jung (1979) suggested stress wave grading techniques for use on veneer sheets used in the manufacture of parallel laminated veneer (PVL). Increases of 30% in allowable bending stress are being realized by sorting PLV strips using stress wave measures of MOE.

Stress wave velocity measurements have also been used to locate the presence and extent of decay in wood members. Hoyle and Pellerin (1978) used a sonic stress wave device to identify areas of decay in two school buildings constructed with glued-laminated timber arches. Areas of decay were successfully mapped out in the wooden arches by comparing stress wave velocity measurements from nondecayed portions of the arches to those from decayed portions. This correlation between perpendicular-to-grain (radial and tangential) stress wave velocities and levels of decay has been further investigated by Rutherford (1987), who found that stress wave velocity measurements could be used to predict about 80% of the variability in MOE for decayed and nondecayed portions of the wood.

Nondestructive tests on a wood barn by Lanius et al. (1981) compared two methods, sonic stress wave velocity measurements and visual grading. Visual-grading estimates of strength and stiffness and those based upon stress wave measurements were compared statistically. It was found that the stress wave method yielded mean material property estimates comparable to the visual grading technique. Additionally, the authors noted the importance of the stress wave measurements in detecting low-strength members.

Acoustically based NDE devices have found widespread use in the electric power industry and are used to monitor the structural condition of wood utility poles subject to the degradative effects of decay and insect attack (Miller et al. 1965, Shaw 1978, Arita and Kuratani 1984). Bulleit and Falk (1985) worked with the power industry to investigate the use of stress wave velocity measurements to distinguish between strength-reducing decay and non-strength-reducing growth ring separations (ringshake). Using the results of field test measurements and finite element model analysis, they concluded that stress wave velocity measurements alone were not sufficient to confidently differentiate between ringshake and decay.

These acoustic NDE devices have also been useful in the inspection of marine pilings subjected to marine borer attack and decay (Walden and Trussel 1963, Pellerin 1978, Agi 1978, Aggour et al. 1984). Using a procedure similar to wood utility pole inspection, increases in wave passage time attributable to damaged wood are compared with a baseline time measurement for undamaged wood.

An ultrasonic stress wave device has been investigated for use in the detection of straight-grained wood in lumber to provide information for cutting decisions (McDonald 1978). Utilizing a pulsed, 500 kHz transducer and a 1 MHz receiver to propagate sound through wood, this patented (McDonald and Cox 1972) through-transmission technique locates all internal and surface defects that are associated with local grain-angle changes and includes location of decay or rotted wood. Using water as a couplant between the wood and the sensor, the device has been shown to function adequately for hardwood and softwood species, rough-sawn or planed, and green or dry lumber of thicknesses up to 4 in.

Application of sound velocity techniques for in-place wood structures has seen limited implementation because of the important coupling requirements necessary for proper signal transmission. Improvements in transducer construction and the development of self-coupling transducers may improve the utility of these techniques.

Acoustic Emission Techniques

Most of the failure modes that occur in wood are fracture-type mechanisms that emit high-frequency sound energy. Each release of sound energy is typically referred to as a burst of acoustic emission (AE). These brittle-type failure processes begin with microscopic fractures, particularly at stress concentrations around naturally occurring defects or those resulting from the manufacturing process. Acoustic emission techniques have been used extensively as NDE tools for materials other than wood. Typically, the material being evaluated is subjected to a proof load while the quantity and location of AEs are measured.

Although most of the research in utilizing AEs from wood has focused on damage occurring during destructive testing, recent research by Groom and Polensek (1987) shows that bending strength, proportional limit load, and ultimate bending deflection of lumber could be predicted with some accuracy through AE measurements below the proportional limit. Porter et al. (1972) evaluated use of AEs to assess quality of finger joints in wood products. It was concluded that the rate of AEs generated just beyond the proportional limit of the material was a reasonable predictor of finger joint strength.

Wood decay within a structural member causes significant loss of strength before one can visually detect the presence of decay. Beall and Wilcox (1987) used AE techniques to quantify the level of incipient decay in wood members by monitoring AEs during a compression test of small blocks of wood loaded in the radial direction. The specimens with early decay were found to begin emitting sound energy at a much lower load than the nondecayed control specimens.

As research progresses toward understanding the mechanisms that generate AEs during destructive tests of wood products, the technology will become more useful for NDE of in-place wood members. For example, many types of chemicals are applied to wood to enhance its behavior for a particular end use. Some chemical treatments, such as those used for fire retardancy, are suspected to have a degrading effect on wood strength if subjected to temperatures in the range of 150 to 180°F. Monitoring AEs while testing the treated wood may provide a useful index of damage resulting from these treatments.

Waveform Analysis Techniques

There is considerably more information available in the acoustic waveform received through a material than is contained in the velocity measurements previously described. High-frequency sound waves (20 to 300 kHz) that pass through a material are influenced by the properties of the material. Acousto-ultrasonics is a technology that combines the AE analysis of the received waveform with ultrasonic pulsing technology. A typical system uses resonant sensors. Acousto-ultrasonics can be used in either a through-transmission or a one-sided mode. Both the time and the frequency domains of the received waveform are evaluated in a relative sense, using measurements such as waveform ringdown counts, measures of average signal level (root mean square, RMS), peak amplitude, and frequency content.

Patton-Mallory et al. (1987) used multiple waveform characteristics to distinguish decayed wood from nondecayed wood under controlled conditions in the laboratory. Dunlop (1983) analyzed the velocity and damping characteristics of longitudinal compression waves to evaluate deterioration in wood utility poles. The technique indicated that wood decay influences attenuation to a greater degree than it influences wave velocity.

Beall (1987) evaluated waveform RMS to successfully monitor glueline curing of a wood adhesive connection and found RMS to increase proportionally with the curing of the adhesive. Lemaster and Dornfeld (1987) used acousto-ultrasonics in an effort to detect knots in lumber, decayed wood, grain orientation, and wood moisture content. They concluded that the laboratory technique looked promising for detecting knots, decayed wood, and a range of grain orientations. The results were inconclusive for detecting moisture content changes. The acousto-ultrasonic technique is still evolving and may have an important role in future evaluation of wood structures after additional research is undertaken.

NONDESTRUCTIVE EVALUATION METHODS BASED ON VIBRATIONAL MEASUREMENTS

While transverse vibrational techniques have been used to assess the material properties of wood members in bending (Pellerin 1965) and have lead to the development of commercially available grading devices, recent efforts have been directed towards a better understanding of the time and frequency domain characteristics of longitudinally induced acoustic signals. Sound wood of a particular geometry has characteristic damping properties that influence the propagation of different frequencies of sound or vibration. The measurement of energy storage or energy loss can be made in the time or frequency domain. Depending on the method used, lower frequency (0 to 20 kHz) or higher frequency (20 to 350 kHz) ranges are analyzed.

Wang (1980) investigated the change in the fundamental mode of frequency for wood specimens exposed to decay fungi by measuring their vibrational characteristics. Increasing levels of decay exposure dramatically reduced the stiffness of the specimens and hence lowered the fundamental mode of frequency.

Modal analysis techniques have also been used for the determination of in-place wood utility pole stiffness and strength (Murphy et al. 1987). This technique monitors the frequency content of the pole and utilizes the fact that stronger and stiffer poles vibrate at higher frequencies than do poles degraded by wood decay. Frequency content can be related to pole condition and hence to expected remaining life.

Additional research is needed in this area to correlate the vibrational characteristics (energy storage and energy dissipation) of the wood member with its stiffness and strength while accounting for the presence of knots, splits, grain deviations, incipient decay, and other defects.

OTHER NONDESTRUCTIVE EVALUATION METHODS APPLIED TO WOOD

In addition to acoustically based NDE methods, a number of other techniques have been applied to the evaluation of wood and wood structures. Many NDE systems have been developed and implemented for the NDE of wood products in a myriad of applications before, during, and after processing from logs to finished wood products. Included are the use of microwaves, x rays, lasers, optics, and digital images. A brief summary of these applications follows.

Microwaves have been used to detect defects in wood products based on wave sensitivity to density changes and grain direction. Using a rotating microwave transmitter, the grain angle at the surface of wood has been measured under laboratory conditions (James et al. 1985). Application of the technique, however, will require additional research to account for the interaction between density and moisture content effects.

The surface grain angle of wood can be measured by two other methods. One uses an electrical capacitance technique to obtain the grain angle based on the dielectric constant of wood, which is about 1.5 times greater along the grain than across the grain (McDonald and Bendtsen 1985). The other is an optical measurement using a laser light source and detectors to measure the variations in the intensity of light reflected from the wood surface to sense grain direction and other defects (Soest 1987).

X-ray tomography, also known as computerized axial tomography (CAT), uses a photon source and an array of receivers to measure the absorption of gamma rays. This technique is capable of determining internal damage in wood (Ellinger et al. 1979, Funt and Bryant 1987) and has found application mainly in the detection and location of internal defects in round wood material, such as logs, poles, and pilings. Defects located include knots, cracks, worm holes, and decay.

Nuclear magnetic resonance (NMR) imaging of logs has been explored (Chang et al. 1987, Hailey and Swanson 1987), with positive results that are predicted to have an impact on future NDE applications of assessing the internal quality of wood. Digital images of slices of logs have shown the internal wood structure and defects in detail. Pattern recognition techniques will be necessary to automate the interpretation of such images for future application. Practical application of these commonly used medical devices for wood evaluation is limited by high cost and by the lack of portability necessary to test in-place structures.

Several optic systems have been developed for sensing and locating surface defects on lumber surfaces for grading processing decisions (Connors et al. 1987, Godinez 1987, Wallis 1987). These systems collect reflected light from the board surface or process analog or digital images from which pattern recognition algorithms obtain the defect information. Surface quality and acceptable blemishes are variables that must be given serious attention with these systems when transferring the technology to the production line.

RESEARCH NEEDS

This discussion indicates there are several nondestructive test methods that are being used for the evaluation of wood products and structures. However, many of these techniques cannot be fully utilized until more information is available to help interpret NDE measurements.

Specific areas of need are in relating NDE parameters to wood material properties and estimates of remaining life. This is especially important in the case of degraded wood. When estimating or predicting yhe strength of a piece of wood, it is known that flaws often cause failure. Techniques that can reliably and economically detect the presence of flaws will be considered useful in evaluating the strength of members. Research is also needed to bridge the gap between the known presence of a flaw, or decayed area, and the calculated strength or lifetime prediction.

Quality assurance and quality control procedures are emerging areas requiring research. Particularly for wood composite products and built-up wood members, reliable methods are needed to determine adhesive-to-wood bond strength and to assess mechanical properties for chemically treated wood materials. The challenge of NDE of wood products is enhanced by the natural variation found in this unique building material that is manufactured by nature.

LITERATURE CITED

Aggour, M. S., Ragab, A. M., White Jr.,E. J., Determination of In-Place Timber Piling Strength, Transportation Research Record 962, Univ. of Maryland, College Park, pp. 69-77, 1984.

Agi, J. J., Nondestructive Testing and Structural Analysis of In-Place Wood Marine Piling, Proceedings of the Fourth Nondestructive Testing of Wood Symposium, Washington State University, Pullman, WA, 1978.

Arita, K., and Kuratani, K., Wooden Pole Tester: For Determining the Strength of Decayed Poles, Japan Telecommunications Review, Nippon Telegraph and Telephone Public Corp., Tokyo, Japan, pp. 167-173, July 1984.

Beall, F. C., Wilcox, W. W., Relationship of Acoustic Emission during Radial Compression to Mass Loss from decay, Forest Prod. J. 37(4): 38-42, 1987.

Beall, F. C., Acousto-Ultrasonic Monitoring of Glueline Curing, Wood Fiber Sci. 19(2): 204-214, 1987.

Bertholf, L. D., Use of Elementary Stress Wave Theory for Prediction of Dynamic Strain in Wood, Bulletin 291, Technical Extension Service, Washington State University, Pullman, WA, 1965.

Bucur, V. and Archer, R. R., Elastic Constants for Wood by an Ultrasonic Method, Wood Sci. Technol. 18: 255-265 (1984).

Bulleit, W. M., and Falk, R. H., Modeling Stress Wave Passage Times in Wood Utility Poles, Wood Sci. Technol. 19: 1983-191, 1985.

Chang, S. J., Wang, P. C., and Olson, J. R., Nuclear Magnetic Resonance Imaging of Hardwood Logs, Proceedings 2nd International Conference on Scanning Technology in Sawmilling, Oakland/Berkeley Hills, CA, IX-1-IX-8, Oct. 1987.

Connors, R. W., Klinkhachorn, P., McMillin, C. W., Franklin, J. P., Ng, C., A Computer Vision System for Grading Hardwood Lumber, Proceedings 2nd International Conference on Scanning Technology in Sawmilling, Oakland/Berkeley Hills, CA, XV-1-XV-7, Oct. 1987.

Dunlop, J. I., Testing of Poles by Acoustic Resonance, Wood Sci. Technol. 17: 31-38, 1983.

Ellinger, H., Morgan, I. L., Klinksiek, R., Hopkins, F., and Thompson, J. N., Imaging Applications for Automated Industrial Inspection and Assembly, Proceedings of the Society of Photo-Optical Instrumentation Engineers, Washington, DC, Apr. 1979.

Funt, B. V., and Bryant, E. C., Detection of Internal Log Defects by Automatic Interpretation of Computer Tomography Images, Forest Prod. J. 37(1): 56-62, 1987.

Galligan, W. C., and Courteau, R. W., Measurement of Elasticity of Lumber With Longitudinal Stress Waves and the Piezoelectric Effect of Wood, Proceedings of the Second Symposium on the Nondestructive Testing of Wood, Washington State University, Pullman, WA, 1965.

Gerhards, C. C., Effect of Cross Grain on Stress Waves in Lumber, Forest Products Laboratory Research Paper FPL 386, U.S. Department of Agriculture, Forest Service, Madison, WI, May 1980.

Groom, L., and Polensek, A., Nondestructive Prediction of Load-Deflection Relations for Lumber, Wood Fiber Sci. 19(3): 298-312, 1987.

Godinez, P. A., Automated Optical Inspection of Wood Grain Patterns and Defects, Proceedings 2nd International Conference on Scanning Technology in Sawmilling, Oakland/Berkeley Hills, CA, XVI-1-XVI-5, Oct. 1987.

Hailey, J. R. T., and Swanson, J. S., Imaging Wood Using Magnetic Resonance, Proceedings 2nd International Conference on Scanning Technology in Sawmilling, Oakland/Berkeley Hills, CA, VII-1-VII-12, Oct. 1987.

Hoyle, R. J., Background to Machine Stress Rating, Forest Prod. J. 18(4):87-97, 1968.

Hoyle, R. J., and Pellerin, R. F., Stress Wave Inspection of a Wood Structure, Proceedings of the Fourth Nondestructive Testing of Wood Symposium, Washington State University, Pullman, WA, 1978.

James, W. L., Boone, R. S., and Galligan, W. L., Using Speed of Sound to Monitor Drying in a Kiln, Forest Prod. J. 32(9): 27-34, 1982.

James, W. L., Yen, Y., and King, R. J., A Microwave Method for Measuring Moisture Content, Density, and Grain Angle of Wood, Forest Products Laboratory Research Note FPL-0250, U.S. Department of Agriculture, Forest Service, Madison, WI, 1985.

Jayne, B. A., Vibrational Properties of Wood as Indices of Quality, Forest Prod. J. 9(11): 413-416, 1959.

Jung, J., Stress-Wave Grading Techniques on Veneer Sheets, Forest Products Laboratory Research Paper FPL 27, U.S. Department of Agriculture, Forest Service, Madison, WI, 1979.

Kaiserlik, J. H., and Pellerin, R. F., Stress Wave Attenuation as an Indicator of Lumber Strength, Forest Prod. J. 27(6): 39-43, 1977.

Lanius, R. M., Tichy, R., and Bulleit, W. M., Strength of Old Wood Joists, J. Struct. Div. ASCE 107(ST12), 1981.

Lemaster, R. L., and Dornfeld, D. A., Preliminary Investigation of the Feasibility of Using Acousto-Ultrasonics to Measure Defects in Lumber, J. Acoust. Emiss. 6(3): 157-165, 1987

Logan, J. D., Machine Stress Rating, Proceedings of the Fourth Nondestructive Testing of Wood Symposium, Washington State University, Pullman, WA, 1978.

McDonald, K. A., Lumber Defect Detection by Ultrasonics, Forest Products Laboratory Research Paper FPL 311, U.S. Department of Agriculture, Forest Service, Madison, WI, 1978.

McDonald, K. A., and Bendtsen, B. A., Localized Slope-of-Grain--Its Importance and Measurement, Proceedings of the Fifth Nondestructive Testing of Wood Symposium, Washington State University, Pullman, WA, pp. 477-489, 1985.

McDonald, K. A., and Cox, R. G., Ultrasonic Detection of Lumber Defects, U.S. Patent No. 3,664,180, 1972.

Miller, B. D., Taylor, F. L., and Popeck, R. A., A Sonic Method for Detecting Decay in Wood Poles, American Wood Preserves Association, Form No. 1351, 1965.

Murphy, M. W., Franklin D. E., and Palylyk, R. A., A Nondestructive Testing Technique for Wood Poles, B.C. Hydro Research and Development, Surrey, BC, 1987.

Patton-Mallory, M., Anderson, K. D, and DeGroot, R. C., An Acousto-Ultrasonic Method for Evaluating Decayed Wood, Proceedings of Sixth Nondestructive Testing of Wood Symposium, Washington State University, Pullman, WA, pp. 167-189, 1987.

Pellerin, R. F., Vibrational Approach to Nondestructive Testing of Structural Lumber, Forest Prod. J. 15(3): 93-101, 1965.

Pellerin, R. F., Nondestructive Testing of Wood--A Possible Method for Timber Piling, Proceedings of the Fourth Nondestructive Testing of Wood Symposium, Washington State University, Pullman, WA, 1978.

Porter, A. W., El-Osta, M. L., and Kusec, D. J., Prediction of Failure of Finger Joints Using Acoustic Emissions, Forest Prod. J. 22(9): 74-82, 1972.

Ross, R. J., and Pellerin, R. F., NDE of Wood-Based Composites With Longitudinal Stress Waves, Forest Prod. J. 38(5): 39-45, 1988.

Rutherford, P. S., Nondestructive Stress Wave Measurement of Incipient Decay in Douglas Fir, M.S. Thesis, Department of Civil and Environmental Engineering, Washington State University, Pullman, WA, 1987.

Shaw, D. A., Bandpass Filter Comparator for Confirming the Absence of Serious Internal Voids in Wood Poles, IEEE Trans. Instrum. Meas. IM-27 (31), 1978.

Soest, J. F., Applications of Optical Measurements of Slope of Grain, Proceedings 2nd International Conference on Scanning Technology in Sawmilling, Oakland/Berkeley Hills, CA, XII-1-XII-10, Oct. 1987.

Vogt, J. J., Evaluation of the Tensile and Flexural Properties and Internal Bond of Medium Density Fiberboard Using Stress Wave Speed and Attenuation, M.S. Thesis, Washington State University, Pullman, WA, 1985.

Walden, C. C., and Trussell, P. C., Sonic Inspection of Marine Piles, Wood Pres. News, Jan. 1963.

Wallis, H., Defect Detection for Wood Classification, Proceedings 2nd International Conference on Scanning Technology in Sawmilling, Oakland/Berkeley Hills, CA, XVIII-1-XVIII-8, Oct. 1987

Wang, S., Suchsland, O., and Hart, J. H., Dynamic Test for Evaluating Decay in Wood, Forest Prod. J. 30(7): 35-37, 1980.

QUALIFICATION OF PARTICLEBOARDS ON THE MILL LINE

A. T. Green
Hartford Steam Boiler Inspection Technologies
Sacramento, CA 95815

INTRODUCTION

Particleboard is a manufactured wood product commonly defined as a composite since it is composed of wood materials and binders. The extensive use of particleboard has required that methods for physically determining its properties be established. One of these measurements, the Internal Bond strength, has been obtained through a destructive physical test procedure on specimens taken from a sampling of boards.

Utilizing a nondestructive test we have been able to develop an, on-line Internal Bond strength measurement. This measurement can be made on each board and continuously along the direction of board travel on the line. The nondestructive measurements have been shown to have a high correlation to the values obtained by the destructive physical test.

This paper described the nondestructive technique, the particleboard product and the equipment now being used to obtain evaluation of this manufactured product within the particleboard industry.

Background

Particleboard products are replacing natural sawn wood products in many applications. Second and third growth trees in many cases do not have the strength properties of virgin forest lumber. Consequently manufactured wood materials whose physical properties may be engineered and designed into the product are gaining wider acceptance. Although these materials can offer significant advantages, they are subject to variations in mechanical strength and integrity.

Frequently, particleboards are manufactured from economically nonproductive wood materials such as; sawdust, shavings and/or chips from low value trees. Because the principal constituents of the boards are renewable resources, (ie: wood and small amounts of glue or resin) and very little energy is used in their manufacture, they are economically attractive. Unlike sawn products, boards can be manufactured in various sheet sizes, thicknesses, surface finishes, and densities containing fire retardants, preservatives and other treatments.

The principal attractiveness in properties is uniformity. Through particle alignment, particle geometry and resin content, particleboards are manufactured that rival, and exceed the structural capability of lumber and plywood. Previous limitations on stiffness, load carrying capability and dimensional stability commonly available with sawn lumber and plywood have been over come.

For the manufacturing process to run smoothly while producing boards with consistent properties from run to run, off-line destructive testing is now required. These tests can only economically be done on a statistical sample basis, moreover considerable substandard product may result during lengthy property measurement procedure. A real-time nondestructive measurement system can provide high economic value to improvement of particleboard manufacturing.

Particleboard

The American Society for Testing and Materials (ASTM) Standard D1554 entitled "Standard Definition of Terms Relating to Wood-Base Fiber and Particle Panel Materials" describes the terms used in the United States. ASTM describes particleboard as a generic term for a panel manufactured from lignocellulosic materials primarily in the form of discrete pieces or particles, as distinquished from fibers, combined with a binder material of synthetic resin and bonded together under heat and pressure in a hot press by a process in which the entire inter particle bond is created by the added binder, and to which other materials may have been added during manufacturing to improve certain properties. When the pressure of manufacturing is applied perpendicular to the outer faces as in the conventional multi-platen hot presses the boards are called 'flat-platten pressed'. When the pressure is applied parallel to the outer faces they are defined as 'extruded' products.

In the flat-platen press the machinery forms the boards from 'mats' of the wood product and resin inserted into multi openings and compressed while heated. The extruded boards are made by forcing the wood/resin mixture between heated dies in a nearly continous process. The dry wood particles have been coated (usually by spraying) with a resin, typically a urea-formaldehyde compound. The resin flows under the heat of the press platen or dies, around 270 degrees F. Figure 1 shows a multi opening flat-platen press in operation. The pressure is applied mechanically by hydraulic cylinders, and the heat by steam circulated through the platens. Figure 2 shows the mat of wood particles and resin prior to pressing.

Within the forest products industry, particleboard usually means a panel made of fine wood particles such as sawdust and or planar shavings. These boards are typically found in applications where structural needs are not high such as under flooring, shelving, siding, furniture, cabinets and similar products.

Where high structural needs are encountered the industry is producing boards specifically designed to meet these requirements. These products are designed with chips, strands and other materials included so as to produce the specific requisite property(ies). Some of these board products are; chipboard, flake board, wafer board, strand board, oriented strand board, and medium density fiber board.

Standards

Voluntary standards for the mat formed wood particleboard products exist under the auspices of the American National Standards Institute (ANSI). The ANSI Standard A208.1 lists the minimum levels for dimensional tolerances, physical and mechanical properties for the various grades. ANSI recognizes two types of panels with Type 2 having a resin binder system producing waterproof qualities required for exterior usage. Within each type they classify three (3) density levels; low at less than 40 lbs/cu. ft. between 40 to 50 lbs/cu. ft. and high at over 50 lbs/cu. ft.

The tests used to determine the physical and mechanical properties are conducted in accordance with methods and conditions prescribed in the American Society for Testing and Materials (ASTM) Standard D1037. The values specific within ANSI A208.1 are used for quality control and certification purposes as they are substantially higher than the design levels as used by engineers/designers.

Manufacturers of particleboard need to produce products conforming to ANSI Standard A208.1 when, amongst other reasons, they represent them as; conforming, required by clients, or by design specifications. At various times in the U.S. and Canada there have been as many as six organizations who administer programs that certify particleboard for individual mills.

The ANSI Standard A208.1 provides for the following strength/material properties: Modulus of Rupture, Modulus of Elasticity, Internal Bond, Hardness, Linear Expansion and Face/Edge Screw Holding.

For Internal Bond strength (IB) determination, tensile strength perpendicular to the surface, a test specimen of 2 in. by 2 in. dimension is pulled apart with the applied force being maintained perpendicular to the faces of the specimen. This measures how well the particleboard, at the location where the specimen was obtained, is glued together. The value at which the specimen fails is noted as the Internal Bond (IB) strength.

Internal Bond values need to be obtained from five (5) boards of any production run. Mills that have certification programs may be inspected at any time and required to establish values for properties of their product. Typically these certifications average approximately one per month. Mills who do not meet the minimum requirements of ANSI 208.1 are not allowed to stamp their products as certified.

Calibration is achieved through the correlation of the A/U and infra-red measurements with the measurement of IB by ASTM methods. Once this has been established further production quantities of boards may be measured and their IB values determined from the calibration. A programable logic controller (PLC) controls an ink spray system which is used to mark those boards having substandard IB values.

Since the A/U measurement is made while the board is still warm the boards surface temperature is measured to allow a correction to be made for IB value at ambient.

IB Measurement System

Figure 5 provides a block diagram of the A/U measurement system. The A/U measurement system consists of three basic subsystems:

1) The measurement subsystem includes the wheeled sensors and associated hardware; remote pulser and preamp modules mounted in a NEMA 13 enclosure with power supply (located near the wheels); infrared pyrometer with air purge cleaning; AET Model 302 Acousto-Ultrasonic Bond Tester for control of the pulser/receiver network and signal measurement; and a dual-channel oscilloscope for visual signal analysis and trouble-shooting (housed in computer enclosure).

2) The control subsystem includes board presence and overlap limit switches; the PLC in a NEMA 13 enclosure; pneumatic cylinders for raising and lowering the wheels (includes solenoids, air cleaners, regulators, air valves); "end-of-press-load" detector (if used); and the ink spray system for marking panels.

3) The computer subsystem includes a locking environmental enclosure with heat exchanger; a PC/AT-compatible computer with EGA color graphics, 20 or 40 MByte hard disk, 1.2 MByte floppy, an externally mounted industrial keyboard and high speed line printer; an opto-isolation interface between the computer and PLC; and an uninterruptible power supply.

Hardware options include a NEMA-rated enclosure with air conditioning for the computer subsystem, rack-mounted industrial PC/AT-compatible computer; serial I/O link to another computer system; remote terminals and keyboard (up to 1000 feet); modem; and custom sensor designs.

The IB measurement system software provides for real-time displays in various formats. The seven (7) most important displays are:
1. IB list by press opening
2. IB histogram for boards in the current press run
3. IB trend information by press load
4. IB trend information by press opening number
5. Averaged board IB profile for any selected press opening
6. Press run statistics
7. System statistics

The software is controlled through menu commands and function keys. All file maintenance and calibration functions are protected and available only to the Quality Control operator. All changes to any of the system files are logged for ease of maintenance.

Calibration requires that the average IB value obtained by the ASTM standard method be manually entered into the system computer to complete the determination of the constants in the calibration equation. Once the basic IB equation has been established for a product, it is maintained in a master file, and can be automatically re-entered into the computer each time that particular product is run. The previous twenty (20) calibration updates are stored for trending changes in system components or the process. Updating the calibration can be performed at any time.

Figure 6 shows the screen copy of the IB values for 5/8in. boards for each press opening. IB trend by press load is shown in Figure 7, and by press opening in Figure 8. During early trial runs the press operator had stated his concern that certain openings were not producing high quality boards. Even without calibration data the A/U measurements clearly showed his suspicion to be correct. The IB histogram display is shown in Figure 9. Press run statistics are shown in the screen copies in Figure 10.

Expected Results

These nondestructive examinations are expected to enable the particleboard mill operator to:

1. Reduce material costs through optimization of the amount of resin used in order to obtain the required IB value.
2. Reduce waste material created during shakedown and startup.
3. Optimize press operation with regard to changes in temperature, time, particle moisture content, or input products.
4. Identify press adjustment problems.
5. Produce consistent IB products.
6. Link into a process control network.

With a typical 400 ton-per-day particle board plant using 7% resin in the product, 56,000 pounds of resin are used per day. Estimating the resin cost at $0.17 per pound provides a yearly resin cost of over $3,000,000. If we assume the on-line acousto-ultrasonic method capable of reducing the amount of resin used yearly we define the following savings.

Reduction of Resin used %	Annual Savings $
0.1	$ 3,000
0.3	$ 9,000
0.5	$15,000
1.0	$30,000
1.5	$45,000

In a similar vein, if we assume the plant to have a reject rate of 5%, the yearly loss is over $900,000. A reduction in the reject rate by 1%, from 5% to 4%, would represent a yearly savings of over $180,000.

While it is too early to have the full savings of the installed system established, we clearly expect the data to show the economic merit of the acousto-ultrasonic measurement.

Actual Results

The data shown in Figure 11 compares the A/U voltage versus IB values obtained during the laboratory experiments. The correlation value of 0.95 has been exceeded in the mill installed system.

ACKNOWLEDGEMENTS:

The development of this acousto-ultrasonic nondestructive examination measurement was accomplished jointly between Weyerhaeuser Company and Hartford Steam Boiler Inspection Technologies (formerly Acoustic Emission Technology Corporation). Participants within Weyerhaeuser Company are Dwayne Shearer, Charles Anthony and Frank Beall. Participants within Hartford Steam Boiler Inspection Technologies are John Rodgers, Mike Shaw, Howard Calkin, Steve Hamilton, Jim Babb and James Leaird. Without their individual and joint efforts this project would not have been successfully accomplished. Our gratitude is extended to each of them.

REFERENCES:

1. Carll, C. "Wood Particleboard and Flakeboard types, grades and uses." USDA Forest Service, Forest Products Lab, General Technical Report FPL-GTR-53, Madison, WI. 1986.

2. American Plywood Association. "Performance Standards and Policies for Structural - Use Panels." APA PRP-108. Dec. 1986.

3. American National Standard. "Mat-Formed Wood Particleboard." ANSI/A208.1 1987.

4. American National Standard. "Medium Density Fiberboard for Interior Use." ANSI/A208.2. 1986.

5. American Plywood Association Product Guide. "Performance Rated Panels." Form No. F4056/revised. Jan. 1988.

6. dos Reis, J.L.M., Bergman, L.A., and Bucksbee, J.H. "Adhesive Bond Strength Quality Assurance using the Acousto-Ultrasonic Technique." The British Journal of Non-Destructive Testing (in print).

7. Maloney, T.M. Data presented at Washington State University Particleboard Symposium, Pullman, Washington. Feb. 1987.

Figure 1 Multi-opening Flat Platen Press in Operation

Figure 2 MAT of Wood Particles and Resin Prior to Pressing

Figure 3 Upper Acousto-Ultrasonic Wheel

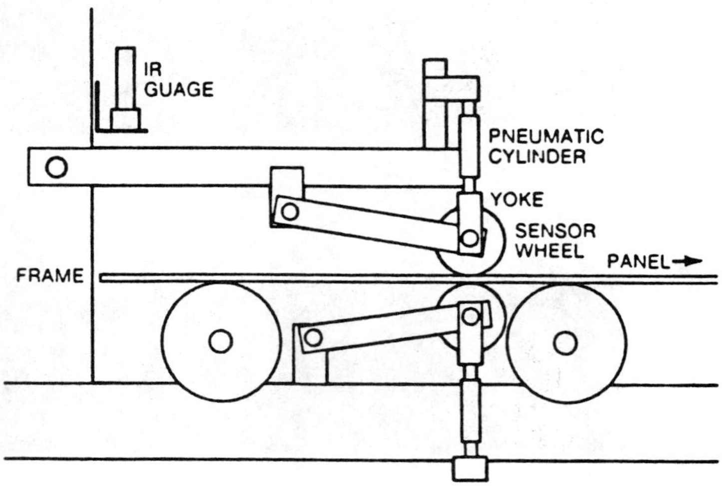

Figure 4 Diagram of the Wheeled Sensor Installation

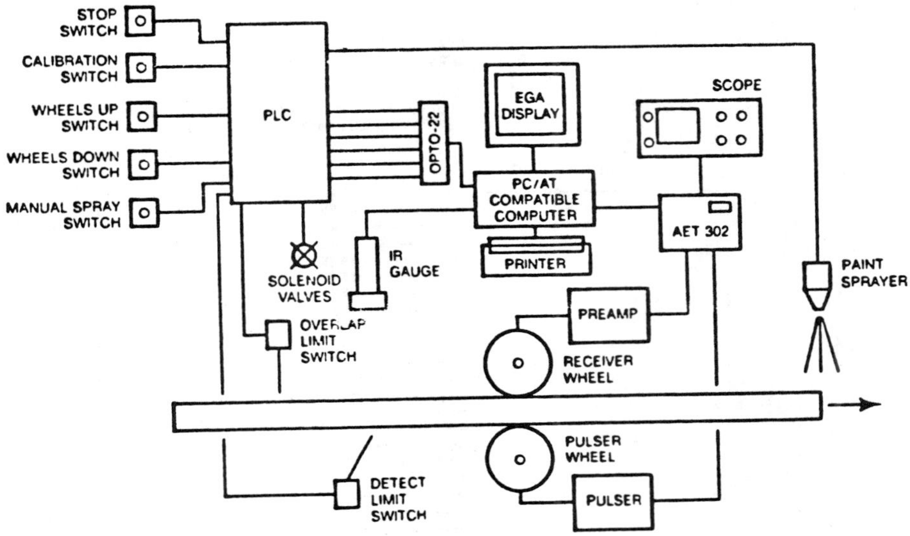

Figure 5 Diagram of the Acousto-Ultrasonic System

PANEL IB VALUES NOT Saving to Disk 11:42:10 05-25-1988

5/8 Timblend for customer xyz
Lot# 123 Operator name: hwc

Target IB = 100 psi Lower limit IB = 92 psi
Current panel = 504 Current press load = 36
Mean IB = 100.3 Standard Deviation = 4.04

	Time	Opening	IB psi	Range		Temperature C	Temperature F	
	14:54	1	97	88	104	115	239	
	14:55	2	102	93	109	113	235	
	14:55	3	99	90	106	112	234	
	14:56	4	101	92	108	111	232	
	14:56	5	99	90	106	110	230	
	14:56	6	100	91	107	109	228	
	14:57	7	99	90	106	108	226	
	14:57	8	97	88	104	107	225	
	14:58	9	102	93	109	106	223	
	14:58	10	104	95	111	105	221	
	14:58	11	99	90	106	104	219	
	14:59	12	96	87	103	103	217	
	14:59	13	97	88	104	102	216	
-->	14:60	14	97	88	104	101	214	<--

Figure 6 IB Values for 5/8in. Boards for Press Opening

IB TREND BY PRESS LOAD NOT Saving to Disk 11:16:02 05-25-1988

5/8 Timblend for customer xyz
Lot# 123 Operator name: hwc

Target IB	=	100 psi	Lower limit IB	=	92 psi
Current panel	=	504	Current press load	=	36
Mean IB	=	100.3	Standard Deviation	=	4.04

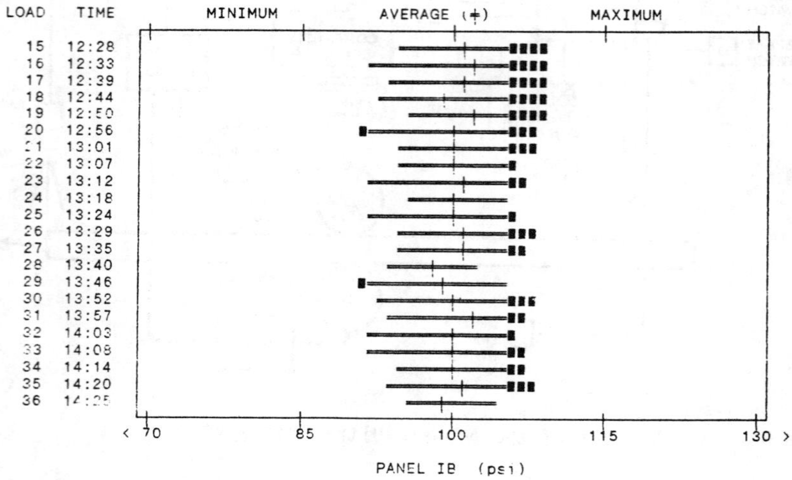

Figure 7 IB Trend by Press Load

IB TREND BY PRESS OPENING NOT Saving to Disk 11:17:23 05-25-1988

5/8 Timblend for customer xyz
Lot# 123 Operator name: hwc

Target IB	=	100 psi	Lower limit IB	=	92 psi
Current panel	=	504	Current press load	=	36
Mean IB	=	100.3	Standard Deviation	=	4.04

Averaged over the last 10 press loads.

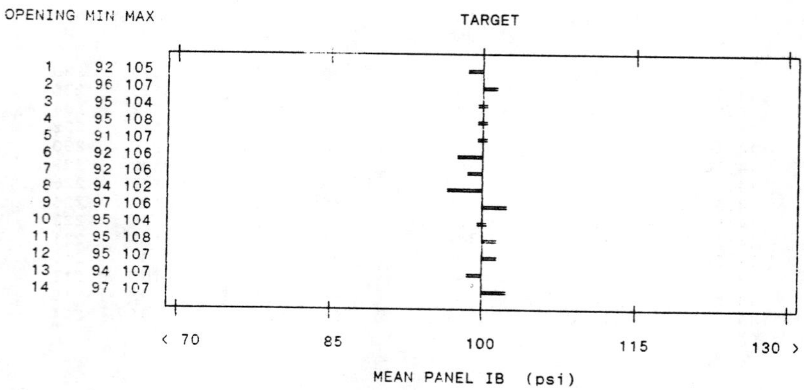

Figure 8 IB Trend by Press Opening

```
IB HISTOGRAM                          NOT Saving to Disk      16:48:54  05-25-1988

5/8  Timblend               for customer xyz
Lot#  123                                 Operator name: hwc

Target IB        =    100 psi             Lower limit IB       =  92  psi
Current panel    =    504                 Current press load  =  36
Mean IB          =    100.3               Standard Deviation  =   4.04
```

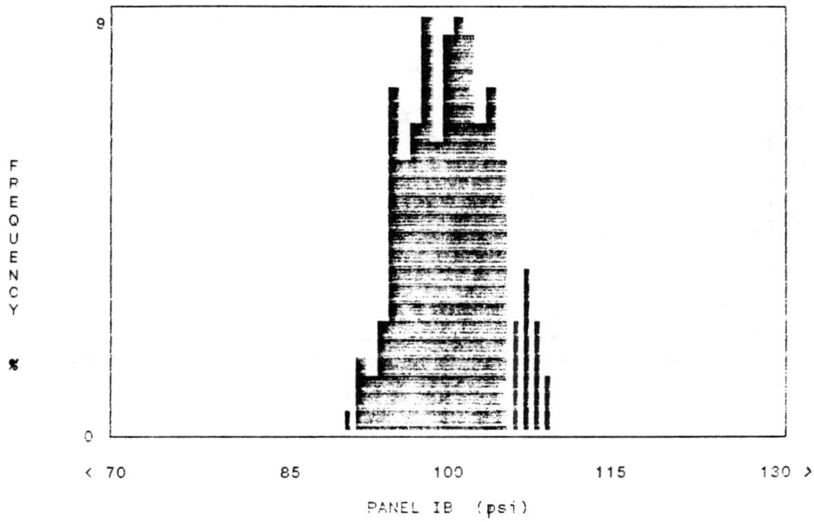

Figure 9 IB Histogram Display

```
PRESS RUN STATISTICS                  NOT Saving to Disk      11:18:28  05-25-1988

5/8  Timblend               for custome- xyz
Lot#  123                                 Operator name: hwc

Target IB        =    100 psi             Lower limit IB       =  92  psi
Current panel    =    504                 Current press load  =  36
Mean IB          =    100.3               Standard Deviation  =   4.04
 10 Load Trend   =     99.9               Loads Per Hour      =   9.9
 14 Panel Trend  =     99.2               Panels <  92 psi    =  20
Missing Panels   =      0

Press Run Start Time  =  05-25-1988  07:39
Elapsed Time          =    3.65 hours

Calibration in progress? YES
Press run file name?     NONE
```

Figure 10 Press Run Statistics

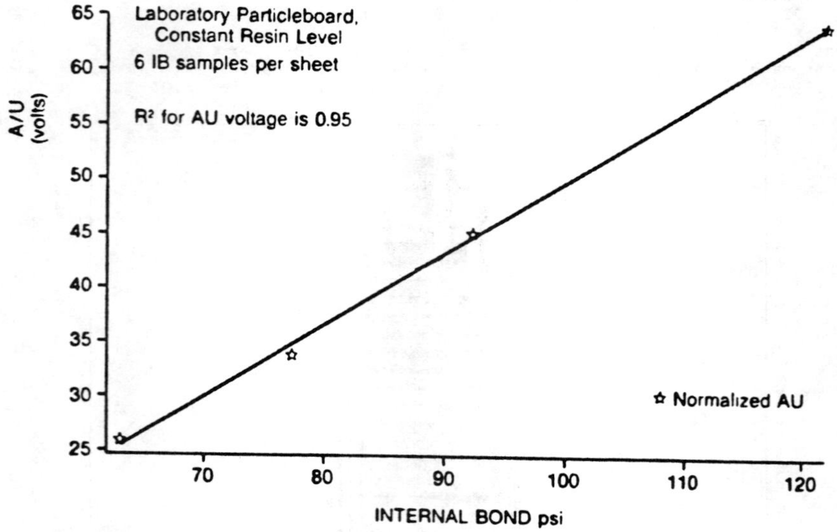

Figure 11 Acousto-Ultrasonic Voltage VS IB Value

DETERMINING VERTICAL DENSITY PROFILES IN WOOD COMPOSITES USING ACOUSTIC EMISSION

R. L. Lemaster
University of California Forest Products Laboratory
Richmond, CA 94804

D. A. Dornfeld
University of Califorrnia, Berkeley
Berkeley, CA 94720

1. INTRODUCTION

The manufacturing of wood composite board and building materials requires the evaluation of many variables. The size of the wood fibers and particles, glue content, pressure, pressing time, temperature, and other factors must be determined and checked in order to produce a board of consistent quality. An accepted measure of consistency is the density profile of the board (Maloney 1977). If the density profile of a composite board with the desired strength characteristics can be reproduced, then the strength characteristics will be maintained.

Several different methods for evaluating density profiles have been used in industry and in the laboratory. The gravimetric method, in which the specimen is planed, measured, and weighed in small increments, is most direct. The density measurements, in this case, can be evaluated approximately every 0.076 cm (0.030 in.) of thickness. Two methods involving radiation sources have also been evaluated. Parker et al. (1980) used x-rays to produce a photographic image of the profile of a composite wood specimen. This method allows for virtually continuous density measurements across the cross-section of the board with correlations made with known standards. Laufenberg (1985) and Winistorfer et al. (1986) used a gamma ray source to scan the thickness of a specimen. The transmitted radiation was then correlated to the material density and a continuous measurement was again possible.

Of the methods discussed, the conventional gravimetric method for density determination is the most time consuming and least resolute. Although the density calculations are the most direct (weight/volume), as the specimen is planed the measurements become more difficult to obtain due to the decreasing thickness. A thicker specimen would be easier to handle, but would also require more time due to the greater number of sections being removed.

The x-ray and gamma ray techniques for density measurement each provide continuous profiles of the specimen, but also require radiation sources and a greater knowledge of equipment operation by the user. The methods are faster than the gravimetric method and provide information about densities at specific points rather than averages over small sections. Both claim to be nondestructive, however this would not be the case in practical applications in industry. The samples being evaluated must be no more than 7.62 cm (3 in.) in width so some part of the manufactured board must be cut. The operation of these densitometers would not only require training, but also incur a substantial initial cost before consistent measurements could be made.

In this paper, acoustic emission (AE) was used to evaluate the density profile of a wood composite. Two methods were investigated. The first method used a 15 cm (6 in.) circular sample prepared from the material to be evaluated. The specimen was mounted on a lathe and rotated at high speed. A piezoelectric transducer coupled to a cutting tool monitored the AE signal produced as the tool was moved across the thickness of the rotating sample. The RMS voltage of the received signal was then correlated to the density of the specimen. This technique although destructive, offers an accurate and repeatable method for density profile determination without the use of radiation or the limitation of sample size.

The second method uses a rotating fly cutter or similar tool. The cutting tool moves across the edge of a wood composite panel while an AE transducer mounted on the panel records the changes in the RMS voltage. This technique while also destructive can be considered a non-destructive technique in the practical sense due to the fact that only a small 2.54 cm (1 in.) wide by .25 cm (.1 in.) thick notch is made in the panel. This notch can be made prior to the final sizing of the panel and hence does not affect the final product.

2. REVIEW OF ACOUSTIC EMISSION

Acoustic emission can be defined as the stress or pressure waves generated during dynamic processes in materials (see Figure 1). Though usually in the ultrasonic range (typically 100 kHz - 1MHz), these waves can sometimes be sensed by the unaided ear as when timber cracks when loaded to near failure.

The AE signal is usually detected by an instrumentation system using sensors (transducers) which, when stimulated by stress waves, transform mechanical excitations into electrical signals which are then amplified and transmitted to an oscilloscope, counter, root mean square (RMS) voltmeter, recorder, or spectrum analyzer, depending on the type of analysis. The most common type of transducer is a piezoelectric ceramic element (lead zirconate titanate).

The greatest problem in the application of AE in process control is the analysis or interpretation of the signals. An actual AE signal is non-periodic, contains many frequencies, and cannot be described by an explicit mathematical relationship (Ono, et al. 1972). This calls for use of statistical tools in analyzing AE data. Among the various methods frequently used to characterize the emission from a material are: the cumulative count, count rate, RMS voltage, spectral analysis, probability density function, auto-correlation function, and amplitude distribution analysis.

The voltage, V, of the signal can be theoretically represented as an exponentially damped sinusoidal wave according to the relationship (Harris, et al. 1972):

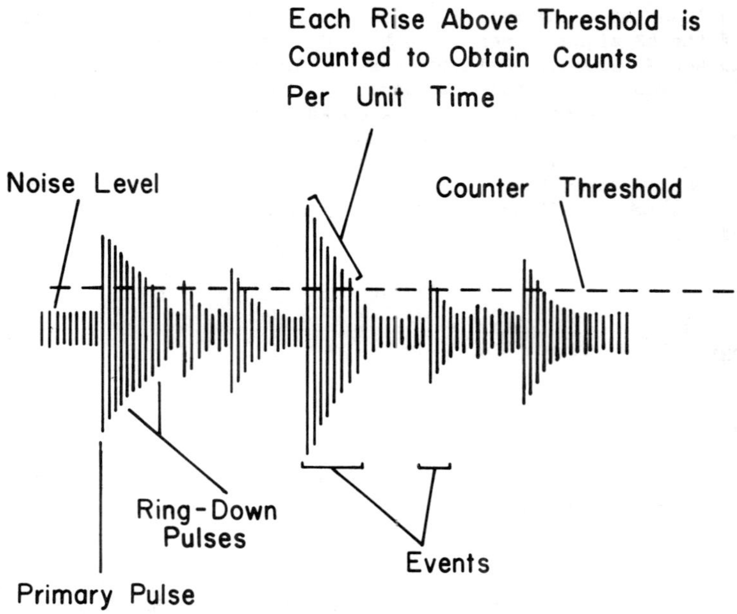

FIGURE 1. Schematic of acoustic emission signal.

$$V = V_0 e^{-\beta t} \cos 2 \pi f t \qquad [1]$$

where

V_0 = threshold voltage

V = initial output voltage of transducer

β = damping factor

f = linear frequency

t = time

This signal is due to an AE event which is defined as a single occurrence of AE activity (Figure 1). A sharp tap on a transducer will register one event while a vigorous shake will register several events. The number of counts resulting from the events is the number of times the signal caused by the events exceeded a preset threshold voltage. This means that the number of counts measured for a given event and a preset threshold voltage depends on the energy contained in the event and the damping factor β. It can be assumed that the output voltage of a transducer is proportional to the square root of the energy released during a given deformation process. This is so as the output voltage of a piezoelectric disk is proportional to the stress on the disk face and because the energy density of an elastic stress wave is proportional to the square of the stress.

The RMS voltage of a continuous AE signal can be used to measure the energy of the AE signal. Here, the term continuous emission (as defined by the 1974 U.S. Acoustic Emission Working Group, refers to a sustained signal level whose characteristics are such that "...the average time interval between the beginning of the emission signals of comparable amplitude is less than or comparable to the duration of the emission signals". The RMS value of the AE signal is that value of a DC signal which, if passed through the same circuit for the same time, would produce the same expenditure of energy as the AE signal. Analytically, it is defined as:

$$RMS = \frac{1}{\Delta T} \int_0^{\Delta T} V\,(t)dt^{\quad 1/2} \qquad [2]$$

where

$V(t)$ = the function of the signal with respect to time.

ΔT = the time period.

In previous studies (Lemaster, Klamecki and Dornfeld 1982; Lemaster, Tee and Dornfeld 1985; and Lemaster and Dornfeld 1985) it was determined that the acoustic emission was generated at the primary deformation zone at the tip of a cutting tool moving through wood. Furthermore the levels of acoustic emission generated were found to be sensitive to the cutting parameters such as feed speed, cutting speed and depth of cut as well as to biological factors such as density. This sensitivity of AE to density gave rise to the idea of correlating AE levels to density profiles in composite wood products.

3. EXPERIMENTAL SET-UP AND TESTS

3.1 Method 1

Experimental Procedure. For the first method, specimens were prepared by cutting a 15.2 cm (6 in.) diameter disk from a board of unknown density. The circular specimen was then secured to a metal face plate which was mounted on a lathe. A piezoelectric transducer (175 kHz resonant frequency) was coupled via a steel mounting bracket and a resinous couplant (AET-SC6) to the carbide-tipped cutting tool. The AE signal received by the sensor was run through a 40 dB pre-amplifier and a 0.125-2 MHz signal filter to an AET 5000 Signal Analyzer. The AET 5000 was used to record the output signal voltage along with many other parameters including signal rise time, ringdown counts, and peak amplitude. This was useful in checking to see if any other signal characteristic could be used to predict material density. The amplified signal was then input to an RMS voltage unit which converted the raw AE signal to a true RMS voltage. This RMS voltage signal was printed out on a high speed strip chart recorder and digitized into an IBM PC/AT for storage and further analysis. Figure 2 shows the experimental lathe set-up for Method 1.

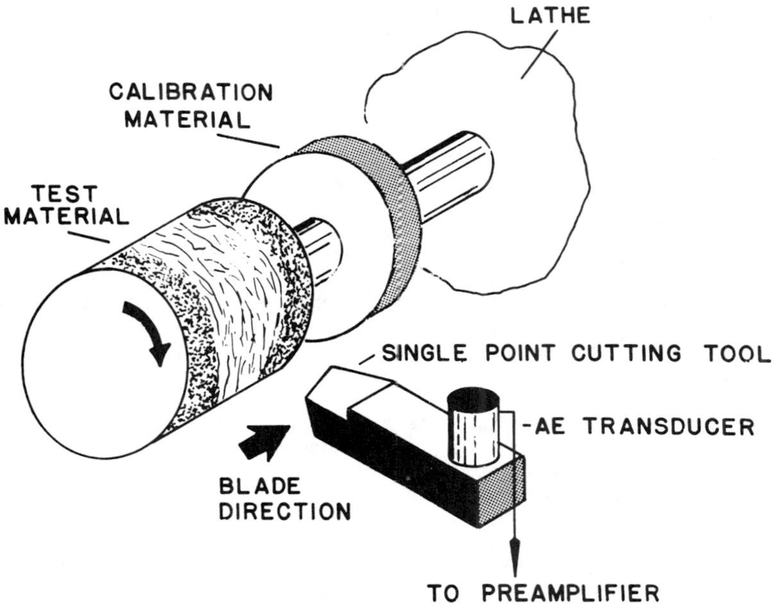

FIGURE 2. Experimental set-up for Method 1.

After preliminary investigations it was determined that the RMS was the best indicator for density. The RMS voltage was therefore used as a comparative measure for material density evaluation for the remainder of the study. The equation:

$$\text{RMS voltage} = f\ (\dot{\varepsilon},\ \text{volume, material property}) \tag{1}$$

was used to relate RMS voltage to known quantities of the experimental setup. In this setup $\dot{\varepsilon}$, the material strain rate (dependent on cutting velocity), remained constant for a given composite wood sample. The volume of material cut as determined by feed speed and depth of cut was also constant for each data acquisition run. The material property of the composite specimen which varies the most is the density. For this experiment therefore, the material property related to the output RMS voltage was density. The above equation can then be simplified to:

$$\text{RMS voltage} = (C)(\text{Density}) \tag{2}$$

where C is a proportionality constant dependent on other aspects of the material being tested and the initial conditions for the data run. Changing such parameters as depth of cut, output signal gain, or frequency of the transducer did affect the magnitude of the RMS voltage, but the effect was constant throughout the data run (as predicted by eq. 1) for depth of cut, for example) and thus could be incorporated into C. From this equation, absolute densities within a sample were predicted using the output signal RMS voltage.

Since absolute densities are not normally required in a quality control application, the repeatability of the output signal voltage produced by testing a specimen was also evaluated. Output signal repeatability for a given specimen would allow for samples to be checked for signal variance from a desired standard without performing the calculations for absolute densities.

Both single density and multiple density materials were tested for consistency of output signal produced over the course of the data runs.

Once the repeatability of this method was verified, a calibration technique was used to allow the prediction of actual density profiles in the composite wood specimens tested. For a given data acquisition run the ratio of RMS voltage to density was constant (eq. 2). Since this ratio was known for the calibration material, eq. 2 was used to predict the density profile of the composite specimen by an analysis of the output RMS voltage signal.

<u>Repeatability Tests</u>. The first phase of this experiment considered the repeatability of the output signals produced when different materials were tested. The repeatability of this density measurement technique was determined by making three separate data acquisition runs for each sample using Method 1 and comparing the shape and magnitudes of the output RMS signals for each run. As can be seen from Figure 3, the shape of the output signal for high density particleboard was maintained, although the magnitudes decreased. This reduction of RMS voltage was principally due to the decreasing radius of the sample, hence reduced cutting speed or strain rate (eq. 1). Any tool wear would also result in a decrease in the RMS levels (Lemaster et al. 1985).

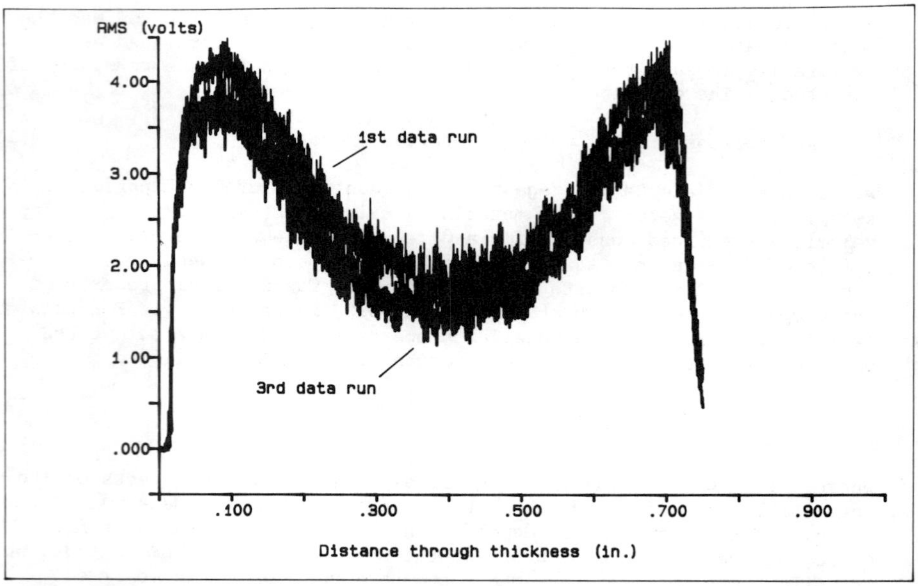

FIGURE 3. Repeatability of the technique (test material - high density particleboard, Method 1).

Single density samples such as solid pine and hardboard were run to determine if a constant RMS voltage would be produced. Results showed that the single density samples had a consistent signal over the thickness of the specimens giving rise to the possibility of developing a calibration technique for determining absolute densities. This technique would allow for densities of unknown samples to be calculated based on the magnitude of the output signal produced by the single density calibration material. This would eliminate variability in the testing procedure caused by variability in the cutting speed, feed speed, cutting depth, and tool sharpness since each test run would be essentially self-calibrating.

Density Calibration Tests. Phase two of this study utilized a calibration material as a reference to predict actual density profiles of different composite samples. The predicted values were compared to actual densities as calculated by a conventional gravimetric method. This method consisted of weighing and measuring a sample of material, shaving off a small layer, and then reweighing and measuring to get a profile. Layers of 0.064 cm (0.025 in.) in thickness were shaved (a finer resolution than typical methods used in industry). The method adopted for this study consisted of mounting a circular sample of the test specimen on a face plate on an engine lathe and securing it in place with a bolt through the center of the sample. The sample was weighed and a face cut was made from the outside diameter of sample to just outside the diameter of the mounting bolt.

This procedure of alternately weighing and face cutting the sample was continued for the entire thickness of the sample. It was believed that this procedure was superior to the planing, sawing or sanding methods used in the past due to the precision of the engine lathe. Although great care was taken to maintain a high level of consistency during these measurements, the resulting densities which were calculated often fluctuated from measurement to measurement. This could be due to the extremely small amount of material being removed which caused small deviations in measurements to be magnified when densities were calculated as well as to actual in board density variations. The lathe version of the gravimetric method also appeared to require a certain amount of expertise on the engine lathe. This was apparent in a visual inspection of the gravimetric samples where some of the earlier samples exhibited evidence of fiber tear out and an uneven stopping point for subsequent cuts. This gravimetric method required approximately two hours to completely evaluate a 1.91 cm (0.75 in.) thick sample.

Specimens from the same boards used for the gravimetric method were then tested using a calibration technique. Hardboard was chosen as the calibration material because of its essentially consistent through thickness density. Almost any material could be used such as brass or plexiglass for calibration as long as the actual density of the material was known. For this experiment, both tempered and untempered hardboards were used as calibration materials with no noticeable difference in results.

The materials to be evaluated using this method were a medium density particleboard (720 kg/m^3 or 45 lb/ft^3), a high density particleboard (1000 kg/m^3 or 65 lb/ft^3) and a medium density fiberboard. The unknown specimen was mounted on the lathe with the hardboard. A rubber washer separated the two specimens. The cutting tool was fed across the hardboard and then the unknown composite sample resulting in an output signal such as shown in Figure 4 for medium density particleboard and Figure 5 for medium density fiberboard. The decreased RMS signal towards the end of the plot in Figure 5 was due to fiber tear out as the blade was leaving the cut. This problem was frequently encountered in this study though it is believed it could be compensated for by making subsequent cuts in the specimen from opposite sides and adding the signals together. Another way of reducing or eliminating the occurrence of tear out is to use a lathe with a faster cutting speed. The 333 rpm used in this study is actually too slow to yield smooth cuts in wood.

Acoustic emission determined densities were calculated from the acquired data and compared to the results of the gravimetric method. Using a signal processing program the output RMS voltage signal was analyzed. The amplitude of the calibration material signal was first averaged over the entire thickness and an average RMS voltage for the hardboard was recorded. The signal produced by the unknown composite was then segmented into corresponding 0.064 cm (0.025 in.) intervals. The average RMS voltage for the interval was calculated and recorded over the entire thickness of the composite sample. The predicted densities were then calculated using the relationship:

$$\frac{\text{calibration density}}{\text{average RMS voltage}} = \frac{\text{unknown density}}{\text{average interval RMS voltage}} \quad (3)$$

A density profile was then generated using the same interval size as the gravimetric method.

The two sets of results were compared and evaluated to determine if the method was consistent for different materials and also for successive data runs. Figure 6 is a graph of the actual material densities as calculated using the gravimetric method and the AE method for the high density particleboard sample. Second order polynomial curves were then fitted to the data points obtained from these two methods such that the squares of the errors were minimized (least squares curve fit). The resulting equations of the form:

$$\text{density} = A\,(\text{distance}) + B\,(\text{distance})^2 + C \quad (4)$$

where A, B, C = coefficients from least squares fit
distance = distance through specimen thickness

allow density to be predicted at any point across the thickness of the specimen. Figure 7 is a graph of RMS versus both the actual gravimetric based density as well as the predicted (fitted) gravimetric based density for the high density particleboard sample.

As can be seen from the previous graphs the source of variability in this study was almost always the gravimetric method. This could be due to the greater sensitivity of the gravimetric method to localized density variation or as is support here, to the inaccuracies of the gravimetric method. As mentioned before the gravimetric method required a degree of experience on the engine lathe to obtain accurate results. This seems to be verified in column II of Table 1 where the r^2 increased with specimens that were machined later. It must be pointed out, however, that once the gravimetric data was smoothed out with a 2nd order polynomial regression, a strong correlation existed between RMS and the fitted data (column IV, Table 1).

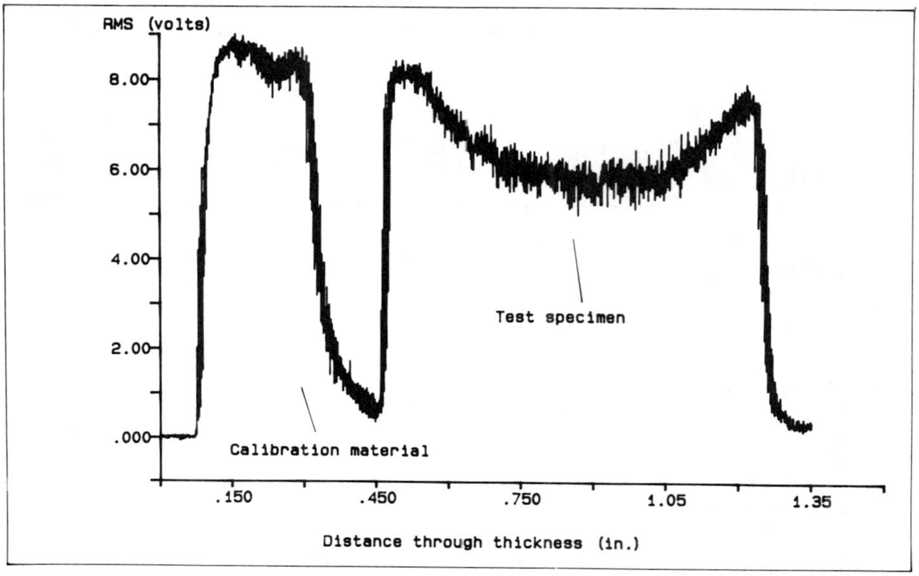

FIGURE 4. RMS output for high density particleboard (Method 1).

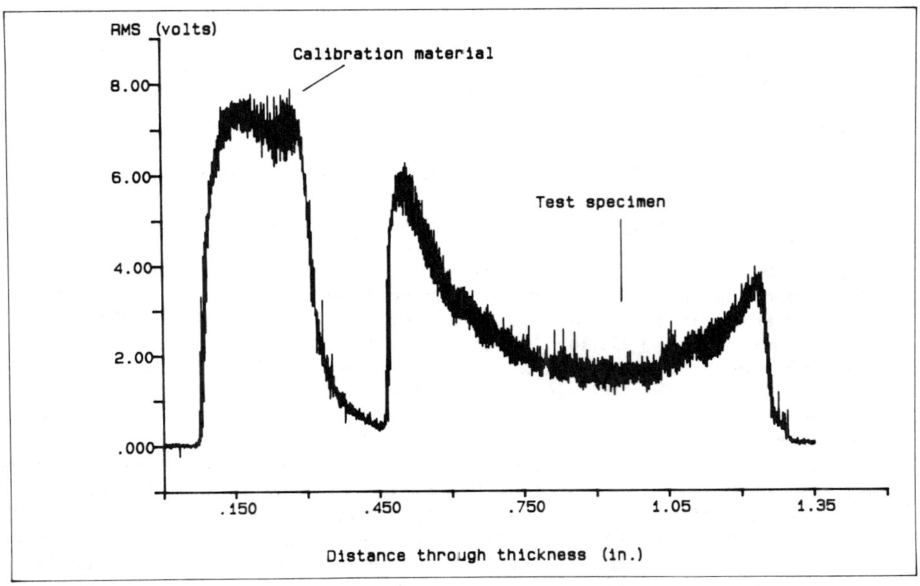

FIGURE 5. RMS output for medium density fiberboard (Method 1).

TABLE 1. Coefficient of Determination (r^2) for Least Square Regression.

Material	AE determined density[1]	Gravimetric density[1]	RMS versus gravimetric density[2]	RMS versus[2] fitted data for gravimetric density method
Medium density fiberboard	0.88	0.32	0.48	0.88
45 lb particleboard	0.96	0.60	0.62	0.95
65 lb particleboard				
Trial 1	0.95	0.63	0.62	0.96
Trial 2	0.96	0.86	0.96	0.89
Trial 3	0.96	0.94	1.00	0.99
Oriented strand board	1.00	.90	0.88	0.95

[1]2nd order polynomial.
[2]1st order polynomial.

3.2 Method 2

Method 2 was devised in an effort to convert this concept of determining density profiles in wood composites into a non-destructive testing technique. As shown in Figure 8, the blade was rotated and the specimen was held stationary. This allows for the possibility of testing full sized panels and consequently an on-line testing procedure is feasible. The need for the calibration material to be tested everytime can be eliminated by the use of wear resistant cutting tools such as carbide or diamond coated tools. This would simplify the testing procedure but would have to be weighed against the higher cost of the cutting tools.

Experimental Procedure. In this study a fly cutter with a carbide single point cutter was used. A cutting arc of 4.44 cm (1.75 in.) rotating at 1200 rpm was used to remove a maximum of .25 cm (.1 in.) from the edge of wood composite specimens. Preliminary work showed that the feed speed selected was critical in obtaining a good density profile. Faster feed speeds could be used though in this study a feed of 1.9 cm/min (.75 in/min) was used. An AET 175 kHz transducer was attached to the specimen, 2.54 cm (1 in.) from the mid-point of the cutter, using a 4 pound constant force spring and a .16 cm (.06 in.) thick black neoprene couplant. The received signal was amplified using an AET 204B instrument having a total gain of 50dB and a .125-2MHz filter. The amplified signal was then sent to a separate true RMS converter with a 25 millisecond response time after which it was digitized and stored in an IBM PC/AT at a rate of 100 samples/second. Since only a one channel acoustic emission system was available the calibration cut and actual test cut had to be done separately as opposed to the method shown in Figure 8. This did not appear to be a severe limitation since the amplitudes between successive

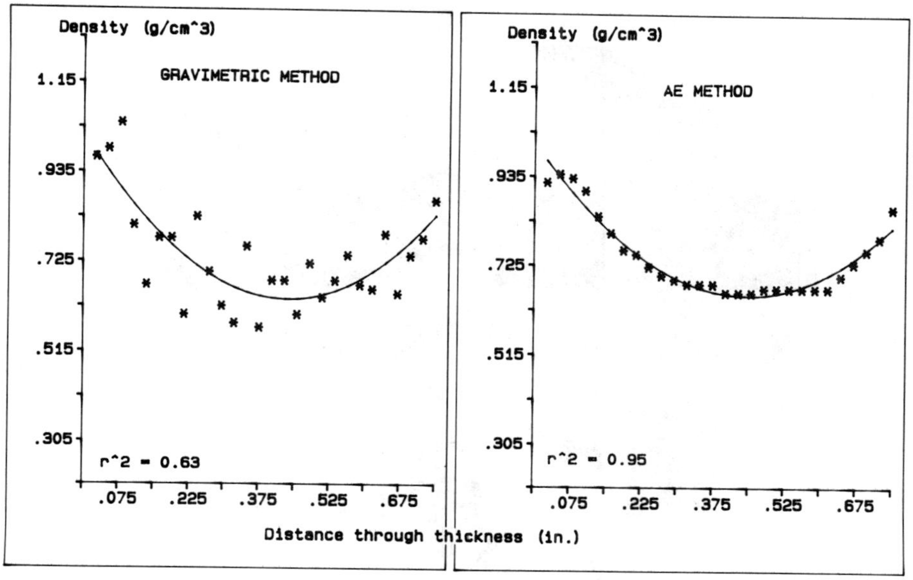

FIGURE 6. Gravimetric and AE determined density profiles for high density particleboard (Method 1).

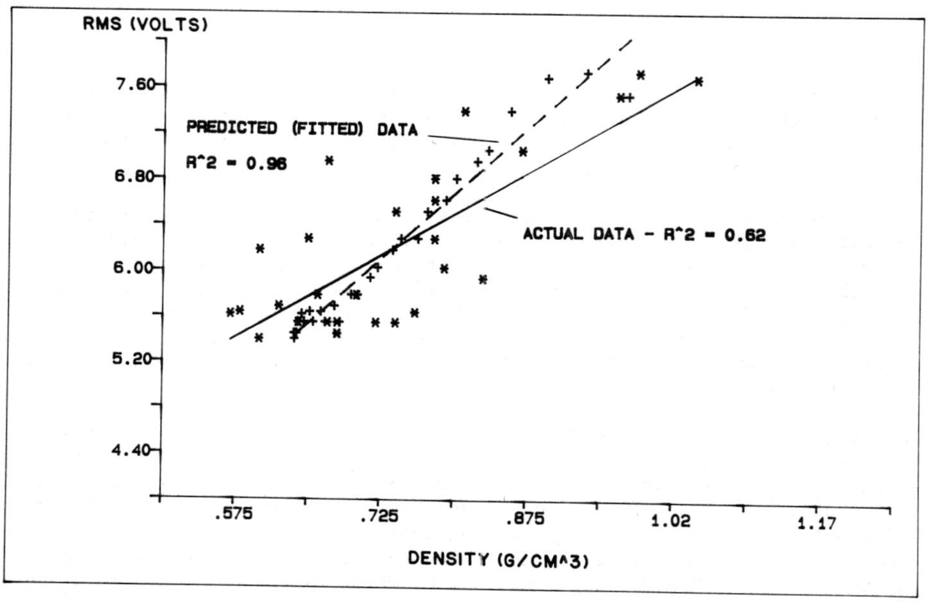

FIGURE 7. RMS versus gravimetric density for high density particleboard (Method 1).

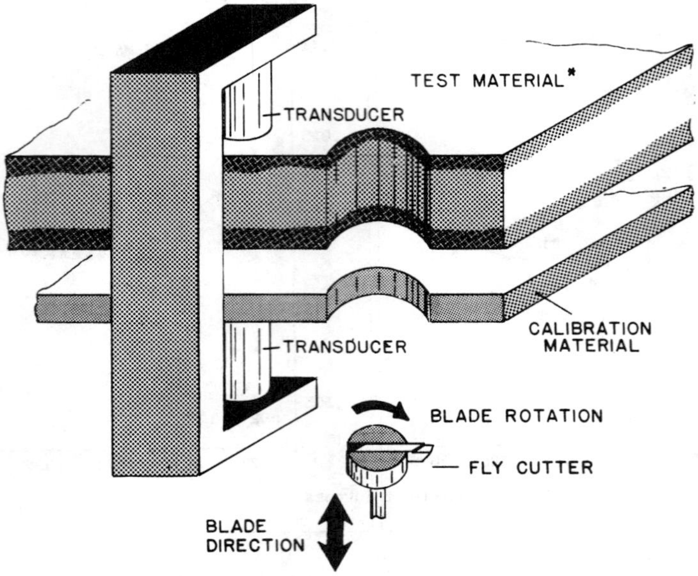

FIGURE 8. Experimental set-up for Method 2.

cuts did not change significantly. This was due, in part, to the fact that a carbide cutter was used as well as the fact that the rate of chip formation did not change between successive cuts as it did in Method 1 when the diameter of the specimen would change.

Figure 9 shows a plot of the RMS data and the corresponding gravimetric density data for a high density industrial (3 layer) particleboard. A second order polynomial was fitted to the data. The RMS data was not as smooth in Method 2 as in Method 1 as reflected by the coefficient determination. This could be due to the intermittent nature of the fly cutter. This could be minimized by using a higher arbor speed (i.e., router) and/or a small diameter cutting tool. Both of these options will be investigated in future research.

One disadvantage of Method 2 which has to be taken into account is that unlike Method 1 the distance between the transducer and the cutting tool changes constantly during testing. Since wood and wood composites are so attenuative this can have a significant effect on the RMS signal. Figure 10 shows a plot of a least squares regression between the smoothed RMS and gravimetric density data. The smoothed data was obtained from the fitted values of the 2nd order polynomial performed on each data set. In Figure 10 the transducer was placed on the front face of the specimen and data collected as the blade moved completely across the thickness of the specimen. As can be seen from the plot of RMS versus gravimetric determined density there was a significant thickness effect. If only the

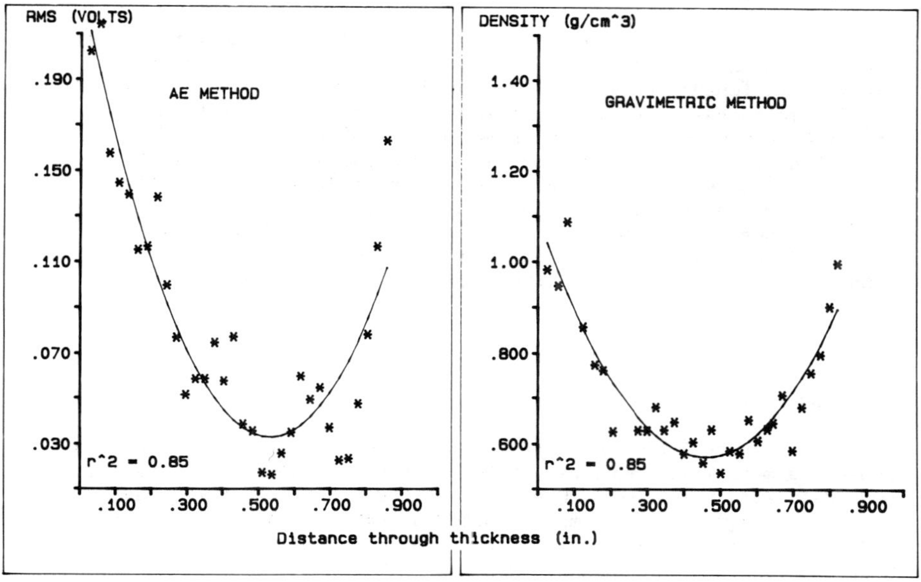

FIGURE 9. RMS and gravimetric density data for high density 3-layer particleboard (Method 2).

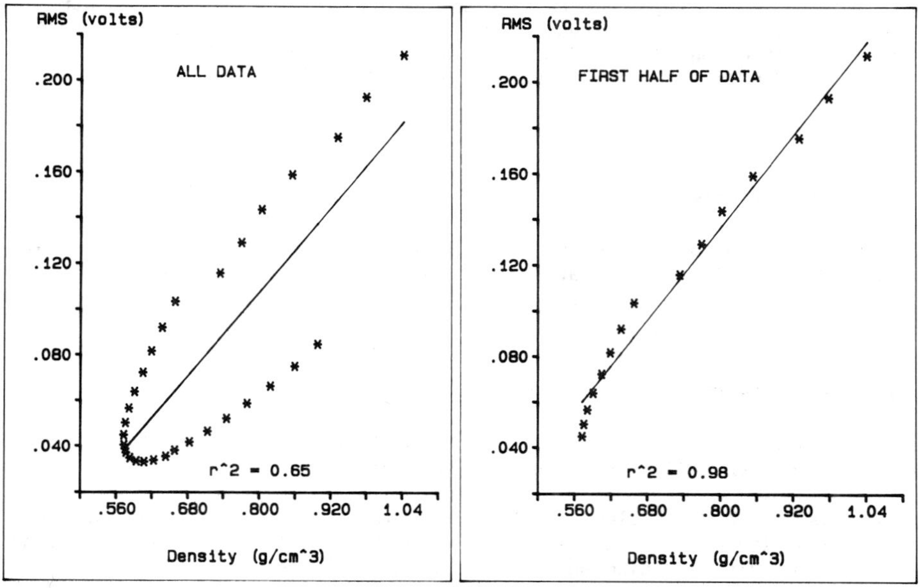

FIGURE 10. RMS versus gravimetric density for high density 3-layer particleboard (Method 2).

first half of the data was used the linear correlation between RMS and density improved dramatically. Subsequent experiments were conducted with the transducer placed on the back face of the specimen. The thickness effect was still obvious but reversed in nature.

By using only the second half of the data (from the midpoint of the specimen to the face where the transducer was mounted) an excellent linear relationship between RMS and gravimetric density was observed. Any attempts that were made to average the entire data set for both front and back face mounted transducer met with limited success. The loop in the curve of RMS versus density was less pronounced but still evident. This was also reflected in the coefficient of determination (r^2) of 0.88 for the combined data. This could be improved upon by taking the front and back face data simultaneously on a two channel AE system. This would reduce the experimental error caused in the experiment by doing two separate cuts and remounting the transducer between each cut.

4. CONCLUSION

The test results from the proposed density measurement techniques have shown the method to be extremely repeatable and accurate for use with composite wood materials. Whether specific densities or relative profiles were required, this method offered results quickly and safely with a relatively small amount of instrumentation, compared to the radiation methods.

Method 2 while more variable in nature than Method 1, still yields an accurate prediction of density. Method 2 also has the additional advantage of being essentially a non-destructive technique. This gives industry the option of using it either as a quick off-line quality control method or an on-line continuous process monitoring system. Additional work will be needed to optimize this concept. (Patent pending).

5. ACKNOWLEDGEMENTS

The authors would like to thank David Leding of the Weyerhaeuser Company for providing the sample material of the industrial particleboard.

6. BIBLIOGRAPHY

1. Cown, D.J. and Clement, B.C. A Wood Densitometer Using Direct Scanning with X-rays, Wood Sci. and Technol., 17:91-99, 1983.

2. Harris, D.O., Tetleman, A.S. and Darwish, F.A.I. Detection of fiber cracking by acoustic emission. Acoustic Emission, ASTM 505, American Society for Testing and Materials, pp. 238-249, 1972.

3. Laufenberg, T.L. Using Gamma Radiation to Measure Density Gradients in Reconstituted Wood Products, Forest Prod. J., 36(2)59-62, 1986.

4. Lemaster, R.L., Klamecki, B.E. and Dornfeld, D.A. Analysis of acoustic emission in slow speed wood cutting. Wood Sci. 15(2):150-160, 1982.

5. Lemaster, R.L, Tee, L.B. and Dornfeld, D.A. Monitoring tool wear during wood machining with acoustic emission. Wear 101:273-282, 1985.

6. Lemaster, R.L. and Dornfeld, D.A. Monitoring the wood cutting process with acoustic emission. Journal of Acoustic Emission 4(2/3):S228-S231, 1985.

7. Lemaster, R.L., Gasick, M.F. and Dornfeld, D.A. Measurement of density profiles in wood composites using acoustic emission. Submitted to the Journal of Acoustic Emission, 1988.

8. Malony, T.M. Modern particleboard and dry-process fiberboard manufacturing. Miller Freeman Publications, 1977.

9. Nearn, W.T. and Bassett, K. X-ray Determination and Use of Surface-to-Surface Density Profile in Fiberboard, Forest Prod. J., 18(1):73-74, 1968.

10. Ranta, L. and May, H.A. Measuring the Density Profile of Particleboard by Gamma Radiation, Holz and Werkstoff 36(12):467-474, 1978.

11. U.S. Acoustic Emission Working Group Subcommittee Report. Chairman, J.R. Frederick, Department of Mechanical Engineering, University of Michigan. August 1974.

12. Winistorfer, P.M., et al. A Direct Scanning Densitometer to Measure Density Profiles in Wood Composite Products, Forest Prod. J., 36(11):82-86, 1986.

CURRENT AE/AU RESEARCH IN SOLID WOOD AND WOOD-BASED COMPOSITES AT THE UNIVERSITY OF CALIFORNIA FOREST PRODUCTS LABORATORY

S. L. Quarles and R. L. Lemaster
University of California Forest Products Laboratory
Richmond, CA 94804

1. INTRODUCTION

Acoustic emission (AE) and acousto-ultrasonic (AU) research at the University of California Forest Products Laboratory (UC FPL) can be divided into two broad areas. The first research area consists of investigations designed to determine the effect of physical and structural properties of solid wood on the propagation of acoustic waves. The second area involves research investigating the potential for using AE/AU technology for wood and wood-based material evaluation and process control systems. The purpose of this paper is to report on the current progress in these two areas.

2. ATTENUATION INVESTIGATIONS

Wood is an anisotropic (orthotropic) material and is highly attenuative. Consequently a significant component of the AE/AU research is concerned with determining the relationship between the physical and structural properties of wood and the propagation of the acoustic wave. The physical properties being investigated include the moisture content, density, and temperature of wood. Structural properties include propagation direction (i.e., grain direction) and within ring growth pattern (i.e., diffuse versus ring porous vessel distribution for hardwoods and abrupt versus gradual transition from early to late wood for softwoods).

Preliminary investigations indicated that attenuation was influenced by moisture content and propagation direction, but the trends were not consistent (Lemaster and Dornfeld 1987, Quarles and Zhou 1987). In fact Lemaster and Dornfeld (1987) reported that if the measurements made on the oven-dry specimens were removed from the analysis, the data showed virtually no relationship between moisture content and signal energy (as measured by RMS voltage). The current studies were conducted to clarify the findings from the preliminary investigations.

2.1 Experimental

Pacific madrone *(Arbutus menziesii)*, a diffuse porous hardwood was used for this portion of the experiment. Specimen material was machined into 2 x 2 x 2 cm cubes. Specimens were processed to obtain one of the four grain angle orientations used to evaluate the effect of propagation direction. Nominal grain angle orientations included 0/90, 15/75, 30/60, 45/45 (Figure 1). By testing on the longitudinal and two transverse surfaces, a total of eight propagation directions could be evaluated. Four replications were used for each nominal grain angle orientation.

Two acoustic sources were used. The first consisted of a burst-type emission generated by a 0.3 mm pencil lead break (simulating AE events). The second consisted of a near continuous signal generated by an electronic pulser (i.e., AU testing). Separate samples were used for each set of measurements associated with a given acoustic source. Five pencil lead breaks were made on each surface, giving a total of 15 breaks per specimen at each testing condition. Three repeated measures were made on each surface during the AU testing.

Attenuation measurements were made at seven moisture contents. These included the fully saturated condition, three equilibrium conditions below the fiber saturation point (at 95, 50 and 10% relative humidity), the oven-dry condition, and at two points between the saturation and 95% relative humidity equilibrium condition. A saturated condition was obtained by submerging the cubes in water and soaking under a series of vacuum and atmospheric pressure cycles for three days. The dry bulb temperature was maintained constant at 70°F. The last two conditions using the near-continuous (AU) signal could not be completed in time to include in this manuscript.

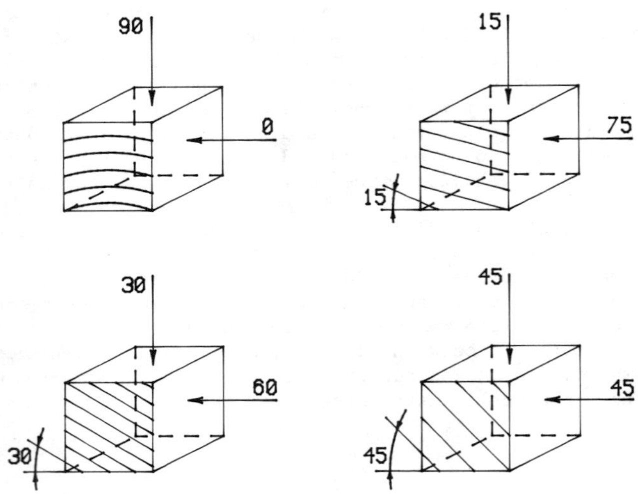

FIGURE 1. Nominal grain angle orientations used in the attenuation investigations.

Peak amplitude and ring-down counts were measured for the AE testing and RMS voltage was measured for the AU testing. An AE International AEI Composite Tester using a 110 kHz transducer with a 40 dB internal preamp was used for the AE experimentation. A hot melt adhesive was used as a couplant. A 110 kHz transducer served as the pulsing transducer for the AU testing. A 175 kHz transducer was used for the receiving transducer (50 dB total system gain). A 250 volt pulser operating at 0.5 kHz was used to excite the pulsing transducer. Black neoprene rubber was used as a couplant.

2.2 Results and Discussion

An analysis of variance was performed on the AE (peak amplitude and counts) and AU (RMS) data. The longitudinal and transverse measurements were analyzed separately. Results of the analysis of variance consistently showed a significant difference ($\alpha = 0.01$), between a given acoustic signal measurement parameter and moisture content and propagation direction (for transverse measurements) and no significant differences in the repeated measurements (p-values were always greater than 0.5). The average coefficient of variation (standard deviation divided by the mean) for the repeated AU measurements was 12.6%. This value tended to be a little lower when the average moisture content was well above 30% (average coefficient of variation of 9.6%) and somewhat higher at lower moisture contents (average coefficient of variation of 15.6%). This trend indicated that coupling was more consistent at the higher moisture contents, when the surface would also tend to be at a higher moisture content. Unfortunately there was generally no consistent trend associated with the significant differences. In addition, some of the observed differences, although statistically significant, were small.

Figure 2 shows the effect of moisture content on the given acoustic parameter. Each plotted value was averaged over the repeated measurements, replications, and transverse propagation directions. Exact moisture contents are not given for the AU (RMS) readings because testing has not been completed, and consequently the specimens have not been oven-dried.

Overall, the peak amplitude and counts measurements were insensitive to changes in moisture content. Peak amplitude and counts were significantly lower at the saturated condition, probably due to greater absorption of acoustic energy at this moisture content. It should be noted that this moisture condition is not typically seen in in-use conditions, and therefore might be considered unrealistic (the same could be said for the oven-dry condition).

The AU measurement parameter (RMS) was more sensitive to changes in moisture content than the AE parameters. The lowest RMS reading was at the saturation condition, and then it gradually increased until the average moisture content reached approximately 30-40% (assuming the AU specimen moisture contents are similar to the AE specimens). At this point there was an abrupt increase in the RMS signal. The basic reasoning for this change is the same as that given for the AE parameters. Between replication (sample) variability (Figure 2) would make this trend difficult to use in any practical nondestructive evaluation program. Past research at the UC FPL has shown that RMS, when used in conjunction with AU testing, was more sensitive to certain physical and structural properties of wood, and therefore the observed trend between moisture content and RMS was not surprising.

As seen in Figure 3, the effect of moisture content on measured acoustic parameters in the longitudinal direction conflicted somewhat with the

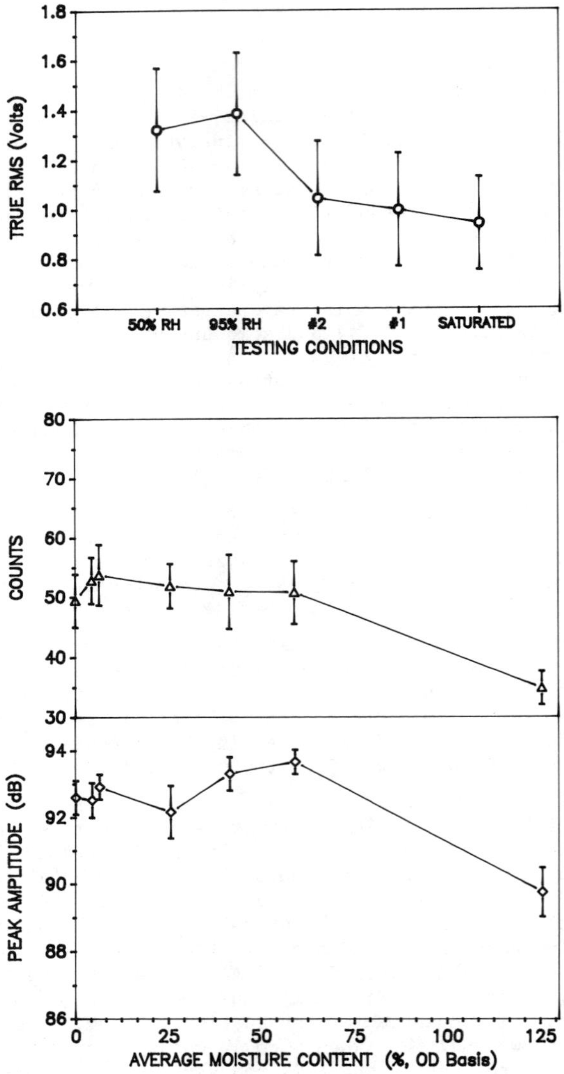

FIGURE 2. The effect of moisture content on the measured acoustic parameters in the transverse grain directions. The 95% confidence intervals are given by the upper and lower bars.

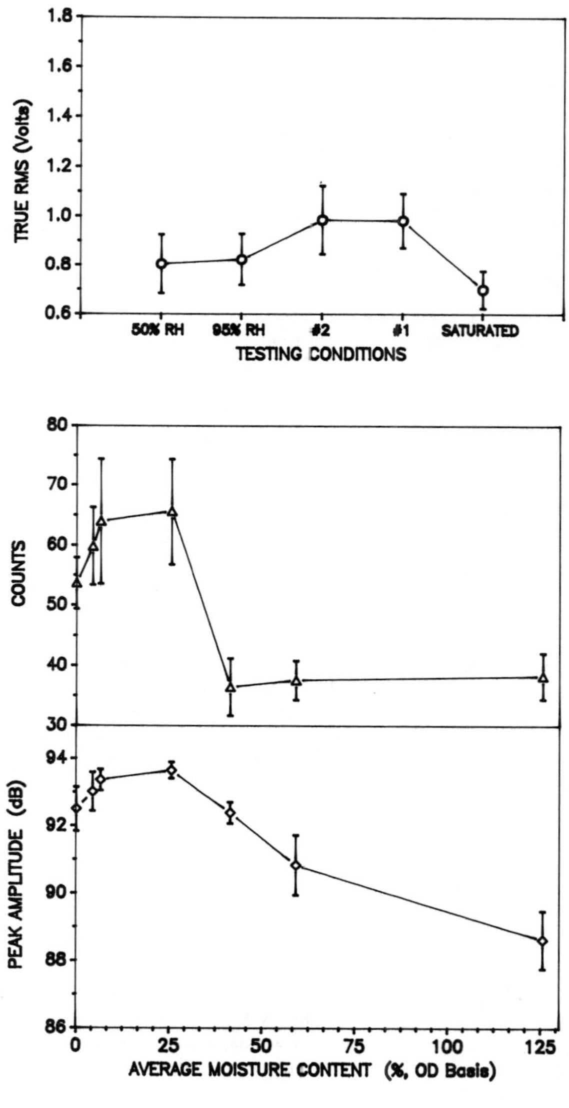

FIGURE 3. The effect of moisture content on the measured acoustic parameters in the longitudinal direction. The 95% confidence intervals are given by the upper and lower bars.

results observed in transverse testing. Average values are also plotted in this figure. There was an inverse relationship between peak amplitude and average moisture content above approximately 30%, but below this point no change was observed. Counts remained fairly constant until the average moisture content reached 30%, and then increased abruptly, exhibiting the same trend as seen with the RMS signal in transverse testing. There was no consistent trend between moisture content and RMS in the longitudinal direction, which contradicted results observed with transverse testing. The reason for this apparent inconsistency is not known, and research is continuing in this area.

The effect of grain angle on the measured acoustic parameters are shown in Figure 4. Plotted values were averaged over repeated measurements, moisture contents, and replications. As was the case for the effect of moisture content on AE parameters measured in the transverse directions, peak amplitudes and counts were insensitive to changes in the propagation direction. It was evident that RMS was more sensitive to changes in propagation direction, however, between replication (sample) variability somewhat masked these differences. This is in general agreement with past indications of the effectiveness of AU testing with regard to its sensitivity to irregularities in physical and structural properties.

It should be noted that the measured grain angles did not always agree with the target nominal angles. This was unfortunate, but somewhat understandable, given the anatomical characteristics of madrone (i.e., diffuse porous, extremely small vessels and high density). This research is continuing, and the anatomical characteristics of the current test species make it quite easy to match target and actual grain angles.

3. OPTIMUM COMBINATION OF ELECTRICAL AND MECHANICAL COMPONENTS FOR AU TESTING

The typical AU system consists of two basic elements. The first element includes the pulse generation circuit, including the pulsing transducer, and the second includes the receiving transducer and the signal measurement circuit. In order to determine the optimal combination of electrical and mechanical components of the AU pulser-receiver circuit, a series of experiments were conducted (Gasick *et al.* 1987). Selection criteria were based on maximizing the output RMS voltage while maintaining repeatable measurements. The experimental variables are given in Table 1. Specimen material included incense cedar *(Libocedrus decurrens)* pencil slats, California black oak *(Quercus kelloggii)* and commercial plywood and industrial (3-layer) particleboard.

3.1 System Optimization Results

Results of this series of experiments showed that certain components in the pulsing-receiving circuit significantly improved the measured RMS signal. The RMS signal was consistently optimized when the 110 kHz transducer was used as the pulser. The 110 kHz transducer was enclosed in an all stainless steel case, whereas all the other transducers had an epoxy base. The AU testing apparatus utilized frustrum-shaped stainless steel wave guides (tapering from 2.36 to 0.97 cm with a height of 1.24 cm) between the pulsing/receiving transducers and the sample, and it was therefore possible that the lack of an impedance mismatch with the 110 kHz transducer improved the pulsed signal. It is also possible that the damping characteristics of the transducers were significantly different. In fact two 110 kHz transducers were utilized during these tests, one with an in-

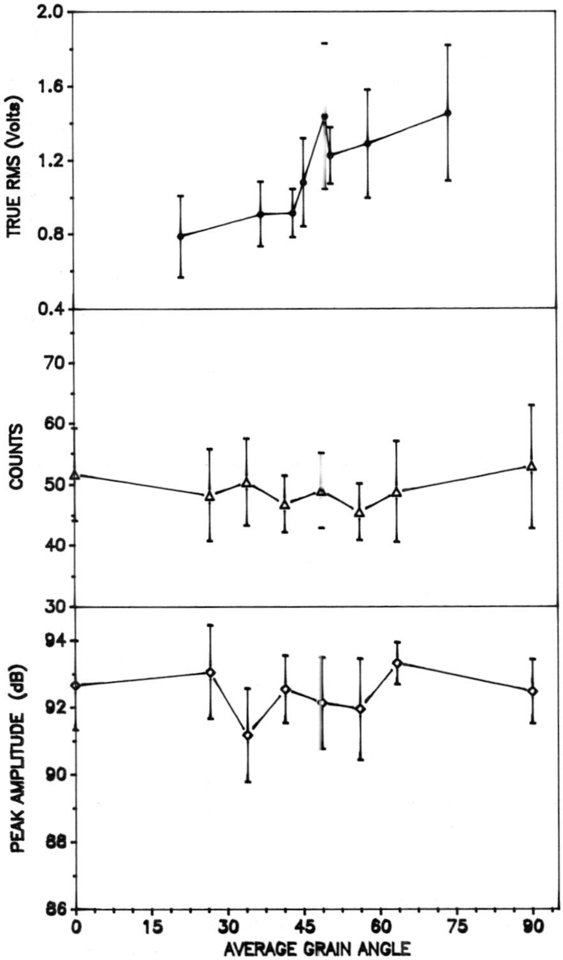

FIGURE 4. The effect of grain angle (propagation direction) as the measured acoustic parameters in the transverse grain directions. The 95% confidence intervals are given by the upper and lower bars.

TABLE 1. Experimental variables investigated in the optimization study.

A. Signal Generation Circuit	B. Signal Measurement Circuit
1. Repetition rate from signal generator	1. Output signal filters
a. 0.5 kHz	a. 0.650 - 125 kHz
b. 5.0 kHz	b. 125 - 250 kHz
c. 50.0 kHz	c. 250 - 500 kHz
	d. 0.125 - 2 MHz
2. Pulsed signal amplitude (input power)	2. Total system gain
a. 100 volts	a. 60 dB
b. 175 volts	b. 70 dB
c. 250 volts	c. 80 dB
	d. 90 dB
	e. 100 dB
3. Pulsing transducers	3. Receiving transducers
a. 75 kHz	a. 75 kHz
b. 110 kHz	b. 140 kHz
c. 140 kHz	c. 175 kHz
d. 375 kHz	d. 375 kHz
e. FAC 500 kHz	

ternal preamp, which was by-passed during testing, and one without. The transducer without the internal preamp was also more damped, and the RMS output when using it was significantly lower than with the RMS transducer with the preamp.

RMS output was less sensitive to changes in the receiving transducer, however the 175 and 375 kHz generally gave the highest readings and maintained a relatively simple output when viewed on the oscilloscope.

The optimum repetition rate from the signal generator was found to be 5 kHz (a period of 200 μs). Whereas this setting did not maximize the RMS output, these readings were consistently the most stable and repeatable. At this rate the received signal showed evidence of heterodyning which would limit some of the signal analysis techniques available.

Previous studies showed that the filter that encompassed the frequency of the pulsing transducer did not always yield the maximum output signal. It was assumed that this was due to the interactions between the transducers, waveguides and coupling medium which tended to change the actual propagating frequencies. In this case, the RMS output was always maximized when the 0.650 - 125 kHz filter was used in conjunction with any other set of components. The 0.650 - 125 kHz filter is not a standard, off-the-shelf, item and consequently the 125 - 250 kHz filter was selected for use in subsequent AU research. The performance of the 0.125 - 2 MHz and 125 -250 kHz filters were very similar, and selection of the 125 - 250 kHz filter was arbitrary.

The input power delivered from the AU pulser and the system gain were adjusted depending on the given material being tested. Adjustments were made such that a high RMS output voltage was obtained without saturating (clipping) the output signal (monitored on the oscilloscope).

The variability in output RMS voltage observed in these experiments indicated the need to standardize components in the AU circuit in order to obtain repeatable and transferable results. Results from these preliminary studies also showed that the optimum coupling, transducer and pulsing combinations for one wood product would not always guarantee optimum performance for another wood product. The need for developing standards for AU testing has been discussed by Leaird and Kingdon (1987).

4. APPLICATIONS TO MATERIAL EVALUATION AND PROCESS CONTROL

Current AE and AU research at the UC FPL includes evaluating the potential effectiveness of each for process monitoring and quality control procedures. AE projects include monitoring 1) stress development in wood during drying and 2) tool wear and feed speed in circular saws during processing, and evaluating vertical density profiles in wood-based composite materials. Only the drying research will be discussed in this paper. Other NDE projects at the UC FPL, using AU technology, include investigating its potential for detecting defects in incense cedar pencil slats and detecting early stages of decay in wood. Only the defect detection research will be discussed in this paper.

4.1 Defect Detection Using AU

AU is one of several nondestructive evaluation techniques that have shown potential for detecting defects in wood. AU is ideal for detecting passive defects throughout the entire volume of material (i.e., it is not necessary for the defect to be visible on the surface). In addition, AU typically operates in a relatively low frequency range (generally less than 500 kHz), whereas ultrasonics technology usually requires operational frequencies greater than 1 MHz. The lower frequencies associated with AU testing are more desirable because high frequencies attenuate rapidly in wood.

<u>Procedure and experimental results</u>. A number of pencil slats (0.5 x 7.2 x 18.4 cm) exhibiting defects such as surface checks (splits from the end grain and cracks), knots, voids, grain deviation, and decay were obtained from California Cedar Products Company, Stockton, California and tested (Rahman *et al.* 1988). Control samples containing no visible defects were also tested. For these types of defects, one would expect an attenuation of the signal when the defect is in a direct path between the pulsing and receiving transducers.

The AU testing apparatus was attached to a milling machine which allowed the pulsing and receiving transducers to move transversely, longitudinally, and vertically relative to the sample. The pulsing and receiving transducers were 110 and 175 kHz respectively, and a 0.5 kHz rep rate and 125-250 kHz filter were also used. A water couplant was used for all testing.

The pencil slats were tested in the longitudinal and transverse directions (Figure 5) at 1 cm and 0.5 cm intervals respectively. RMS output voltage was the measured parameter. The results presented here represent the average of three measurements made at each location on each pencil slat.

Results from the control slats showed that there was virtually no change in the AU signal level across the slat. Results from the slats containing defects showed that knots, surface checks, and decay could readily be detected by an abrupt drop in the output RMS voltage. A typical example of this is shown in Figure 6, which shows the longitudinal and transverse RMS output profiles across the width and length of the slat. The longitudinal profile clearly detected the defect, and whereas the reduction in RMS output as seen in the transverse profile was not as dramatic, it still decreased by a factor of two in the region of the check. The apparent effectiveness of the longitudinal profile in this instance was most likely caused by a reduction in wood-couplant contact in the void area created by the end check, and was not due to a greater sensitivity to this type of defect. All other defects were detected with a similar decrease in output RMS voltage in either the longitudinal or transverse profile.

Future research will concentrate on increasing the throughput of slats beneath the detector and developing defect classification schemes using pattern recognition techniques.

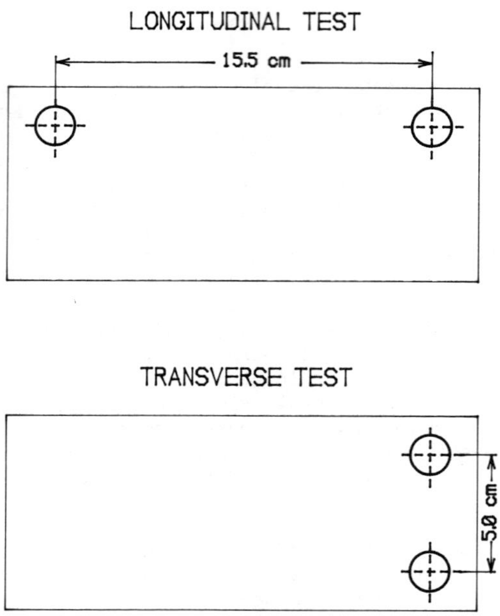

FIGURE 5. Test configurations.

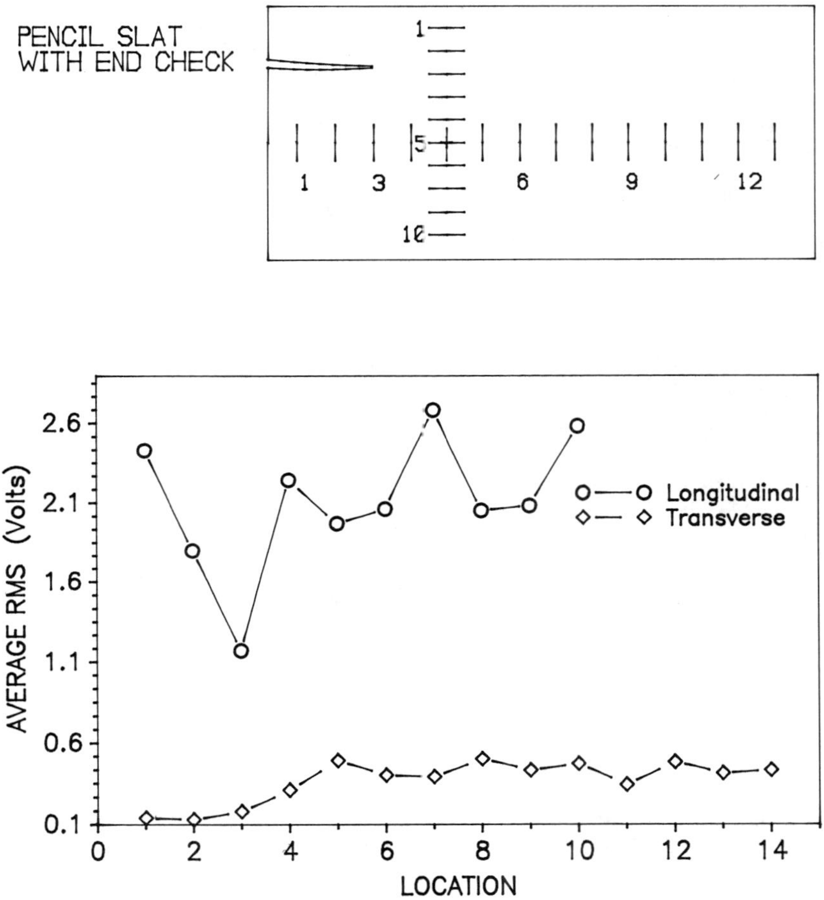

FIGURE 6. Schematic of pencil slat with end check and accompanying longitudinal and transverse AU test results.

4.2 Drying Related Research Using AE

The central goal of lumber drying is to remove the bulk of the water in wood as rapidly as possible while minimizing degrade. Since moisture content can be readily determined or estimated, drying schedules have traditionally been based on empirical relationships developed between moisture content and defect development. In recent years, considerable research effort has been directed towards developing drying control systems in order to maximize drying time while concurrently minimizing degrade. The ideal control system would be reactive, signaling for conditions of increasing or decreasing severity based on direct measurements of stress development in the drying wood. One such control system which shows promise is based on AE technology, and is the basis for a significant portion of the drying research at the UC FPL.

Results of preliminary investigations at this institution have been reported (Quarles and Zhou 1987). Research in the drying area since this publication have focused two critical issues, the first being to determine the effect of wood properties on the attenuation of the acoustic wave and the second to determine the relationship between AE parameters and surface check development. Progress in the first area has been reported in this paper (see Section 2. Attenuation Investigations).

Work towards meeting the second objective has concentrated on developing the appropriate methodology to be able to consistently position the transducer in close proximity to the propagating surface check. This is necessary in order to establish the relationship between AE parameters and a given type of failure (in this case surface checking).

In order to predict the location of the surface check, a small razor blade cut will be initiated on the planed (i.e., smooth) surface at the interface of a ray and fiber. Previous research has shown that presurfacing refractory hardwoods can significantly reduce the amount of surface checking (Gaby 1963). The reduction in surface checking was attributed to the removal of a damaged surface layer originally caused by microfailures occurring during the primary breakdown. This procedure will, in effect, selectively simulate the small fractures which occur during the primary breakdown, and consequently it will be possible to accurately place the AE transducers on the board surface. The drying surface will be recorded using a video camera, and check growth will be quantified by analyzing sequential video footage using an image analysis system. Work in this area of drying is continuing.

REFERENCES

1. Lemaster, R.L. and Dornfeld, D.A. Preliminary investigation of the feasibility of using acousto-ultrasonics to measure defects in lumber. Journal of Acoustic Emission 6(3):157-165. 1987.

2. Quarles, Stephen L. and Zhou, L. Use of acoustic emissions to detect drying defects: a preliminary report. Proceedings Sixth Symposium on the Nondestructive Testing of Wood, Washington State University, Pullman, WA, pp. 95-111. 1987.

3. Gasick, Michael F., Lemaster, R.L. and Dornfeld, D.A. Optimal transducer pulser-receiver combinations for use in acousto-ultrasonic testing of wood and wood composites. Technical Report No. 35.01.442, Progress Report No. 1, University of California, Forest Products Laboratory, Richmond, CA, 27 pp. 1987.

4. Leaird, James D. and Kingdon, M.C. Considerations for developing standards for acousto-ultrasonic inspections. Proceedings Conference on Acousto Ultrasonics: Theory and Applications (in press). 1987.

5. Rahman, Babu S., Lemaster, R.L. and Dornfeld, D.A. Defect detection of pencil slats using acousto-ultrasonics. Proposal No. 34.04.103, University of California, Forest Products Laboratory, Richmond, CA, 34 pp. 1988.

6. Gaby, L.I. Surface checking of white oak as related to mechanical processing. Forest Products Journal 13(12):529-532. 1963.

ACOUSTIC EMISSION EVALUATION FOR TIMBER INCLUDING KNOTS

K. Sato and M. Fushitani
Tokyo University of Agriculture and Technology
Tokyo, Japan

INTRODUCTION

Since wood is produced by a growing tree, it is heterogeneous and its mechanical properties, such as the modulus of rupture (MOR) and the modulus of elasticity (MOE), are variable. Furthermore, the existance of defects like knots, which is inevitable, makes the distribution of the strength of the timber much scattered. The effect of the knot on the evaluation of structural timber is one of the most important problems in the timber engineering.

There are two methods for evaluating timber, namely, visual grading and mechanical grading. The former is most generally performed in the wood products factory and the sawmill. Defects in timbers which reduce strength are visually detected by the sorter. Kunesh et.al.[1] have reported the correlation between the knot size and the tensile strength of dimension lumber of Douglas-fir. The Japanese Agricultural Standard prescribes the knot size or the knot ratio as standards for grading timber. The latter measures the MOE by either static loading or vibration or the velocity of the sound wave, and is based on the principle of a high correlation between MOR and MOE. Though the latter can be determined automatically, the MOE is only a nominal average quality of the material, and thus it is also necessary to visually evaluate the timber.

Since acoustic emission (AE) is sensitive to the stress concentration around a defect, it is considered to be an effective method to evaluate wood for knots. In a previous report[2], many AE signals were generated even at a low bending stress level in a specimen of ezomatsu containing a knot or compression failure. Therefore, the AE can be used to inspect the defect. In paticular, it is effective on compression failures which cannot be detected by the MOE.

Hence, the inspection of the defect in the timber which depends on visual grading may be performed automatically using AE, and grading by measuring both AE and MOE will make the timber evaluation more reliable. In this work, specimens of sugi including knot were tested in bending, and the evaluation using both AE and MOE was discussed.

EXPERIMENTAL PROCEDURES

Specimens (20(R)X20(T)X320(L)mm) were sawn from sugi (*Cryptomeria japonica*). Ninety five specimens with a knot located on the plain-sawn surface, and ten defect-free specimens for control were tested. The bend testing was performed at 20°C and 65RH using a Shimazu Servopulser EHF-EA5. The stress rate was 100kgf/cm min. A system selected was the center loading on quarter-sawn surface of the beam with a span of 280mm. AE signals were detected with a PZT type transducer (NAIS F-217M, 200kHz) attached to the quarter-sawn surface with silicon grease and held in place with a rubber band that was on the tension side of specimen at 25mm from the center of the span. Detected AE signals were magnified by a preamplifier (60dB) and an amplifier (41dB), and the number of AE events whose amplitude were greater than the threshold (0.8V) were measured using a one channel AE analyzer (PAC 4300).

RESULTS AND DISCUSSION

Relationship between MOR and MOE

The mechanical grading method used for estimating the strength of the timber is based on the high correlation which exists between the MOR and the MOE. In the case of static loading, the load less than the proportional limit is given to obtain the deflection, and the timber is graded by its calculated MOE. FIGURE 1 shows the relationship between the MOR and the MOE. The correlation is significant at 1% level (correlation coefficient 0.747), but it is not reliable enough to estimate the strength because there are some specimens with a high MOE which have low rupture stresses. For example, specimens with an MOE of about 60000kgf/cm² exhibited MOR's which ranged from 100-600 kgf/cm².

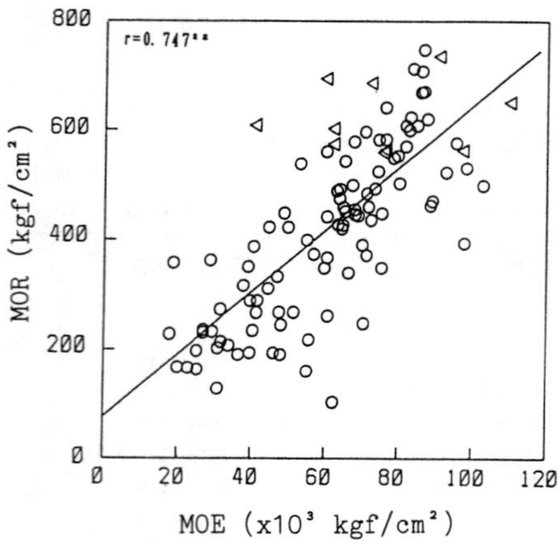

FIGURE 1. Relationship between MOR and MOE.

△: defect-free specimen
○: specimen including knot

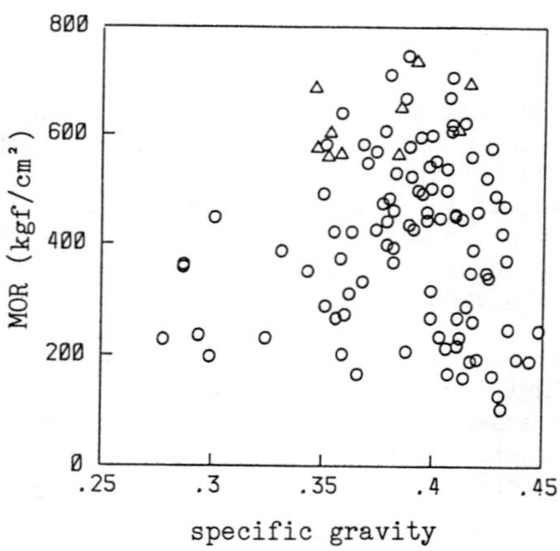

FIGURE 2. Relationship between MOR and specific gravity.

△: defect-free specimen
○: specimen including knot

Relationship between MOR and Specific Gravity

A physical property of wood which is used in addition to the MOE is the specific gravity. In general, the specific gravity relates to the MOR of clear specimens which means the quantity of wood substance present in a certain volume. However no correlation was found between MOR and specific gravity for the specimens with knots, as is shown in FIGURE 2.

Evaluation with Knot Ratio

A knot ratio was defined as the ratio of the width of the face knot to its width on the plain-sawn surface. FIGURE 3 shows the relationship between the MOR and the knot ratio. The correlation coefficiet of -0.755 is similar to that between the MOR and the MOE. Specimens whose knot ratios are greater than about 30% exhibit MOR values lower than 400kgf/cm^2. This contributes to the high correlation. The reduction of MOR caused by the knot is thought to be a result of the sloping and crossing grain around the knot and the check existing in the knot.

The MOE can evaluate the slope of grain to a certain degree, but it cannot be expected to evaluate local stress concentration around a knot since it is a parameter of nominal average quality of the material. Though the correlation between the knot ratio and the MOE in FIGURE 4 is significant at the 1% level, the correlation coefficient of -0.488 is relatively small. This means that the information included in each of them is different.

FIGURE 5 shows that the plot of the relationship between the MOR and the MOE of specimens which includes knots (FIGURE 1) can be distinguished by the knot ratio. The specimens whose MOR is greater than 400kgf/cm^2 can be selected by choosing those specimens with an MOE greater than 50000kgf/cm^2 and a knot ratio less than 20%. It is, therefore, necessary to establish a standard such as the knot ratio to directly evaluate a defect.

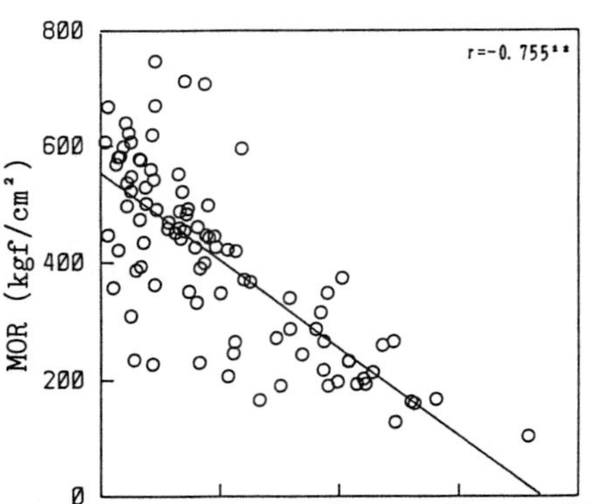

FIGURE 3. Relationship between MOR and knot ratio.

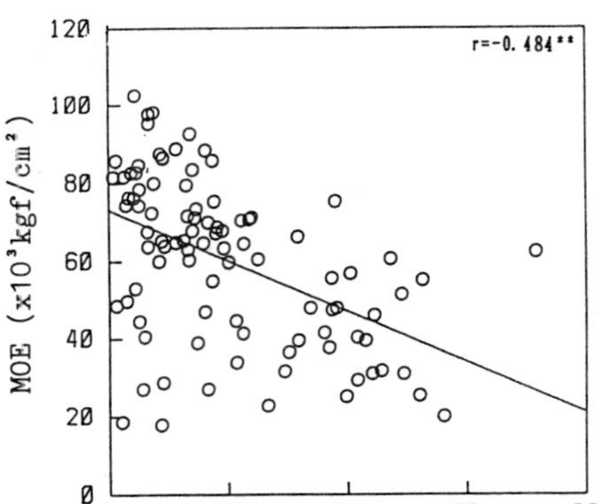

FIGURE 4. Relationship between MOE and knot ratio.

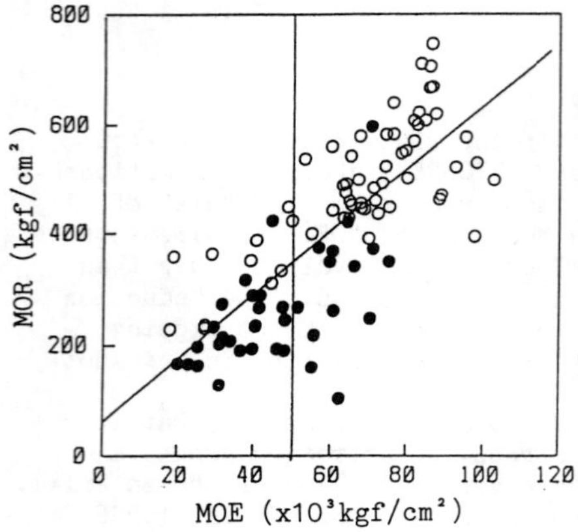

FIGURE 5. Evaluation with MOE and knot ratio.

○: Knot ratio <20%
●: Knot ratio ≧20%

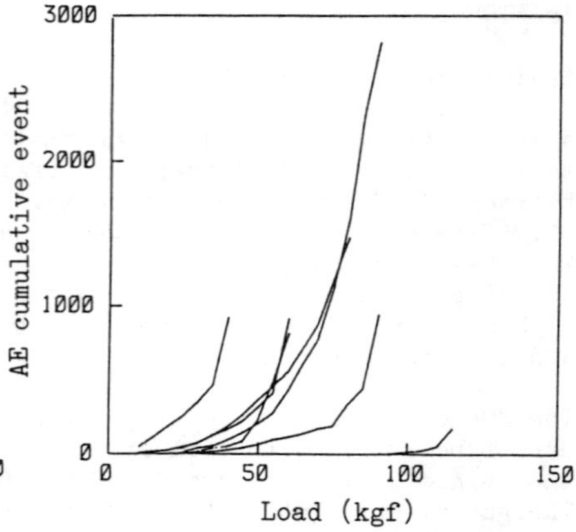

FIGURE 6. Load-AE cumulative event diagrams for specimens with MOE range of 55000-65000 kgf/cm².

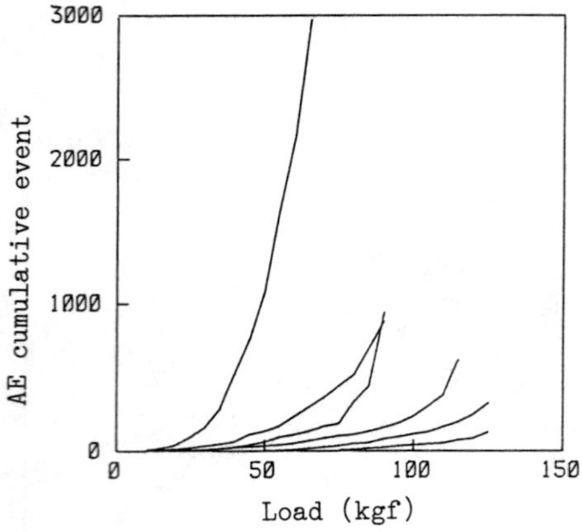

FIGURE 7. Load-AE cumulative event diagrams of specimens with knot ratio range of 8-12%.

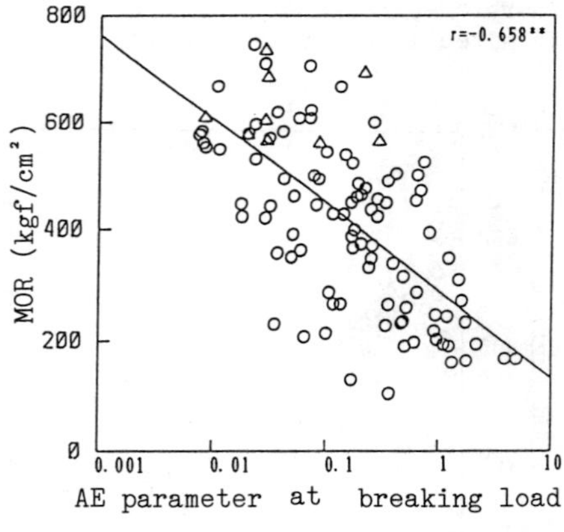

FIGURE 8. Relationship between MOR and AE parameter regressed at breaking load.

In a wood products factory or a sawmill, timber selection is currently made by visual inspection besed on the observed knot ratio. However, in this case there is a limit to the volume of timber which can be inspected, and the reliability of graded timber is low. Thus it is expected that a new method of automatically evaluating defects can be developed which will replace the knot ratio.

AE Generation from Knot

Typical relationships between the load and AE cumulative events for two groups of specimens, one with an MOE of 55000-65000kgf/cm², and the other with a knot ratio of 8-12%, are shown in FIGURE 6 and 7, respectively. Since the specimens which had small values of MOR began to generate AE signals at a low stress level, the AE count rate was high. It is possible to estimate the MOR using the AE. The AE is considered to give information about defects which cannot be detected by MOE measurement because the AE is sensitive to stress concentrations around the defect. The AE method is expected to be used as a new automatic quality control system for timber. An important advantage of this system is that MOE and AE can be measured simultaneously when used in conjunction with the current mechanical grading systems which measure MOE.

Evaluation Using AE

AE parameter used was the regression coefficient of AE cumulative events with respect to the load squared as given in Equation (1).

$$N = A P^2 + B \quad (1)$$

where N is the AE cumulative event, P is the load, A is a regression coefficient used for an AE parameter and B is a coefficient relating to the Kaiser effect. This equation was effective for evaluating defects in ezomatsu[2]. FIGURE 8, 9 and 10 show the relationship

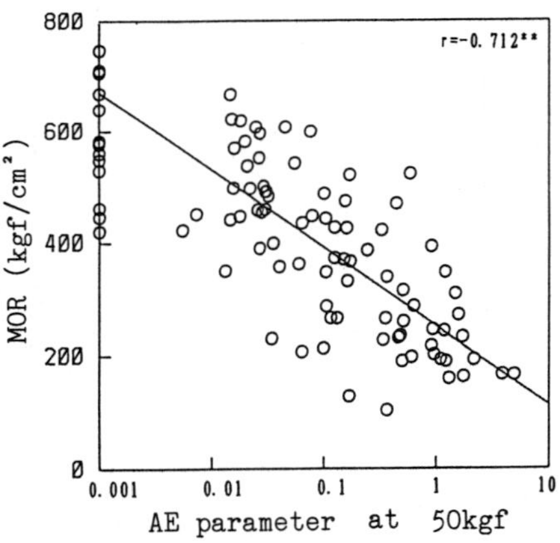

FIGURE 9. Relationship between MOR and AE parameter regressed at load of 50kgf.

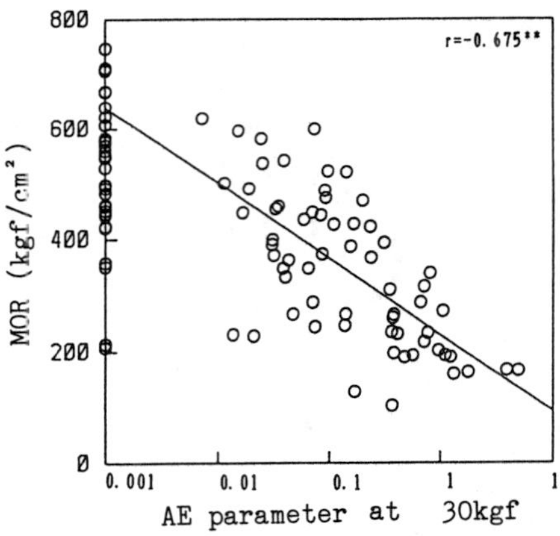

FIGURE 10. Relationship between MOR and AE parameter regressed at load of 30kgf.

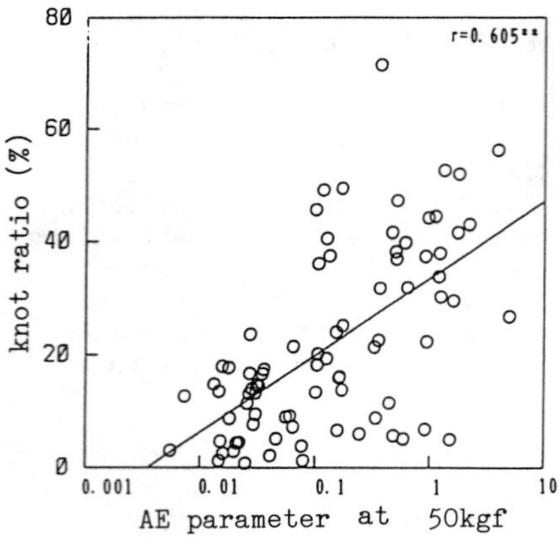

FIGURE 11. Relationship between knot ratio and AE parameter regressed at load of 50kgf.

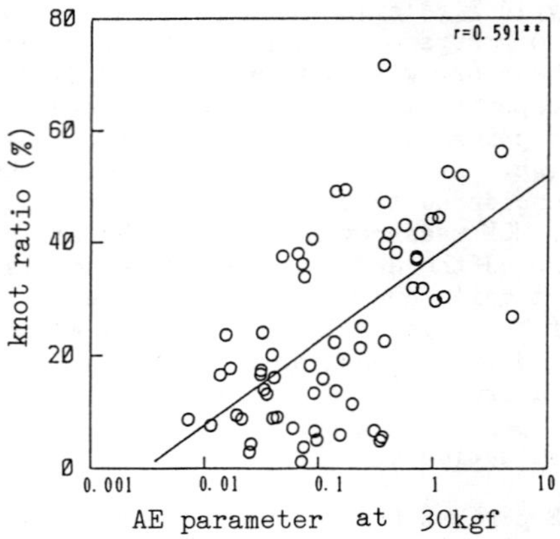

FIGURE 12. Relationship between knot ratio and AE parameter regressed at load of 30kgf.

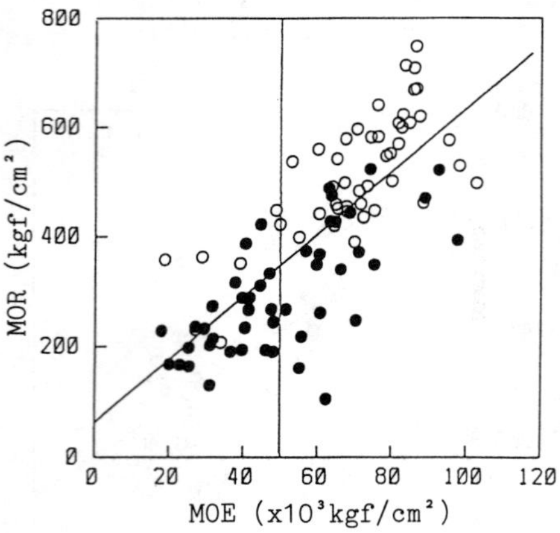

FIGURE 13. Evaluation with MOE and AE parameter regressed at load of 50kgf.

○: AE parameter <0.1
●: AE parameter ≧0.1

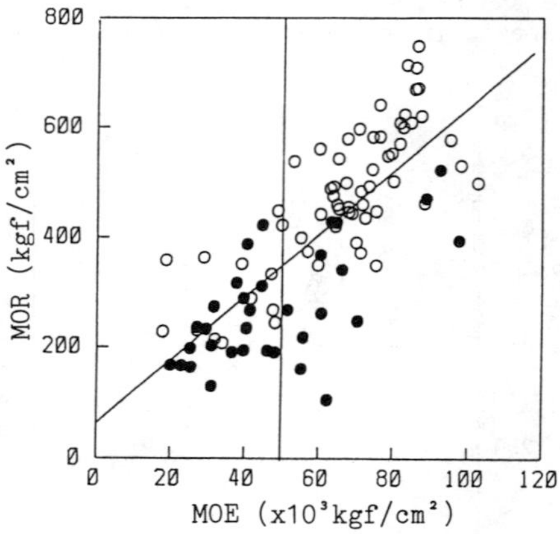

FIGURE 14. Evaluation with MOE and AE parameter regressed at load of 30kgf.

○: AE parameter <0.1
●: AE parameter ≧0.1

between MOR and the AE parameter calculated using Equation (1) for values of P obtained at the breaking load, 50kgf (43% of the breaking load of control), and 30kgf (26% of it), respectively. The AE parameter for specimens which did not generate AE until after the selected load was assumed to be 0.001. A negative correlation was observed between the AE parameter and MOR.

There was a correlation between the knot ratio and the AE parameter as shown in FIGURE 11 and 12, and therefore the evaluation based on AE can replace the one based on the knot ratio.

FIGURE 13 and 14 show that the plots of the relationship between the MOR and the MOE of specimens with knots in FIGURE 1 are distinguished by the AE parameter of 0.1. Specimens with values of AE parameter less than 0.1 have MOR greater than 400kgf/cm² in a way similar to that shown in FIGURE 5. In general, loads used in non-destructive mechanical grading should be as small as possible. These results indicated that a 30kgf load (26% of the breaking load of the control specimens) was all that necessary to obtain adequate results. This loading level is low enough to prevent the failure of most specimens during testing.

These results suggest that measuring both MOE and AE would produce a reliable mechanical grading system for timber.

CONCLUSION

The AE of specimens with knots was measured during bending tests, and an evaluation of timber based on the AE parameter was discussed. Since the MOE is only measure of the nominal average quality of the timber, it is necessary to use the MOE jointly with an index that can offer information on the effect of the knot as a defect. The evaluation using AE is similar to that with the knot ratio, and it can be measured automatically even using a slight load. A new reliable mechanical grading system for timber by measuring the MOE and AE is expected. Future research will address development of an in-line system and determination of the appropriate AE parameters.

REFERENCES

1. Kunesh, R. H., Johnson J. W., Effect of Single Knots on Tensile Strength of 2- by 8-Inch Douglas-Fir Dimension Lumber, Forest Products Journal, vol. 22, no. 1, pp. 32-36, 1972.

2. Sato, K., Okano, T., Asano, I., Fushitani, M., Application of AE to Mechanical Testing of Wood, Proc. 2nd Int. Conf. on Acoustic Emission, Lake Tahoe, pp. 240-243, 1985.

NONDESTRUCTIVE EVALUATION/CHARACTERIZATION OF ADHESIVE BONDED CONNECTIONS IN WOOD STRUCTURES

H. L. M. dos Reis
University of Illinois
Urbana, IL 61801

F. C. Beall
Forest Products Laboratory
Richmond, CA 94804

J. V. Carnahan, M. J. Chica, K. A. Miller, and V. M. Klick
University of Illinois
Urbana, IL 61801

INTRODUCTION

The feasibility of nondestructive evaluation/characterization of laminated wood products has significant economic ramifications. Currently, destructive methods of testing require that the performance of a sample be evaluated in order to characterize the larger group represented by the sample. If variation in bond strength is significant, it is possible that perfectly acceptable materials in the larger group may be rejected as a result of the statistical analysis of destructive tests of the smaller sample. Such needless waste can be avoided through nondestructive testing, possibly in the form of an on-line quality monitoring program.

Analytical ultrasonics implies the measurement of material microstructure and associated factors that govern mechanical properties and dynamic response. It goes beyond flaw detection, flaw imaging and defect characterization and includes assessing the inherent properties of material environments in which the flaws reside.

Acousto-ultrasonics is an analytical ultrasonic NDE technique which measures the relative efficiency of energy transmission in the specimen [1,2]. An ultrasonic pulse is injected with a transmitting transducer mounted on the surface of the specimen.More damage (i.e., flaws, changes in the microstructure, etc.) in the specimen tends to result in higher signal attenuation, resulting in lower stress wave factor (SWF) readings. Traditionally, the SWF has been evaluated as the number of oscillations with amplitude greater than some chosen threshold in the ring down oscillations in the output signal from the receiving transducer. The SWF has already been correlated with the mechanical strength of composite materials by Vary and Lark [3], Williams and Lampert [4], Kautz [5], and Govada, et al. [6]. SWF measurements have also been correlated, by the author of this study, with damage in wire rope [7], with swelling of wood products [8], and with the adhesive bond strength between rubber and steel [9,10]. A good review of analytical ultrasonics in materials research and testing is given in references [1,2].

In this study, the feasibility of using the acousto-ultrasonic (AU) technique to characterize the adhesive bond strength in laminated wood beams and in adhesive bonded connections is investigated. Various stress wave parameters, i.e., stress wave factors, were used and correlated to the shear strength of the adhesive bonded connections. Stress wave factor measurements were obtained both before and after the wood beam samples were subjected to temperature and humidity conditioning to simulate aging or wear of the glueline. A number of stress wave factors in the time and frequency domain exhibited excellent correlation with the shear strength of the adhesive bond. In addition, the stress wave factor measurements before and after conditioning correlated well with one another. Therefore, the results of this study show promise not only for eventual implementation in on-line quality control but also for evaluating the deterioration of the bond line integrity in the field.

EXPERIMENTAL PROCEDURE

To examine the feasibility of using the acousto-ultrasonic (AU) approach to evaluate the adhesive bond strength in wood connections, eighteen laminated wood beam specimens were manufactured. The specimens consisted of two 2.5 mm x 127 mm x 203 mm (1" x 5" x 8") blocks of wood laminated together. To obtain specimens with different levels of adhesive bond strength, the samples were made with variations in the gluelines as indicated in Table 1. These variations were chosen to represent common faults that occur in the glueline during manufacturing.

Figure 1 represents the configuration of the acousto-ultrasonic data acquisition system used in this study. The system consisted of the portable Acoustic Emission Technology (AET) model 206 AU instrument, two transducers, an AST computer equipped with an STR*825 analog-to-digital board and Digiscope software developed by Sonotek, Inc. Both the transmitting and receiving transducers were AET AC-175 L having an approximate sensitivity of -66 dB (relative to 1V/ μ bar) at a resonance frequency of 175 kHz. The transmitting transducer was excited by the Acoustic Emission Technology (AET) model 206 AU instrument which was set at a pulsing rate of 250 pulses/s. The output signal of the receiving transducer was amplified 40 dB by a preamplifier (AET model 140B) with a plug in filter having a passband between 30 kHz and 2 MHz. The signal was further amplified another 45 dB in the model 206 AU instrument, given a total gain of 85 dB. The analog signal from the AET 206 AU instrument was then digitized using an A/D board (Sonotec, model STR*825) and the digitized signal (i.e., waveform) stored on a floppy diskette in the 286 AST computer. Silicone rubber disks with 25.4 mm (1") in diameter and a thickness of 1.6 mm (1/16") were used as a dry couplant medium between the wood board specimens and the transducers. The rubber disks were attached to the transducers using a silicone based adhesive and the contact pressure between the wood specimens and transducers was such that the saturation pressure was exceeded. The scanning of the wood block specimens was done using "straight through transmission" with the receiving transducer being placed directly below the sending transducer as shown in Figure 2. The orientation of the wood grain in the specimens with respect to the transducers was held constant during the stress wave factor measurements.

To obtain SWF measurements which were representative of the entire glueline, a 13 mm (1/2") square grid was sketched on the top surface of each laminated block, as shown in Figure 2. The sending transducer was centered in each of these squares and a waveform was stored. To avoid edge effects, waveforms were not taken for the squares on the grid at the edges of the blocks.

Once the SWF for the undamaged laminated wood specimens had been obtained, the laminated wood block specimens were subjected to conditioning by placing the blocks in 90% relative humidity for 16 days at 20°C. The humidity causing the wood to swell created a state of stress on the glueline. This state of stress caused damage in the glueline and therefore simulated age and wear. After conditioning, the wood specimens were scanned in the same manner as before conditioning, the waveforms were sampled and stored, and the respective SWF claculated.

Once all the stress wave factor measurements were complete, the wood block samples were prepared for mechanical testing of the respective adhesive bond strength. Each of the eighteen blocks were cut into six 51 mm x 51 mm (2" x 2") squares for a total of 108 shear test specimens as shown in Figure 3. These 108 shear test specimens were offset to encourage failure to occur at the glueline as shown in Figure 4.

EXPERIMENTAL RESULTS

Acousto-ultrasonic stress wave factor measurements have been conducted on laminated wood beam samples manufactured with controlled adhesive bond strength. Stress wave factor measurements were recorded before and after conditioning where the laminated wood block samples were subjected to 90% relative humidity for sixteen days at 20 C.

The stress wave factor does not yet have a standard definition. In this study the stress wave factor is assumed to be any useful ultrasonic parameter in any domain such as the time and frequency domains. A variety of stress wave factors is provided by Chica, et al. [11] and the complete set of experimental results is given there. The discussion of experimental results given here will focus on the stress wave factors having the highest correlation with the adhesive shear bond strength.

Figures 5 and 6 show the retified area in the time domain, chosen as a stress wave factor, as a function of the shear break force for the test specimens before and after conditioning, respectively . Similarly, Figures 7 and 8 show the largest peak amplitude in the time domain, another possible stress wave factor, as a function of the shear break force for the test specimens before and after conditioning, respectively. Each shear break force value plotted represents the average of 18 measurements (one for each shear test specimen). Recall that for each glueline variation, there were 3 sample blocks prepared; each sample block was cut into 6 smaller squares for the shear test. Sixteen stress wave factor measurements were collected from each of the equal eighteen square shear test specimens. Therefore, each SWF value plotted represents the average of 288 measurements (16 measurements for each of the 18 shear test samples for each glueline variation). The average stress wave factor value was regressed on the average shear break force and the results are provided on each figure. The correlation coefficient is reported as a qualitative measure of the linear relationship between the average stress wave factor and the average shear break force. It was observed that in general the correlation between the mean values of the stress wave factors and the shear break force is somewhat higher for stress wave factors obtained after than before conditioning. Figures 9 and 10 show typical correlations that exist between stress wave factors before and after conditioning, respectively.

CONCLUSIONS

The acousto-ultrasonic approach shows great promise for evaluating the adhesive bond strength in laminated wood connections. Several stress wave factors, particularly the peak amplitude and the rectified area in the time domain, show high correlation with shear bond strength.The results indicate that the approach could be used for quality control, eliminating the need for a costly destructive testing program. In addition, there is potential application of this approach for evaluation of bond strength and possible delamination of wood connections already in the field. Finally, and most importantly, because of the existing correlation between the SWF before and after conditioning, there is the possibility of predicting the adhesive bond strength based on the SWF measured immediately after manufacture, eliminating the need for conditioning the test specimens.

There are several avenues of further investigation which could be pursued. An analysis of variance could be performed in order to study the contributions due to random and deliberate variations in manufacturing, edge effects, and conditioning. A study could be done to choose a statistical model to predict shear strength from the SWF. It is possible that one SWF might emerge as the best in terms of having minimum error of prediction.

ACKNOWLEDGEMENTS

The authors appreciate all the support provided by Weyerhaeuser towards the successful development of this project.

REFERENCES

1. Vary, A., "The Acousto-Ultrasonic Approach," in *Acousto-Ultrasonics: Theory and Aplications*, J. C. Duke, Jr., Ed., Plenum Press, New York, N.Y. 1988,pp. 1-21.

2. "Analytical Ultrasonics in Materials Research and Testing," NASA Conference Publication 2383, A. Vary, Ed., NASA Lewis Research Center, Cleveland, Ohio, November 13-14, 1984.

3. Vary, A., and Lark, F. R., "Correlation of Fiber Tensile Strength and the Ultrasonic Stress Wave Factor," NASA Lewis Research Center, Cleveland, Ohio, NASA TM 78846, 1978.

4. Williams, J. H., Jr., and Lampert, N. R., "Ultrasonic Evaluation of Impact-Damaged Graphite Fiber Composite," *Materials Evaluation*,Vol.38, No. 12, 1980, pp. 68-72.

5. Kautz, H. E., "Ultrasonic Evaluation of Mechanical Properties of Thick, Multilayered, Filament Wound Composites," NASA Lewis Research Center, Cleveland, Ohio, NASA TM 87088, 1985.

6. Govada, A. K., Duke, J. C., Jr., Henneke II, E. G., and Stinchcomb, W. W., "A Study of the Stress Wave Factor Technique for the Characterization of Composite Materials," NASA Lewis Research Center, Cleveland, Ohio, NASA CR 174870, 1985.

7. Reis, H.L.M. dos, and McFarland, D. M., "On the Acousto-Ultrasonic Non-Destructive Evaluation of Wire Rope Using the Stress Wave Factor Technique," *The British Journal of Non-Destructive Testing*, Vol. 28, No. 3, May 1985, pp. 155-156.

8. Reis, H. L. M. dos, and McFarland, D. M., "On the Acousto-Ultrasonic Characterization of Wood Fiber Hardboard," *Journal of Acoustic Emission*, Vol. 5, No. 2, April-June 1986, pp. 67-70.

9. Reis, H. L. M. dos, Bergman, O. A., and Bucksbee, J. H., "Adhesive Bond Strength Quality Assurance Using the Acousto-Ultrasonic Technique," *The British Journal of Non-Destructive Testing*, Vol. 28, November 1986, pp. 357-358.

10. Reis , H . L. M. dos, and Kautz, H., "Nondestructive Evaluation of Adhesive Bond Strength Using the Stress Wave Factor Technique," *Journal of Acoustic Emission*, Vol. 5, No. 4, 1986, pp. 144-147.

11. Chica, M. J., Klick, V. M. and Miller, K. A., "Software Development for Nondestructive Testing and Evaluation of Wood Products," Department of General Engineering, University of Illinois at Urbana-Champaign, Urbana, Illinois, 1988.

TABLE 1. Controlled glueline variations.

SPECIMEN LABEL	NUMBER OF BOARDS	DESCRIPTION
A	3	NORMAL- CONSIDERED TO BE THE CONTROL GROUP
B	3	GLUELINE THICKNESS CONTROLLED BY 0.0104" SHIMS IN THE ENDS
C	3	GLUELINE THICKNESS CONTROLLED BY 0.0364" SHIMS IN THE ENDS
D	3	UNDER-SPREAD : HALF OF NORMAL
E	3	PARTIALLY CURED AS COMPARED TO NORMAL
F	3	TAPERED - GLUELINE VARIES FROM NO SHIM IN ONE END TO A 0.0581" SHIM IN THE OTHER END

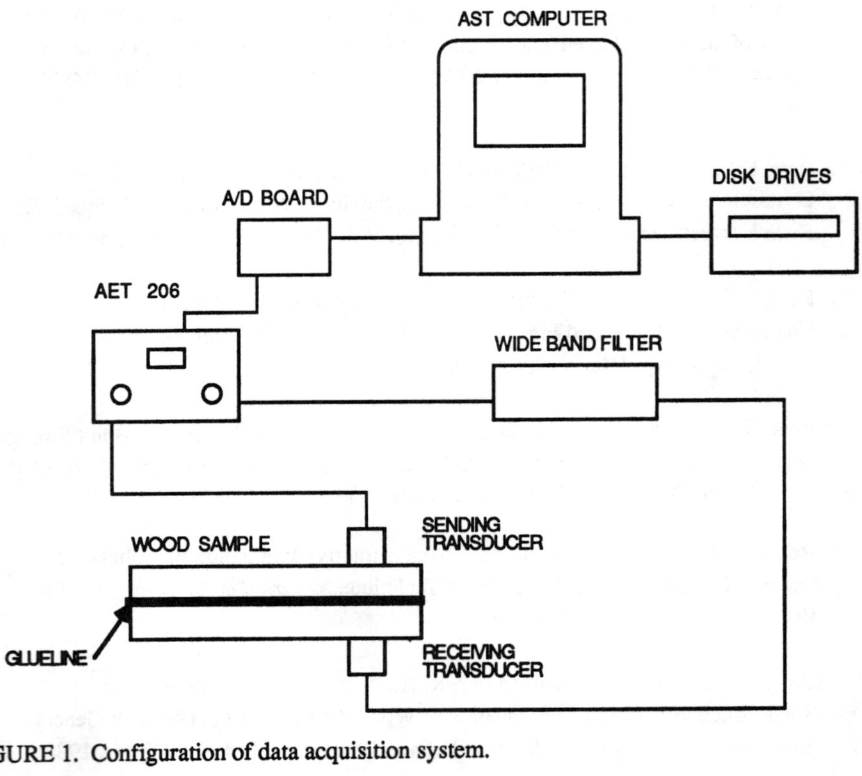

FIGURE 1. Configuration of data acquisition system.

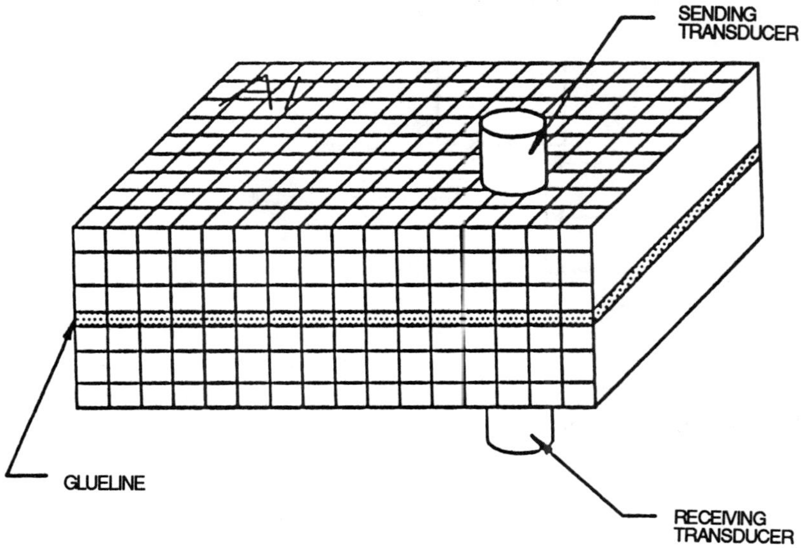

FIGURE 2. Transducer positions for measurement of straight-through transmission.

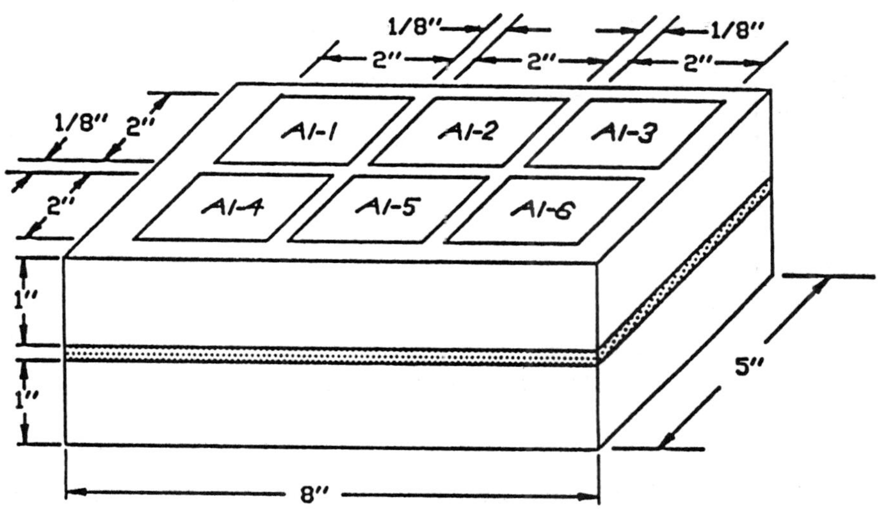

FIGURE 3. Cutting pattern for samples to be tested for shear strength.

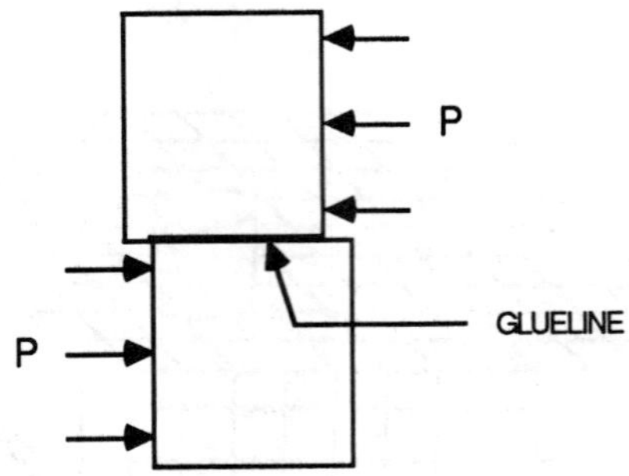

FIGURE 4. Shear test configuration.

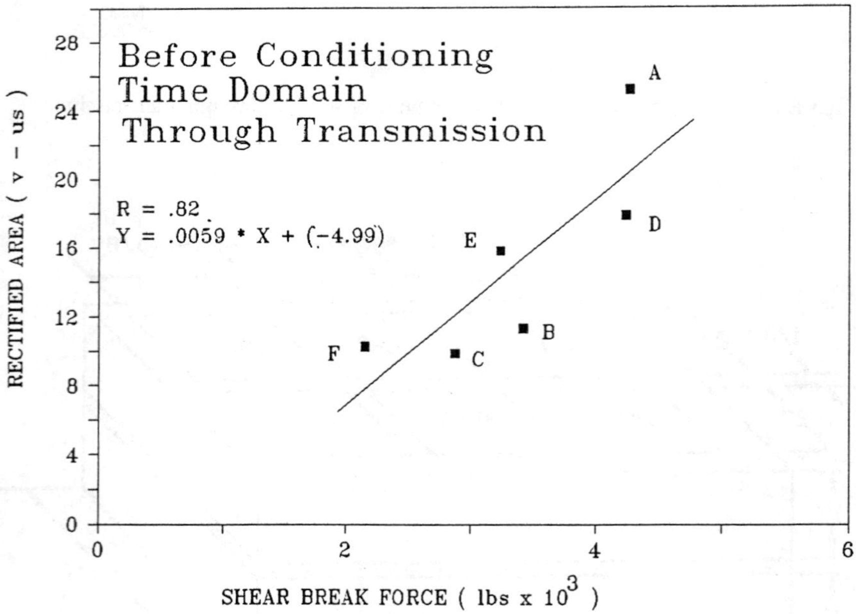

FIGURE 5. Rectified area in the time domain vs. shear break force for specimens before conditioning.

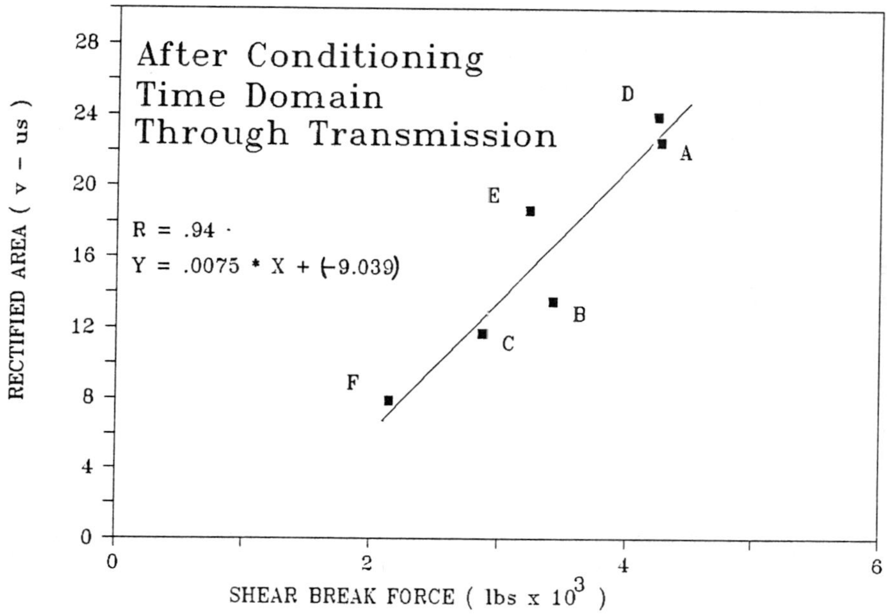

FIGURE 6. Rectified area in the time domain vs. shear break force for specimens after conditioning.

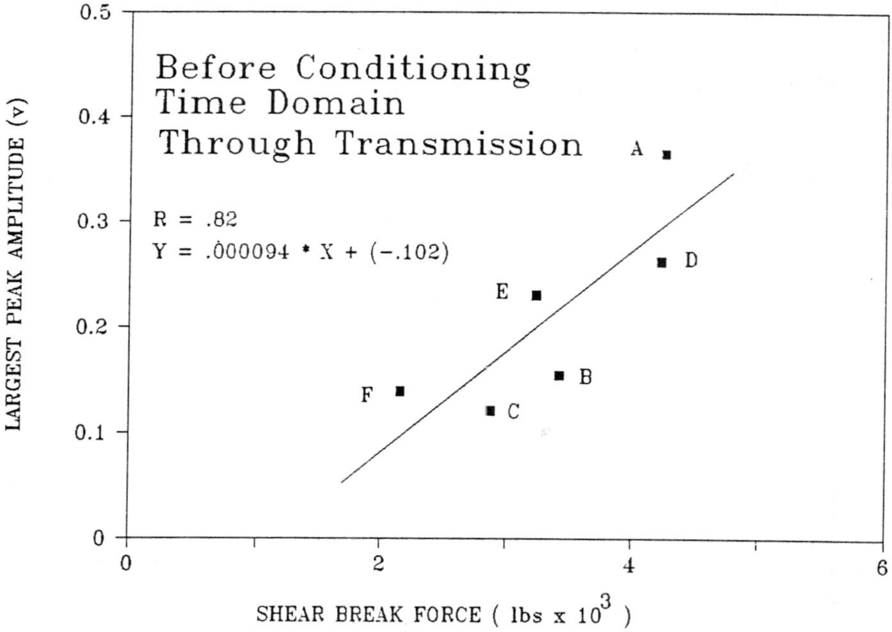

FIGURE 7. Largest peak amplitude in the time domain vs. shear break force for specimens before conditioning.

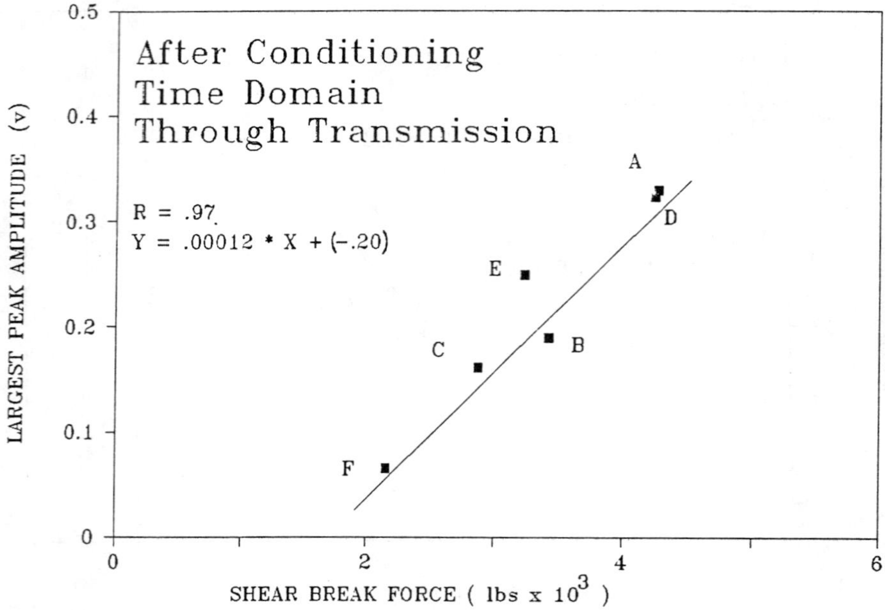

FIGURE 8. Largest peak amplitude in the time domain vs. shear break force for specimens after conditioning.

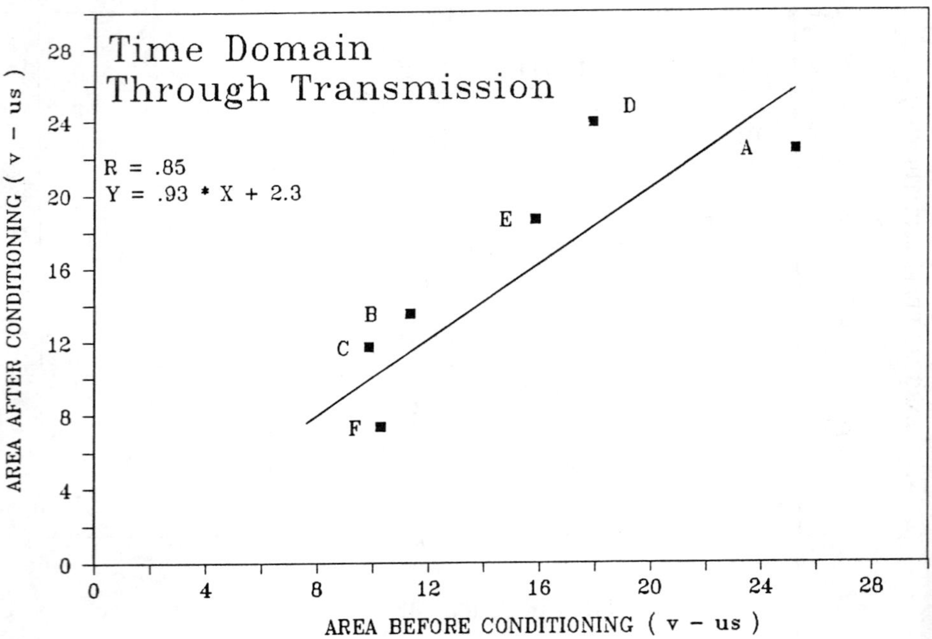

FIGURE 9. Rectified area after conditioning vs. rectified area before conditioning.

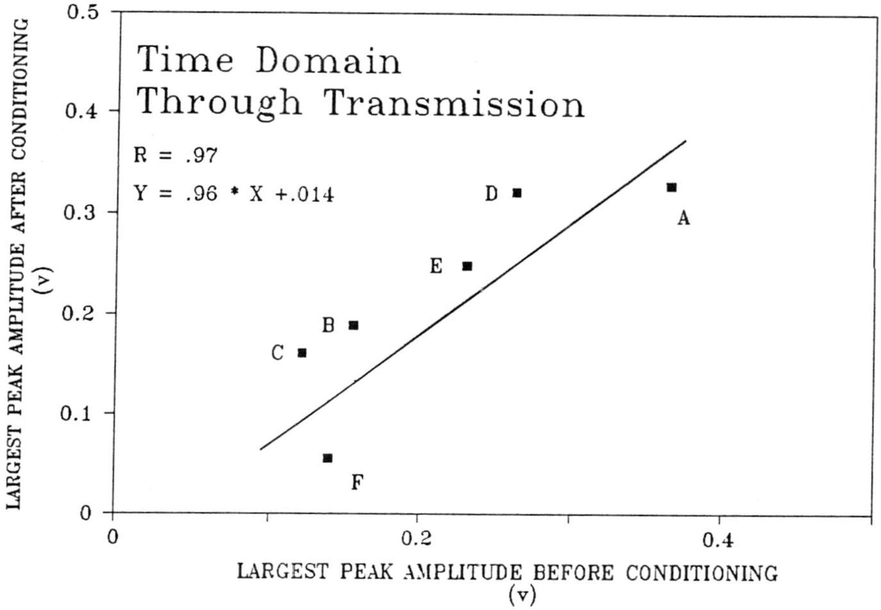

FIGURE 10. Largest peak amplitude after conditioning vs. largest peak amplitude before conditioning.

IMPACT-ECHO: A NEW METHOD FOR INSPECTING CONSTRUCTION MATERIALS

N. J. Carino
National Bureau of Standards
Gaithersburg, MD 20899

M. Sansalone
Cornell University
Ithaca, NY 14853

INTRODUCTION

The development of nondestructive evaluation (NDE) techniques for non-metallic, heterogeneous construction materials, such as concrete, masonry, and timber, has lagged behind the progress made in the development of techniques for relatively homogeneous materials like metals and ceramics. Most of the NDE methods that have been developed are based upon monitoring the interactions of some form of radiation (electromagnetic or stress waves) with the material properties or defects within a test object. The characteristics of non-metallic construction materials make the inspection by common forms of radiation impossible or extremely difficult. For example, concrete and masonry are non-magnetic, poor electrical conductors, heterogeneous, and contain macroscopic defects which may not be detrimental to performance.

Ultrasonic stress wave propagation methods are some of the most highly developed procedures for nondestructive evaluation of materials. They can be used to locate internal defects and to characterize elastic properties. The two common applications of ultrasonic waves are "through transmission" and "echo" methods. In the through transmission method, one measures the time it takes for a pulse to travel from a transmitting transducer located on one surface of the test object to a receiving transducer located on the opposite surface. If the distance between the transducers is known, the wave speed can be computed, and inferences can be made about the condition of the material between the transducers. In the echo methods, the transmitter and receiver are located on the same surface, and one measures the time it takes for the pulse to travel from the transmitter to a reflecting interface and back to the receiver. If the wave speed is known, the depth of the reflector can be determined.

Applications of stress wave methods for evaluation of concrete structures have recently been reviewed [1,2]. The implementation of ultrasonic techniques for evaluation of concrete has been hampered by the heterogeneous nature of concrete which leads to scattering of stress waves and limits penetration of the waves into a test object. In the past, success has been achieved in using the through transmission method to measure ultrasonic pulse velocity. In contrast, there has been limited success in the development of reliable echo methods for concrete structures. There have been some achievements in the testing of long slender structures, such as piles, but there is no generally accepted procedure for flaw detection in other types of members.

The National Bureau of Standards (NBS) has undertaken a long-term study to develop suitable procedures for flaw detection in concrete and other construction materials. One of the techniques that has been developed is based on transient stress wave propagation and is known as "impact-echo" [3]. The research program has involved analytical studies of transient wave propagation [4-7], laboratory testing of specimens with known defects [8-11], and field studies [12]. This paper provides a brief explanation of the impact-echo method and presents results which illustrate the capabilities of the method to locate a variety of flaws in plain and reinforced concrete.

PRINCIPLE OF THE METHOD

The principle of the impact-echo technique is illustrated in Fig. 1. A transient stress pulse is introduced into a test object by mechanical impact on the surface. The stress pulse propagates into the object along spherical wavefronts as compression (P) and shear (S) waves [4]. In addition, a surface (R) wave travels along the surface away from the impact point. The P- and S-waves are reflected by internal cracks and voids and the boundaries of the test object. The arrival of these reflected waves at the surface where the impact was generated produces displacements which are measured by a receiving transducer. If the receiver is placed close to the impact point, the displacement waveform is dominated by the displacements caused by P-wave arrivals [4]. The displacement waveform can be used to determine the time from the initiation of the pulse to the arrival of the first P-wave reflection. If the P-wave speed in the test object is known, the distance, T, to the reflecting interface can be determined.

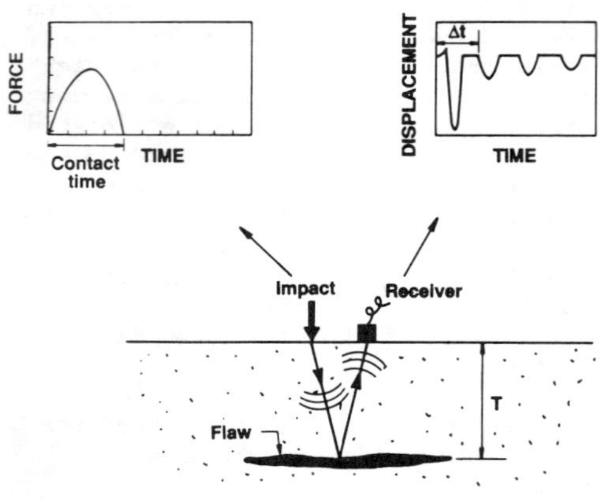

FIGURE 1. Principle of the impact-echo method.

Theoretical Response

Analytical studies have been carried out to gain an understanding of the expected response of an object subjected to impact at a point. Theoretical surface displacement waveforms have been calculated for infinite, elastic plates using the Green's function approach [3,4,9,13] and for plates containing defects using the finite element method [3,4-7,10]. As an example, Fig. 2 shows the computed vertical displacement (using Green's function) at a point 0.10 m away from the impact point. The plate is 0.5 m thick, the P-wave speed is 3500 m/s, and the duration of the impact is 0.25 of the time for the P-wave to travel from the top surface, to the bottom surface, and back to the top surface. The large downward displacement beginning at the point labeled "R" is produced by the arrival of the R-wave. The duration of the R-wave disturbance is equal to the duration of the impact. After the R-wave passes by, there is no displacement until the arrival of the P-wave which has reflected from the bottom surface. The arrival of the P-wave, which is indicated as "2P"[1], causes a downward displacement. Note that subsequent arrivals of the P-wave, after undergoing multiple reflections between the top and bottom surfaces, also produce downward displacements. The pattern shown in Fig. 2 is characteristic of the impact-echo response of either a plate or a large planar void within a solid.

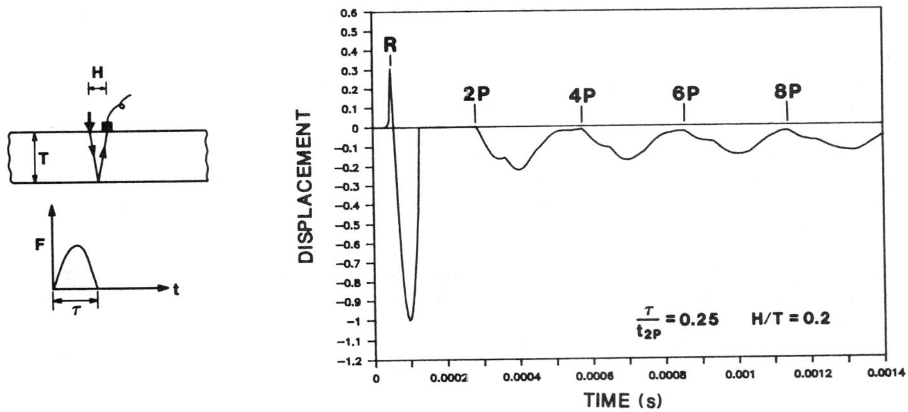

FIGURE 2. Computed vertical displacement 0.1 m away from the point of impact on a 0.5-m thick plate.

Instrumentation

An impact-echo test system is composed of three components: an impact source; a receiving transducer; and a waveform analyzer which is used to capture the transient output of the transducer, store the digitized waveforms, and perform signal analysis. Currently there is no commercially available test system; however, a suitable system can be readily assembled from off-the-shelf components.

Impact source. The selection of the impact source is a critical aspect of a successful impact-echo test system. The force-time history of an elastic impact may be approximated as a half-sine curve, and the duration of the impact is the "contact time." The contact time determines the frequency content of the stress pulse [13]. Impacts with short contact times contain the higher frequency (shorter wavelength) components that are needed for testing thin objects and for detecting small flaws. The

[1]The labels "2P", "4P", "6P", etc. in Fig. 2 refer to the number of plate thicknesses that have been traversed by the P-wave when it arrives at the top surface.

highest frequency component with significant amplitude has a frequency value that is approximately equal to the inverse of the contact time [13]. The contact time of an impact depends primarily on the size and mass of the impactor and on the condition of the concrete surface. An impact on a smooth, hard surface has a short contact time relative to that of the same impact on a rough surface. Small diameter (4-12 mm) steel ball bearings and spring-loaded, spherically-tipped impactors have been used successfully as impact sources. The contact times produced by these sources range from about 20 to 80 microseconds. The approximate value of the contact time in a test can be determined from the recorded waveform by measuring the duration of the downward displacement of the surface due to the R-wave.

Receiver. The receiver that has been found to produce successful results is a conically-tipped, broadband, displacement transducer, originally developed at the National Bureau of Standards as a secondary reference standard for calibrating acoustic emission transducers [14]. A special housing was built to hold the transducer so that it could be used on vertical surfaces. A thin lead strip is used to provide acoustic coupling between the piezoelectric tip of the transducer and the test surface. No coupling fluids or special surface preparations are required.

Waveform analyzer. The output from the receiver is captured using a digital oscilloscope. The appropriate sampling rate depends upon the geometry of the test object and the depth of the reflecting interfaces. In current studies, slab-like specimens with thicknesses less than 1 m have been tested and a sampling rate of 500 kHz has usually been used. As will be discussed in the next section, the system should be capable of performing spectral (frequency) analysis of the recorded waveforms. An alternative to a waveform analyzer is a personal computer with an A/D (analog to digital) converter and software to perform spectral analysis.

Data Analysis

In early work [8,9], interpretation of the recorded displacement waveforms was performed in the time domain. This required establishing the time of impact initiation and the arrival time of the first P-wave reflection. This was time-consuming and only feasible for simple test geometries. A more efficient approach is to analyze displacement waveforms in the frequency domain [3,13].

Frequency analysis. The principle of frequency analysis can be understood by referring to the impact-echo response shown in Fig. 2. The stress pulse generated by the impact propagates back and forth between the top and bottom surfaces of the plate. Therefore, a resonance condition is created. Each time the pulse arrives at the top surface it produces a characteristic downward displacement. Thus the displacement waveform is periodic. For a point close to the impact point, the travel path is approximately equal to twice the thickness of the plate, and the period is equal to the travel path (2T) divided by the P-wave speed (C). Since frequency is the inverse of the period, the frequency, f, of the characteristic displacement pattern is:

$$f = \frac{C}{2\,T} \qquad (1)$$

In practice, the frequency content of the digitally recorded waveforms is obtained using the fast Fourier transform (FFT) technique. This technique is based on the principle that any waveform can be represented as a sum of sine waves, each with a particular amplitude, frequency, and phase shift. Figure 3 shows the amplitude versus the frequency for each component sine wave in the theoretical waveform shown in Fig. 2 [13]. There is a single large amplitude peak at a frequency of 3.42 kHz which corresponds to multiple reflections between the bottom and top surfaces of the slab [6]. The frequency associated with reflections between the top and bottom surfaces is called the "thickness" frequency.

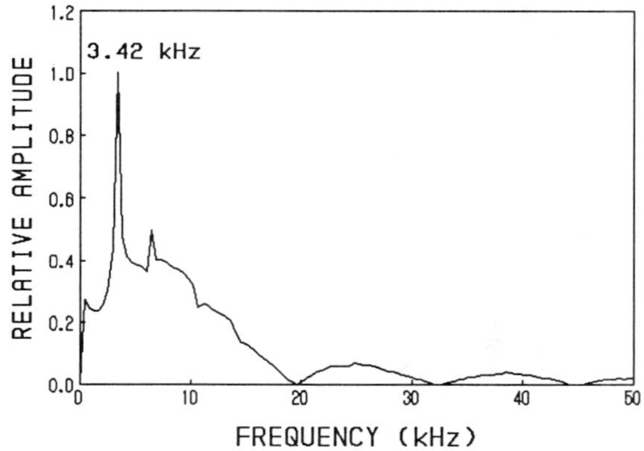

FIGURE 3. Amplitude spectrum of the displacement waveform in Fig. 2.

Equation (1) can be rewritten as follows:

$$T = \frac{C}{2\ f} \tag{2}$$

Thus, if the P-wave speed is known, the thickness of the plate (or the distance to a reflecting interface) can be determined from the predominant frequency in the displacement waveform. Equation (2) is the fundamental relationship for interpreting impact-echo results using frequency analysis.

The resolution in the digital amplitude spectrum, that is, the frequency difference between adjacent points, is equal to the sampling frequency for the waveform divided by the number of points in the waveform record. This affects the resolution of the depth calculated according to Eq. (2). Because frequency is inversely related to depth, for a given resolution in the amplitude spectrum, the resolution of the calculated depth improves as the thickness frequency increases, that is, as depth decreases. Typically, waveforms are sampled at a rate of 500 kHz and contain 1024 points; therefore, the resulting digital spectra have a resolution of 500/1024 = 0.488 kHz.

Figure 4 illustrates the application of frequency analysis. Figure 4(a) shows the amplitude spectrum for an impact-echo test performed over a solid portion of a 0.5-m thick concrete slab. There is a thickness frequency peak at 3.42 kHz. Using Eq. (1) and solving for C, the calculated P-wave speed is 3420 m/s. Figure 4(b) shows the amplitude spectrum obtained from a test over a portion of the slab containing a disk-shaped void. The peak at 7.32 kHz corresponds to multiple reflections between the top of the plate and the void. Using Eq. (2), the calculated depth of the void is 3420/(2 * 7320) = 0.23 m, which compares favorably with the known distance of 0.25 m.

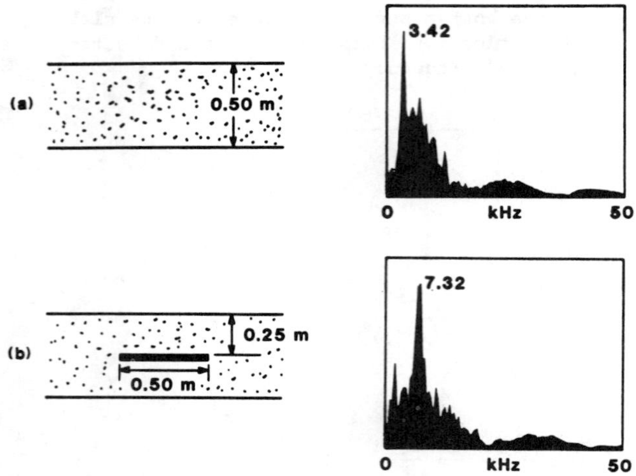

FIGURE 4. Amplitude spectra: (a) over solid portion of 0.5-m slab; (b) over a 0.25-m deep void.

Spectral peak plotting. In using the impact-echo method to determine the locations of flaws within an object, tests are performed at regularly spaced points along "scan" lines marked on the surface. Examination of the amplitude spectra from these scans will reveal the depth and approximate size of defects which may be present. One technique for visualization of the location of defects along a given scan is to create a "waterfall plot" of the individual spectra. In a waterfall plot, the spectra are plotted consecutively in the order corresponding to the test location along the scan line. Figure 5(a) is an example of a waterfall plot. In this case, a scan was performed along the top surface of the laboratory floor, which is a reinforced concrete slab on grade. Each of the spectra contains two significant peaks: a low frequency peak at approximately 1.5 kHz which corresponds to resonance of the transducer assembly; and a frequency peak at approximately 8 kHz which corresponds to the thickness frequency of the slab. The absence of other large peaks indicates that there are no significant defects within the slab along this scan. A close study of the waterfall plot shows that the thickness frequency peaks do not have the same frequency values. Thus the floor slab does not have a constant thickness. Because of the inverse relationship between depth and frequency, as indicated by Eq. (2), points with higher frequency values represent locations of shallower depth.

The information contained in a waterfall plot can be further processed to produce a cross-sectional view of the test object along the scan line. The technique that has been developed is called "spectral peak plotting," and it involves the following operations for each spectrum in the scan:

1. Locate the peak in the spectrum which has the highest amplitude;
2. Normalize the amplitude values by dividing by the highest amplitude;
3. Select a threshold value for the relative amplitude;

4. Locate all peaks in the spectrum whose relative amplitudes exceed the threshold value;
5. For each peak selected in step 4, compute the depth (T) corresponding to the frequency values (f) using the P-wave speed (C) and Eq. (2); and
6. Using the depth calculated in step 5, plot the depths along the scan line.

The above procedure has been incorporated into a computer program which uses the individual spectra as input and generates a picture of the cross section along the scan line. Spectral peak plotting permits a rapid assessment of impact-echo results by allowing the user to "see" inside the test object.

The spectra in Fig. 5(a) were processed using the spectral peak plotting computer program, and the resulting cross section is shown in Fig. 5(b). The P-wave speed was 4000 m/s. For this case, the maximum depth for the plot was chosen as 0.5 m; therefore, each tick mark along the depth axis represents 0.05 m. The tick marks along the top surface represent the individual test locations. The depth calculated at each test point is plotted as a "dash" between the midpoints of adjacent test points. The irregularity in the thickness of the slab can be clearly seen.

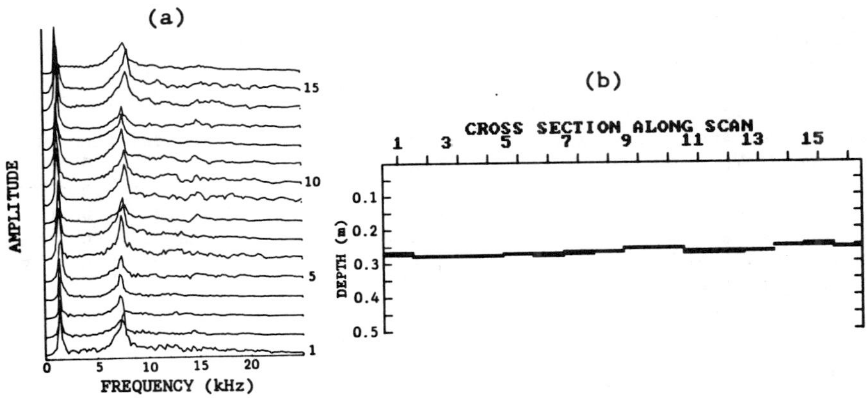

FIGURE 5. (a) Waterfall plot of amplitude spectra for scan of floor slab; (b) results form spectral peak plotting of amplitude spectra.

APPLICATIONS

The impact-echo method has been used successfully in the laboratory for detecting a variety of defects in concrete structures including: cracks and voids in plain and reinforced concrete [3,8,12]; the depth of surface-opening cracks [10]; voids in prestressing ducts [10]; honeycombed concrete [10]; delaminations in slabs with and without asphalt concrete overlays [11].

Most of the experimental studies have been controlled-flaw studies where specimens containing flaws at known locations have been constructed and tested. However, two "blind" studies have been carried out which involved detecting delaminations in a reinforced concrete slab [11] and locating voids in the grouting of prestressing ducts in a 1-m thick, reinforced concrete specimen which was built to simulate the wall of an off-shore structure [15]. In both of these studies, the locations of the flaws were not known by the authors prior to testing. Only after the impact-echo results were submitted to the project coordinators were the actual locations of the flaws revealed. In both cases, the impact-echo method was used successfully to detect the flaws. The following sections highlight the results from some of these studies; details can be found in the listed references.

Voids in Concrete Slabs

Laboratory study. One of the first laboratory specimens constructed at NBS was a 0.5-m thick concrete slab having plan dimensions of 2.0 by 4.5 m. When the slab was cast, 25-mm thick polyurethane disks of various diameters were embedded at different elevations. Because of their low density, the disks simulated voids. This specimen was used to establish the basic experimental procedures for the impact-echo method and to validate the results of analytical studies [3,9]. The spectra shown in Fig. 4 are examples of the results obtained from this slab.

One of the 0.5-m diameter disks embedded in the slab was inclined at an angle of 30 deg to the bottom surface. Figure 6(a) is a cross section of the slab showing the inclined disk. A scan was performed over the disk using a 50-mm spacing between test points. A mechanical impactor was used to generate the impact. The location of the test points with respect to the disk are shown in Fig. 6(a). The disk was located approximately between points 5 and 13. Figure 6(b) shows the cross section as obtained using the spectral peak plotting program. The presence of the inclined disk is indicated below test point 3 through 13. Note that for the tests over the disk, the apparent thickness of the slab is greater than 0.5 m. This is due to the increased path length of rays which are diffracted by the disk boundary and reflected by the slab surfaces [3,12]. The boundary of the disk is indicated by the distinctive pattern of regularly spaced lines at test points 5 and 14. This pattern is believed to be caused by diffraction effects at the boundary of the void.

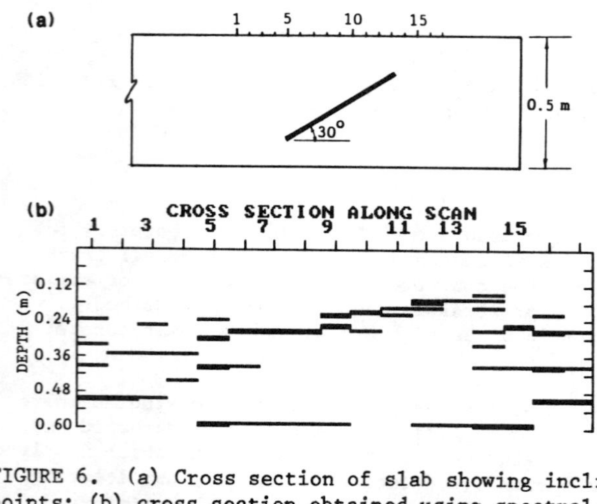

FIGURE 6. (a) Cross section of slab showing inclined disk and test points; (b) cross section obtained using spectral peak plotting program.

Field Study. A field study was performed to demonstrate the capability of the impact-echo method for detecting pockets of unconsolidated concrete in a concrete slab for an ice-skating rink [12]. The 150-mm thick slab contained cooling pipes at a depth of approximately 80 mm. In one-half of the rink, leaks developed in the pipes as a result of corrosion, and it was believed that the corrosion was linked to the presence of voids around the pipes. The voids, in turn, were probably a result of improper consolidation of the concrete due to mechanical difficulties with the vibrators during the placement of a major portion of the slab.

The pockets of unconsolidated concrete were believed to be about 50 to 100 mm in length and located below the cooling pipes. Prior to field testing, a laboratory mock-up of the slab configuration was prepared to develop the appropriate testing procedure for detecting the small, shallow voids. A 150-mm thick (0.5 by 1 m in plan) slab was cast in which reinforcing bars and a cooling pipe were embedded. As shown in Fig. 7(a), a small void was created below the cooling pipe using polyurethane foam. The mock-up slab rested on insulation board to simulate actual field conditions.

A 5-mm steel ball bearing was used to produce an impact with a duration of approximately 20 microseconds. The resulting short duration stress pulse had the high frequency components which are needed to detect the small, shallow void. Figure 7(b) shows the cross-section obtained from spectral peak plotting of the amplitude spectra for the test points shown in Fig. 7(a). The void was located approximately between test points 6 and 11, and the depth of the void is indicated by the reference line drawn at a depth of 0.8 m. The presence of the void is indicated below test points 6, 7, 8, and 11. In addition to producing frequency peaks associated with the top surface of the void, the void also produces changes in the frequency peaks associated with the bottom of the slab. Based on these results it was concluded that the impact-echo method could detect the small pockets of unconsolidated concrete in the skating rink slab.

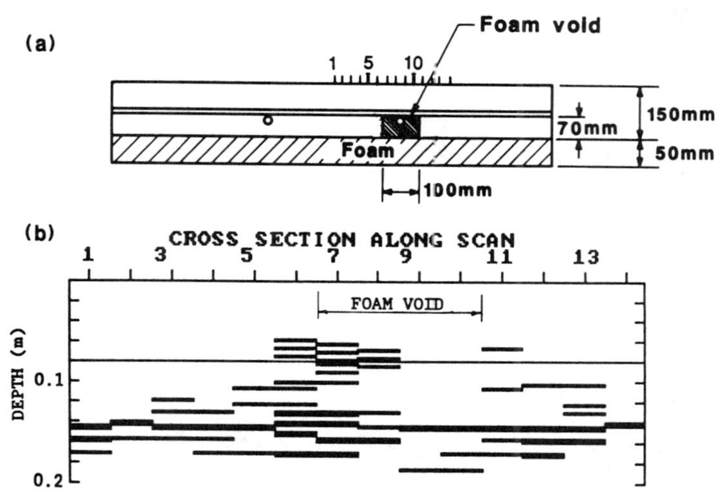

FIGURE 7. (a) Cross section of mock-up of skating rink slab; (b) cross section obtained using spectral peak plotting.

During the field testing, three regions of the ice-skating rink slab were selected for testing: a region believed to contain sound concrete; and, two regions believed to contain voids. A P-wave of 4000 m/s was determined by performing impact-echo tests adjacent to core holes previously drilled in the slab. As in the laboratory study, impacts were generated by dropping a 5-mm diameter steel ball; the contact time was approximately 20 microseconds. Scans were performed using a spacing of 100 mm between test points.

Figure 8(a) shows the cross section obtained using spectral peak plotting of the amplitude spectra along one scan line. The results indicate that this scan was carried out over generally sound concrete because, except for test point 10, the cross section shows only the bottom of the slab. In contrast, Fig. 8(b) shows the cross section obtained along a scan line in an area of the slab believed to contain voids. The presence of voids is indicated by lines within the slab and by the multiple lines associated with the bottom of the slab. Note that this portion of the slab is considerably thinner than that shown in Fig. 8(a). Thus this field study demonstrated that, by using an appropriate short duration impact, the impact-echo method could identify the presence of small voids in a thin slab.

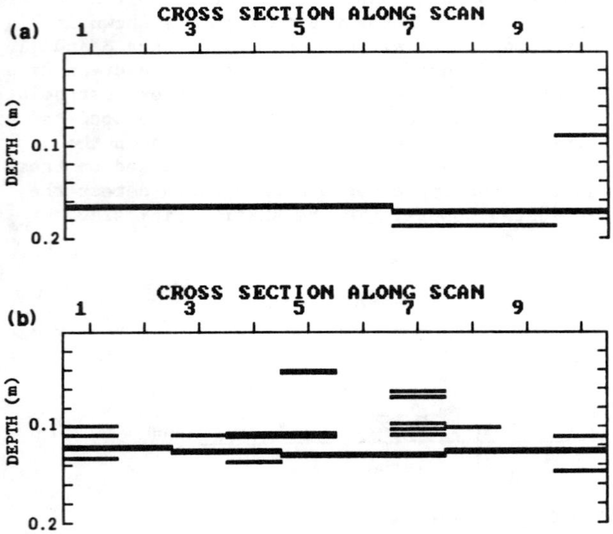

FIGURE 8. (a) Cross section in portion of skating rink slab believed to be sound concrete; (b) cross section in portion believed to contain voids.

Delaminations in Concrete Slab

A critical need in the nondestructive evaluation of concrete structures is a method to detect delaminations caused by the disruptive stresses arising from corrosion of reinforcing steel. While progress has been made in the development of techniques based upon acoustic impact, ground penetrating radar, and infrared thermography [11], there is still a need for a simple and highly reliable method.

The authors participated in a "blind" study designed to demonstrate the feasibility of wave propagation methods for rapid detection of delaminations in parking garage slabs [11]. A 200-mm thick reinforced concrete slab, 3 m x 1.5 m in plan, was cast by consultants who were in charge of the study. Thin sheets of plastic were embedded in the slab to simulate delaminations. Figure 9(a) shows the intended locations of the plastic sheets. Some sheets were located on the bottom layer of wire mesh (approximately 140 mm from the top surface), and some sheets were located on the top layer of wire mesh (approximately 40 mm below the top surface). These artificial delaminations could not be detected by sounding (chain drag method). At the time of testing, the authors were advised that the delaminations were located along two lines 440 mm distant from the long sides of the slab. However, the number, sizes, or exact locations of the simulated delaminations were not disclosed.

Before describing the results, the expected response for an impact-echo test over a delaminated portion of a slab is discussed. The amplitude spectrum would be expected to contain two major frequency components: (1) a thickness frequency peak corresponding to multiple P-wave reflections between the top surface and the delamination; and, (2) a low frequency peak corresponding to flexural vibration of the delaminated portion of the slab [6,11]. To be able to measure the thickness frequency associated with the depth of the delamination, the stress pulse must contain significant frequency components which are higher than the thickness frequency. Therefore, according to Eq. (1), as the depth of a delamination decreases, shorter contact times are required for the thickness frequency peak to be present in the spectrum. If the contact time is too long, the depth of the delamination can not be measured, but the presence of the delamination will be indicated by the low frequency peaks associated with flexural vibrations. These low frequency peaks are absent in solid portions of the slab.

Initial tests were performed at 200-mm intervals along two scan lines to determine the approximate locations of delaminations. The two scan lines are labelled "A" and "B" in Fig. 9(b), and the test points (location of receiver) are numbered from 0 to 14. Subsequent tests were performed at a 100-mm spacing and along orthogonal scan lines to determine the extent of each delamination.

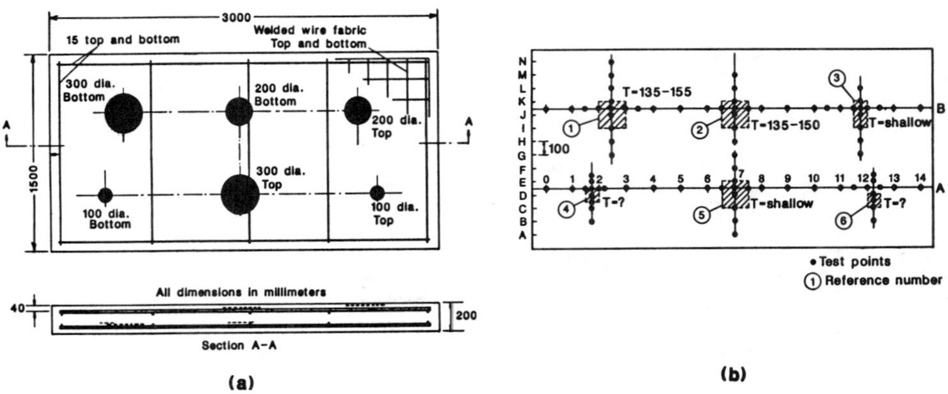

FIGURE 9. (a) Planned locations of simulated delaminations; (b) impact-echo test points and detected locations of delaminations.

The amplitude spectrum obtained from a test carried out over a solid portion of the slab had a single large amplitude peak at 9.76 kHz; this is the thickness frequency [11]. Using Eq. (1) and the known thickness of 200 mm, this frequency value gives a calculated P-wave speed of 3900 m/s. Because the concrete was tested when it was only four days old, the hardness of the surface was low and the contact time of the impact was approximately 40 to 50 microseconds. Therefore, the stress pulse did not contain significant frequency components above 25 kHz. Thus, according to Eq. (2), it was not possible to determine the depths of delaminations shallower than approximately 80 mm.

Figure 10(a) shows the spectral peak plotting results for the initial scan along line B. It is evident that there is a delamination below test points 2 and 3 and a delamination below test point 7. There are two clues which indicate the presence of these delaminations: (1) lines indicating the depth of the delaminations; and (2) lines at depths greater than the slab thickness. Note that at test point 12 there is no line to indicate the depth of a delamination nor is there a line corresponding to the depth of the slab. It was concluded that there is a shallow delamination below point 12. The reasoning for this conclusion is as follows. Because of the limited frequency content of the impact, the thickness frequency associated with the shallow delamination could not be measured. However, the low frequencies associated with the flexural vibration of the delaminated section were measured, and these frequencies result in lines at fictitious depths greater than the slab thickness of 200 mm.

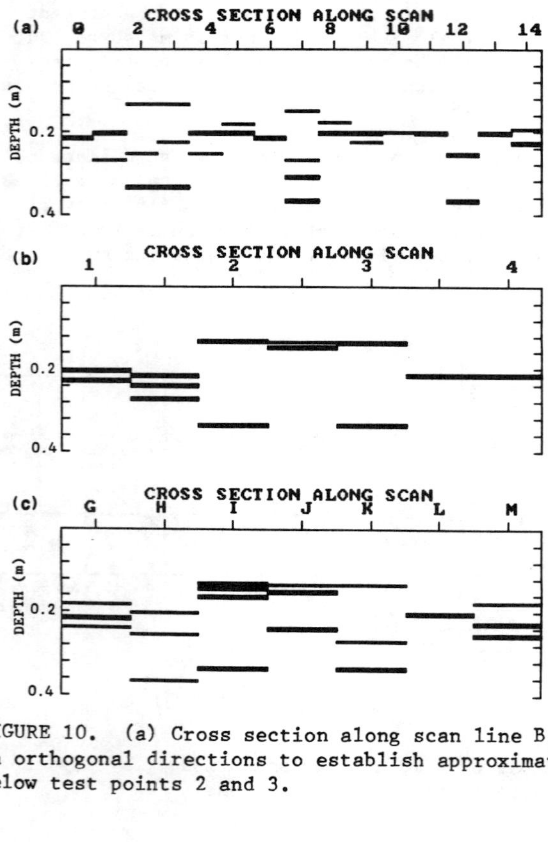

FIGURE 10. (a) Cross section along scan line B; (b) and (c) cross section in orthogonal directions to establish approximate size of delamination below test points 2 and 3.

Figures 10(b) and 10(c) show the results for two orthogonal scans which were performed to establish the approximate size of the delamination below test points 2 and 3 on scan line B. From these results it was concluded that the size of delamination was approximately 300 mm in each direction.

Figure 9(b) shows the approximate boundaries of the delaminations that were detected using the impact-echo method. Delaminations numbered 1 and 2 along line B were clearly identified to be at the depth of the bottom layer of wire mesh. The spectra obtained from delamination number 3 on line B and number 5 on line A contained only very low frequency peaks indicating that these delaminations were shallow. Delamination number 4 on line A was detected, but its depth could not be determined because of its small size. The presence of delamination number 6 on line A was detected because the thickness frequency associated with the depth of the slab shifted to a slightly lower value, indicating that the stress waves were being diffracted around the defect. However, the depth of this delamination could not be determined because it was too shallow. A comparison of Figs. 9(a) and 9(b) shows that all the delaminations were detected. Approximate sizes were correct, and where depths could be determined, they were accurate.

SUMMARY

A technique called impact-echo has been developed for nondestructive testing of concrete structures. The testing technique is based on a simple concept: mechanical impact is generated on the surface of the test object, and the surface displacement close to the impact point is measured. The stress waves which propagate into the object undergo multiple reflections between the test surface and internal defects or the opposite boundary of the test object. By monitoring the surface displacements caused by the arrival of the multiple reflections (echoes), the depth of the reflecting interface can be determined. While a similar approach has been used successfully by others for the evaluation of one-dimensional structures (piles and piers), the impact-echo method has proven to be successful for three-dimensional structures.

Implementation of the impact-echo concept has been successful because of three novel approaches: (1) the use of short duration impacts with well defined force-time histories; (2) the use of a broadband, high fidelity, displacement transducer to monitor surface displacement at a point; and, (3) the use of frequency analysis for interpretation of recorded displacement waveforms.

The contact time (duration) of the impact is critical in determining the depth and size of internal defects which can be detected. As a general rule, the inverse of the contact time has to be at least as great as the thickness frequency which is intended to be measured, and the shortest wavelength (wave speed divided by maximum frequency) has to be approximately equal to or less than the dimension of the smallest flaw to be detected.

The receiving transducer must be broadband so that it will respond over a wide range of frequencies, and it should be high-fidelity so that its output is directly proportional to the surface motion. In the studies which have been performed to date, a displacement transducer has been used so that test results could be compared directly with analytical results. Broadband accelerometers or velocity transducers may also be suitable; however, these devices have not yet been investigated by the authors.

Traditionally, the results from echo testing have been analyzed in the time domain, that is, the depth of the reflector has been determined by measuring the time from the start of the pulse to the arrival of the echo. The presence of surface waves and the complex displacement waveforms produced by impact-echo tests make time domain analysis impractical. Therefore, frequency analysis is the recommended method for interpreting the results of impact-echo tests. Peaks in the amplitude spectrum can be readily converted to the depth of reflecting interfaces. The spectral peak plotting technique that has been developed can be used to construct an "image" of the interior of the test object.

Presently, an impact-echo test system is not commercially available. However, a suitable system can be readily assembled from commercially available components.

Experimental and numerical work on the impact-echo method continues. Up to now, investigations have been limited to relatively thin (less than 1 m) concrete structures. However, plans are underway to investigate the feasibility of the technique for massive concrete structures and to explore the feasibility of applying the method to masonry and timber components. The results reviewed in this paper and the results of future studies will form the basis for a reliable and versatile method for the nondestructive evaluation of heterogeneous construction materials.

ACKNOWLEDGMENTS

The authors thank Dr. Nelson N. Hsu at NBS for his contributions toward the successful development of the impact-echo method.

REFERENCES

1. Carino, N.J., and Sansalone, M., "Pulse-Echo Method for Flaw Detection in Concrete," Technical Note 1199, National Bureau of Standards, July, 1984.

2. Thornton, H.T, and Alexander, A.M., "Development of Nondestructive Testing Systems for In Situ Evaluation of Concrete Structures," Tech. Report REMR-CS-10, Department of the Army, Waterways Experiment Station, Dec. 1987.

3. Sansalone, M., and Carino, N.J., "Impact-Echo: A Method for Flaw Detection in Concrete Using Transient Stress Waves," NBSIR 86-3452, National Bureau of Standards, Sept., 1986, 222 pp. (NTIS PB #87-104444/AS)

4. Sansalone, M., Carino, N.J., and Hsu, N.N., "A Finite Element Study of Transient Wave Propagation in Plates," Journal of Research of the National Bureau of Standards, Vol 92, No. 4, July-August, 1987, pp. 267-278.

5. Sansalone, M., Carino, N.J.,and Hsu, N.N., "A Finite Element Study of the Interaction of Transient Stress Waves with Planar Flaws," Journal of Research of the National Bureau of Standards, Vol 92, No. 4, July-August, 1987, pp. 279-290.

6. Sansalone, M., and Carino, N.J., "Transient Impact Response of Thick Circular Plates," Journal of Research of the National Bureau of Standards, Vol 92, No. 6, Nov.-Dec., 1987, pp. 355-367

7. Sansalone, M., and Carino, N.J., "Transient Impact Response of Plates Containing Flaws," Journal of Research of the National Bureau of Standards, Vol 92, No. 6, Nov.-Dec., 1987, pp. 369-381.

8. Carino, N.J., "Laboratory Study of Flaw Detection in Concrete by the Pulse-Echo Method," in In Situ/Nondestructive Testing of Concrete, V.M. Malhotra, Editor, ACI SP-82, American Concrete Institute, 1984, pp. 557-579.

9. Carino, N.J., Sansalone, M., and Hsu, N.N., "A Point Source - Point Receiver Technique for Flaw Detection in Concrete," Journal of the American Concrete Institute, Vol. 83, No. 2, April, 1986, pp. 199-208.

10. Sansalone, M., and Carino, N. J., "Impact-Echo Method: Detecting Honeycombing, the Depth of Surface-opening Cracks, and Ungrouted Ducts," Concrete International, Vol. 10, No. 4, April 1988, pp. 38-46.

11. Sansalone, M. and Carino, N.J., "Detecting Delaminations in Concrete Slabs With and Without Overlays Using the Impact-Echo Method," To be published in the ACI Materials Journal, 1988.

12 Sansalone, M., and Carino, N.J., "Laboratory and Field Study of the Impact-Echo Method for Flaw Detection in Concrete," to be published in a Special Publication by the American Concrete Institute on Nondestructive Testing of Concrete, 1988.

13. Carino, N.J., Sansalone, M., and Hsu, N.N., "Flaw Detection in Concrete by Frequency Spectrum Analysis of Impact-Echo Waveforms," International Advances in Nondestructive Testing, 12th Edition, W.J. McGonnagle, Ed., Gordon & Breach Science Publishers, New York, 1986, pp. 117-146.

14. Proctor, T. M., Jr., "Some Details of the NBS Conical Transducer," Journal of Acoustic Emission, Vol. 1, No. 3, 1982, pp. 173-178.

15. Cumming, N. and Seabrook, P., "Evaluation of Pulse-Echo Technology for Detection of Voids in Grouted Ducts," Vol. 3, Prepared for Energy Mines Resources Canada, Contract No. 23SQ-23440-4-9212, January 1988.

HIGH RESOLUTION IMPULSE RADAR IMAGING - CIVIL ENGINEERING APPLICATIONS

P. A. Okamoto and A. T. Ciolko
Construction Technology Laboratories, Inc.
Skokie, IL 60077

1.0 INTRODUCTION

Considering the significance attached to assessing as-built condition and presence of deterioration and defects in engineered structures, the civil engineering research community has, for the last 30 years, attempted to develop nondestructive imaging techniques for concrete structures. When compared to nondestructive examination in metals and other fine-grained solids, development of imaging techniques for concrete has been hampered by inherent non-homogeneity of building materials, such as concrete, and the large variability in geometric configurations and physical dimensions of structural elements. Generally, the candidate testing technologies included echo-sounding or through-transmission methods utilizing low-frequency sonic and electromagnetic energy. These rapid, nondestructive methods were intended to reduce dependence on destructive, expensive sampling techniques such as coring and methods of questionable accuracy such as audible sounding for evaluating depth of concrete deterioration. In the last decade, the need for nondestructive imaging techniques has increased in connection with rehabilitation of the nation's infrastructure.

This paper presents the underlying principles and measurement techniques of impulse, or short-pulse radar, a nondestructive imaging technique for use in civil engineering applications. The authors' experience related to evaluating as-built characteristics and presence of flaws in concrete, masonry, and bituminous concrete are presented.

2.0 BASIS, ORIGIN AND OPERATION

Man-made and naturally-occurring materials are categorized as metallic or dielectric. Metallic substances are generally good conductors, both electrically and thermally, whereas dielectric substances are poor conductors. Electromagnetic waves readily penetrate and propagate through dielectric materials. On the other hand, metallic materials totally reflect electromagnetic waves. An electromagnetic wave emitted from an airfield radar transmitter, for example, propagates through the air (a dielectric medium), strikes the metallic skin of aircraft, and is reflected back to the radar receiver. Similarly, an electromagnetic wave emitted by impulse radar into a dielectric material such as a pavement structure propagates through the material until it intercepts a discontinuity.(1) A discontinuity is any boundary between materials of differing dielectric properties, such as concrete to steel, concrete or asphalt to subbase, concrete or asphalt to air, soil to rock, etc. At each boundary, a portion of the incident energy is reflected and a portion is transmitted through the boundary. Reflected waves are detected and electronically processed to permit nondestructive imaging of layered systems.

2.1 Origin

Commercial impulse radar equipment was developed in the late 1960's for geophysical applications. Initial applications included recording geological profiles, locating bedrock, and finding buried objects, including pipe and tanks.

In general, the impulse radar apparatus uses a single transducer for transmitting and receiving, together with signal processing and recording hardware. The radar signal is directed into the material under test and patterns created by the reflected waves are observed. The radar receiver measures the time it takes the transmitted pulse to travel to a dielectric boundary and for reflected pulses to return.

In air, a radar pulse travels approximately 305 mm every billionth of a second. In dielectric solids, the velocity of propagation is reduced by the square root of the dielectric medium's relative permittivity. In concrete, for example, a radar pulse travels approximately 38 to 64 mm every billionth of a second. The time delay of a radar echo from a dielectric boundary is directly related to its depth.

Electromagnetic energy in the 100 to 1000 MHz frequency range is selected for radar operation depending on the penetration depth and resolution desired. The penetrating range of impulse radar ranges from 0.50 to more than 30 m, depending on antenna frequency and bandwidth. For a specific radar system configuration, resolution or ability to differentiate closely spaced objects is a function of the antenna's signal frequency and bandwidth. Very closely spaced changes in dielectric properties require a higher frequency and wider bandwidth, whereas lower frequency and narrow bandwidth provides greater depth of penetration with less resolution.(2)

For applications such as geophysical mapping, low frequency radar is used. For high resolution applications such as flaw detection and evaluation of as-built characteristics of concrete, masonry, and bituminous concrete elements, high frequency radar is used.

2.2 Methodology and Instrumentation

The equipment is designated Subsurface Interface Radar (SIR) System and is manufactured by Geophysical Survey Systems, Inc. (GSSI) of Hudson, N.H. The impulse radar system radiates repetitive short-time-duration electromagnetic pulses into the materials from a broad band width antenna placed on, and electromagnetically coupled to, the material's surface. The equipment functions as an echo sounding system using electromagnetic pulses of a few nanosecond duration. The system is capable of detecting location and measuring depth of reflecting discontinuities.

This high resolution equipment consists of radar control circuitry, a line scan graphic recorder and a high frequency transmitting-receiving antenna. Radar control circuitry maintains system operation. The control electronics generate and send power (+12 volts DC, -12 volts DC and +150 volts DC) and a synchronizing signal to the impulse generator in the antenna. Whenever the transducer detects a reflected radar pulse, this electromagnetic signal is fed to the receiver. The receiver converts the pulse, about a nanosecond in duration and sends this signal back to the control electronics. The signal is then processed and prepared to be fed to the printing electronics.

In practice, impulse radar imaging is performed by moving the hand-held antenna/transducer on the surface of the material to be scanned. The transducer, control electronics and graphic recorder are shown in Fig. 1. The control electronics furnishes a regulated DC voltage, called a trigger pulse, to the transmitting section of the transducer, which when triggered, generates a transmit pulse of approximately 0.75 nanoseconds in time duration. The radiated signal is a brief electromagnetic transient having a center frequency of approximately 1000 MHz. The transmitter pulse repetition rate is 50 KHz, peak power is 2 watts, and average power is 0.058 milliwatts. Reflected signals from dielectric constant boundaries are detected by the receiving section in the hand held portion of the transducer and fed through a control cable to be processed by the control electronics and printed on the graphic recorder.

2.3 Data Analysis

Data is presented using a line scan graphic recorder. This graphic recorder accepts the signal from the control electronics and produces a continuous, permanent recording on electro-sensitive paper. Reflected analog signals are plotted as vertical scans at rates of up to 32 scans per second, and a continuous profile is developed in real time, showing position and profile of dielectric constant boundaries.

The single channel mechanical line scan graphic recorder system prints shades of gray lines that are proportional to input analog voltage levels. The recorder is an intensity-modulated device, with the sweep of the stylus across the chart paper synchronized with the pulse transmitter trigger. Signal amplitudes above a preset threshold level are printed as black but weaker signals remain white. The paper moves as the stylus is swept, producing a profile such as that shown in Fig. 2. In Fig. 2, the dark bands occuring at hyperbolic signal peaks represent position of metallic embedments in a concrete slab. Since the paper vertical axis is

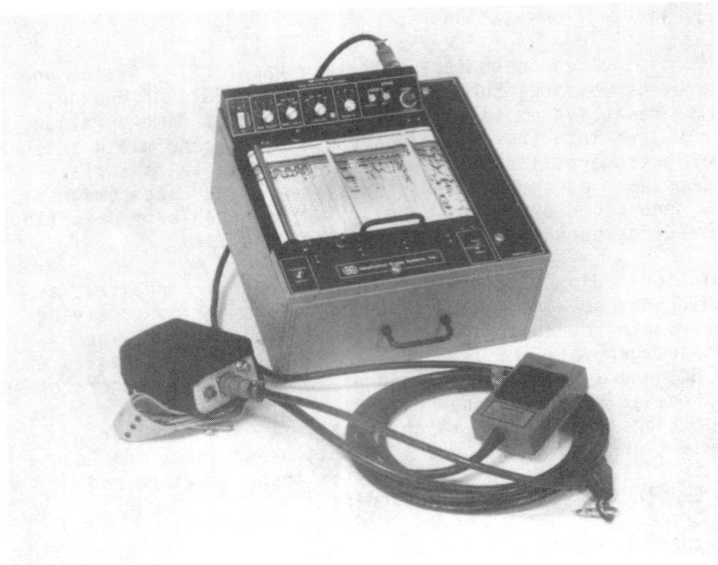

FIGURE 1. Impulse radar equipment

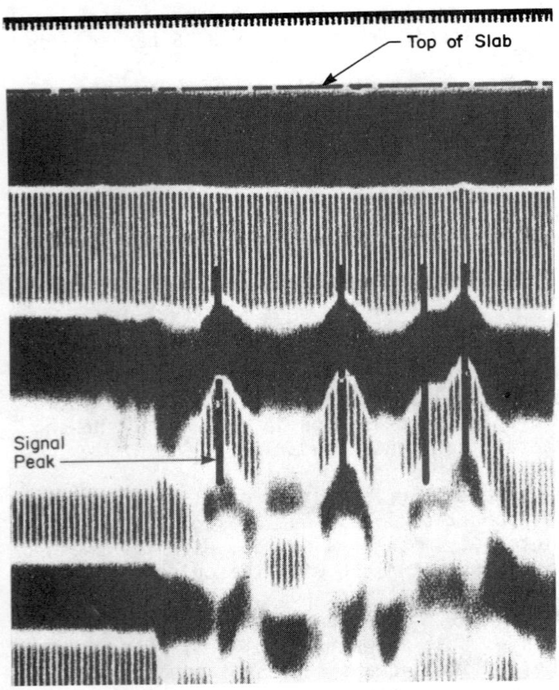

FIGURE 2. Radar trace depicting position of metallic embedments in concrete

calibrated in time in nanoseconds, the depth of a target in the material can be estimated once the material's velocity of propagation is established through calibration.

3.0 APPLICATIONS

In recent years, the use of high resolution impulse radar testing has increased to cover a broad spectrum of civil engineering applications. The application of high resolution radar can be classified into fields of characterizing as-built structures or flaw detection. The use of radar as a nondestructive test method eliminates the need for extensive, time consuming and expensive destructive testing. Nondestructive testing can replace expensive concrete coring, and methods of questionable accuracy such as audible sounding for evaluating the depth and extent of concrete deterioration. The continuous real-time graphics radar output also eliminates some of the uncertainty associated with noncontinuous sampling techniques.

3.1 As-Built Characteristics of Structures

Impulse radar imaging has been used to determine characteristics of structural components of buildings, bridges, and pavements. Radar testing can be used to generate thickness profiles and locate reinforcing bars, strand, conduits and other components embedded in concrete structures. As the transducer is moved along the concrete surface, a real-time graphic recording is automatically generated, showing the location and depths of all embedments along the transect line.

Since electromagnetic waves traveling through concrete are totally reflected by steel reinforcing bars, a strong signal peak is received by the antenna and processed by the control unit. The characteristic output from a steel reinforcing bar is the three-banded hyperbolic feature labelled "Signal Peak" on the output trace presented in Fig. 2. The antenna has a fairly broad radiation pattern. The electromagnetic energy transmitted is conical in shape with the apex located at the center of the antenna. The conical pattern forms an angle of approximately a 0.8 to 1.0 radians to the horizontal. As the antenna is passed over the bar, the time of travel for the radar pulse and return reflection is minimized, and the resulting output signal reaches a hyperbolic peak. As the antenna is moved away from the embedment and the pulse travel distance increases the output profile depth increases.

The horizontal scale (relative location) of the embedment is dependent on the speed at which the transducer is moved along the concrete surface (as well as the paper feed rate of the graphic recorder). To facilitate accurate determination of an embedment's position, the equipment has a reference marking system. Reference marks are printed on the graphic output when manually triggered by the radar operator. These marks can be references to specific surface features or preselected marks on the concrete. As the antenna centroid passes the mark or feature, the marking system is triggered. By interpolating between reference marks on the output, the embedment can be easily located relative to the mark or feature on the concrete surface.

To determine depth of steel reinforcing bars, strand, pipe, or other embedded components in concrete, the vertical scale on the graphics output must be converted from a time scale to a depth scale. Vertical position on the output is both a function of range setting, which adjusts the print window graphics output, and a function of the velocity of propagation of the pulse in concrete. For a given radar range setting, as the velocity of propagation increases, the relative depth of reflected signals decreases. The velocity of propagation is a function of dielectric constant of the concrete. As the dielectric constant increases, the velocity of propagation decreases. The vertical time scale can be directly converted to a depth scale if the dielectric constant of concrete is known. Since the dielectric constant can change within the concrete structure due to impurities, changing water contents, and aggregate type; the easiest way to determine feature depth on the graphic output is to indirectly calibrate the radar output.

Cores or known steel bar depths are used to calibrate the graphic recordings to obtain actual embedded depths of the steel bars. A calibration constant is calculated which converts units of graphic output depth to units of concrete cover over the steel. Frequently, the calibration data suggests that the calibration constant is a function of output chart depth. If enough data is available, a least-squares linear regression equation can be developed as shown in Fig. 3. The actual depths to top of steel, determined from coring, are regressed on measured chart output depths to increase the degree of radar accuracy when estimating steel depths.

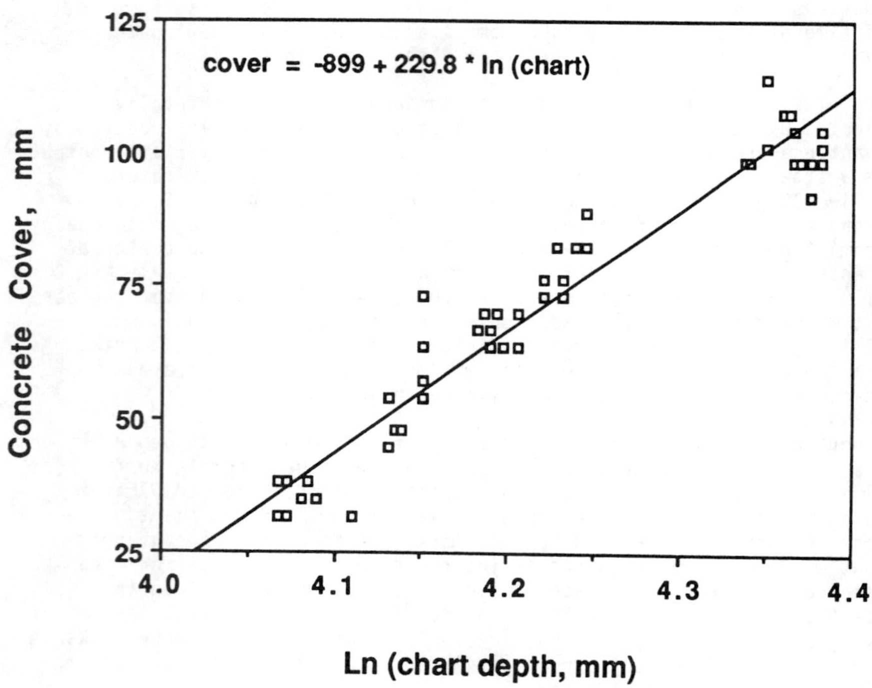

FIGURE 3. Radar calibration curve to determine reinforcement depth

A similar procedure is used to determine thickness profiles of structural components including structural concrete slabs, asphalt pavements, and concrete pavements. The radar output provides a continuous profile showing the change in thickness with horizontal travel. The reference marking system can be used to identify horizontal position on the output. Many internal discontinuities and boundaries can exist within the structural member such as cold joints between concrete pours and overlays over existing pavements, which reflects energy back to the radar control unit. These boundaries which show up on the radar output can make identification of the slab bottom difficult. The easiest way to identify which band on the output corresponds to the slab bottom or other desired feature, is to run the transducer over an area where a distinctive feature will cause a strong surface reflection. This is often accomplished by placing a steel rod or reinforcing bar under the pavement layer and identifying the reflective signal on the output trace corresponding to the top of steel and bottom of pavement layer. For a structural concrete slab where access to the underside is possible, a steel plate or bar is positioned at the slab bottom. Once the band corresponding to the slab bottom is identified, the profile can be quickly determined. To convert profiles to actual depths, the chart output depths at several locations can be calibrated as previously discussed.

Radar imaging has also been used in profiling geological strata, river and lake bottoms, ice thickness, and railroad beds. The continuous output capability of radar is ideal for many other profiling applications.

3.2 Flaw Detection

Impulse radar imaging has also been used to detect and locate flaws in structural members including grouted hollow-core masonry walls, pavements, and bridge decks. Types of flaws which are detected using radar include ungrouted cavities in hollow-core concrete blocks, large voids under concrete pavements, degree of dowel bar misalignment at concrete pavement joints, and concrete bridge deck delaminations.

Locating ungrouted cavities in hollow-core concrete block walls is relatively easy since the interface reflection between concrete and air is distinctively different than the resulting reflection between concrete block and sand-cement grout. The hollow cores within concrete blocks are grouted during construction to act as structural columns. The absence of grout can significantly reduce the capability of the wall to transfer loads from beams and girders to building foundations.

The resulting output of a grouted block wall is a distinctive band from reflected energy at the concrete-air interface. When the hollow cores are grouted the interface becomes harder to distinguish since the dielectric constant of concrete and sand-cement grout are similar.

Impulse radar has also been used to locate misaligned dowels at concrete pavement joints. Dowel bars are placed at mid-depth of concrete slabs to reduce joint deflections due to heavy loads. Dowels are placed at joints using wire basket assemblies or inserted directly into the plastic concrete by an automatic dowel bar inserter mounted at the back of a concrete slip form paver. The dowels should be placed as parallel as practical to the longitudinal axis and the horizontal plane of the pavement to prevent the joint from locking up as adjacent slabs move due to both temperature change and shrinkage of concrete. Dowels with excessive misalignment prevent the joint from easily opening and closing and can cause joint spalling, joint cracking, corner cracking, and a reduction in transfer capability.

To detect dowels with misalignment exceeding allowable specified levels, a radar pass is made 152 mm on each side of the joint. Dowels are located in both the horizontal and vertical plane on each side of the joint using the same technique as that for locating steel reinforcing bars. Using data from both sides of the joint, misalignment in both the horizontal and vertical direction can be nondestructively evaluated for construction specification compliance.

Radar imaging can also be used to detect voids beneath pavements and slabs-on-grade. Voids can occur due to pumping action of slabs at joints and cracks. If joints and cracks have inadequate load transfer capability, excessive deflections can cause the underlying subgrade and subbase material to erode in the presence of water. Loss of subgrade support can lead to premature slab cracking and joint faulting.

A radar pass is made, usually in the outer wheel path, so a continuous profile is generated. Data can be collected at a rate of approximately 1.3 meters per second. Since the velocity of propagation is higher in air than in concrete, the reflected energy at concrete-air interfaces will have a shorter travel time than reflected energy at concrete-subbase interfaces(3). Voids under pavements appear as either positive signal peaks or multiple interfaces in the graphics output. The amplitude of the reflected electromagnetic energy is also larger at the concrete-air interface thereby producing a darker band on the graphics display, similar to that found at ungrouted hollow-core concrete block. Using the reference marking system, voids can be located in the horizontal plane. The size of voids found with radar ranges from 6 mm deep to over 50 mm deep(4). Radar is not only used to detect voids but also used as a quality control device to check that voids are filled during subsealing restoration operations.

Radar profiling has also been used to detect concrete bridge deck deterioration(5). Continuous profiles are generated as a radar scan is made across the width of decks at rates of approximately 0.1 meters per second. The reference marking system is used to locate the areas of deterioration in the horizontal plane.

As bridge deck concrete cracks and water is allowed to penetrate to the level of steel reinforcement, corrosion can occur. Use of deicing salts to prevent decks from icing over during the winter months can add to the degree and rate of corrosion. As the steel corrodes, the volume expands and if enough expansion occurs the resulting constraining forces can cause surrounding concrete to crack.

Areas of concrete delaminations produce irregularities in the radar output reflection band near the reinforcing bar locations. The presence of delaminations are associated with distinctive depressions in the resulting return reflective energy signal on the radar output. Reflection profiles and delamination features may vary appreciably among decks, depending on concrete condition, presence of overlays, and depth of steel reinforcing bars.

Radar imaging has similarly been used to detect joint deterioration at concrete pavement joints[4]. Radar passes can be made in wheel paths at speeds of approximately one meter per second. Joint deterioration may not always be visible from the surface. Radar can be used to nondestructively evaluate subsurface joint conditions. The distinguishing feature on the radar output corresponding to joints exhibiting a significant loss of cross-section was confirmed with a series of cores. The results demonstrated that joints could be quickly and accurately evaluated using radar imaging.

4.0 CASE STUDIES

The following cases illustrate how the use of nondestructive radar testing eliminates the need for traditionally used and time consuming destructive testing.

4.1 Dowel Placement

Radar nondestructive testing for dowel bar alignment at concrete pavement joints has been successfully demonstrated. Radar has been recently used to evaluate the effectiveness of new automatic dowel bar inserter equipment for the Federal Highway Administration (FHWA) and numerous state highway agencies[6].

Dowel bar alignment at pavement joints was measured in sections where a new automatic dowel bar inserter was used and in sections where dowels were anchored to the subbase with conventional dowel basket assemblies. The dowel bar inserter eliminates the use of expensive basket assemblies and associated intensive labor costs by vibrating dowels into the plastic concrete behind slipform pavers. Concern was raised as to whether the equipment could properly place dowels parallel to the longitudinal axis of the pavement in both the vertical and horizontal plane.

Dowel bar alignment was evaluated in two states where the inserter and conventional basket assemblies were used. A 9.7 km long 254 mm thick section of Interstate 45 near Dallas, Texas was evaluated in 1987 for the Federal Highway Administration Demonstration Projects Division. To compare the inserter and dowel basket placement procedures, 51 and 52 joints were evaluated in the basket and inserter project portions, respectively. Dowels were 560 mm long, 32 mm in diameter and spaced at 305 mm along the 8.2 m long joint. Approximately 2800 m of each type of dowel placement was evaluated in four days of testing performed by one engineer and one technician. Equipment was carried, along with a portable generator, in the back of a pickup truck as shown in Fig. 4.

The evaluation procedure at each selected joint was done by making a pass with the radar transducer 152 mm away from each side of the joint. The output for each pass was recorded graphically for future interpretation. The transducer was passed on top of a 9.5 mm thick plexiglass guide. The use of the plastic guide served several purposes including:

1. allowing dowel depths to be determined at a constant offset distance from the joint

FIGURE 4. Nondestructively evaluating dowel bar alignment at concrete pavement joints.

2. facilitating uniform travel rate rates along the joint surface to allow for horizontal skewness estimation

3. providing a template for triggering reference marks on radar output.

To estimate horizontal dowel skewness, a triggering mechanism was developed which printed a reference mark on the radar recording. Metal rivets mounted on the plastic guide at uniform 102 mm spacings triggered the marker. The guide was consistently placed in the same position on each side of the joint to allow for estimation of horizontal misalignment levels.

When the transducer passed very near or over the end of the dowel, a weak return signal was received resulting in a lighter recording. In such instances, to eliminate additional subjective variance in interpolating output, the test was repeated at a 102 mm joint offset. This repetition also allowed for qualitative verification that the dowel embedment lengths were at least 102 mm.

During the radar evaluation, 32 cores were obtained at locations of dowel bars identified by the radar. Cores were obtained from the section with the dowel baskets as well as from sections where the inserter was used. These cores and core holes were used to measure the depth of dowel bars from the pavement surface. From the core data, the graphic recordings were calibrated to obtain the actual dowel bar depths (top of steel). A simple least squares linear regression analysis was performed on the core data. The depth on each side of the joint for each dowel was determined by measuring the distance on the graphic output from slab surface to peak of dowel signal. This output depth was input into the regression equation to determine estimated dowel depth.

Using data obtained from both sides of the joint the depth, vertical misalignment, horizontal misalignment, longitudinal displacement, and horizontal displacement was determined for each dowel.

A similar analysis, sponsored by the Wisconsin Department of Transportation was done in the summer of 1987. An 8 km pavement section of Interstate 90 was evaluated near Janesville, Wisconsin. Fifty joints were evaluated at joints where an automatic dowel bar inserter was used and 30 joints evaluated where basket assemblies were used. The 7.3 m wide 254 mm thick concrete pavement had 460 mm long by 32 mm thick dowels spaced at 305 mm.

Test results from both test sites indicate the dowel placement with the dowel bar inserter is as good as basket assembly construction. The use of radar nondestructive testing has eased some of the uncertainty associated with the development of new equipment and construction practices. The acceptance of these new dowel bar inserters should improve paving production rates and reduce costs.

4.2 Masonry Voids

Radar has also been used to accurately pinpoint randomly located voids in load-bearing grouted concrete masonry walls. Hollow cores at specified locations are grouted during construction to act as load bearing columns. The absence of grout can significantly reduce the capability of the wall to transfer loads from the beams and girders to the building foundation.

During a recent investigation to identify unfilled grout cavities, engineers discovered that numerous voids occurred at random locations in the walls. To be absolutely sure that all voids would be discovered and repaired, impulse radar nondestructive testing was used to scan each core area designated to contain grout and identify suspect locations.

Radar passes at all specified grouting locations were vertically made at rates of approximately 75 mm/s. The transducer was located so it ran along the block cores. The output was analyzed and ungrouted cells immediately identified. To verify radar findings, 13 mm-diameter holes were drilled at suspected void locations. A fiber optic borescope was used to view the ungrouted cavities one meter in each direction of the drilled holes The reliability of the radar to identify ungrouted cavities was over 95%. Due to conservatism in interpretating output, most errors resulted in identifying grouted cells as ungrouted. The size of voids ranged from a 3 mm cavity parallel to the face shell of a block to an entire core cell 2.5 m high. The use of radar results, supplemented by the borescope analysis, enabled the size and shape of the unfilled cavities to be accurately laid out. Once the extent of ungrouted cells was mapped out, 25 mm diameter grout injection ports were located to assure that the voids were completely filled. A portland cement grout was then injected into the cavities to bring the load-bearing capacity of the wall back up to design specifications.

5.0 CONCLUSIONS

Impulse radar imaging, also referred to as ground penetrating radar, was initially developed for geophysical applications to provide continuous profiles of subsurface conditions. Since its development, applicability of impulse radar has increased to include nondestructive testing of several other types of construction materials, including reinforced concrete, masonry, and asphalt.

6.0 ACKNOWLEDGMENTS

The authors acknowledge Joseph V. Rosetta, Jr. and Sam Heald of Geophysical Survey Systems, Inc. for their help in developing some of the radar applications described in this paper.

The opinions and findings expressed or implied in this paper are those of the authors. They are not necessarily those of Geophysical Survey Systems, Inc. or the Federal Highway Administration.

7.0 REFERENCES

1. Alongi, A. V., Cantor, T. R., Kneeter, C. P. and Alongi, A., Concrete Evaluation by Radar Theoretical Analysis, Transportation Research Record 853, pp. 37-42, 1982.

2. Cantor, T. R., Review of Penetrating Radar as Applied to Nondestructive Evaluation of Concrete, In-Situ/Nondestructive Testing of Concrete, SP-82, American Concrete Institute, Detroit, pp. 581-591, 1984.

3. Kovacs, A., and Morey, R. M., Detection of Cavities Under Concrete Pavement, US Army Corp of Engineers CRREL Report 83-18, July 1983.

4. Holt, F. B. and Eales, J. W., Nondestructive Evaluation of Pavements, Concrete International, vol. 9, no. 6, pp. 41-45, June 1987.

5. Clemeña, G. G., Nondestructive Inspection of Overlaid Bridge Decks with Ground-Penetrating Radar, Transportation Research Record 899, pp. 21-32, 1983.

6. Okamoto, P. A., Field Evaluation of Dowel Placement Along a Section of I-45 in Texas, 67th Annual Meeting, Transportation Research Board, Washington, D.C., January 1988.

EVALUATION OF IN - PLACE STRENGTH OF CONCRETE BY THE BREAK - OFF METHOD

T. R. Naik, Z. Salameh, and A. Hassaballah
The University of Wisconsin - Milwaukee
Milwaukee, WI 53201

1.0 INTRODUCTION

The Break-Off test is one of the recent developments for testing in-place concrete strength. In the Break-Off method, the force required to break-off an in-place concrete cylinder, 2.17 in. (55 mm) diameter and 2-3/4 in. (70 mm) long, is related to the compressive strength of the concrete from a predetermined calibration curve developed for the particular concrete mix.

This method was developed in 1977, by researchers at the Norwegian Technical University (NTH), and the Research Institute for Cement and Concrete in Norway [1]. The method is now used worldwide and is standardized in several countries [2,3,4].

This research was undertaken to investigate the reliability of the Break-Off test as a measure of the in-place strength of 4000 psi concrete. Concrete of this strength is generally used for concrete pavements and floors in North America. To achieve this objective, the following parameters were investigated: (1) aggregate shape; (2) slab thickness; and, (3) method of obtaining cylindrical Break-Off test specimens, either by inserting a sleeve in the fresh concrete or by drilling a cylindrical core after the concrete has hardened.

2.0 PRINCIPLE OF THE BREAK-OFF TEST METHOD

The Break-Off test method measures the flexural strength of the in-place concrete with a rupture plane located at 2-3/4 in. (70 mm) from the concrete surface. The test principle is shown in Fig. 2.1. It involves breaking off a cylindrical test specimen formed in the concrete by applying a force at the top of the test specimen. The test specimen is formed either by removing a tubular sleeve which was inserted into the fresh concrete, or by drilling a core in the hardened concrete at the time of the Break-Off test. The core is drilled with a special diamond tipped drill bit. The force is applied through a load cell placed at the top of the specimen. Approximately one stroke per second (about 70 psi per second) is applied by the use of a manual hydraulic pump attached to the load cell. The pumping is continued until the break-off specimen snaps off. The pressure required to break-off the test specimen is measured by a mechanically operated manometer.

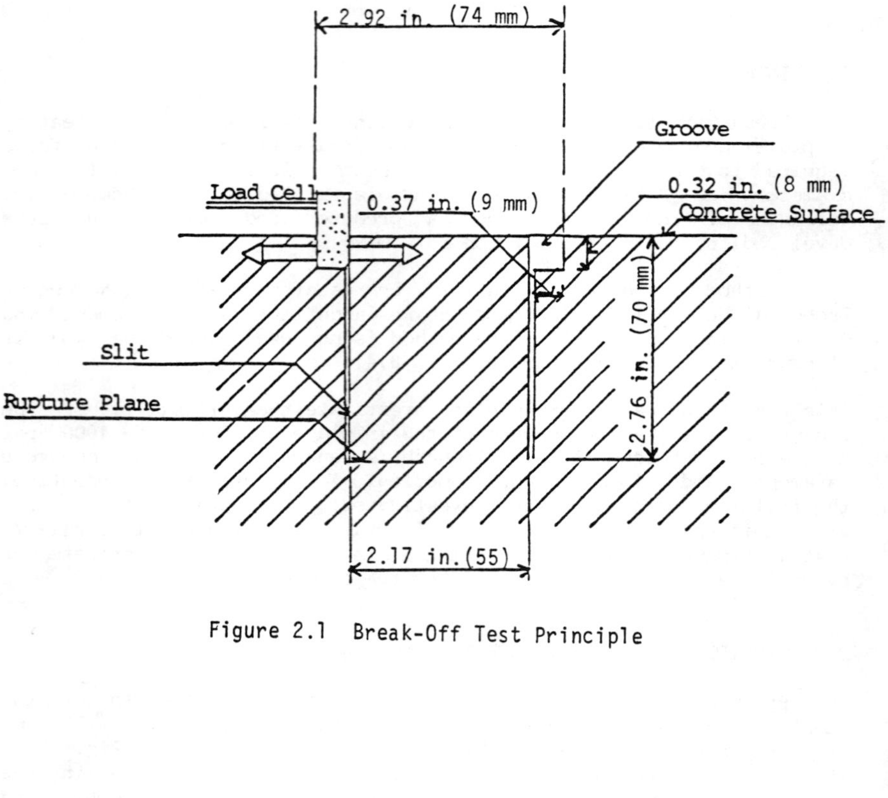

Figure 2.1 Break-Off Test Principle

3.0 CASTING OF CONCRETE SLABS AND TEST SPECIMENS

Two different concrete mixes were used, Mix 4R which had 3/4 in. (20 mm) maximum size rounded aggregates, and Mix 4C which had 3/4 in. (20 mm) maximum size crushed limestone aggregates. Mix proportions per cu. yd. of concrete for the two mixes are given in Table 3.1. The fresh concrete test results are given in Table 3.2.

Table 3.1. Concrete Mix Proportions

	Mix 4R*	Mix 4C
Nominal 28-Day Compressive Strength, psi	4000	4000
Cement, Type I, lbs	595	423
Fly Ash, Type C, lbs	320	100
Water, lbs	280	308
Ratio of Water to Cementitious Material	0.34	0.59
Sand, SSD, lbs	1130	1600
3/4" Max. Aggregates, SSD, lbs	1810	1710
Airentraining Agent, liq. oz. (Daravair-R)	9	none

Note: 1 psi = 6.9 KN/m^2 ; 1 lb = 0.453 Kgf

* This mix was originally proportioned to give a compressive strength of 8000 psi. Due to a mistake at the batch plant, air entraining agent was added instead of the water reducing agent. The air-content was 6%, Table 3.2. This decreased the compressive strength to approximately 4000 psi. Therefore, this mix was considered as the nominal 4,000 psi concrete.

Table 3.2. Fresh Concrete Test Results

Mix	Air Content, %	Slump, in.	Unit Weight, $lb./ft.^3$
4R	6.0	5-1/2	145.9
4C	1.8	4-1/2	148.9

Note: 1 in. = 25.4 mm; 1 lb/ft^3 = 16.18 kgf/m^3

For each mix, two 4 X 4 ft. (1.2 X 1.2 m) slabs were cast at the UWM Concrete Research Laboratory. One slab was 5 in. (125 mm) thick and the other was 7 in. (175 mm) thick.

In each slab, 30 greased tubular shaped sleeves, which create the test specimens, were inserted at a distance of 6 $\pm$ 1/4 in. (150 $\pm$ 5 mm) center to center between sleeves. An edge distance of 6 in. (150 mm) was also maintained. Figure 3.1 shows the position of sleeves in the slab. The remaining area of the slab was for locating drilled specimens. After finishing operations, the slabs were covered with plastic sheets. Beginning the next day after casting of slabs, they were soaked at least once every day with water. The plastic sheets were removed at 7 days age but wetting was continued up to 28 days.

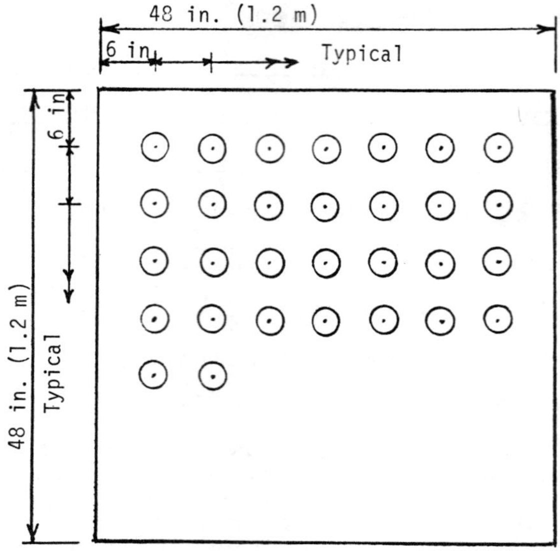

Figure 3.1 Position of Sleeves

From each concrete mix, seventeen 6 X 12 in. cylinders were cast according to ASTM Test Designation C-192. The tops of the cylinders were covered with plastic bags to protect them from drying out. These specimens were allowed to harden in the laboratory for 24 hours, at approximately 75° F and 50% ± 10% R.H., then de-molded, at which time they were placed in a lime saturated curing tank at approximately 75° F until the time of the test.

4.0 TEST PROGRAM

For each slab, all Break-Off test locations were chosen randomly. Six sleeve specimens were tested at ages of 1 and 3 days, while five sleeve specimens were tested at ages of 5, 7, and 28 days. This was done because of the expected higher variability of concrete strength at early ages, i.e., 1 and 3 days age. For drilled specimens, five specimens were tested at ages of 5, 7, and 28 days. Also, at each of these test ages, and for each mix, 3 cylinders were tested to determine the standard cylinder compressive strength of concrete. These test ages were chosen to give a complete picture of the strength variation with respect to time. The Break-Off and the standard cured cylinder test results are given in Table 4.1 for Mix 4R and in Table 4.2 for Mix 4C. These tables show concrete cylinder compressive strength, and Break-Off readings for sleeve and drill specimens for 5 in. and 7 in. slabs.

Table 4.1. Results of Standard Cylinder and Break-Off Tests for Mix 4R

Test Age, days	Cylinder Compressive Strength, psi		Break-Off Readings, bar							
			Sleeve Specimen Slab Thickness, in.				Drill Specimen Slab Thickness, in.			
			5		7		5		7	
	Actual	Average	Actual	Ave.	Actual	Ave.	Actual	Ave.	Actual	Ave.
1	1570 1380 1330	1425	55,49 53,53 58,54	54	49,42 50,50 62,54	51	-	-	-	-
3	2650 2650 2705	2670	62,50 56,54 80,66	63	78,80 80,72 61,57	71	-	-	-	-
5	3180 3165 3180	3175	77,71 69,59 81	71	57,70 83,66 63	68	77,82 82,73 69	77	76,63 67,62 70	68
7	3360 3250 3325	3310	77,72 71,70 62	70	77,75 86,85 85	82	87,78 75,78 79	79	77,85 88,81 75	81
28	3990 3890 4280	4055	95,95 79,90 91	90	94,61 83,95 89	84	95,100 109,97 82	97	80,87 84,95 100	89

Note: 1 bar = 0.1 Mpa; 1 psi = 6.9 KNf/m²

Table 4.2. Results of Standard Cylinder and Break-Off Tests for Mix 4C

Test Age, days	Compressive Strength, psi		Break-Off Readings, bar							
			Sleeve Specimen Slab Thickness, in.				Drilled Specimen Slab Thickness, in.			
			5		7		5		7	
	Actual	Average	Actual	Ave.	Actual	Ave.	Actual	Ave.	Actual	Ave.
1	2120 2210 2210	2180	41,50, 59,57, 57,64	55	59,59 57,56 62,61	59	-	-	-	-
3	3360 3380 3200	3315	76,70, 70,76, 70,77	73	76,72 73,70 77,74	74	-	-	-	-
5	4085 3925 3800	3935	78,75, 80,83, 80	79	70,86 77,79 81	79	104,94 91,98 100	97	91,87 89,94 80	88
7	4155 4070 4070	4100	67,75, 73,80, 72	73	86,82 79,79 85	82	95,84 80,88 89,	87	93,84 80,90 83	86
28	5040 4950 4880	4955	104,92 83,96, 96	94	102 90,91 98,98	96	123,113 105,105 112	112	118,120 118,112 120	118

Note: 1 bar = 0.1 Mpa; 1 psi = 6.9 KNf/m²

None of the test data reported in this paper were rejected as outliers. The objective of this research was to test the reliability of the Break-Off method under field conditions by which the results will be reported as they are and conclusions will be drawn based on such results.

5.0 DISCUSSION OF RESULTS

In order to draw broader conclusions based upon this laboratory investigation, data collected from all sets of test specimens were analyzed using appropriate statistical procedures.

5.1 Within Test Coefficient of Variation

The degree of variability of the Break-Off test readings with respect to the different variables was determined by analyzing the coefficient of variation (V). Table 5.1.1 shows the within test standard deviation (σ) and the coefficient of variation for compressive strengths and Break-Off readings for Mixes 4R and 4C. Table 5.1.2 shows the average coefficient of variation for each mix, as well as the overall average coefficient of variation for both mixes, for the standard cylinder tests, and the Break-Off tests. From Table 5.1.2, the average coefficient of variation for all the Break-Off readings is calculated to be 7.7 percent, versus only 3.2 percent for standard cylinders. These data, therefore, indicate that somewhat higher coefficient of variation can be expected for the Break-Off test results than that for the standard cylinder test results. However, the Break-Off test results cannot be considered highly variable compared to other nondestructive tests such as the pull-out and the impact hammer tests. The coefficient of variation for the Break-Off test obtained in this study agrees with the findings of Barker and Ramirez (5) and Nishikawa (6).

Table 5.1.1. Standard Deviation (σ), and Coefficient of Variation (V) for Cylinder Tests and Break-Off Tests for Mixes 4R and 4C

Mix No.	Test Age, Days	Standard Cured Cylinder Compressive Strength Test		Break-Off Test							
				Sleeve Specimen Slab Thickness				Drill Specimen Slab Thickness			
				5 in.		7 in.		5 in.		7 in.	
		σ, psi	V%	bar	V%	bar	V%	bar	V%	bar	V%
4R	1	131	9.2	2.9	5.4	6.6	12.9	-	-	-	-
	3	31	1.1	9.3	14.8	10.1	14.2	-	-	-	-
	5	10	0.3	8.3	11.7	9.7	14.2	5.7	7.4	5.7	8.4
	7	54	1.6	5.5	7.9	5.1	6.2	4.5	5.7	5.4	6.7
	28	277	6.9	6.5	7.2	13.9	16.5	9.7	10.7	8.2	9.2
4C	1	51	2.3	8.0	14.8	2.3	3.9	-	-	-	-
	3	79	2.9	3.5	4.8	2.6	3.5	-	-	-	-
	5	142	3.6	3.0	3.7	5.9	7.5	5.1	5.2	5.0	5.7
	7	51	1.3	4.7	6.4	3.3	4.0	5.6	6.5	5.3	6.2
	28	80	1.6	7.6	8.1	5.1	5.3	7.4	6.6	3.3	2.8

Note: 1 in. = 25.4 mm; 1 bar = 0.1 Mpa; 1 psi = 6.9 KNf/m²

Table 5.1.2. Average Within test Coefficient of Variation, V, Percent

Mix	Coefficient of Variation, V, Percent				
	Standard Cylinder Tests	Break-Off Tests			
		Sleeve Specimens Slab Thickness		Drill Specimens Slab Thickness	
		5 in.	7 in.	5 in.	7 in.
4R	3.8	9.4	12.8	7.7	8.1
4C	2.5	7.6	4.8	6.1	4.9
Ave.	3.2	8.5	8.8	6.9	6.5

Note: 1 in. = 25.4 mm

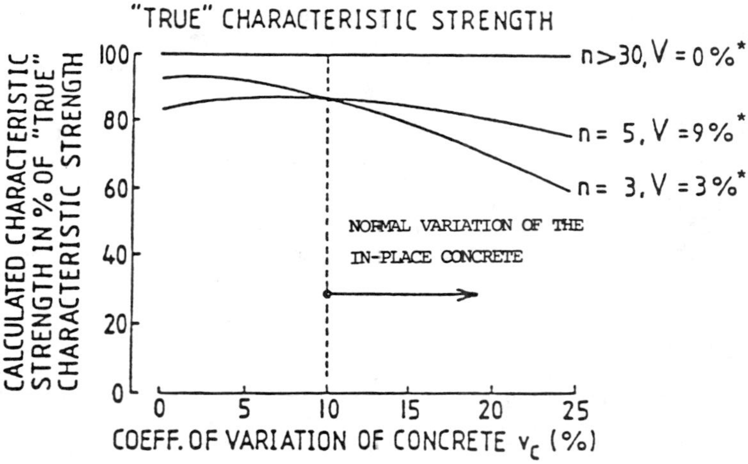

Figure 5.1.1 Coefficient of Variation of Concrete Vs. Percentage of "True" Strength

* n = Number of Test Specimens; and,

V = Within Test Coefficient of Variation for the Test Method

Fig. 5.1.1 shows in general how to calculate the accuracy of a test method, based upon the coefficient of variation of the test concrete as the number of test specimens per test and within-test coefficient of variation for the test method changes (1). If the number of test specimens per test, n, for any test method is larger than 30, n > 30, and the within test variation for this test method is zero (V = 0%), then the strength of the concrete can be predicted on a 100% accuracy level by this test method. This is regardless of the coefficient of variation for the test concrete. Of course, test methods available in real situations are less accurate (i.e., they have a higher coefficient of variation). For example, as shown in Fig. 5.1.1, a test method with V = 3% and n = 3, will predict the strength of the test concrete on a 60 to 85% accuracy level depending on the coefficient of variation for the concrete. Similarly, Fig. 5.1.1 also shows that if a test method has the coefficient of variation of 9% and the number of test specimens per test is 5, then the strength of the test concrete can be predicted on an 87% accuracy if the coefficient of variation for the concrete being tested, e.g., the in-place concrete, is 10 percent. This indicates that with the average coefficient of variation for the Break-Off Method being 7.7%, Table 5.1.2, and with five Break-Off readings for a set of tests, the "true" in-place strength is within less than 13% of the predicted in-place strength of the test concrete. Johansen (1) obtained similar results in his research.

The average coefficient of variation for Break-Off tests for 5 and 7 in. slabs, Table 5.1.2, shows that the 7 in. slab test data has about the same degree of variability as that for the 5 in. slab (V = 7.7%). This shows that the accuracy of the Break-Off test results are not slab thickness dependent. No similar research has been performed by others with respect to this variable.

From Table 5.1.2, the average coefficient of variation for the rounded aggregate concrete, Mix 4R, was 9.5% compared to 5.9% for the crushed aggregate concrete, Mix 4C. This clearly indicates that the variability of the Break-Off test data is much higher for rounded aggregate concrete than for crushed aggregate concrete. Aggregate type, therefore, clearly affects the consistency of the Break-Off test results. Barker and Ramirez (7) had similar findings.

Table 5.1.2 also shows that the sleeve specimen test data has a higher variability (V=8.7%) than that for the drilled specimen test data (V=6.7%). This also shows that the variability of the Break-Off test results are dependent on the specimen type, sleeve or drilled. The sleeve specimens are expected to give higher variability because of the fact that the in-place concrete is disturbed by the insertion process of the sleeves compared to the undisturbed concrete in the case of drilled specimen (7,8). No other research has been reported by others with respect to this variable.

5.2 Variance Analysis

To identify the variables that might have significant effect on the Break-Off test readings, one way variance analysis was done with respect to: (1) the slab thickness; (2) the aggregate shape; and, (3) the method of obtaining the Break-Off test specimen.

One way variance analysis compares two or more sets of data and gives sufficient evidence to indicate if there is a significant difference between the means of these data. For example, the mean value for all the Break-Off readings of the drilled specimens is 90 bars compared to 81 bars for the sleeve specimen. This suggests that drilled specimen Break-Off readings are higher than those for sleeve specimens. Such a conclusion may not be accurate. The difference can be attributed to chance. Variance analysis tests the data and estimates to which level, or percentage, the difference is attributed to chance. If the difference between the means is less than 5% attributed to chance, the difference is significant and meaningful, otherwise it is insignificant. For a more detailed discussion on variance analysis the reader is encouraged to refer to Reference 9.

Using a statistical package, MINITAB, the data with respect to different variables were analyzed by one way variance analysis.

The variance analysis for these data indicate that there is no significant difference between the test data means of the 5 in. and 7 in. slab thickness. It is concluded, therefore, that the slab thicknesses of 5 in. and 7 in. do not have any significant effect on the Break-Off readings. No similar research with respect to this variable is reported in the literature.

The rounded aggregates Break-Off test data is compared by the variance analysis to the crushed aggregates data. The Break-Off test data used for this analysis consists only of concrete of equal compressive strength. Although the Break-Off test data mean for the rounded aggregates is lower than that for the crushed aggregates data, the variance analysis indicates no significant difference, (i.e., at the 5 percent level, F=1.79 which is less than F=4.96 for the variance analysis between crushed aggregates and rounded aggregates data; therefore, it is concluded that there is no significant difference between crushed aggregates and rounded aggregates data). This also clearly indicates that the type of aggregates will not affect the Break-off test results. Similar observations were reported by other researchers (5,6).

The last variable that was analyzed using the variance analysis was the sleeve specimens data versus the drilled specimen data. The analysis indicates also no significant difference between the sleeve specimen data and the drilled specimen data, (i.e., at the 5 percent level, F=3.53 which is less than F=4.30 for the variance analysis between sleeve specimens and drilled specimens data). It is concluded, therefore, that the method of obtaining the Break-Off specimens does not affect the Break-Off readings. No similar research with respect to this variable is reported in the literature.

5.3 Regression Analysis

Considering the Break-Off readings as an independent variable and the cylinder compressive strengths (f'c) as a dependent variable, linear regression analysis was performed. From the variance analysis it was shown that the slab thickness of 5 in. and 7 in. did not have a significant effect on the Break-Off readings. Therefore, the Break-Off data for 5 in. and 7 in. slabs were combined, and then the regression analysis was performed. Fig. 5.3.1 and 5.3.2 show plots for the regression equations for Mixes 4R and 4C for sleeve specimens and drilled specimens, respectively. The regression equation, the standard deviation of error (S.E.), and the correlation coefficient (C.C.), are also shown for each mix in these figures. The S.E. shown in the Fig. 5.3.1 and 5.3.2 is the estimated standard deviation of error of the actual cylinder test data about the regression line. It is equal to the square root of the mean square of the residual error.

The regression equations shown in Fig. 5.3.1 and 5.3.2 indicate that the S.E. is very low. The average S.E. from the sleeve tests is calculated to be 255 psi for estimating the standard cured cylinder compressive strength for Mix 4R and 4C. Similarly, the average S.E. from the drill specimens is 210 psi. It is concluded that the calculated standard deviation of error (S.E.) for the compressive strength obtained from the Break-Off test is very low. Therefore,

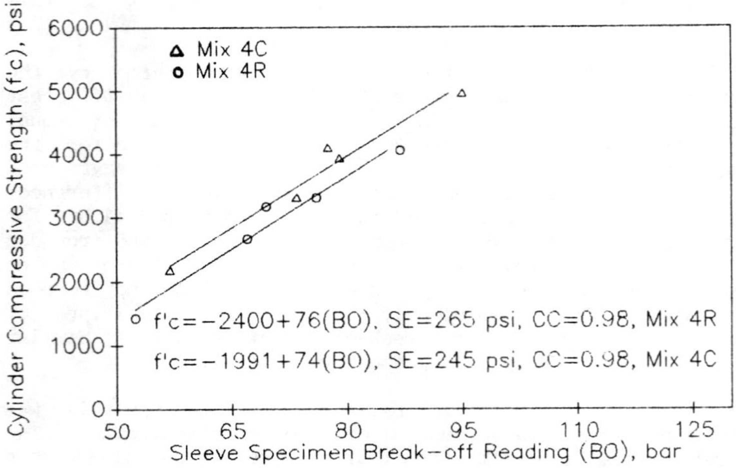

Figure 5.3.1 Sleevs Specimen Break-off Vs. Cylinder Compressive Strength

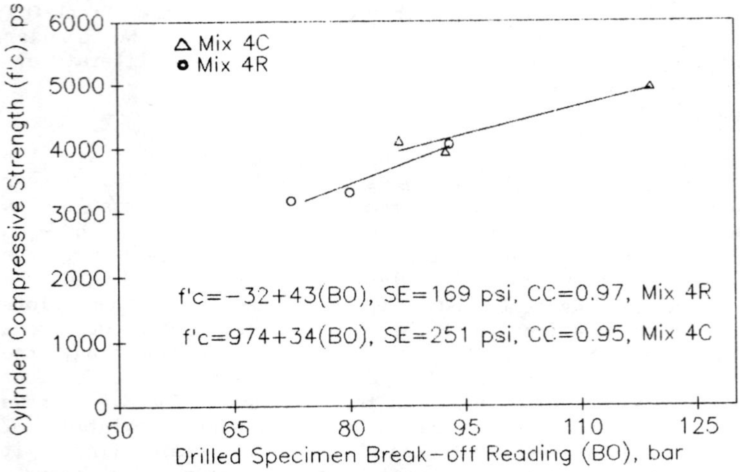

Figure 5.3.2 Drilled Specimen Break-off Vs. Cylinder Compressive strength

the Break-Off test can very accurately determine the in-place compressive strength of concrete. Johansen [4] concluded that the correlation of the Break-Off test reading with that of the standard cube compressive strength was very satisfactory. The standard deviation of error for his results were also on the average about 150 psi for concrete cube strength ranging from 2000 to 8000 psi. Other researchers have also reported similar results (10, 11, 12)

6.0 CONCLUSIONS

The following conclusions can be drawn from this research reported here.

(1) The Break-Off test readings show higher variability than the standard cylinder test. This higher variability, however, can be easily offset by taking 5 Break-Off test readings.

(2) Slab thickness does not affect the accuracy of the Break-Off test results.

(3) The variability of the rounded aggregates Break-Off test results is higher than that for the crushed aggregates test results.

(4) Sleeve specimen Break-Off test results have a higher variability than that for the drilled specimen test results.

(5) Slab thickness, type of aggregate and method of obtaining Break-Off test specimens did not have any significant effect on the Break-Off readings.

(6) From the regression equations, it is concluded that a high degree of correlation exists between the Break-Off readings and the compressive strength of the in-place concrete. These equations show a high correlation coefficient between the standard cylinder tests and the Break-Off tests. The calculated standard deviation of error for the Break-Off test was very low. In summary, the Break-Off test can very accurately determine the in-place compressive strength of concrete.

REFERENCES

1. Johansen, R., "In-Situ Strength Evaluation of Concrete - The Break-Off Method", Concrete International: Design and Construction, V. 1, No. 9, September 1979, pp. 45-51.

2. Nordtest, "Nordtest Method NT BUILD 212," Edition 2, Approved 1984-05, 5 pages.

3. Swedish Standard, SS137239, Approved 1983-06, 3 pages.

4. British Standard, "Break-Off Test," BS 1881: Part 201, 1986, P. 17.

5. Barker, G. and J.A. Ramirez, "Determination of Concrete Strengths with Break-off Tester", ACI Material Journal, V. 85, No. 4, July - August 1988, pp. 221-228.

6. Nishikawa, A., "A Non-destructive Testing Procedure for In-Place Evaluation of Flexural Strength of Concrete," Report No. JHRP-83-10, Joint Highway Research Project, School of Civil Engineering, Purdue University, West Lafayette, Indiana, 1983.

7. Salameh, Z., "Evaluation of In-Place Rounded Aggregates Concrete by the Break-Off Method," Department of Civil Engineering and Mechanics, the University of Wisconsin-Milwaukee, December 1987.

8. Hassaballah, A., "Evaluation of In-Place crushed Aggregates Concrete by the Break-Off Method," Department of Civil Engineering and Mechanics, the University of Wisconsin-Milwaukee, December 1987.

9. Kennedy, B.J. and A.M. Neville, "Basic Statistical Methods for Engineers and Scientists", 2nd ed., T.Y. Crowell Co., New York, 1976, 490 pp.

10. Byfors, J., "Plain Concrete at Early Ages", CBI Research No. 3:80, Swedish Cement and Concrete Research Institute, Stockholm, March 1980, pp. 329-330.

11. Dahl-Jorgensen, E., "In-Situ Strength of Concrete, Laboratory and Field Tests", Report No. STF 65A82032, Foundation of Scientific and Industrial Research, Norwegian Institute of Technology, Trondheim, 1982, 57 pages.

12. Smith, L.M., "Evaluation of the Scancem Break-Off Tester", Central Laboratories Report No. M4.8616, Ministry of Works and Development, Lower Hutt, June 1986, 10 pages.

ACOUSTIC EMISSION FROM PLAIN CONCRETE AND CONCRETE WITH POLYPROPYLENE FIBER REINFORCEMENT

J. D. Leaird
Hartford Steam Boiler Inspection Technologies
Sacramento, CA 95815

M. A. Taylor
University of California, Davis
Davis, CA 95616

ABSTRACT

There is substantial benefit to be gained from finding a cost effective method of increasing the crack resistance of structural concrete. One suggested method is to add chopped polypropylene fibers to the concrete. Fibers are currently used for toughness, plastic shrinkage, and impact resistance. A series of beams were manufactured with fibers added and a series without the fibers. These beams were tested in four point loading. The acoustic emission activity from each beam was measured during the load test. The fiber reinforced beams had 1/2 to 1/3 the acoustic emission activity of the beams without fibers. This suggests that the addition of fibers will indeed increase the crack resistance of concrete. Crack data was obtained from the beams at various load levels. The crack width information indicated smaller widths in the fiber reinforced concrete than in the plain concrete at all loads which correlates well with the acoustic emission data.

INTRODUCTION

Acoustic Emission Testing

It may be helpful to briefly describe the technology of acoustic emission (AE). Acoustic emission is the name of both a physical phenomena and a technology. An acoustic emission is defined (ASTM E610-82) as "..the class of phenomena whereby transient elastic waves are generated by the rapid release of energy from localized sources within a material, or the transient elastic waves so generated." Therefore, practitioners refer to the source mechanism (crack initiation or extension) as an acoustic emission and the resultant stress wave as an acoustic emission. These are different but related phenomena, each having its own set of physical laws. A most important point is that each different type of acoustic emission source generates a stress wave that has distinctive properties. This enables a test engineer to identify the sources thereby eliminating unwanted sources from the data. An earthquake is a classical acoustic emission. It has often been suggested that a more appropriate name for the technology would be structural seismology.

One of the key attributes of AE is the ability to do detection, remote from the source, and the complimentry ability of source location. Source location was not used in this program but the ability to detect remote sources is critical (i.e. noninvasive volumetric monitoring). Without this ability the technique would only detect sources in direct contact with the sensor.

Fiber Reinforced Concrete

Concrete is an essentially brittle material; this contributes to some of the following problems facing designers.

1. Cracking. Cracking occurs both in the plastic state and in hardened concrete. In hardened concrete such cracking is architectually unsightly and may allow water to penetrate the concrete and stimulate corrosion of the reinforcing steel. This is especially dangerous for members in corrosive environments and for prestressed beams (where the wires are under high stress and often of small cross-section).

In the plastic state, rapid hydration and the evaporation effects of winds can cause much unsightly cracking. The cracks remain throughout the life of the structure, with the attendant problems mentioned above.

2. Toughness (Ability to absorb energy at a given displacement). Concrete is strong but brittle in compression and weak and brittle in tension. Hence, steel (ASTM A 305 "..any deformed steel bar intended for use as reinforcement in reinforced concrete construction" ..commonly referred to as rebar) is required to provide any tension resistance and impart some ductility to the component material. Plain concrete is a very rare structural material. Toughness is of special concern in regions where earthquakes occur. This is because earthquakes can cause unexpected loads and unexpected reversals of load.

Present day philosophy of "crack control" in concrete is to limit the width of cracks to a limit that is presumed to be "safe". If cracks are held below this limit then their number is insignificant. This philosophy underlies the use of deformed rebar in reinforced concrete. The ribs on the rebars create a large number of narrow cracks which is much preferred over a small number of wide cracks which form when smooth rebar is used. Similarly a large number of small diameter rebar are preferred over a few large diameter rebars if cracking is a major concern.

Present day methods of designing for the risks (ie. cracking and toughness) often includes so much steel that it is difficult or impossible to place the concrete. This makes the problem worse instead of better. Obviously any system which could alleviate the above mentioned problems of cracking or toughness would be of great benefit to construction.

The effects of including quantities of short, straight fibers in concrete has been studied for almost 30 years. The first fibers proposed were brass coated steel wires (a few thousandths of an inch in diameter and from one-half to several inches long). The advantage predicted by inventors included; a) improved tensile strength, b) improved toughness, and c) improved crack control.

In recent years synthetic polypropylene fibers have also been introduced. The advantages of plastic fibers include; 1) lighter weight, 2) lower cost, and 3) easier mixing. In common with steel fibers these newer fibers are believed to function either by retarding cracks or preventing them (or both).

This report presents results of tests on beams which were arranged to have a constant moment region reinforced with FIBERMESH (trademark name of FIBERMESH Corp. Campbell, CA.) fibers. One objective of the test was to observe and measure the width and extent of cracking during bending. Acoustic emission data were taken as the load was applied and the acoustic emission characteristics were correlated with the cracking behavior.

ACOUSTIC EMISSION SYSTEM

The typical acoustic emission test system is composed of three elements. These are: 1) a sensor to detect the acoustic emission, 2) an amplifier to boost the signal energy for transmission, and 3) a processor to detect and quantify the signal. All the above elements are interconnected by shielded cabling.

1. The sensor. The sensors used on this test have a sensitivity of approximately -65dB referred to 1V per microbar. The sensors selected were the AET AC175L which are resonant at 175 kHz. These sensors were originally chosen because they are high enough in frequency to be immune to almost all ambient noise but low enough to detect signals from a significant volume of concrete. The beams were brushed off and any extra spill-over was broken off prior to attaching the sensors. The sensors were acoustically coupled to the beam with silicone grease. The sensors were mechanically restrained with constant force clamps.

2. The amplifier. The second part of the system is an amplifier, to boost the signal for transmission from the sensor to the quantification system. The preamplifier used in this work was the AET model 160B utilized 60dB (x1000) of gain. A secondary function of the preamplifier is to bandpass the signal for optimal signal-to-noise characteristics. The bandpass filtering will be determined by the resonant frequency of the sensor used. The filter used bandpassed the signal from 125kHz to 250kHz.

3. The processor. The processor detects and quantifies the acoustic emission signal. This process is keyed to the amplitude output of the sensor/preamp combination, any post amplification and the threshold selected. Total system gain was 80dB. The amplified AE signal was compared to a 0.5 volt threshold. Once the system is triggered the signal is measured for all of the acoustic emission parameters (e.g.; duration, counts, rise-time, peak amplitude, slope, and energy). A detailed explanation of this measuring process is beyond the scope of this paper. Suffice it to say that the short duration acoustic emission event is measured with a digitizing rate of 8 MHz. The processor used for the principal investigation is the AET Model 5000B. This system includes a microcomputer as an operator controlled interactive display. Also included in the system is the ability to digitally record characteristics of the AE signal for post-test analysis. The AET 5000B system has software graphics capabilities to display the acoustic emission signal characteristics and data trends.

Additional components may be added to the basic system outlined above. These may include an oscilloscope to view the acoustic emissions, an audio unit to listen to the acoustic emissions, and a hard-copy-unit to make a permanent real-time record of the acoustic emissions. All these were used during the study.

The authors believe that acoustic emission should be used on a statistical basis, as opposed to giving to much meaning to individual acoustic emission events. Acoustic emission instruments are capable of detecting, processing and recording many hundreds to many thousands of events per second. In a test there are generally many events per hour to many per second thus giving the engineer many data points and a statistically sound chance to detect and interpret the sources of interest. Placing too much emphasis on single events increases the chance of false interpretation. The experimentation reported below indicates that the plain and reinforced concrete can generate many acoustic emissions per minute minimizing the risk of false intrepretation.

THE CONCRETE BEAMS

Cracking in beams results from bending, from shear, from de-bonding and from interaction of these mechanisms. For simplicity this test was arranged for the central region of the beams to develop pure bending so as to avoid any complications from interaction with shear.

Beam molds were formed from 16 gage metal and coated with a releasing agent (i.e. Formkote 7). Reinforcing bars and stirrups were wired together then placed in the forms. Top width was maintained by use of metal hairpins placed across the forms at about a 4 foot spacing. Rebar cages were also hung from these hairpins. (Note, beam dimensions and load data are given in English units and crack data are reported in metric.)

The beams were 10 feet long with a slightly tapered cross section. The taper was from 6.5 inches at tension face and 8.5 inch at compression face. Depth was 13 inches.

Stirrups were placed in the end regions (2 feet long) but the central 4 foot region (to be the constant moment region) had no stirrups. The main reinforcement was two #5 (5/8 inch) deformed rebar; two #4 (4/8 inch) rebars were used in the compression zone to hold the stirrups.

The concrete was designed and inspected by FIBERMESH engineers. The plain mix used 1 inch maximum size aggregate normal-weight concrete. The FIBERMESH concrete used 7.5 pounds per cubic yard of fibrillated fibers 1/2 inch long. This quantity of fibers represents approximately one-half (1/2) percent by volume of the concrete or 0.2 percent by weight. The water and cement contents of the two mixes were held constant. The reinforcing steel was tested to have a yield at 65 ksi.

Three beams were constructed using regular concrete and three were constructed with fibers added. For a more detailed account of the design please refer to the UC Davis report of investigation, reference 1.

The beams were cast on March 15, 1988. Companion cylinders were cast for the two mixes to determine strength and unit weight. Concrete was placed in two horizontal layers. An internal vibrator was used to consolidate each layer. This resulted in beams with virtually no surface defects. Slump tests were conducted to assure constancy of concrete.

After casting, the beams were covered with a sheet of plastic. After initial set (approximately 4 hours) the beams were sprayed with water 3 to 4 times per day for 4 days. Then the water spray was reduced to once per day until 2 days before testing.

Acoustic emission data is reported on two of the plain concrete beams (P2 and P3), and on two of the fiber added beams (F2 and F3). There is crack information on both of the two plain concrete beams. Crack information is also available on three fiber reinforced beams (F1, F2 and F3). Beam P1 was broken in an uncontrolled load experiment, and beam F1 was tested prior to start-up of the acoustic emission testing. For these reasons there is no acoustic emission data on either P1 or F1, but there is crack data on F1.

Crack Dimension

The tension faces of the beam were 6.5 inches wide. A pencil line was drawn down the center of this face through the full 48 inch test section. Lines parallel to the center line were drawn 3/4 inches in from the side of the beam. Crack widths were measured at these locations. The largest value is reported as the maximum crack width. Generally the widest crack was located at the sides not along the centerline.

Crack widths were measured with a transparent template gage. Two people did all the crack dimension measurements. An effort was made to cross check the accuracy, consistancy and repeatability of these two people. The effort consisted of having both people read a set of cracks before, during and after the test. This effort reduced the reading error inherent in this type measurement.

The penetration of the cracks toward the compression face was measured on both sides of the beam. This value is reported in centimeters of crack height and in percentage of beam thickness.

The sum of all the crack widths in the constant moment test section (48 inches long) are also reported.

Cracks (in the test section) were counted and the number of cracks divided into 48 inches (121.92 cm) to obtain an "average crack spacing." However, at some load levels new cracks would appear. Cracks are identified in the table (Table 5) as either "PRE" or "NEW". The line labeled "PRE" indicates cracks that existed before the new load level was obtained and the line labeled "NEW" indicates cracks that formed during the current load increment.

Experimental Method

The spreader beam (used to apply point loads at the ends of the beams-see Figure 1) was placed on the concrete beam, and an MTS controlled hydraulic ram was lowered into position against it. A straight rising load was applied, until the first hold period. The load ramp was manually controlled. The hold points were at 5, 10, 15, 17.5, 20, 22.5, 25, and 27.5 kips. The 17.5 kip hold point was not used on every beam. The duration of the hold was controlled by the amount of time it took to physically measure height, width and length of every crack in the beam within the region of interest. This hold period varied from 9 minutes for beam F2 at 15 kip to less than a minute for most beams at 5 kip.

The acoustic emission tests were usually started just after the load ramp started. This minimized noise generated by the contact of the load and reaction beams with the concrete beam. A short preliminary experiment indicated that noise was not transmitted from the RAM or top of the load beam to the AE sensors. The one exception to the above start time was beam P3. The AE start time on P3 was immediately prior to the application of load. This was done to check the AE of the concrete beam with just the load of the steel beam on it and also at 500 lbs of RAM load. All AE tests were stopped when the beam lost load or the test was terminated by the test engineer.

An attenuation study was conducted on one of each beam type, e.g.; plain and fiber reinforced. The study utilized 175kHz sensors bandpassed from 125 to 250kHz. A 40dB preamplifier was used with a threshold set at 1.0 volt. The sensor was attached in the same fashion as for the above test. The simulated source was a mechanical pencil lead break. A 0.5mm diameter HB lead was extended approximately 3mm. The pencil was inclined to the concrete beam surface at 30 degrees, and load was applied until the lead snapped. This was done at the sensor and at 6 inch intervals from the sensor. All breaks were done with the source and the detecting sensor on the same side of the beam.

DISCUSSION

Attenuation was slightly higher in the fiber reinforced beam than in the plain beam. This attenuation information is plotted in Figure 2. The authors selected the attenuation difference between 6 inches and 12 inches to represent the attenuation of the concrete, because, 0 inches is susceptible to saturation hence incorrect amplitude measurement. The average attenuation produced is 19dB per 6 inches in the fiber reinforced beams and 16dB per 6 inches in the plain beam.

The plain concrete (e.g.; P2 and P3) was more emittive than the fiber added concrete (e.g.; F2 and F3). The fiber added concrete had fewer acoustic emission events than the plain concrete.

The sum of all valid events for each beam is indicated in Table 1:

TABLE 1

BEAM	TOTAL VALID EVTS.	TEST TIME(min)	EVTS/MIN
P2	29090	35.43	820
P3	41324	45.23	910
F2	11254	35.48	320
F3	11320	45.53	250

TABLE 1A

P2/F2 RATIO	2.6
P3/F3 RATIO	3.6

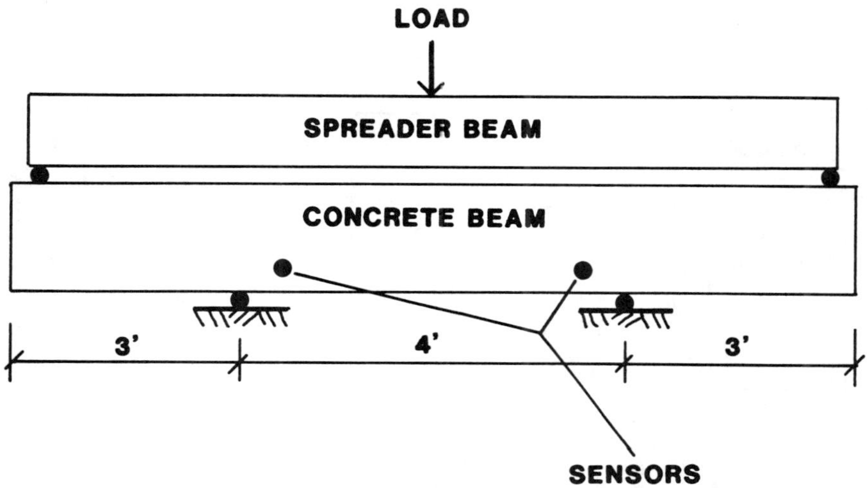

FIGURE 1 EXPERIMENTAL SET-UP

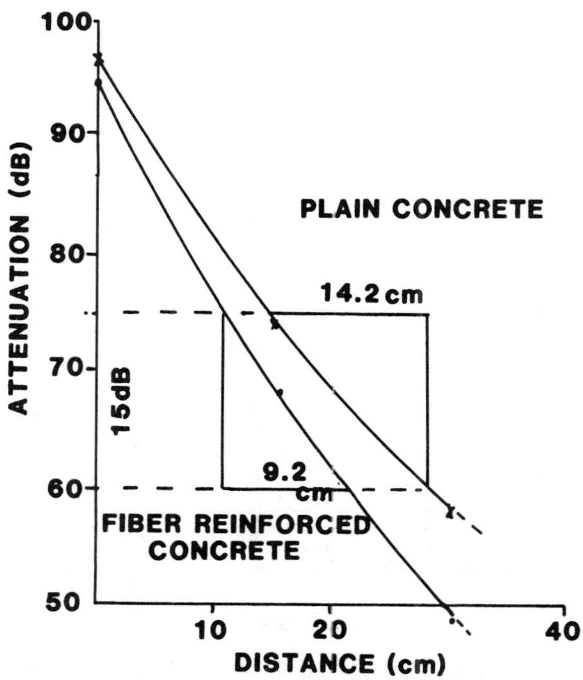

FIGURE 2 ATTENUATION vs. DISTANCE

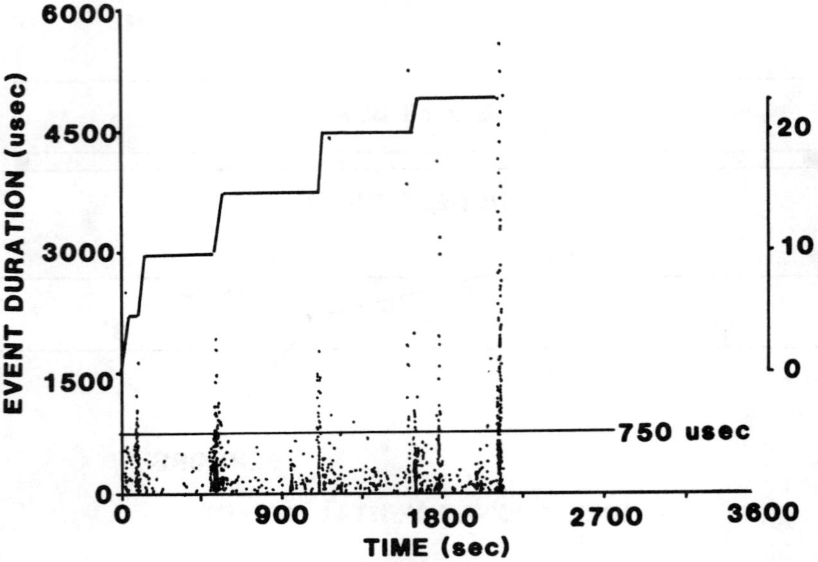

FIGURE 3 EVENT DURATION and LOAD vs. TIME SENSOR 2 BEAM F2

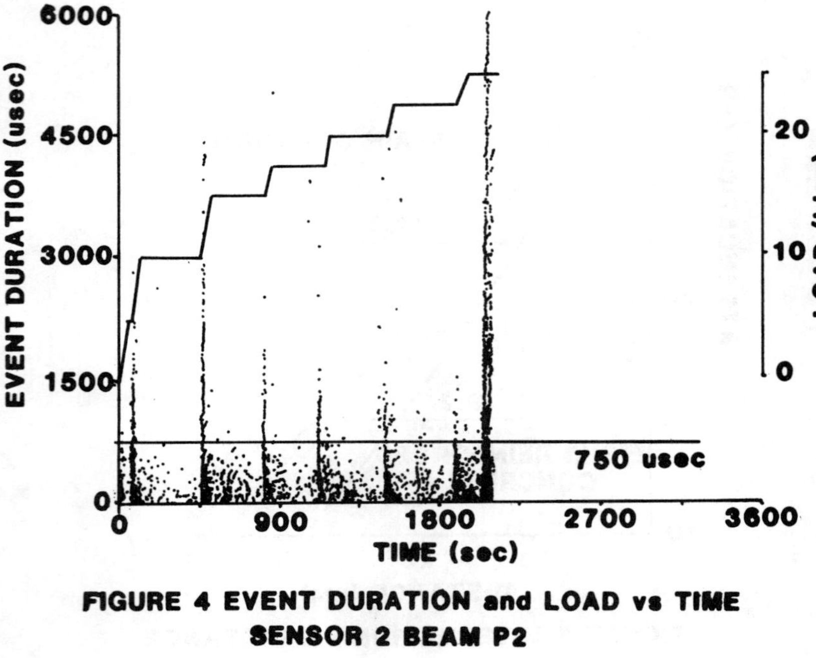

FIGURE 4 EVENT DURATION and LOAD vs TIME SENSOR 2 BEAM P2

In the above table a valid event is defined as one that had 2 or more counts and less than 4090 counts. This criterion was established to delete those events that could be interpreted as noise. An event with 1 count could be caused by electrical noise, while one with more than 4090 counts exceeds the range of the counter.

Comparing the data above indicates that there was 2.6 to 3.6 times as much event activity in the plain concrete as in the fiber reinforced concrete. These ratios are obtained by comparing long term test to long term test and short term test to short term test (e.g.; P3/F3 and P2/F2). All things are equal in the mix design and manufacture of the concrete except for the addition of the fibers. This would tend to attribute the increased activity in the plain concrete to decreased fracture resistance caused by lack of fibers.

If the fibers are retarding the crack extension this should be indicated by AE event duration and energy (e.g.; counts). Intuitively one would expect shorter duration signals and lower energy signal from the fiber added/reinforced concrete. Indeed, shown in Figure 3 and 4 are the event durations of the events detected in beams F2 sensor 2 and P2 sensor 2. 4.4 percent of the events have a duration greater than 750 microseconds (an arbitrarily chosen level) in beam F2 as compared to 12.3 percent in beam P2.

This same relationship holds true in the matter of counts as well. Figures 5 and 6 are the counts versus time plots for beams F2 sensor 2 and P2 sensor 2. Fiber reinforced beam F2 has 1.6 percent of the events with greater than 100 counts (an arbitrarily chosen value). Plain concrete beam P2 has 4.4 percent of the event greater than 100 counts. Both of these indicators strongly suggests that the fibers are indeed impeding the growth of cracking in the concrete.

The crack width measurement data is contained in Table 2. There is a very narrow range (although there are only two samples) of average crack in the plain concrete beam. This range is 0.26mm to 0.27mm at 22.5 Kips load, which is the maximum load that all beams were subjected to. The fiber reinforced beams had crack width values at 22.5 Kips of 0.16mm, 0.18mm and 27.5mm.

Crack height measurements are considered at 22.5 Kips load for the reason stated above. The plain concrete beams had crack heights of 18.4cm and 19.9cm. The fiber reinforced beams had crack heights of 16.0cm, 16.5cm and 17.5cm. This information is shown in Table 3.

Table 4 shows the data from the total crack opening in the test section of the beams. The plain concrete beams had total crack openings of 2.63mm and 2.99mm. The fiber reinforced beams had total crack openings of 1.79mm, 2.37mm and 2.49mm.

The number of cracks that existed in the beams at the end of the 22.5 kips load cycle are shown in Table 5. The fiber reinforced beams had the most cracks. This set of beams had 11, 13, and 14 cracks in the test section. The plain concrete beams both had 11 cracks in the test section.

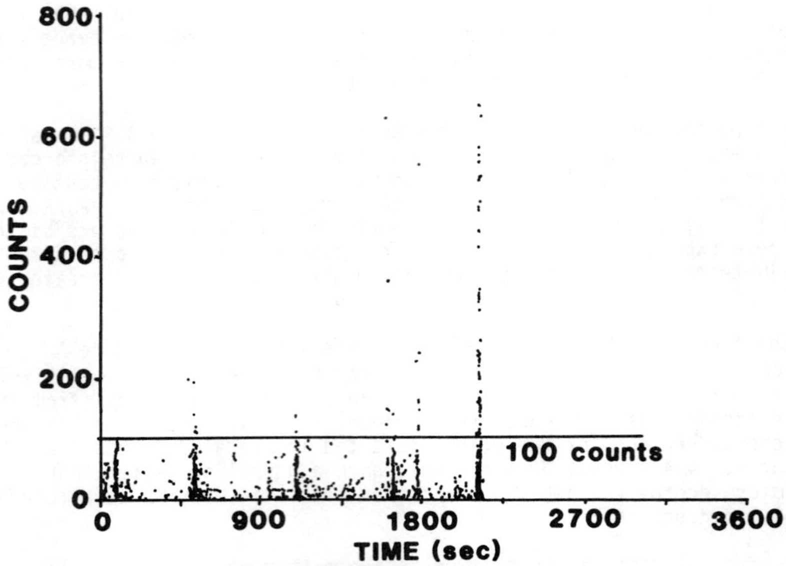

FIGURE 5 COUNTS vs TIME
SENSOR 2 BEAM F2

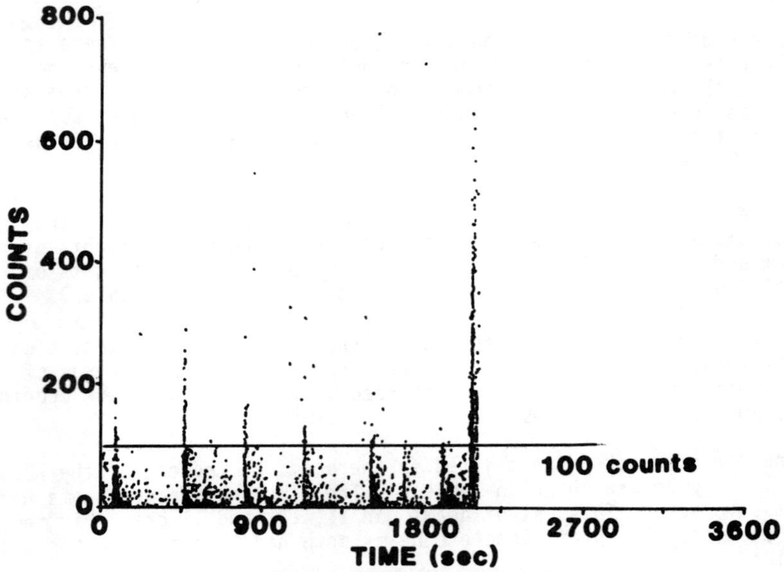

FIGURE 6 COUNTS vs TIME
SENSOR 2 BEAM P2

TABLE 2
CRACK WIDTH MEASUREMENT

	PLAIN CONCRETE							
BEAM	FACTOR/LOAD	10.0	15.0	17.5	20.0	22.5	25.0	27.0
P2	MAX CR WID (mm)	.13	.25	.33	.40	.40		
	AVG CR WID (mm)	.06	.13	.18	.20	.27		
P3	MAX CR WID (mm)	.10	.20	.25	.33	.33		
	AVG CR WID (mm)	.07	.14	.19	.23	.26	.30	.36
	FIBER REINFORCED CONCRETE							
BEAM	FACTOR/LOAD	10.0	15.0	17.5	20.0	22.5	25.0	27.0
F1	MAX CR WID (mm)	.10	.18		.25	.25	.33	
	AVG CR WID (mm)	.08	.11		.11	.16	.23	
F2	MAX CR WID (mm)	.08	.20		.30	.30		
	AVG CR WID (mm)	.06	.09		.13	.18		
F3	MAX CR WID (mm)	.08	.20	.25	.25	.33	.40	
	AVG CR WID (mm)	.05	.11	.15	.20	.27	.28	

TABLE 3
CRACK HEIGHT MEASUREMENT

	PLAIN CONCRETE							
BEAM	FACTOR/LOAD	10.0	15.0	17.5	20.0	22.5	25.0	27.0
P2	HEIGHT (cm)	10.6	14.5	15.5	17.3	18.4		
P3	HEIGHT (cm)	10.8	15.0	16.8	18.5	19.9	19.9	20.4
	FIBER REINFORCED CONCRETE							
BEAM	FACTOR/LOAD	10.0	15.0	17.5	20.0	22.5	25.0	27.0
F1	HEIGHT (cm)	5.4	11.4		14.4	16.0	18.4	
F2	HEIGHT (cm)	6.3	11.5		15.8	17.5		
F3	HEIGHT (cm)	9.4	13.5	14.5	15.0	16.5	17.3	

TABLE 4
TOTAL CRACK OPENING IN 48 INCH REGION

BEAM	PLAIN CONCRETE FACTOR/LOAD	10.0	15.0	17.5	20.0	22.5	25.0	27.0
P2	OPENING (mm)	.50	1.20	1.98	2.19	2.99		
P3	OPENING (mm)	.55	1.30	1.90	2.17	2.63	3.01	3.56
BEAM	FIBER REINFORCED CONCRETE FACTOR/LOAD	10.0	15.0	17.5	20.0	22.5	25.0	27.0
F1	OPENING (mm)	.31	1.02		1.29	1.79	2.30	
F2	OPENING (mm)	.41	.95		1.86	2.49		
F3	OPENING (mm)	.47	.92	1.46	2.06	2.37	3.07	

TABLE 5
CRACK SPACING MEASUREMENT

BEAM	PLAIN CONCRETE FACTOR/LOAD	10.0	15.0	17.5	20.0	22.5	25.0	27.0
	PRE	-	7	8	10	11		
P2	NEW	7	1	2	1	-		
	SPACING (cm)	17.4	15.3	12.2	11.1	11.1		
	PRE	-	7	9	9	10	11	12
P3	NEW	7	2	-	1	1	1	-
	SPACING (cm)	17.4	13.6	13.6	12.2	11.1	10.2	10.2
BEAM	FIBER REINFORCED CONCRETE FACTOR/LOAD	10.0	15.0	17.5	20.0	22.5	25.0	27.0
	PRE	3	4		9	12	13	
F1	NEW	1	5		3	1	-	
	SPACING (cm)	30.5	13.6		10.2	9.4	9.4	
	PRE	-	6		11	14		
F2	NEW	6	5		3	-		
	SPACING (cm)	20.3	10.2		8.7	8.7		
	PRE	-	6	8	10	11		
F3	NEW	6	2	2	1	-		
	SPACING (cm)	20.3	15.3	12.2	11.1	11.1		

The authors acknowledge that the attenuation difference between the two types of concrete (fiber reinforced concrete and plain concrete) may account for the differences noted in detected signal. To ensure that this possibility is taken into account a calculation of the events per crack within a given attenuation range is made here. This calculation will consider beam F2 and P2 which represent the most conservative case. Attenuation is plotted on Figure 2. Assume that the system only detects a signal within a 15dB range. For the fiber reinforced concrete beam (F2) 15dB represents 9.2cm, while in the plain concrete (P2) 15dB represents 14.2cm. Further assume that the beam is a two dimensional plane and that the detecting sensor is placed so that it will detect events at double the indicated distance (e.g.; 18.4cm in the fiber reinforced beam and 28.4cm in the plain concrete beam). At 22.5 kips load the crack spacing in P2 is 11.1cm and in F2 the crack spacing is 8.7cm. Dividing the crack spacing into the effective monitoring area indicates that the sensors on P2 are responding to 2.6 cracks while the sensors on beam F2 are responding to 2.1 cracks. Since each beam was monitored with two sensors divide the total number of valid events detected in either beam (P2 or F2) by 2. This gives 14,500 events for beam P2 and 5,700 for beam F2. Then divide the resultant values by the number of cracks in the effective monitoring area. That is 14,500 events / 2.6 cracks for beam P2 and 5,700 events / 2.1 cracks for beam F2. This indicates that the cracks in the plain concrete beam generated 5,600 events per crack while the cracks in the fiber reinforced beam generated 2700 events per crack. The authors admit that this events per crack calculation is rudimentary, but believe that it reflects the ability of the fibers to retard the cracking.

CONCLUSION

Acoustic emission data indicates that crack growth is retarded by the addition of FIBERMESH fibers. There are substantially fewer events in the fiber reinforced beams. Further, the events that do exist are of shorter duration and lower in energy. The attenuation of the fiber reinforced beam is slightly greater than the plain beam. This will account for some, but not all, of the difference in the detected AE activity. Indeed a simplified calculation indicates that the plain beam generated approximately 2700 to 3400 events per crack.

The highest load that was common to all beams reported was 22.5 Kips. At this load the fiber reinforced beams had more cracks and the cracks were finer (narrower). The plain concrete beams generally had wider openings, longer cracks and more openings but fewer cracks in the test section. The only exception to this generalization is beam F3 which had one crack that was as wide as the plain concrete beam cracks.

This study represents the first stage of tests to be conducted. In the future the cracking due to earthquake generated loading of the fiber reinforced concrete will be studied.

ACKNOWLEDGEMENTS

The authors wish to thank Gary Vondran, Director of Research and Development, FIBERMESH Corporation, Campbell, CA. for his (and their) support of this research. We also wish to thank Ms. Guo, visiting scholar from Bejing, China and Mr. Bill Sluis, Development Engineer, UC Davis for their assistance with the testing.

REFERENCES

1. "CRACKING BEHAVIOR OF FIBERMESH CONCRETE", Final Report to FIBERMESH Company, Michael Taylor, University of California-Davis, Civil Engineering Dept. June 1988.

SEISMIC, SONIC, AND VIBRATION METHODS FOR QUALITY ASSURANCE AND FORENSIC INVESTIGATION OF GEOTECHNICAL, PAVEMENT AND STRUCTURAL SYSTEMS

L. D. Olson and C. C. Wright
Olson-Church, Inc.
Denver, CO 80218

INTRODUCTION

Case histories are presented to illustrate quality assurance (QA) and forensic applications of nondestructive testing and evaluation (NDT&E) for the construction industry. The sonic, seismic and vibration methods described are based on the measurement of wave propagation behavior. The measurements provide data on dynamic properties that are directly and indirectly related to physical conditicns of the system under test. They typically provide engineers with critical data that cannot be economically obtained with traditional destructive tests.

The paper begins with descriptions of three NDT&E methods for deep foundations that have been under active research and development since the 1960's. Building on this experience with deep foundations, three other similar NDT&E techniques are described for testing slabs and pavements, structural systems, and the subsurface.

DEEP FOUNDATIONS

A number of NDT&E methods exist for deep foundations. There are routine tests for QA and special tests for forensic investigations as reviewed by Stokoe, et.al. (1987). Two of the more common tests are the Seismic Echo (SE) and Impulse Response (IR) methods. A third ultrasonic test, Crosshole Sonic Logging (CSL) is used less commonly, but is valuable for QA of drilled shafts for difficult subsurface conditions.

The SE and IR tests are conducted from the surface as illustrated in Fig. 1. Typical test equipment includes an impulse hammer, accelerometer and geophone receivers, and a dynamic signal analyzer. The impulse hammer is instrumented with a load cell to measure the force of an impact. Both methods involve hitting the top of a foundation with the hammer to generate a compression (P) wave which travels down the foundation. The wave is reflected off the bottom of a sound shaft or an irregularity. Receivers at the surface and embedded in the shafts track the wave travel. The embedded receivers are installed during shaft construction for QA investigations.

Example SE and IR records are presented and described below for two drilled shaft foundations. The first is a microwave tower shaft 10.7 m long and 1.5 m in diameter, where the tests were conducted as part of a QA investigation in New Jersey. The second is a research shaft 15.2 m long and 0.9 m in diameter in Texas with a defect at 9.8 m. The planned defect consists of a 30 cm thick lense of clay.

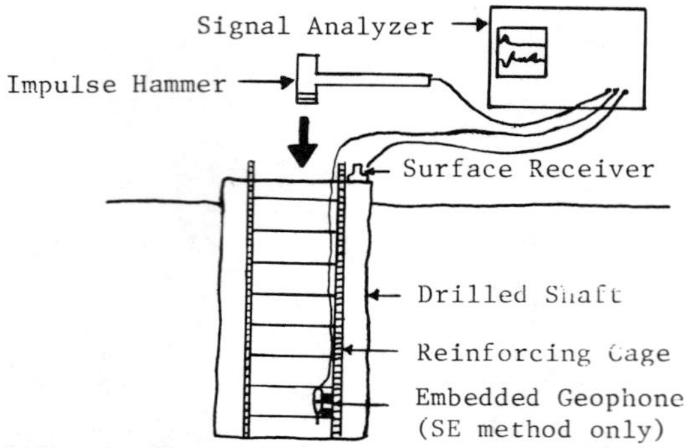

FIGURE 1. Seismic Echo and Impulse Response Methods.

Seismic Echo

A 2 channel SE test record is presented for the tower shaft in Fig. 2. The dashed lines illustrate compression wave travel through the shaft. The top trace is from an accelerometer receiver (A) at the surface. It represents the large initial acceleration of the shaft top immediately after the shaft is hit with a 3 or 12 lb hammer. The surface wave energy begins to damp out, while the compression wave travels down the shaft. The bottom trace is from an embedded geophone (G) at 7.3 m. The direct arrival time of the compression wave energy at 1.87 milliseconds (ms) is indicated on the bottom trace. A direct compression wave velocity of 3,900 mps is calculated by

$$Vc = Gd / t \tag{1}$$

where Vc = compression wave velocity, Gd = embedded geophone depth, and t = time. The 3,900 mps (meters per second) velocity is indicative of good quality concrete (Harrel and Stokoe, 1984).

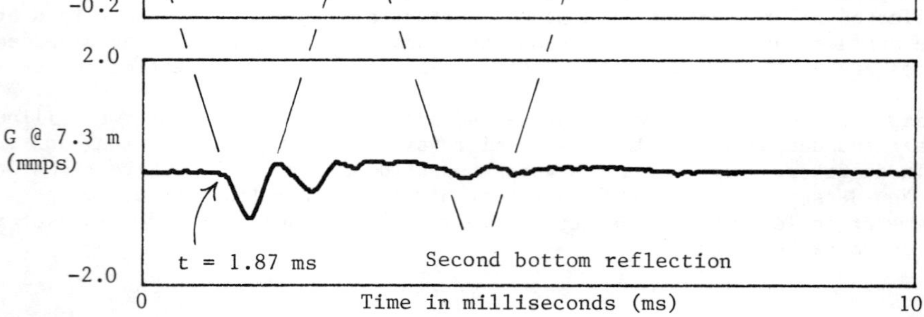

FIGURE 2. Seismic Echo Test record of New Jersey tower shaft

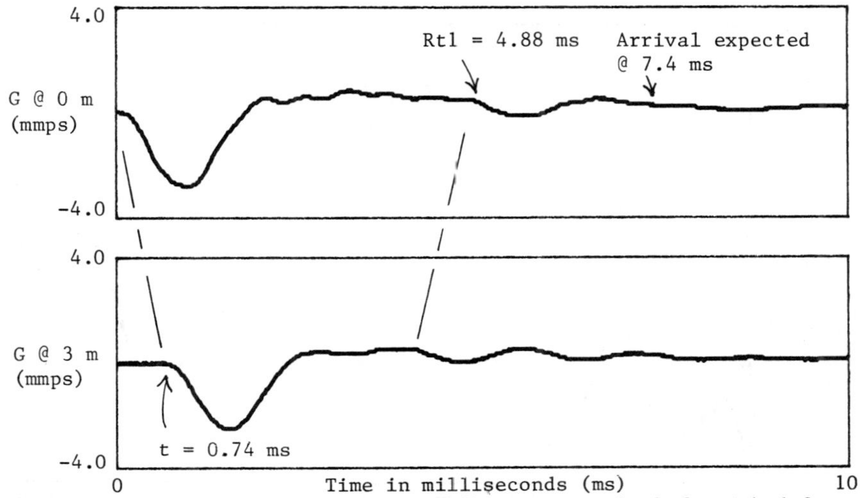

FIGURE 3. Seismic Echo test record of Texas research shaft with defect.

Moving back up to the top trace of Fig. 2, an increase in vibration acceleration is apparent at a time of 5.7 ms, Rtl. The event at 5.7 ms is identified as an arrival of reflected tension P-wave energy. No earlier arrival of reflected compression wave energy is identified. A depth, D, of 11.1 m is calculated for the reflected energy according to the equation

$$D = (Vc * Rt1) / 2 \qquad (2)$$

The estimated SE reflector depth of 11.1 m compares favorably with the actual length of 10.7 m. The event at 5.7 ms is identified as a reflection from the bottom of the shaft. A second event at 11.4 ms is identified as a second reflection, Rt2, from the bottom of the shaft. The wave energy is reflected off the bottom, off the top, and off the bottom a second time as illustrated on Fig. 2. In summary, results of the SE tests indicate the shaft concrete is of good quality and that the shaft is sound to full depth.

Texas Research Site. In comparison, a 2 channel SE record is presented in Fig. 3 of the research shaft with the defect at 9.8 m. A wave arrival at the top surface is detected at 4.88 ms. According to Equations (1) and (2), the reflector depth is 10.0 m which agrees with the defect depth. No bottom echo is apparent at the expected arrival time for a sound 15.2 m long shaft.

Impulse Response

Results for the IR test are analyzed in the frequency domain instead of in time as for the SE test and only a surface receiver is used (Stain, 1982).

New Jersey Site. An IR mobility plot for the tower shaft is presented on Fig. 4. It is a plot of the vibration response at the shaft top, normalized to the impulse force from the hammer blow, as a function of frequency. The frequency, df=178 Hz, of the spacing between resonant peaks and the velocity of 3900 mps identifies a reflector depth of 11.0 m according to the equation

$$D = Vc / (2 * df) \qquad (3)$$

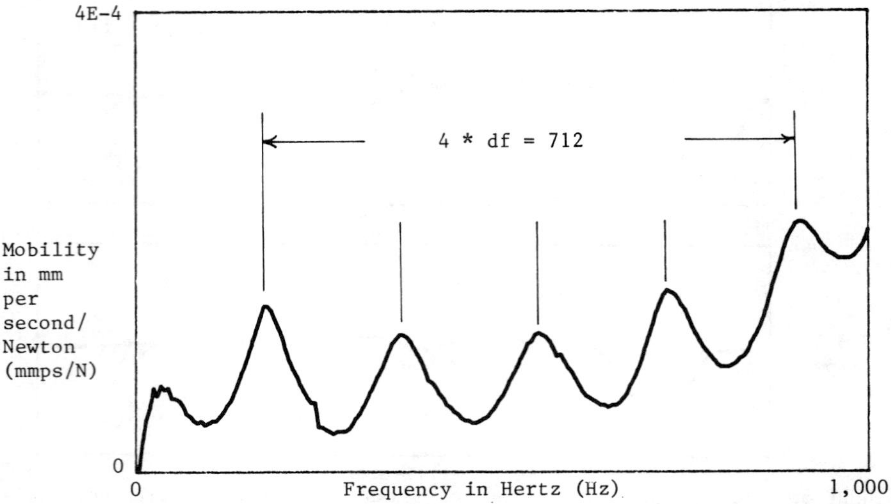

FIGURE 4. Impulse Response mobility plot for New Jersey Tower Shaft.

The estimated IR reflector depth of 11.0 m agrees with the actual length of 10.7 m for the tower shaft.

The IR test also measures the dynamic stiffness at the head of a shaft, an indication of soil-foundation interaction conditions. In a simple sense, dynamic stiffness is analagous to placing a weight on a spring and measuring the deflection of the spring to calculate the static spring stiffness (Newtons force per millimeter). Hence, dynamic stiffness can be correlated with the static stiffness measured with a load test. Dynamic stiffness is typically from 1 to 2 times the low strain static stiffness for deep foundations. Comparatively lower dynamic stiffnesses are measured for foundations with defects such as weak soil-foundation interaction, soil inclusions, breaks and necking. Dynamic stiffnesses should be compared only for similar sized shafts (length and diameter) in similar soils and not used to predict foundation capacity.

Dynamic stiffness is measured at low frequencies where the soil-foundation system may be idealized as a linear spring and a straight-line response is produced over frequencies below 100 Hertz (Hz). The slope of this straight line defines the flexibility, or millimeters displacement per Newtons force, of the foundation head. Dynamic stiffness is the inverse of flexibility. The dynamic stiffness of the tower shaft head is 2.2 MN/mm.

<u>Texas Research Site.</u> An IR mobility plot is presented as Fig. 5 for the research shaft with a defect at 9.8 m. Resonant peaks identify the shallow defect at 10.5 m. The dynamic stiffness of the defective shaft is 1.7 MN/mm, compared with 2.2 MN/mm for a similar sized sound shaft at the site.

Downhole Test

Crosshole Sonic Logging (CSL) is a downhole method for quality assurance testing of drilled shaft foundations. Access tubes, typically PVC plastic or steel, must be cast-in-place during construction to permit logging as

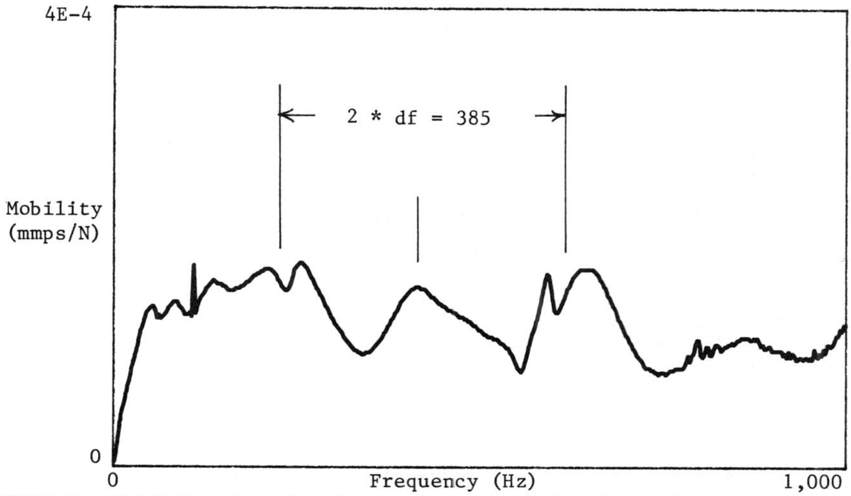

FIGURE 5. Mobility plot for Texas research shaft with defect.

illustrated in Fig. 6. The logging involves passing an ultra-sonic pulse through the concrete between source and receiver probes in a water-filled tube pair. Data from the receiver probe is recorded and processed by a computerized CSL system. The logging is conducted as the probes are simultaneously withdrawn from the tubes. The probes are withdrawn from the tubes at a rate such that measurements are conducted at 3 to 6 cm intervals coming up the shafts.

Analyses to evaluate the integrity of the concrete include measurement of wave travel times between the source and receiver, calculation of corresponding wave velocities, and receiver response energy. Longer travel times and corresponding slower velocities are indicative of irregularities in the concrete between tubes. The complete loss of signal is indicative of a defect in the concrete between tubes provided there is direct contact between the concrete and tubes.

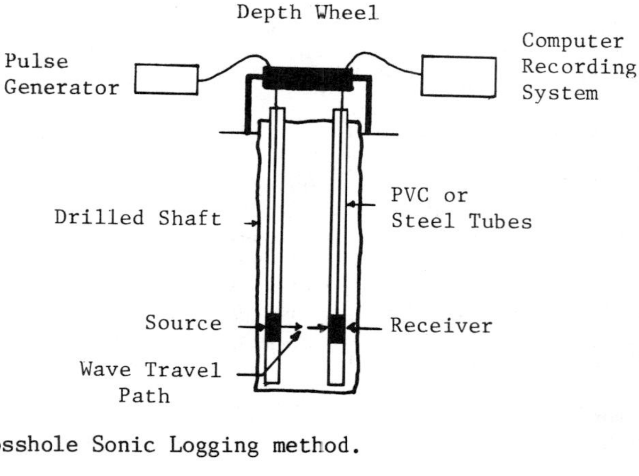

FIGURE 6. Crosshole Sonic Logging method.

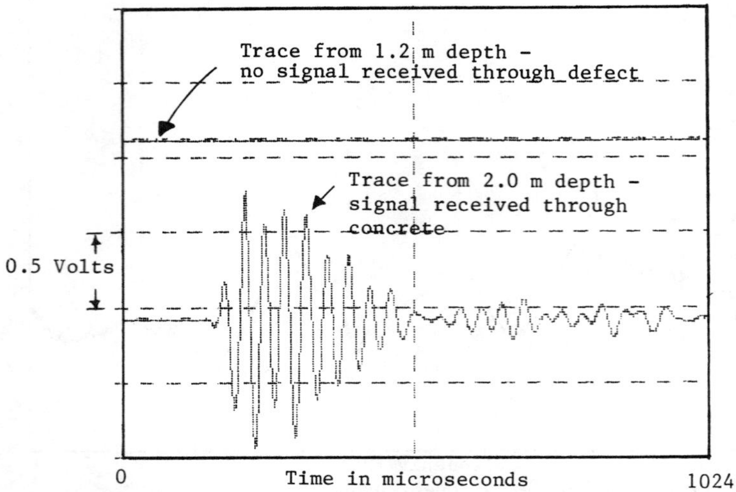

FIGURE 7. Example individual receiver response records from Crosshole Sonic Log of Utah research drilled shaft.

An example log is described below of a Utah research shaft drilled to a depth of 2.7 m and 2.4 m in diameter. The research shaft contains a lense of clayey sand to simulate a defect between 0.9 and 1.5 m and a gravel lense below 2.4 m to the bottom of the drill hole where ground water occurred.

Two example individual receiver response traces are presented on Fig. 7 from between a pair of tubes in the research shaft. The tubes are spaced a distance, d, of 54 cm apart. The upper trace is from a depth of 1.2 m in the soil defect zone and the lower trace is from a depth of 2.0 m below the top of the shaft in a sound concrete zone. On the lower trace, the receiver is quiet until the pulse from the source arrives at a time, t, of 150 microseconds. This corresponds to a sonic compression wave velocity, Vc, of 3,600 mps according to the equation

$$Vc = d / t \tag{4}$$

The 3,600 fps sonic wave velocity is indicative of sound, good quality concrete. On the upper trace from 1.2 m, no signal is received. The signal was highly attenuated and delayed by the clayey sand defect.

Individual receiver response records at incremental depths can be reviewed because the computer stores each record on a magnetic disk during logging. For general report presentation of CSL test results, however, the hundreds of individual receiver response records are not presented. Instead, computer software processes the data, determines times for the wave to travel from the source to the receiver based on a threshold amplitude selected by the operator, and plots wave arrival times vs depth. This plot is the crosshole sonic log. It is a one page presentation of arrival times vs depth from the hundreds of receiver response records covering the length of the shaft between a pair of tubes.

A crosshole sonic log between a tube pair of the research shaft is shown on Fig. 8. The horizontal axis of the log is travel time in microseconds (us).

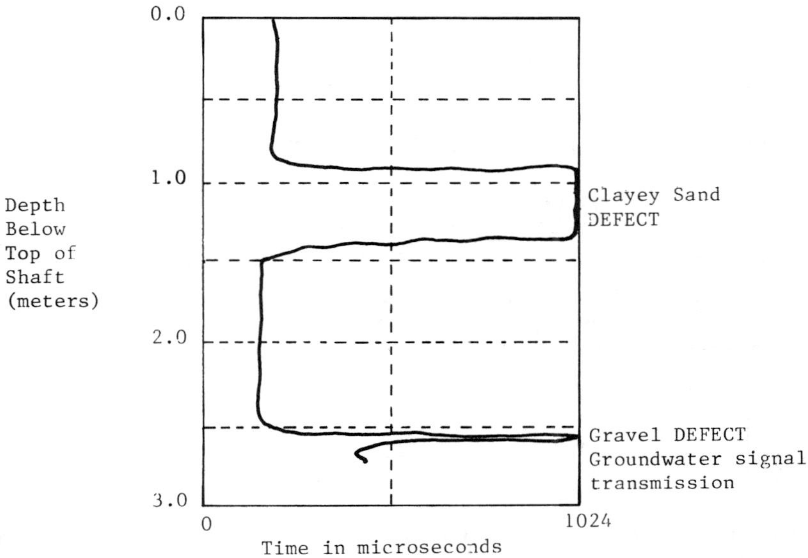

FIGURE 8. Crosshole Sonic Log.

Time 0 uS corresponds to when the pulse generator excited the source. The vertical axis is distance below the top of the shaft in meters. Depth 0 m corresponds to the top of the shaft. The defects between 0.9 and 1.5 m and below 2.5 m are identified by the absence of a signal arrival in time. The slower transmission of the signal through the groundwater in the deepest gravel is also apparent at the bottom of the shaft in Fig. 8.

Discussion of Deep Foundation NDT&E Methods

The types of deep foundations that can be tested with the surface methods include concrete drilled shafts and driven or auger cast piles. Advantages of the surface tests include speed and cost. Under good test conditions, surface tests can be conducted in a matter of minutes and little preparation is required. In forensic investigations, embedded geophones are not necessary (Olson and Thompson, 1985).

The SE and IR methods complement each other. Reflector depths and wave velocities are more accurately estimated with the SE method, while the IR method also measures soil-foundation interaction conditions. Disadvantages include wave attenuation with depth and the lack of bottom reflections from foundations in concrete-like bedrock.

An advantage of the downhole CSL test is that it can detect defects at any depth. Disadvantages include the lack of data on soil-foundation interaction conditions, the necessity of installation of PVC or steel pipe, and the requirement that the tubes be well bonded to the concrete.

SLABS AND PAVEMENTS

The Impulse Respone (IR) method for integrity testing of deep foundations has been adapted for subgrade support condition evaluation of concrete slabs

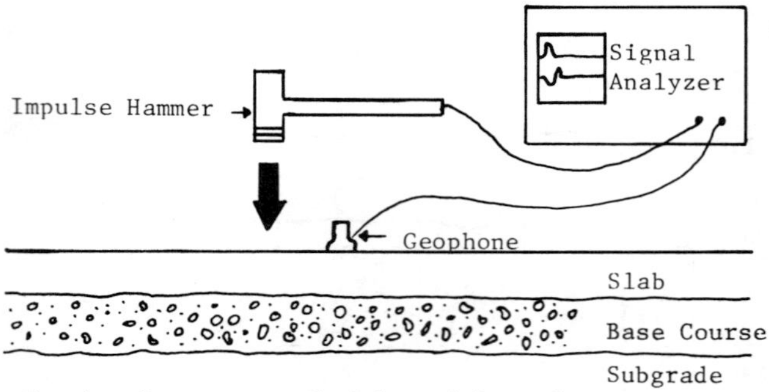

FIGURE 9. Impulse Response method for slabs and pavements.

and pavements (Pederson and Senkowski, 1986). Typical equipment for IR testing of slabs is illustrated on Fig. 9. Analysis of a mobility plot produced with a 3 lb hammer and geophone, similar to an IR test record from a deep foundation, provides information on the subgrade support conditions within a radius of 15 to 30 cm of the test point depending on slab thickness. A case history slab investigation is described below.

Case History

Subgrade support conditions were evaluated with the IR method at proposed column locations on a slab for an electronic clean room manufacturing area. The investigation was conducted in response to concern that previous flooding from a sprinkler line break nearby may have created voids below the slab. The slab is approximately 15 cm thick. The subgrade consists of 8 to 10 cm of bedding sand with a plastic vapor barrier. The bedding sand is underlain by 0.3 to 2.5 m of structural fill, underlain by residual soils and bedrock. Of the 295 locations tested, IR test results indicated void to weak support conditions at 91 points and good support at the remaining 204 points. Subsequent grouting operations confirmed the results. Representative IR test records are presented and described below for void versus good subgrade support conditions, and void before and after grouting.

Void versus Good Subgrade Support. Mobility plots are presented on Fig. 10 for void versus good subgrade support conditions. Two characteristics of a mobility plot are indicative of support conditions in particular: the inverse of the slope of the straight line portion of a plot, or dynamic stiffness, and the mobility at higher frequencies. The mobility plot of a test point with good support indicates a dynamic stiffness of 0.16 MN/mm, a flat response from about 200 to 600 Hz, and an increasing response thereafter. The mobility plot of a test point over void indicates a dynamic stiffness of 0.03 MN/mm and an irregular response thereafter. The slab underlain by void is more flexible and mobile than the slab with good subgrade support.

Void Before and After Grouting. Slab support conditions can be checked after repair with the IR method. For example, mobility plots are presented on Fig. 11 for a test point over void before and after grouting. The mobility plot over void before grouting indicates a dynamic stiffness of 0.03 MN/mm and an irregular response thereafter. The mobility plot after grouting indicates a dynamic stiffness of 0.09 MN/mm and a somewhat less

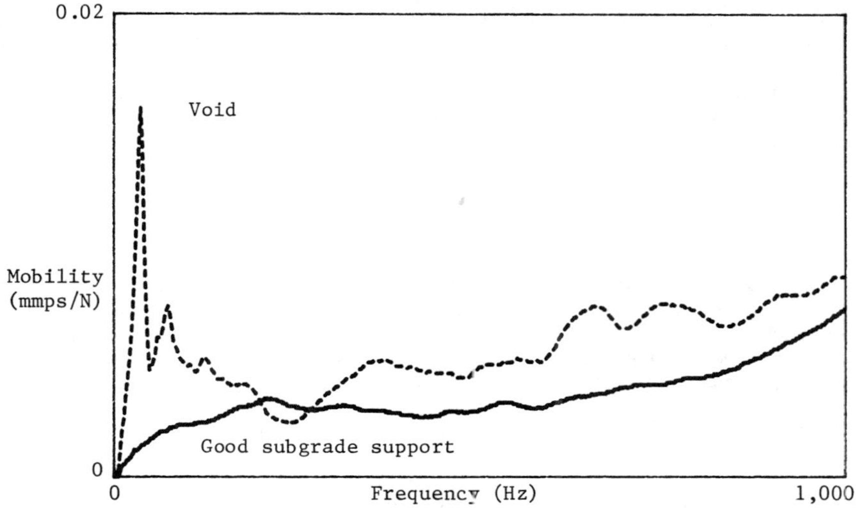

FIGURE 10. Mobility of slabs with void and good subgrade support.

irregular response thereafter. Results indicate the grout substantially improved support conditions at this test point.

Discussion of IR Slab Testing

The types of slabs one can test with IR method include concrete slabs, pavements, runways, spillways, pond and pool bottoms, and tunnel linings. The tests can be performed on concrete or any other rigid, concrete-like

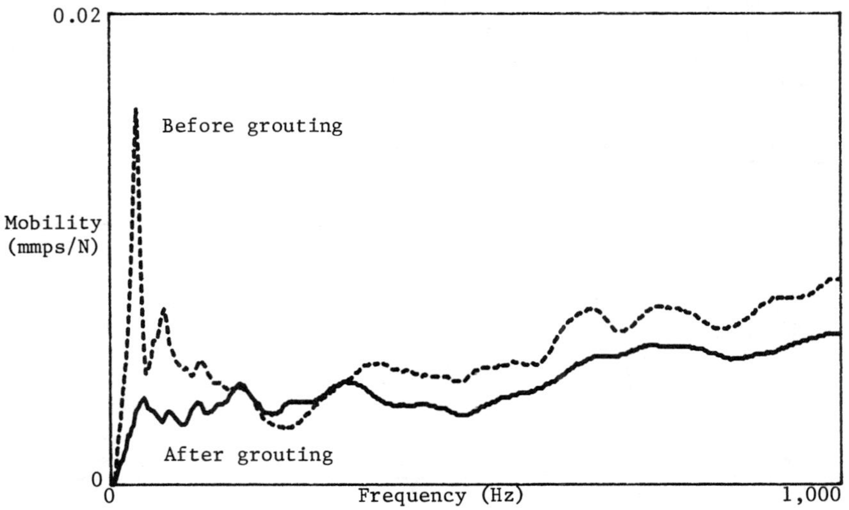

FIGURE 11. Mobility plots for slab before and after grouting of void

material. In forensic investigations, areas of void below slabs can generally be identified and mapped out at the time of testing, although the thickness of a void cannot be determined by an IR test. As quality assurance testing after construction or repair operations, the method can identify areas of potential problems. Checking the effectiveness of grout behind a precast tunnel liner is an ideal application, for example. As part of a planned maintenance program to improve long term slab performance, the method has the potential to become an integral tool.

STRUCTURAL SYSTEMS

The Impact Echo (IE) method is a sonic test for evaluation of concrete member integrity and thickness. The nondestructive test is conducted from one surface in a fashion similar to IR tests of slabs. The IE method has been researched by Drs. Nicholas Carino and Mary Sansalone at the National Bureau of Standards since the early 1980's (Carino and Sansalone, 1986).

The IE test involves impacting the concrete at a point. A surface receiver measures the vibration of the concrete. Analysis of the frequency contents of the test results provides information on the condition of the concrete as described in the following case history investigation.

Parking Garage Deck

The investigation was conducted with the IE method in response to concern regarding spalling and delamination of concrete due to the corrosion of reinforcing steel in the decks. The deck slab is nominally 20 cm thick and is reinforced with #5 steel bars. The thickness of the asphalt overlay ranges from 0 to 3.8 cm and averages 2.5 cm.

Dragging chains and listening for hollow sounds to locate voids was not feasible for the parking decks because of the asphalt overlay. Instead, IE tests were conducted from the undersides of the decks with a small impulse hammer and accelerometer. Test results indicate a variety of conditions including sound concrete/asphalt deck, top delamination, debonded concrete/asphalt, and bottom delamination areas as illustrated on Fig. 12. Test records of the first two conditions are presented and described below.

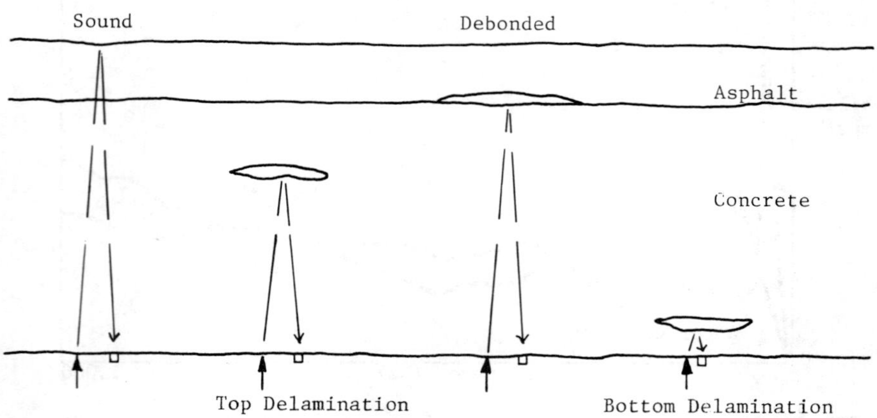

FIGURE 12. Deck schematic of Impact Echo test results.

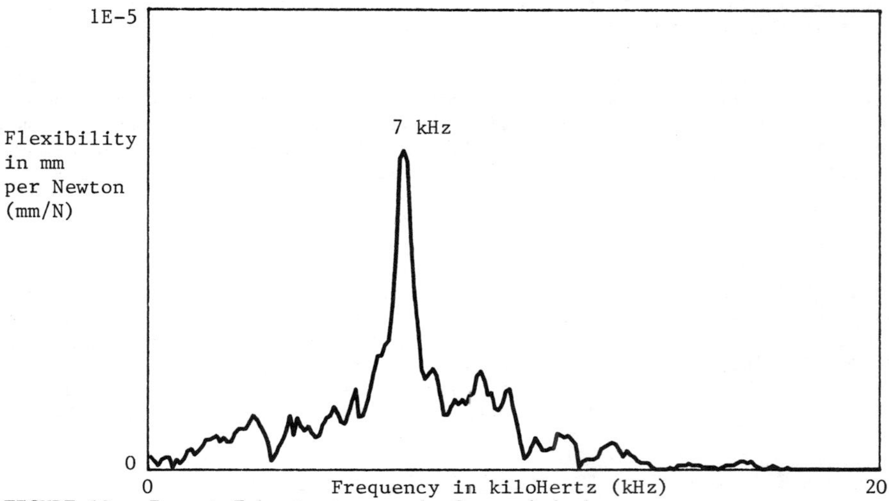

FIGURE 13. Impact Echo test record of sound deck.

An IE test record is presented on Fig. 13 for sound concrete/asphalt. The record is a plot of vibration displacenent (integrated from the accelerometer output) per force (mm/N) or flexibility as a function of frequency. The dominant frequency peak at 7 kHz corresponds to the periodic compression wave reflections between the top and bottom surfaces of the deck. The 3,400 mps velocity of the compression wave through the 24 cm thick deck can be calculated according to the equation

$$Vc = 2 * T * f \tag{5}$$

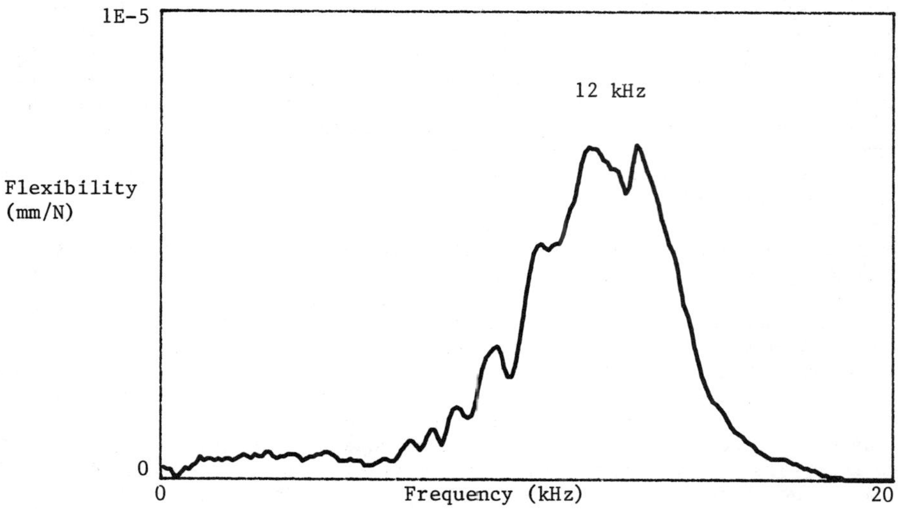

FIGURE 14. Impact Echo test record of deck with top delamination.

where Vc is the compression wave velocity, T is the thickness of the deck, and f is the frequency peak.

An IE test record is presented on Fig. 14 for a top delamination. The 12 kHz dominant frequency peak corresponds to reflection off a delamination 15 cm above the test point. The higher frequency peak indicates the shallower reflecting depth according to Equation (5) if a similar velocity is assumed. Excellent confirmation of the IE test results was provided by 15 cores.

Discussion of IE Testing

Impact Echo (IE) tests can be conducted from one surface on structures such as slabs, walls, bridge decks, dams, and tunnel linings as well as parking garage decks. Flaws such as void, honeycomb, cracking, and delamination are detected. Thicknesses can be estimated. Fire and frost damage can be evaluated. The method can also be used for QA purposes and prediction of early age concrete strength if correlated with destructive test results.

SUBSURFACE SEISMIC VELOCITY PROFILING FROM THE SURFACE

A new seismic technique known as Spectral Analysis of Surface Waves (SASW) has been developed for the determination of shear modulus profiles with depth at soil sites and Young's modulus profiles at pavement sites (Nazarian and Stokoe, 1985 and 1986). The SASW method and results of current research on its potential for Arctic permafrost regions are discussed below.

SASW Method

The method is based upon the theory of stress waves propagating in layered elastic media. The ratio of surface wave velocity to shear wave velocity varies slightly with Poisson's ratio, but can be assumed to be equal to 0.90 with an error of less than five percent (Stokoe and Nazarian, 1985). Measurement of the surface wave velocity of the SASW method similarly allows calculation of compression wave velocity. Knowledge of the seismic wave velocities and mass density of the material layers allows calculation of shear and Young's moduli for low strain amplitudes. Surface wave (Rayleigh - R wave) velocity varies with frequency in a layered system with velocity contrasts, and this frequency dependence of velocity is termed dispersion. A plot of surface wave velocity versus wavelength is called a dispersion curve. The SASW tests and analyses are performed in three phases: (1) collection of data in situ; (2) construction of a dispersion curve; and, (3) inversion of the dispersion curve.

The SASW field test consists of impacting the ground surface to generate surface wave energy at various frequencies that is transmitted through the medium. Two vertical receivers located on the surface monitor the passage of the surface wave energy as illustrated on Fig. 15. To obtain more representative data, several tests with different receiver spacings are performed by doubling the distance between the receivers about the imaginary centerline between the receivers. A signal analyzer digitizes the analog receiver outputs and records the signals for spectral (frequency) analyses to determine the phase information of the cross power spectrum between the two receivers for each frequency. The dispersion curve is developed by knowing the phase ($\emptyset$) at a given frequency (f) and then calculating the travel time (t) between receivers of that frequency/wavelength by:

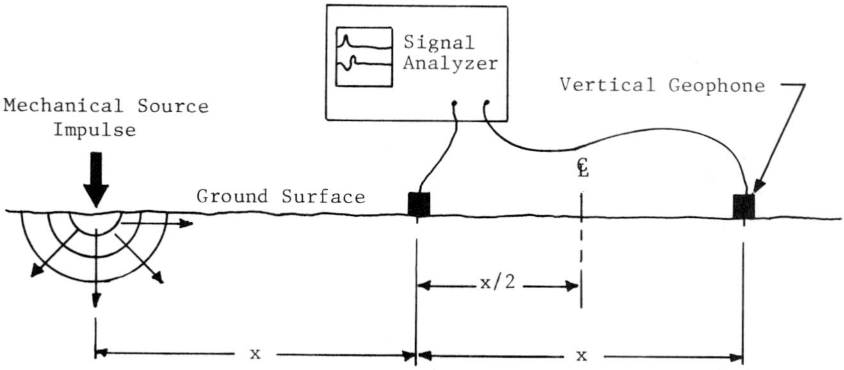

FIGURE 15. Spectral Analysis of Surface Waves (SASW) survey method.

$$t = \emptyset / 360f \tag{6}$$

Surface wave velocity (Vr) is obtained by dividing the receiver spacing (X) by the travel time at each frequency (Stokoe and Nazarian, 1985):

$$Vr = X / t \tag{7}$$

The wavelength (Lr) is related to the phase velocity and frequency by:

$$Lr = Vr / f \tag{8}$$

By repeating the above procedure for every frequency, the surface wave velocity corresponding to each wavelength is evaluated, and the dispersion curve is determined.

Inversion is the process of determining the "true" shear wave velocity profile from the "apparent" velocity of the dispersion curve. The inversion process is iterative and involves assuming a shear wave velocity profile and constructing a theoretical dispersion curve. The experimental and theoretical curves are compared, and the assumed shear wave velocity profile adjusted until the two curves match. An interactive computer algorithm, INVERT, has been developed to compute a theoretical dispersion curve based upon an assumed shear wave velocity profile (Nazarian, 1984).

Permafrost Research with the SASW Method

Facilities affected by freeze-thaw damage in permafrost regions include structures and their foundations, pipelines, highways, runways and utility distribution poles. A nondestructive method is needed to determine the extent and thicknesses of frozen and thawed ground to supplement soil test borings. Results of a feasibility research study of the potential for use of the SASW method to delineate thawed (softer) and frozen (stiffer) ground are discussed below.

In addition to SASW tests at three sites in Fairbanks, Alaska, a soil boring was drilled at each of the sites and downhole seismic tests were performed

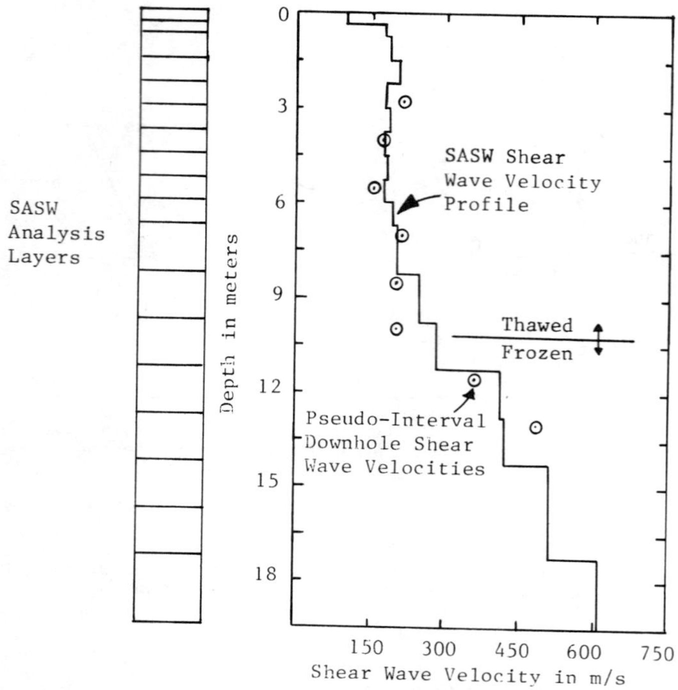

FIGURE 16. Shear Wave Velocity Profiles at the Univ. of Alaska hayfield.

in the borings at two of the sites for comparison with the SASW results. Results from a sandy silt and silty sand site in a hay field at the University of Alaska campus are presented below.

Shear wave velocity profiles determined with the SASW and downhole seismic tests at the hay field site are presented on Fig. 16. The shear wave velocity profiles determined by the two methods are in good agreement. A significant increase in shear wave velocity occurs between the depths of 10 to 12 meters for both methods. A substantial increase in the penetration resistance of the driven soil sampler was also noted over this depth interval. Examination of the soil samples indicated that the soil was frozen below a depth of 10.5 m. The research findings indicate the SASW method accurately identified the increase in velocity (stiffness) associated with the frozen ground without drilling a boring. The method is believed to have excellent potential in permafrost and more research is proposed.

CLOSURE

Nondestructive methods have several advantages versus destructive methods.

1. Economical test coverage of large areas.
2. Definition of limits and nature of defects;
3. Portable equipment and ability to test in confined areas;
4. Data on intact structures rather than laboratory specimens;
5. Little or no damage due to testing;

The principal advantages of destructive versus nondestructive testing are:

1. Direct measurement of material properties like strength and density;
2. Accepted standards for quality control of construction;
3. Tests are familiar to engineers, architects, and contractors.

Test results from the seismic, sonic, and vibration methods described in this paper have been correlated with destructive test results many times to best meet engineering investigation goals in the authors experience.

The authors would like to thank Dr. Kenneth H. Stokoe, II and Dr. Soheil Nazarian of the University of Texas for their active participation as technical consultants for the SASW research in permafrost. The Phase I Study was funded by the Small Business Innovation Research program of the National Science Foundation whose support is gratefully acknowledged.

REFERENCES

Sansalone, M., and Carino, N.J., "Impact-Echo: A Method for Flaw Detection in Concrete Using Transient Stress Waves," National Bureau of Standards Report NBSIR 86-3452, Gaithersburg, Maryland, September, 1986.

Harrel, S. and Stokoe, K.H., II, "Integrity Evaluation of Drilled Piers by Stress Waves," Research Report 257-1F, Center for Transportation Research, The University of Texas at Austin, January, 1984.

Nazarian, S., "In Situ Determination of Elastic Moduli of Soil Deposits and Pavement Systems by Spectral-Analysis-of-Surface-Waves Method," Doctoral Dissertation, The University of Texas at Austin, 1984.

Nazarian, S. and Stokoe, K.H., II, "In Situ Determination of Elastic Moduli of Pavement Systems by SASW Method (Practical Aspects)," Report 368-1F, Center For Transportation Research, The University of Texas at Austin, 1985.

Nazarian, S. and Stokoe, K.H., II, "In Situ Determination of Elastic Moduli of Pavement Systems by SASW Method (Theoretical Aspects)," Report 437-2, Center For Transportation Research, The University of Texas at Austin, 1986.

Olson, L.D., and Thompson, R.W., "Case Histories Evaluation of Drilled Pier Integrity by the Stress Wave Propagation Method," Session Proceedings Drilled Piers and Caissons II, American Society of Civil Engineers (ASCE) National Spring Convention, Denver, Colorado, May, 1985.

Pederson, C.M. and Senkowski, L.J., "Slab Stabilization of PCC Pavements," a paper presented at the Transportation Research Board Meeting in Washington D.C. by the Research and Development Division of the Oklahoma Department of Transportation, January, 1986.

Stain, R.T., "Integrity Testing," Civil Engineering, April, 1982, pp. 53-73.

Stokoe, K.H., II, and Nazarian, S., "Use of Rayleigh Waves in Liquefaction Studies," Measurement and Use of Shear Wave Velocity for Evaluating Dynamic Soil Properties, ASCE Publication National Convention, Denver CO, May, 1985.

Stokoe, K.H., II, Olson, L.D., and Reese, L.C., "Integrity Testing of Drilled Shafts," Geotechnical Engineering Faculty Workshop Proceedings, International Association of Drilled Shaft Contractors, Fort Collins, Colorado, July, 1987.

FIELD PROVEN NONDESTRUCTIVE TESTING METHODS FOR EVALUATING DAMAGE AND REPAIR OF CONCRETE

R. A. Muenow
Muenow and Associates, Inc.
Charlotte, NC 28226

INTRODUCTION

Nondestructive testing (NDT), or testing that does not require dismantling or destructive sampling of an object, has been performed for centuries. Early Italian bellmakers used NDT methods when they thumped a newly completed bell with a mallet to see if it "rang pure."

Concrete testing has employed the same type of auditory method - rapping a concrete member with a hammer or archaeological pick to test its tone. At best, an experienced investigator can determine only if the damage is cosmetic or structural by listening to the ring. Location, type, or degree of damage cannot be ascertained.

NDT remained in this primitive form until the technological advances of the 20th century. In the early 1930s Harvey Firestone developed the reflectoscope, which has been used widely and with confidence, for the non-destructive testing of steel.

The reflectoscope works on the principal that stress waves propagate in steel until there is a density change, and then a certain percentage reflects. At a steel/air interface, 100 percent of the stress waves reflect. Equipped with a cathode ray tube, the reflectoscope is able to graphically display this reflection. An experienced operator can determine with extreme accuracy where a discontinuity in steel is located.

With the advent of computerized signal enhancement, sonic methods of NDT for in-place concrete - plastic or hardened - have improved markedly with respect to definition and location of discontinuities.

But these methods are still not as widely utilized as destructive testing - destructive because it requires the taking of multitudinous core samples. These samples are then taken back to a laboratory and tested to determine the nature, and if possible, the extent of the problem.

In some cases the problem can be readily identified with destructive testing. In other cases, the answer is not so apparent. The process is time-consuming, and to be certain of the problem's extent requires hundreds of cores.

This paper examines the various methods of field proven NDT, particularly impulse radar and pulse echo, for the purpose of structural evaluation. Case histories that shows field applications for these fast and highly accurate, testing techniques are discussed.

Resonant Frequency

This method is currently used to test relatively low cost materials such as granite, stone, or pre-cast concrete panels. A sensor is mounted on the panel's surface, and the panel is then rapped with a hammer. If it rings pure, it is acceptable. If it thuds, it is usually discarded because it is cheaper to buy another than it is to fix the panel. The method depends upon the consistency and regularity of a member concerning size, shape and weight. In general, pure modes of vibration are measured during longitudinal, flexural and to small degree torsional. This is not a diagnostic method, but rather a comparator.

Magnetic Field

Magnetic testing devices, which have a probe to create a magnetic field, are used to detect and measure the depth of reinforcing steel in concrete.

By bringing the magnetic field into proximity with ferrous materials, one measures disturbances in the field caused by the presence of the ferrous material. This is a precise method, usually $\pm$ 1/4 inch, for locating and measuring depth of cover of reinforcing bars in concrete. In order to replicate the survey with a high degree of accuracy, repeated calibration of the instrumentation is necessary with known standards. The limiting factor of use states that the concrete cover (depth to bar) must be less than the horizontal spacing of reinforcement bars.

Pulse Velocity

The pulse velocity method is a through transmission double transducer testing technique utilizing low frequency (20khz) stress waves to evaluate insitu concrete. Repeated pulses of stress waves are introduced to one side of a member. These pulses pass through that member and are received by a second transducer on the opposite face of the member. Propagation times, amplitude, frequency and rise time characteristics for the wave propagation are recorded for each test location. From the recorded data comparative compressive strengths are calculated, crack presence and orientation are estimated. A measure of pulse velocity is correlated to concrete quality and uniformity.

There are limitations to structural members for evaluation depending upon the internal condition of the concrete. Two opposite surfaces of the member must be available, and member thickness must be known.

Pulse Echo

The final method to be discussed is pulse echo, by far the most versatile testing technique for the inspection of insitu concrete.

While the pulse velocity method requires access to both sides of a member, the enormous advantage of pulse echo is that testing can be done with access to only one surface of a member.

A spring activated device transmits a low frequency mechanical stress wave into the concrete. Wave energy reflections occur at discontinuities within the member and at the front and back surfaces of the member. All reflections are received by a piezoelectric transducer placed on the concrete surface; electrical signals are presented to a processor and a computer enhanced display of the data is shown on a storage CRT.

Data recorded include time between reflections, frequency of reflections, attenuation, and decay time for the wave energy. Analysis of the signals allows an experienced operator to determine transit times, compressive strength, voiding, honeycombing, cracking, loss of bond, modulus of elasticity, Poisson's ratio, deterioration of cement matrix, and generally the location, orientation, and extent of most internal discontinuities and cracks.

Impulse Radar

Radar has been used for years for geologic mapping and shallow subsurface foundation mapping. The method utilizes electromagnetic wave energy or radio waves to detect and locate metallic embedments by reflected energy in non-metallic construction materials, other applications are useful, but do not produce as positive data as R/F identification.

The principle of operation is based on measuring reflected energy at changes in the dielectric constants, within a material. It is particularly useful in detecting and locating reinforcing steel bars and other concrete embedments because there is a substantial change in dialectric characteristics between steel and concrete.

Radar is also useful in determining the location of both water and air-filled voids beneath slabs, pavements and runways. The effectiveness of the radar method to detect and map voids is made possible and greatly enhanced when test data can be correlated to a known condition.

CASE HISTORY

Use of Radar for Parking Deck and Office Project

A medium to large structure in the Florida panhandle, including a parking deck and office condominimum project, was built on a rather speculative basis. There was a minimum of construction supervision and inspection.

Several electrical and plumbing problems were detected when the building was about to be occupied. During the repair, cracks were discovered in the concrete frame. To determine the extent of the structural problems, the architect requested NDT so that a scientific, rather than an emotional analysis, could be made. This project typifies what can be done with impulse radar using an IGH antenna.

First, all perimeter foundation footings and two rows of interior footings were inspected. The spread footings had been designed to be squares of four to six feet, 14 - 18 inches thick, with light reinforcements.

Next, the deck slab thickness was measured. Core samples showed an average thickness of five inches, with variations from three and one-half to seven inches.

Also, there were indications that the lentils above the doors as well as the door framing were cracking because of an absence of reinforcing steel. In addition, the masonry block walls were cracked because the cells in the concrete blocks had apparently not been filled with concrete, nor had vertical steel been placed as specified.

Finally, it was felt that the reinforcing durawall, a layer of steel mesh at each third course of block, had been omitted.

The size of the footings was investigated first. The holes for the spread footings had been dug in sand, and the holes had apparently been left open too long before placing the concrete. The sides of the excavation had sloughed toward the center. Instead of a flat footing, the shape approached that of an arrowhead.

Radar showed the top surfaces of the footings were as large or larger than designed, but indicated that the load bearing surfaces were considerably smaller than the top surfaces. To verify this, concrete and asphalt cover was removed, and visually examined to expose the drastic angling of the footing sides.

Radar was also used to measure thickness of the footings. They varied from less than 10 inches thick to a maximum of 20 inches thick. In some cases the design was not met. In others it was exceeded, indicating a lack of uniformity.

The footing reinforcement pattern was also examined with radar. There was no real pattern. In two-thirds of the footings only 50 percent of the designed amount of steel was present, in others the designed amount was in place.

Next, radar was used to measure slab thickness. The results indicated that the thickness along the edge was the designed five inches, but the interior areas were much thinner. This indicated uneven grading prior to pouring the slab.

About two-thirds of the slab met the nominal five-inch thickness requirements. Of the remaining one-third, 80 percent was less than four inches thick, while 20 percent exceeded five inches. The radar allowed us to map large problem areas that could then be removed piecemeal, regraded, and repoured without having to replace the entire slab.

The radar also detected the 4x4 reinforcing mesh. It was located mostly in the bottom of the slab or even pressed into the sand underneath the slab.

Again employing radar, it was discovered that a high percentage of the lentils had less than the specified amount of steel, while another 30 percent had been installed upside down with the stell on top of the lentil.

Radar was used to test the walls for cavity filling, horizontal reinforcement, and for placement of the vertical reinforcing bars inside the cavities.

The specifications called for the first two levels of the deck wall cavities to be completely filled with concrete. Radar indicated that less than two-thirds of the cavities were completely filled, and that most of the remaining cavities were less than one-half filled. Horizontal reinforcing every third course of concrete block was specified in the architectural plan. Radar proved that the placement of reinforcing was completely non-uniform. In no place did we find reinforcing as specified.

Vertical reinforcing was specified for every fifth vertical cavity from the floor to ceiling. Again, radar showed the placement of steel to be non-uniform.
These findings were verified on a statistical basis using visual observations and sampling because of possible litigation. NDT was used to show how widespread the problems were and to determine the exact location of problems. This allowed building construction supervisors to make a rational, non-emotional decision about how to repair the problems.

Pulse Echo Applications

<u>Measure in-place compressive strength</u>. Pulse echo can be used to find the location of the lowest, the highest, and average velocity in a concrete structure. Core samples are taken at each location; they are broken in compression, and from this information a velocity versus strength relationship is established. Once the velocity and the compression curve are known, you can determine compressive strength to an accuracy of plus or minus four percent, a majority of the time, without having to take hundreds of samples.

The vast amount of pusle echo testing done over the past 20 years has been correlated into velocity-strength relationships for concretes made with hundreds of different mix designs and aggregate.

<u>Measure thickness of members</u>. Let's take a slab for a shopping center that is a quarter million square feet. If the thickness can be reduced 10 percent, a significant amount of money can be saved. But the concrete thickness may not be adequate from a load-bearing standpoint and premature failures will occur.

Pulse echo testing is so fast - each test taking roughly 15 seconds once the equipment is in place - that you can cover the entire slab quickly.

For example, on a flat slab in the time it takes to perform five core samples, you can take 500 pulse echo readings. The savings multiplies quickly when walls, columns or beams need to be cored.

In a statistical sense there is no basis for taking five core samples in a quarter million square feet - you would need many more than that to determine average slab thickness. But 500 NDT readings would be statistically significant.

As a result of pulse echo testing in numerous shopping centers, owners have had revised pay outs and have requested extended guarantees because of insufficient slab thicknesses.

Locating voids under slabs, roadways, and runways. Voiding is a significant problem that often leads to the premature failure of slabs, roadways, and runways. Detection with radar or vibratory methods is limited by the thickness of the slab and the extent of a void.

Pulse echo has been proven by core sampling to delineate voiding beneath slabs whose thickness ranges from fractions of an inch to several feet. Void depth can be as small as one-fourth inch and up to 1 or 2 inches. Pulse echo records voids in excess of 2 feet as infinite in depth, when the void is filled with a material such as water, mud, slurry or loosely compacted subbase stone and sand.

This type of testing does not take the road, slab or runway out of service for any extended period of time, such as would occur with coring.

Detection of honeycombing. This is an indicator of a severe lack of consolidation in concrete that can be readily identified with pulse echo. The typical response on the CRT is a saw-tooth configuration denoting a multi-face internal discontinuity. Data from this response allows for both aerial extent and thickness estimates to be made of the internal discontinuity.

Cracking. A common problem in concrete, most cracking is of minor importance from a structural standpoint. There are times, however, when knowledge of the depth and orientation of a crack is essential in analyzing the structural response of columns, beams, slabs and walls.

Concrete does not crack on an absolutely flat plane, but pulse echo will detect and locate cracks on virtually any plane, due to the extremely rough surfaces and irregularity of crack planes which serve as internal reflectors.

SUMMARY

There are many commonly-used destructive methods for determining damage to concrete, and while these methods are useful, they generally yield limited information. Use of this information is severely limited in assessing the structural impact of the damage because it does not define the type, specific locations, extent, and degree of damage.

In contrast, NDT techniques are readily available and have been field proven and accepted by the Nuclear Regulatory Commission, the Navy, and the Corps of Engineers.

These methods rapidly detect the type, extent, and specific location of damage - precisely the information a qualified engineer needs to assess repair strategies for the damaged concrete.

The information pulse echo can give is so precise that the experienced operator can specify the amount of material needed to fill a void to within five or six percent. The pulse echo technique also serves the indispensable function of evaluating the performed repairs - bonding of new concrete to parent material and to reinforcing steel for example.

These NDT methods provide an accurate, cost-saving, comprehensive, and rapid system for performing a damage survey.

EXAMINATION OF CONCRETE STRUCTURES USING COMPTON SCATTERING

T. Whynot and E. M. A. Hussein
University of New Brunswick
Fredericton, N. B., Canada E3B 5A3

INTRODUCTION

A nondestructive evaluation (NDE) application which entails great demand is the inspection of concrete structures for defective reinforcing components and/or void cavities. Some of the existing NDE methods, such as ultrasonics, eddy current and acoustic emission may be employed for this purpose [1]. Ultrasonic testing is, perhaps, one of the best methods for determining the uniformity of in situ concrete, however, its flaw detection capabilities are limited [1,2]. Eddy current tests are primarily used for detecting the existence of steel rebar located near the surface [1,3]. Acoustic emission, on the other hand, is a passive monitoring technique which, rather than requiring complete volumetric scanning (as in ultrasonics and eddy current methods), employs remote sensors for inspection. Signal processing, however, usually requires complex analysis to retrieve useful information [4].

Conventional nuclear radiation methods, such as radiography, have not been widely exploited for use in the inspection of concrete structures. This is despite the penetration ability of radiation in matter and its sensitivity to changes in the material of a structure. Perhaps the main reason for the lack of radiation applications to this area is the large thicknesses commonly encountered. Customary NDE radiation techniques rely on the measurement of the attenuation of transmitted radiation. Large thicknesses require, therefore, the utilization of high energy and high intensity radiation sources which are difficult to manage in field applications.

This paper focuses on the development of a NDE method which utilizes backscattered gamma radiation. This allows application of the technique to extended structures where access to only one side is possible. The concept is based on the Compton interaction between the incident photons and the electrons of matter. The scattered signal, therefore, provides an indication of the electron density of the material comprising the inspected volume. Hence, the method enables the detection of local defects and the discrimination between materials of different density and composition, such as concrete, void and steel. Compton scattering has been successfully employed for many industrial and medical applications including the inspection of aircraft turbine blades for casting defects, local void fraction measurement in complex two-phase flow geometries, and lung and bone density analysis [5,6].

A feasibility study for utilizing Compton interaction to detect iron rebar and air voids contained in a concrete medium has been previously reported by Gautam et al., [7]. This study was conducted using the 448 keV energy of a large (25 Ci) Iridium-192 gamma source. The use of such a strong source imposes the employment of bulky and heavy radiation shielding which, in turn, limits the mobility and flexibility of the inspection device. In the present work, use is made of smaller sources, in the mCi range, to examine concrete structures. A 300 mCi Cesium-137 source (662 keV in energy) and a 200 mCi Cobalt-60 source (1250 keV) were employed in this study. The amount of shielding required for these sources is not too large to inhibit the design of a portable device.

The paper describes the underlying principles of Compton scattering and the implications concerned with distinguishing steel and void from concrete. The detection capabilities of the method is demonstrated using Monte Carlo simulations and confirmed with laboratory experiments.

PHYSICAL PRINCIPLES

Compton interaction is a scattering event in which a photon collides with a "free" electron, loses a portion of its energy and is deflected from its original direction. Compton scattering is the dominant mode of interaction in most materials if gamma ray energies lie between 30 keV and 2 MeV [6]. The electron binding energies then become negligible in comparison with the incident photon energy [8].

The kinematics of a single Compton scattering event is described by the relationship:

$$E' = E / [1 + (E/mc^2) (1 - \cos(\theta)] \quad (1)$$

where E and E' represent, respectively, the incident and scattered photon energies while mc^2 is the electron's rest mass energy (511 keV).

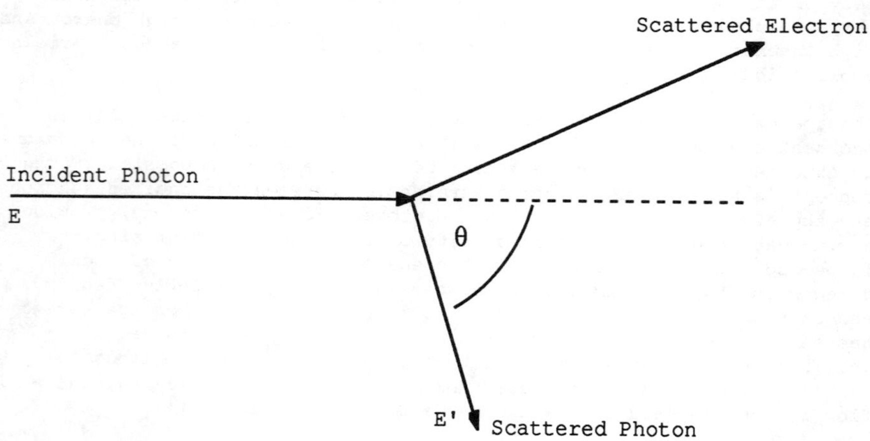

FIGURE 1. Schematic representation of Compton scattering.

The Compton angle of scattering, θ, is depicted in Figure 1. The above equation shows a unique correlation between the energy and angle of scattering. This has proved, as will be shown later, to be a useful feature to exploit in Compton scattering inspection.

The well known Klein-Nishina formula governs the angular scattering probability of photons [8]. According to this relationship, the probability of scattering, at a given angle, decreases as the energy increases. The microscopic Compton cross section represents the scattering probability of a photon interacting with the material electrons. It is obtained by integrating the Klein-Nishina formula over all possible scattering angles. These cross sections, and the density and composition of the medium, enable the calculation of attenuation coefficients, for a particular medium.

COMPTON SCATTERING INSPECTION

The Compton effect is a viable tool for inspecting material since it is an interaction which is strongly dependent on the electron density of the scattering medium. One can visualize an inspecting system, as depicted in Figure 2, that consists of a collimated source together with a detector, equipped with an adequate collimator. The inspection or "sensitive" volume is defined by the intersection of the source and detector view angles. The intensity of the scattered radiation is then expected to strongly depend on the electron density of the material comprising the sensitive volume. By positioning the source and detector on one side of the inspected object, to detect backscattered radiation, scanning of extended structures becomes possible.

The Compton scattering technique relies on changes in electron density for flaw detection. The flaws, however, are usually small, in comparison with the surrounding medium. It is, therefore, essential to focus the incident and scattered beam such that corresponding changes in electron density significantly alter the backscattered signal. This can be accomplished by utilizing metallic or "hard" collimators to define

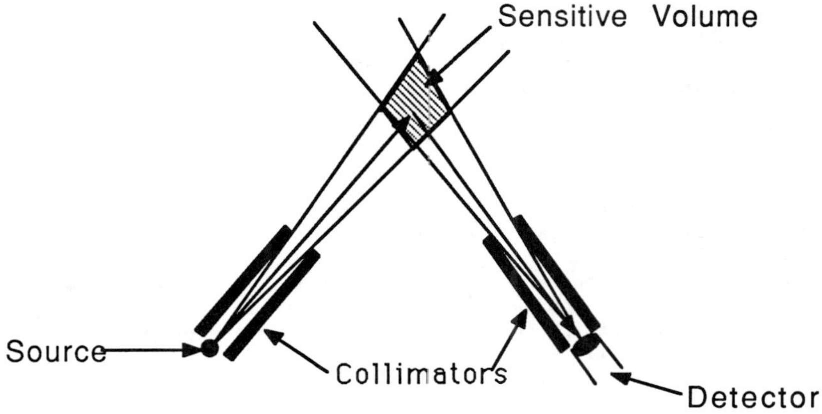

FIGURE 2. Compton scattering inspection.

the view angles of both the source and detector. The size of the sensitive volume can be then controlled by varying the aperture dimensions of the collimators.

The use of hard collimation does not, however, prevent the detection of photons originating outside the sensitive volume as a result of multiple scattering. Detection of only singly scattered photons, originating primarily within the sensitive volume, can eliminate this effect. This can be made possible by utilizing the unique energy-angle relationship defined by Eq. (1). This requires the definition of an energy window, centered around the energy corresponding to the Compton scattering angle. Photons experiencing multiple collisions possess energy, primarily, lower than that defined by the energy window and, therefore, would not significantly interfere with the detected signal. This process of energy discrimination provides an indirect form of detector collimation, since only a particular direction of scattering is considered, and is termed here as "soft" collimation. A combination of soft and hard collimation should enable more adequate definition of the sensitive volume.

Preliminary experiments were conducted to confirm the applicability of the above concept for detecting the presence of steel rebar or air cavities in concrete structures, using sources of strengths in the millicurie range. These experiments are not reported here since no satisfactory results were obtained. Realizing, however, that many complicating factors are associated with the proposed concept, it was decided that a Monte Carlo simulation of the technique should be developed. Such a simulation should provide insight into the physical processes encountered and, in turn, enable better understanding of the phenomena. Monte Carlo simulations also eliminate many of the hardware problems encountered in laboratory experiments. A study of the individual effects can then be performed and methods can, consequently, be devised to eliminate, or compensate for, any undesirable effects.

MONTE CARLO SIMULATIONS

A Monte Carlo computer program was developed to simulate the Compton interaction of photons with matter. In this program, photons were followed, one at a time, by using random sampling techniques, governed by the probability laws that determine the particle's behavior [9]. For simplicity, each particle began with the same energy and the photon's initial direction lied randomly within a cone of user-specified dimensions.

A concrete brick was modelled with cylindrical iron or void inclusions, as shown in Figure 3. The particle trajectory through the medium was terminated upon escape from the system or when the photon energy fell below a preset value. The minimum allowable energy for these studies was 50 keV. A point detector was positioned at a specified distance from the face of the brick, on the same side as the source. Rectangular detector collimation was implemented with user specified dimensions.

Upon each collision within the brick, the program ascertained whether the collision point fell within the view angle of the detector. If so, the contribution to the detector response was calculated using the next

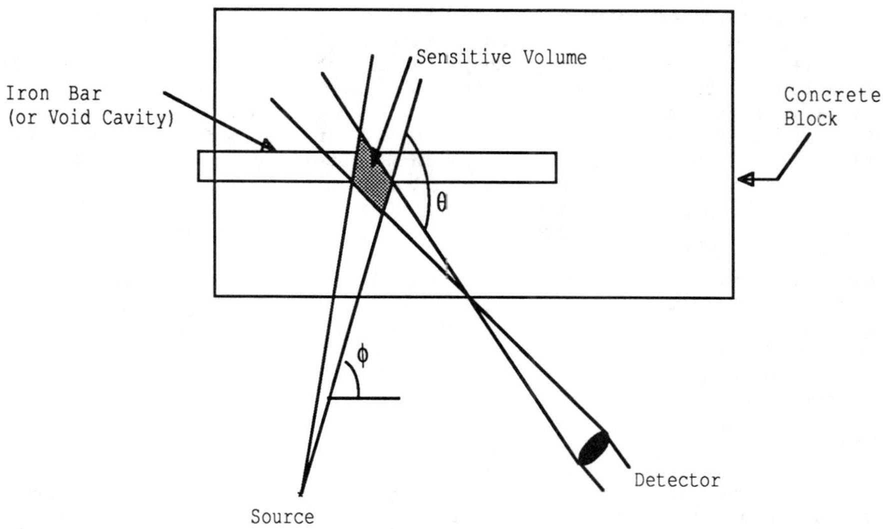

FIGURE 3. Geometry used in the Monte Carlo simulations, θ is the Compton angle and ϕ is the inclination angle.

event estimator [10].

$$S = W \exp(-ARG) P / R^2 \qquad (2)$$

where ARG and R are, respectively, the number of mean free paths and distance from the collision point to the detector. P is the probability associated with the photon scattering at an angle such that it collides at the detector site, as determined by the Klein-Nishina formula. The survival weight W, which is updated after each collision, is simply the probability that the particle has survived up to this point of scattering.

The above estimator represents the probability of the next collision being at the detector site. Each collision, occurring within the confines of the detector collimation, contributes to the detector response. Hard collimators, for both the source and detector, were simulated in the program. When collimation was not employed, photons were detected from collisions occurring everywhere within the brick. The program provided a summary of the detector response, as a function of energy, thus enabling the implementation of soft collimation. A copy of this program can be made available upon request.

The Monte Carlo simulation allowed the determination of the detector response to either singly or multiple scattered photons. The former was obtained by allowing the particle to suffer only one collision before termination. This enabled a direct evaluation of the multiple scattering effect and, hence, its contribution to the detected signal. In addition, the computer model proved to be an excellent testing aid, as it permitted the use of various source energies, depths and flaw sizes. Different collimator arrangements were also incorporated into the simulation.

The results presented in Figure 4 show a scan of a small concrete brick for iron and void. In each scan, the sensitive volume was located at a different position within the inspected object. Each position along the scan represents one Monte Carlo run of twenty thousand particles. A 662 keV source energy, equivalent to that of Cesium-137, was employed. The source and detector collimators were 14 and 20 mm in diameter. The flaw was 110 mm in length and 21 mm in diameter and located in a brick which was 150 x 50 x 90 mm. When an idealized scan using single scattering only was performed, no significant change was observed indicating that adequate use was made of soft and hard collimators.

Similar results were predicted for void detection, as demonstrated in Figure 4. The transition from steel and void into concrete is clear and abrupt. This indicated that the attenuation of the incident and scattered radiation in the concrete block was not significant enough to overshadow the effect of the small sensitive volume, at least for the source energy used.

The Monte Carlo simulation demonstrated that the concept of the inspection technique is theoretically sound. It also indicated that, at least for the source energy and geometry utilized, multiple scattering and attenuation effects do not impose a significant influence on the ability of the method to discriminate against the presence of steel and void. With a combination of both "hard" and "soft" collimation, the model demonstrated that the boundaries of the sensitive volume can be adequately confined to the region of interest. The success of the Monte Carlo simulation then prompted experimental studies to verify the practical feasibility of the proposed inspection method.

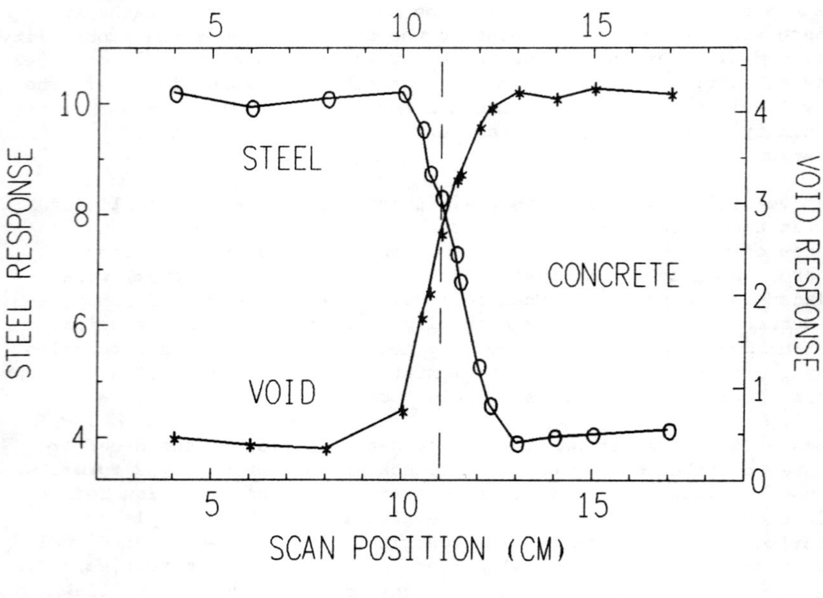

FIGURE 4. Detector response for a Monte Carlo scan of the small brick for E = 662 keV, ϕ = 42°, θ = 104° and E' = 254 ± 4 keV.

EXPERIMENTS

A 300 mCi Cesium-137 source was employed for most of the experimentation. The source was contained in a lead block approximately 60 x 47.5 x 47.5 mm in dimensions. An opening, drilled in one end of the container, allowed insertion of lead plugs, with small apertures, producing the desired collimation. The source collimator, used for most of the experimental work, was 10 mm in diameter.

Two concrete test objects of different sizes (760 x 150 x 150 mm and 190 x 50 x 90 mm) were employed and are referred to here as the large and small bricks. An iron bar was inserted in one end of the brick to simulate reinforced concrete. The bar was 126 mm long and 21 mm in diameter. An air cavity was represented by removal of the iron rod.

The detection system consisted of a NaI(Tl) scintillation detector and Tennelec Pulse Height Multichannel Analyzer (MCA). The NaI(Tl) crystal was 51 mm in length and diameter. Collimation of the measured radiation was achieved by placement of brass cylinders between the test object and detector. Two collimators were tested; each were 300 mm long and their apertures were produced by drilling 10 and 20 mm holes through their centers. The pertinent data was transferred from the MCA to floppy diskette by means of an Apple 2e computer interface.

An axial scan of the test object was obtained by traversing the sensitive volume across the brick. This procedure was greatly facilitated by lateral movement of the test object, as opposed to the source-detector system.

The feasibility of the Compton Scattering method was confirmed with an axial scan of the small brick, shown in Figure 5. The spatial resolution at this depth was as good as could be reasonably achieved through manual movement of the test object, roughly 2.5 mm. The detector collimator, for this scan, had an aperture of 20 mm in diameter. An improvement in resolution could, perhaps, be obtained with finer collimation. However, when the aperture of the detector collimator was reduced, the strength of the Cesium-137 source was insufficient to provide enough deflections within the sensitive volume to provide adequate counting statistics. A stronger source would undoubtedly enable the use of more precise collimation and, hence, an improvement in resolution. Referring to Figure 5, the respective detector responses differ in magnitude because of the counting intervals required for clear distinction. The scans required 70 and 240 second counting periods for void and steel, respectively, indicating that void is more easily detectable with this configuration.

Scanning the large brick presented, however, some difficulties. The higher associated depth decreased the count rate due to radiation attenuation. Positioning the collimators as close as possible to the brick surface, and increasing the counting period helped to improve the statistics. Also, the energy window for soft collimation was increased from 3 keV, for the small brick, to 9 and 11 keV, for steel and void in the large brick. A slightly higher energy window was used for void in order to obtain comparable statistics to that of steel.

Figure 6 shows the scans obtained for the large brick. As expected, the contrast was significantly reduced, as compared to that of the small brick. The spatial resolution obtained was about 5 mm.

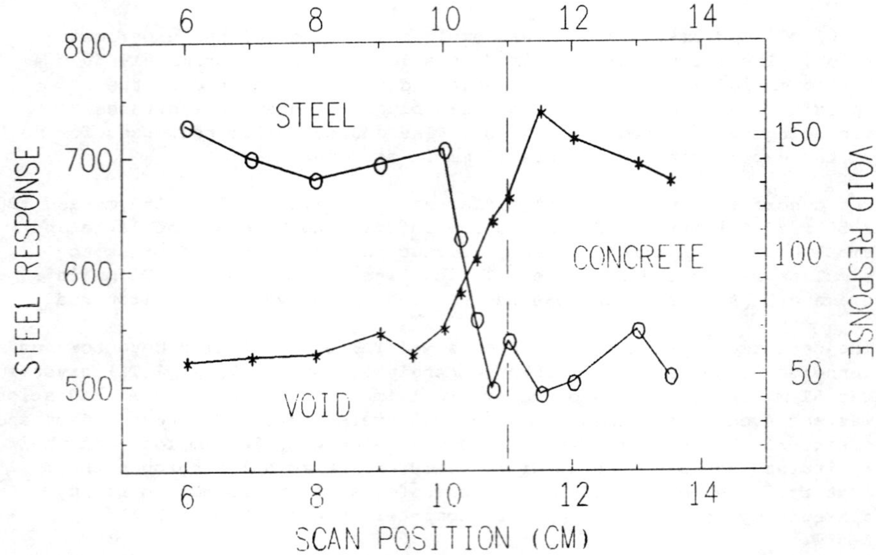

FIGURE 5. Detector response for an experimental scan of the small brick for E = 662 keV, ϕ = 42°, θ = 104° and E' = 254 ± 1 keV.

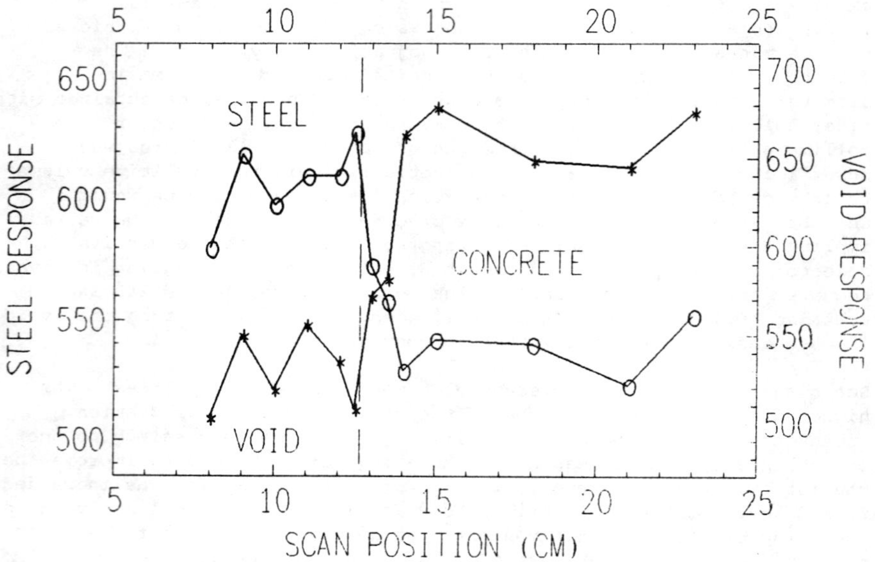

FIGURE 6. Detector response for an experimental scan of the large brick for E = 662 keV, ϕ = 42°, θ = 104° and E' = 254 (± 4 keV for steel; ± 5 keV for void).

The above experiments indicate that the Compton scattering technique is practically feasible provided that sufficient attention is given to the formation of the sensitive volume. This was achieved by utilizing both "hard" and "soft" collimation, as predicted by the Monte Carlo simulation, and by proper allignment of the source and detector. Finally, it should be emphasized that adequate shielding, between the source and detector, was necessary in order to reduce background radiation.

Given the encouraging results of both the Monte Carlo simulation and laboratory experiments, it was then desirable to study, in more detail, the different parameters which might affect the sensitivity and resolution of the method.

PARAMETRIC STUDY

The principal factors which control the sensitivity of the technique were recognized as being the sensitive volume, photon attenuation and the probability of scattering. The variables which determine the significance of these effects were ascertained to be collimation, energy, inclination angle and the Compton scattering angle.

The size of the sensitive volume is established by the dimensions of the collimators (length and aperture) and the depth of the medium which must be traversed to reach the inspection volume. The Monte Carlo simulation had indicated that a combination of both soft and hard collimation was necessary to reduce the effect of multiple scattering and for adequately defining the sensitive volume. Assuming that background radiation can be eliminated, the spatial resolution then becomes dependent on the degree of collimation attainable. As the depth of the flaw is extended, the collimator aperture must be reduced and the source strength (or counting interval) must be increased.

The effectiveness of an inspection volume is not solely dependent on its size. The geometry which defines its location and the energy of the source photons must also be considered. An example of the energy effect is presented in Figure 7, which is a scan of the small brick using a 200 mCi Cobalt-60 source, which has an energy of 1.25 MeV. The same geometry was used, as that used with the Cesium-137 source presented in Figure 4. For this particular depth, an increase in source energy has resulted in a reduced detector response. This resulted in poorer counting statistics and, hence, reduced contrast. This was attributed to the decrease in scattering probability due to the increase in source energy, as indicated by the Klein-Nishina relationship.

The geometry and energy determine the attenuation to and from the inspection volume. The Compton scattering angle, defined by the geometry, plays two important roles. An increase in angle decreases the scattering probability and reduces the energy of the deflected photon. The former effect immediately increases the statistical interference and reduces the contrast. Since the multiple scattering effect predominates at low energies, a reduction in singly scattered photon energy, due to large Compton angles, will allow an increased contribution from multiple scattering. This further reduces the sensitivity of the method. However, decreasing the Compton angle creates two additional problems. The attenuation from the inspection volume will increase and reduce the contrast. In addition, the size of the sensitive volume will increase, thus, reducing the spatial resolution. If one decides to increase the

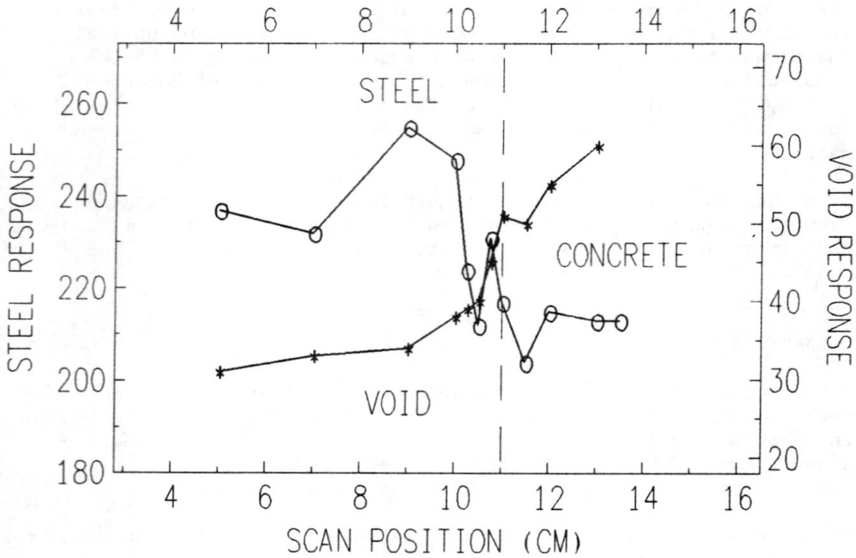

FIGURE 7. Detector response for an experimental scan of the small brick for E = 1250 keV, ϕ = 45°, θ = 103° and E' = 313 ± 1 keV.

energy to overcome the attenuation, the scattering probability, as governed by the Klein-Nishina formula, will be reduced and statistical interference will result.

Determination of an optimum combination of inclination and Compton angles is cumbersome to perform experimentally. In addition, the previous discussion suggests that an optimum energy exists for a particular depth and geometrical configuration. An attempt to achieve such an optimum arrangement is discussed in the following Section.

OPTIMIZATION

A single scattering analytical model was developed for optimization purposes. It calculated the flux of photons reaching the detector as a result of single scatters at the inspection point. This allowed an accurate specification of geometry and enabled a direct correlation between scattering probability and attenuation.

The model is based on the following relationship:

$$F = \left(1/R_i^2\right) \exp\left(-\mu_i \, t_i\right) P \left(1/R_o^2\right) \exp\left(-\mu_o \, t_o\right) \tag{3}$$

F is the probability of photons reaching the detector site per unit area. R_i and R_o are the distances from the source to the inspection point and detector to inspection point, respectively; μ_i and μ_o are the attenuation coefficients of the medium at the source and scattered energies, respectively; t_i and t_o are the corresponding thicknesses of medium to be traversed in and out of the medium; P is the Klein-Nishina probability of scattering as a function of source energy and Compton angle.

The first and last terms of the above relationship represent the divergence of the incident and scattered photons with distance. The second term represents the incident attenuation which is directly related to the inclination angle and source energy. As the source energy intensifies, the attenuation coefficient (μ_i) decreases, which results in a higher probability of photons reaching the inspection point. As the inclination angle is decreased, the incident attenuation increases because of the larger t_i. The Klein-Nishina probability diminishes with increases in energy and Compton angle. The scattered attenuation, however, is reduced with increases in energy and Compton angle.

Optimization aims to minimize attenuation and maximize scattering probability while maintaining adequate spatial resolution. This requires manipulation of source energy (E), inclination angle (ϕ) and the Compton angle of scattering (θ), such that the detector response is maximized. A reinforced concrete brick was modelled with the inspection point positioned at the center of the flaw. In the model, the reinforced concrete medium was made infinitely long such that the only variables which were changing were attenuation and scattering probability. Geometry manipulation was facilitated by changing the detector coordinates, for Compton angle variations, and source coordinates, for inclination angle variations.

At first, simultaneous optimization of all three variables was performed. The results, however, were impractical since the optimum setup would position the source and detector at the same location to minimize attenuation. The optimum energy was very low, in order to increase the scattering probability. When the source energy was increased to 662 keV and 1250 keV and the geometry optimized, the results were, again, impractical. The model predicted the optimum angles to be very small, which makes it difficult to perform measurement on the same side of the object.

The practical use of the model is, evidently, the specification of source energy and one angle while optimization is performed on the other angle. An inclination angle of 45° and a source energy of 662 keV produced optimum Compton angles of 61° and 81° for the small and large bricks, respectively. A source energy of 1.25 MeV resulted in optimum Compton angles of 59° and 80° for the small and large bricks, respectively. The increase in Compton angle with depth demonstrates the predominance of attenuation along the scattered path. An increase in source energy reduces the required scattering angles. At higher source energies the attenuation becomes less significant and the Klein-Nishina probability predominates.

When the Compton angle was held constant, at 90°, the optimum inclination angle was determined to be approximately 45°, regardless of the source energy. With the distance from the collision point to the detector being held constant, the maximum flux occurs when the distances to and from the flaw become equal. This was confirmed by testing other Compton angles, thus demonstrating the predominance of scattering probability over incident attenuation at high source energies.

The conclusion which may be drawn from the preceding analysis is the importance of the Compton angle to the sensitivity of this method. At low depths of penetration, it is probably best to use an inclination angle of 45° and a Compton angle of 90°, because the optimum low angle geometry will improve the sensitivity. However, as the depth is increased, the attenuation of the scattered radiation becomes significant and the Compton angle should be increased to maintain the method's sensitivity. This will also ensure the best possible resolution.

If a wide range of source energies are available, optimization should be performed for energy. Since the inclination angle appears to be the least singificant parameter, it was held constant at 45°. The optimum combination of energy and Compton angle was predicted to be 220 keV and 70°, for the small brick, and 200 keV and 135°, for the large brick. This indicates that as the depth is increased, attenuation predominates over the scattering probability. The influence of attenuation can be alleviated, however, by manipulating the geometry instead of the energy.

The results in Figure 8 represent a Monte Carlo scan of the small brick using the optimum setup predicted by the analytical model. Upon comparing these results with those of Figure 4, the response is clearly higher with the optimum setup. The contrast was better for steel, but not void, with the optimum configuration.

The use of a low energy source produced a peculiar trend in the respone of steel along the scan, as can be seen in Figure 8. As the steel bar entered the sensitive volume, the response intensified due to the increased electron density. When the steel continued to fill the inspection volume, however, the response dropped and then levelled out.

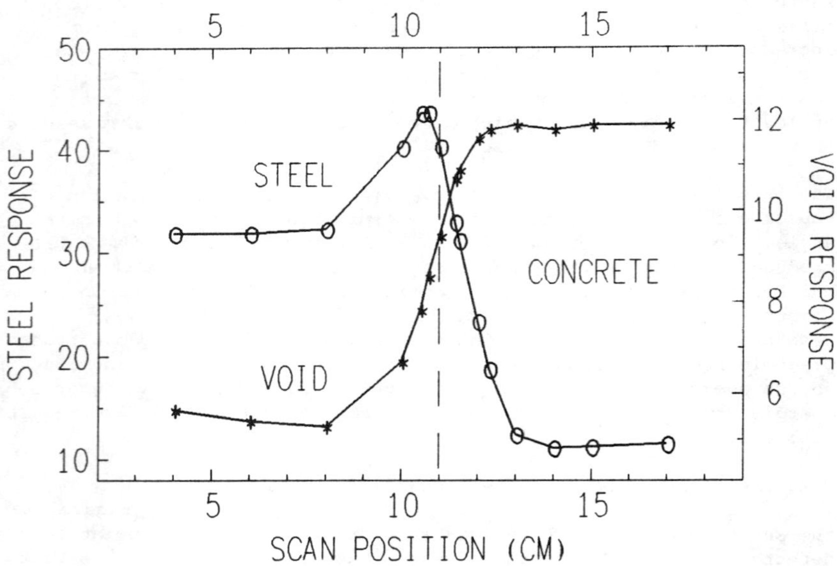

FIGURE 8: Detector response for a Monte Carlo scan of the small brick for E = 220 keV, ϕ = 45°, θ = 70° and E' = 171 + 4 keV.

This was attributed to self-attenuation within the steel at low scattering energies. The energy of the incident photon beam is sufficient to penetrate the steel bar, however, the scattering energy is too low to overcome the attenuation along the path to the detector.

In summary, the best combination of both geometry and energy does not appear to be clearly defined. The inclination angle is the easiest choice because of the low importance of incident attenuation. A 45° angle is probably suitable in most instances. The Compton angle, however, is dependent on the depth which must be achieved. As the penetration increases, the Compton angle should be increased because the scattering attenuation will predominate over the Klein-Nishina probability of interaction.

The source energy should, however, be higher than the predicted optimum for a number of reasons. Firstly, the energy must be high enough to avoid self-attenuation, within the sensitive volume. For high depths of penetration, which require large Compton angles, the energy must be sufficiently high to reduce the multiple scattering effect which predominates at low scattering energies. Finally, in general, the energy must be large enough to overcome the scattered attenuation.

CONCLUSIONS

The proposed Compton scattering inspection method was confirmed to be theoretically sound by Monte Carlo simulations. Laboratory investigations, which followed, verified the practical feasibility of the technique. In addition, a method was presented to optimize the configurations of the device such that its sensitivity to changes in density can be obtained. Further work is being carried out to determine the capabilities of the technique, in terms of the best attainable spatial resolution and maximum detectable depth.

ACKNOWLEDGEMENTS

This work was funded by a grant from the Natural Sciences and Engineering Research Council of Canada.

REFERENCES

1. Clifton, J. and Carino, N., Nondestructive Evaluation Methods for Quality Acceptance of Installed Building Materials, National Bureau of Standards, Washington, D.C., 1982.

2. Krautkramer, J. and Krautkramer, H., Ultrasonic Testing of Materials, 3rd ed., Springer-Verlag Berlin Heidelberg, New York, 1983.

3. Libby, H., Introduction to Electromagnetic Nondestructive Test Methods, Wiley-Interscience, New York, 1971.

4. Matthews, J., Acoustic Emission, Gordon and Breach Science Publishers, New York, 1983.

5. Hussein, E., Radiation Scattering Methods for Nondestructive Testing and Imaging, Int. Advances in Nondestructive Testing, McGonnagle, Ed., Gordon and Breach, New York, in press.

6. Hussein, E. and Kondic, N., Compton Scatter Imaging Systems, in Bioelectronic Systems, CRC Press, D.L. Wise Ed., in press.

7. Gautam, S., Hopkins, F., Klinsksiek, J.K. and Morgan, I., Compton Interaction Tomography I. Feasibility Studies for Applications in Earthquake Engineering, IEEE Trans. Nucl. Sci., NS-30, pp. 1680-1684, 1983.

8. Evans, R., The Atomic Nucleus, Massachusetts Institute of Technology, McGraw-Hill Book Company, Inc., 1955.

9. Wood, J., Computational Methods in Reactor Shielding, Univ. of London, Pergamon Press, 1982.

10. Cain, V., SAMBO, A Collision Analysis Package for Monte Carlo Codes, ORNL-TM-3203, 1979.

RECENT GEOTECHNICAL APPLICATIONS OF THE ACOUSTIC EMISSION/MICROSEISMIC TECHNIQUE

H. R. Hardy, Jr.
The Pennsylvania State University
University Park, PA 16802

ABSTRACT

Since the early 1970s there has been an accelerating interest in the use of acoustic emission/microseismic (AE/MS) techniques for the laboratory and field investigation of geologic materials and structures. Basic AE/MS research and the application of AE/MS techniques to a variety of geotechnical problems has been underway in the Penn State Rock Mechanics Laboratory since 1965. To date, studies have involved a wide variety of field topics, including: cavern stability, stability of natural gas storage reservoirs, strata control in longwall coal mines, rock bolt stability monitoring, effects of blasting, mine and tunnel roof stability, subsidence monitoring, rock slope stability evaluation, and remote monitoring of rock drilling. A considerable number of basic studies have also been underway including those associated with the development of AE/MS field and laboratory monitoring facilities, transducer array design, AE/MS data processing, mechanical waveguide development and the Kaiser Effect.

To provide a more general insite into the subject, this paper will first briefly review the literature relative to the use of AE/MS techniques to study the mechanical behavior of both rock and soil. This will be followed by a brief review of a number of recent AE/MS studies carried out at Penn State and a more detailed discussion of a number of ongoing and proposed geotechnically oriented AE/MS research studies.

INTRODUCTION

It is generally accepted today that most solids emit low-level seismic signals when they are stressed or deformed. This phenomenon will be referred to throughout this paper as acoustic emission/microseismic (AE/MS) activity.

In recent years the application of AE/MS techniques in the general area of geotechnical engineering has rapidly increased. At present such techniques are in use, or under evaluation, for stability monitoring of underground structures such as mines, natural gas and petroleum storage caverns, radioactive waste repositories, and geothermal reservoirs. The technique, originally developed in an attempt to predict and reduce the incidence of rock bursts in hardrock and coal mines, has now found extensive application in soils and soft rocks such as salt. Furthermore, besides their application in large scale field studies, AE/MS techniques are being successfully utilized in an increasing number of laboratory-scale studies both of a basic and an applied nature.

In essence the measurement of AE/MS activity in a field structure or laboratory specimen is relatively simple. A suitable transducer is attached to the structure or specimen, the output of the transducer is connected to a suitable monitoring system, and the seismic signals occurring in the structure or specimen, due to internal or external stress or deformation, are suitably detected, processed and recorded. Figure 1 illustrates a typical example where a number of transducers (an array) are installed to monitor the stability of an underground fluid storage cavern.

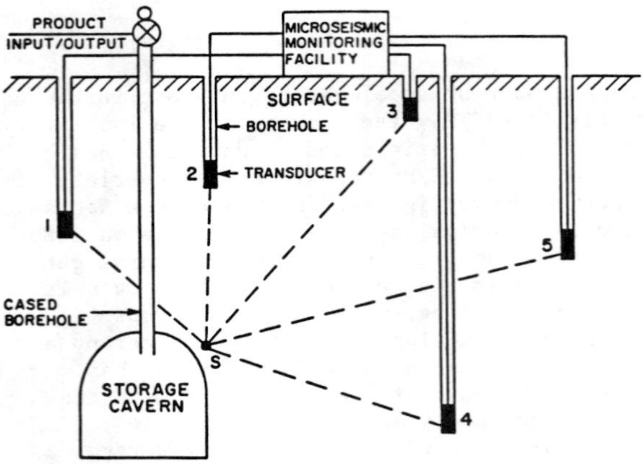

FIGURE 1. AE/MS transducer array installed to monitor the stability of an underground fluid storage cavern. (S denotes the assumed location of an AE/MS event.)

Detailed discussions relative to the fundamentals of the AE/MS technique are presented elsewhere (1,2). In review a number of the more important AE/MS concepts are as follows:

(1) The AE/MS technique is a passive, indirect technique.

(2) AE/MS activity originates as an elastic stress wave at locations where the material is mechanically unstable.

(3) The associated stress wave propagates through the surrounding material undergoing attenuation as it moves away from the source.

(4) With suitable instrumentation AE/MS activity may be detected at locations a considerable distance from its source.

(5) The useable spatial range of the technique is dependent on the frequency content of the source and the characteristics of the media and the monitoring facility.

(6) The character of the observed AE/MS signals provide indirect evidence of the type and degree of the associated instability.

(7) Analysis of data obtained from a number of transducers (array) make it possible to determine the actual location (i.e., spatial coordinates) of the source.

AE/MS STUDIES IN THE GEOTECHNICAL AREA

There is little doubt that the initial development of the modern AE/MS technique was associated with studies in the general geotechnical field, namely mining (3,4). Incorrectly many persons in material science associate the birth of the subject with the work of Kaiser (5), Mason, et al. (6) and others.* Although, in the years following the initial mining related AE/MS studies, research in the geotechnical areas continued, many of the major developments, occurred within the disciplines of geophysics and material science. Unfortunately, due to the wide differences in the general interests and the materials under study in the different areas, and the development of separate and somewhat unique nomenclature, for a number of years the interaction between the various disciplines involved in AE/MS research was relatively limited.

In the last 20 years, due in part to a resurgence of interest in the method by those in the geotechnical field, it has become increasingly clear that a common area of interest exists; and that studies in various disciplines, in most cases, mainly differ in respect to the scale of the specific problem under investigation, and the frequency range of the associated AE/MS activity. In general, AE/MS studies in the geotechnical field overlap at low frequencies with seismology and at high frequencies with material science related AE/MS studies. This overlap presents a number of advantages since specialized instrumentation and experimental techniques have been under development for a number of years in both areas.

Considering the development of the AE/MS technique in the geotechnical area, as noted earlier, studies relative to mining related problems began with the pioneering work of Obert and Duvall in the late 1930s. Obert comments briefly on these studies in an earlier paper (4). The historical development of the technique relative to geotechnical field applications has been described in considerable detail in recent papers by Hardy (1,8). Further details on both the laboratory and field developments in the subject are available in papers by Hardy (9), Koerner et al., (10), and Lord and Koerner (11); in the proceedings of the four Penn State AE/MS Conferences (12-15); and in the proceedings of the eight International Acoustic Emission Symposia held in Japan (e.g. 16-18). A comprehensive bibliography on the general subject of AE/MS activity by Drouillard (7) is also available, which includes approximately 2000 entries as well as a relatively detailed historical review of the general development of the subject. Updates on this bibliography are available in various volumes of the Journal of Acoustic Emission, published since 1982 by the Acoustic Emission Group, University of California in Los Angeles.

The author recently prepared a comprehensive review of the field-related geotechnical AE/MS studies underway during the period 1980-1987 (19). A very limited number of earlier papers were also included for completeness. The review included research and application studies

---

* A detailed history of acoustic emission is presented in the recently published ASNT handbook (2). It is interesting to note that the historical review neglects any mention of activities in the geotechnical area even through these were clearly described in an earlier historical review by Drouillard (7).

underway in a total of 19 countries. More than 225 references to recent and ongoing research were included. A detailed bibliography of earlier research (some 500 entries) is included in the proceedings of a recent AE/MS conference (14).

AE/MS STUDIES AT PENN STATE

General

During the past 22 years, the Penn State Rock Mechanics Laboratory has been actively involved in both the laboratory and field aspects of acoustic emission/microseismic (AE/MS) research. In this period, research funds well in excess of one million dollars have been expended and 13 graduate students have completed M.S. and Ph.D. theses on AE/MS related topics. At present, six graduate students are involved in AE/MS related field research. AE/MS research, to date, has included studies associated with longwall coal mining, underground natural gas storage, effects of underground blasting on surface stability, scenic cavern safety, stability of rock slopes, sinkhole development, earthquake prediction, mine and tunnel roof stability, and a variety of instrumentation and fundamental topics. Over 50 scientific papers, proceedings, monographs and technical reports have been published on these activities.

The present paper will briefly outline a number of the recent and on-going AE/MS field oriented research in the geotechnical area. Further details of these and other studies are available in published papers and reports.

Recent Studies

Cavern stability. In recent years, AE/MS techniques have been applied to an increasing number of geotechnical situations in the area of public and industrial safety. In the rock mechanics field, studies have been undertaken relative to the underground storage of nuclear wastes, crude oil, petroleum products and pressurized gases, and the structural stability of scenic caverns. In 1978 AE/MS studies were carried out by the author in a large scenic cavern located in central USA (1). Since the area had been closed to the public for some years, and since minor apparent structural instabilities had been noted in the past at this site, it was felt that an evaluation of the safety of the area should be undertaken as a preliminary step in the reactivation of the area for public use. It was decided to utilize AE/MS techniques to evaluate the stability of the area due to the size and complexity of the cavern structure, and the fact that the results of earlier AE/MS studies had influenced the decision to close the area to public use.

At the cavern test site, commercially available geophones were utilized as transducers. Measurements were made using the Penn State Mark I mobile AE/MS monitoring facility. Notch filters were used for all channels to minimize 60-Hz electrical noise, and active bandpass filters (1-50 Hz) were also used to reduce both the 60-Hz noise and other high-frequency noises. As a consequence of the extensive filtering and the system configuration employed, very high monitoring system sensitivities were attained. In general, it was possible to sense ground velocities as small as 5×10^{-8} to 25×10^{-8} m/s.

During the four-day monitoring period, only two basic types of discrete AE/MS events were commonly observed. These were denoted as Type I and Type II events. Type I events were the most numerous and characteristically had a low amplitude (<$83x10^{-8}$ m/s), often barely exceeding the generally low background activity, low frequency content (7.5-15 Hz), short durations (<6 s) and were normally detected by all transducers. The dominant feature of the Type II event was its long duration (lasting a minimum of 18 seconds to often more than two minutes).

In general, the data indicated that there appeared to be an excellent correlation between Type I and Type II event rate curves, and that the event rates were relatively high near midday and extremely low near midnight. Based on these results, it was apparent that vibrations from tour buses and other culturally-generated events were responsible for the majority of the events observed. Furthermore, the ground motions at the site due to even the largest Type II events recorded rarely exceed $12.5x10^{-6}$ m/s, which is considerably less than the $5x10^{-2}$ m/s value generally accepted as the lower limit for induced structural change. It was concluded therefore that during the monitoring period, the site, from a mechanical point-of-view, was highly stable. In recent years the site was renovated and is now open to the public. No signs of structural instability have been observed.

Stability of natural gas storage reservoirs. During the period 1966-1982 the Penn State Rock Mechanics Laboratory has been involved in research associated with the mechanical stability of underground gas storage reservoirs. Such reservoirs are regions of highly porous and permeable rock, located at depth, which are utilized by the natural gas industry for storage of compressed natural gas for use during periods of peak demand. During Phase I of the research, laboratory and theoretical studies were conducted. In 1971 an extensive AE/MS field program (Phase II and III) was initiated to evaluate the mechanical stability of full-scale operating reservoirs. Funds for the Penn State reservoir storage study have been provided by the Pipeline Research Committee of the American Gas Association (A.G.A.).

In the Phase III study, initiated in 1975, a detailed AE/MS investigation was undertaken at the New Haven underground gas storage site in central Michigan using refined techniques and experimental facilities developed on the basis of the experience gained during the earlier Phase II studies. Figures 2-4 illustrate various aspects of the associated instrumentation. In order to rapidly and objectively analyze the large volume of AE/MS data obtained during the New Haven studies (some 70,000 m of multichannel tape), it was necessary to develop a suitable computer-based analysis system (see Figure 10). With this system it has been possible to edit, manipulate and store large volumes of AE/MS field data. The data was analyzed to evaluate possible correlations between observed AE/MS activity and various storage parameters. Analysis was also carried out to investigate the approximate source of the observed AE/MS signals. During the overall study many thousands of small AE/MS events were monitored at the New Haven site, however only 26 events of sufficient amplitude and signal quality were observed which were suitable for detailed source location analysis. Details of the overall study have been presented in a two-volume A.G.A. monograph (20,21).

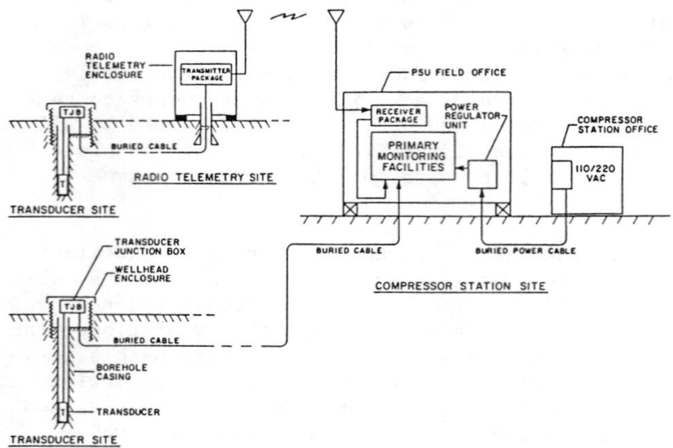

FIGURE 2. General outline of overall AE/MS monitoring system at New Haven gas storage [after Hardy et al. (21)].

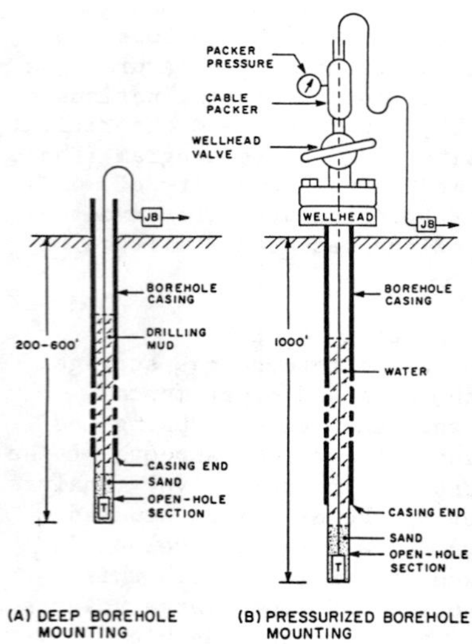

FIGURE 3. Deep transducer installation techniques [after Hardy et al. (21)].

FIGURE 4. HS-10-1B-HP type transducers for deep installation.

Longwall coal mine studies. During the period 1970-1977 the Penn State Rock Mechanics Laboratory was involved in a research project, sponsored by the U.S. Bureau of Mines, associated with the application of AE/MS techniques to coal mine safety. During this project, field studies were undertaken at a central Pennsylvania longwall coal mine to investigate the feasibility of using AE/MS techniques to locate potential zones of instability around underground coal mine workings. This study was unique in the fact that, as illustrated in Figure 5, measurements were made from the surface rather than underground. When this project was initiated in 1970, little or no scientific evidence existed to substantiate the fact that AE/MS activity generated underground, due to routine longwall mining activities, could be successfully detected by near-surface transducers.

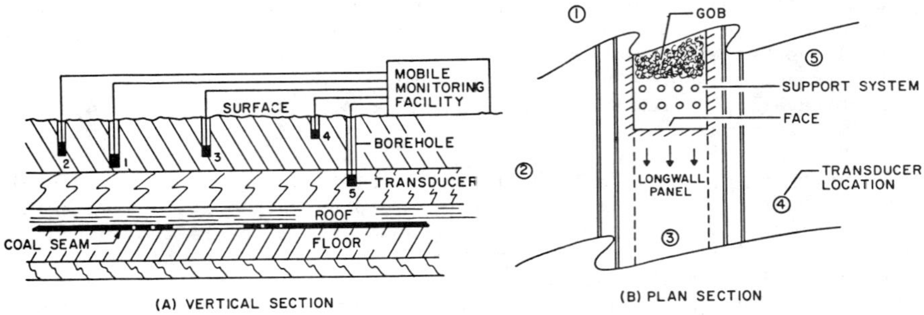

FIGURE 5. Experimental arrangement for AE/MS monitoring of an underground longwall coal mine [after Hardy et al. (23)].

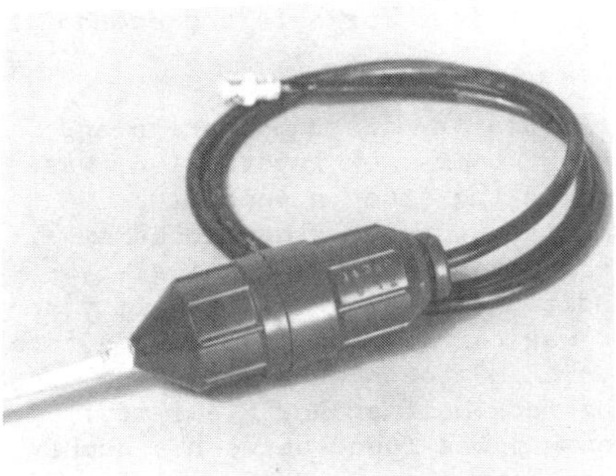

FIGURE 6. AE/MS transducers used at coal mine site.

FIGURE 7. Electronics rack from Mark I mobile AE/MS monitoring facility.

Some four months prior to initiation of the final monitoring program, 15 transducer sites were selected directly over an active longwall panel located some 200 m below surface. At this time, the longwall face itself was some considerable distance from the selected transducer sites. Suitable holes for transducer installation were augered, where possible down to bedrock. Transducers (geophones), as shown in Figure 6, were cemented in place immediately following hole completion. The Mark I mobile AE/MS monitoring facility (see Figure 7) developed earlier was utilized to process and record the field data. In this study a maximum of seven transducers were employed during any one monitoring session. Over a seven-month period, the longwall mining operation progressed some 250 m, and AE/MS data was monitored and recorded on magnetic during some 17 major field trips made to the site.

In general, some thousands of well-defined events were recorded during the study. These were classified into three levels, with level-3 being the largest. To date, source location analysis has been only applied to those data where at least five transducers in the array exhibited level-3 events. In general, sources were found to be clustered along the actual longwall face; the majority lying within (±20 m of the face line). This result is what would be expected from a knowledge of the longwall mining process. The source location data obtained also included information on the depth of the source, however, since the transducer array utilized was almost planar, with maximum depth variations between transducers being less than 3 m, results obtained with respect to source depth were generally poor. A three-volume final report, available from NTIS, was prepared on these studies (22-24). A preliminary paper on the study is available in the open literature (25).

Rock bolt monitoring. In both mining and civil engineering, artificial support of rock structures is often provided by rock bolts. Such bolts are a major means of support in underground coal mines in the United States. However, two of the most crucial problems facing the mining engineer today are that of finding a well defined and standard method for testing the anchoring ability of rock bolts, and of selecting the most suitable type of rock bolt and bolting pattern for a particular mine or a specific mine section. Recent studies have shown the inadequacy of present methods used in selecting and testing of rock bolts and the lack of clearly defined guidelines for safe and economical design of bolting patterns.

A research study, aimed at developing laboratory instrumentation and associated techniques for the evalution of bolt-rock interaction, was carried out at Penn State (26,27). During the study a special test facility was designed and experimental procedures developed in order to investigate the behavior of expansion-shell, mechanical-type rock bolts under a variety of test conditions. Using this facility an overall analysis of rock bolt behavior was carried out involving a total of seven anchorage-testing parameters. AE/MS techniques were also used for the first time in the evaluation of rock bolt anchor stability. During the study a significant relationship was found between a number of the anchorage-testing parameters and the observed AE/MS activity. This result reinforces the possibility of developing an early warning system for monitoring the stability of bolted underground structures, such as mine or tunnel roofs, using AE/MS techniques.

AE/MS field instrumentation. In the early 1970s, the Penn State Rock mechanics Laboratory assembled the Mark I mobile microseismic monitoring facility (28). The system (see Figure 7) allowed for seven channel monitoring at AE/MS field sites. Data was recorded on a multi-channel analog magnetic tape recorder. The monitoring facility was housed in a small van, and operating power was supplied from the mains or from a portable 5 kW motor generator unit. Tests on the completed system using 110 VAC building power indicated that for a passband of 10 Hz to 250 Hz and a system gain of 126 dB ($2\text{x}10^6$ voltage gain) the average system noise level was 1.1 volts peak to peak. This is equivalent to a system noise level of ne = $8.9\text{x}10^{-9}$ volts RMS/Hz, or in terms of ground motion for a typical geophone (sensitivity = 0.6 volts/in./sec) ng = $1.5x10^{-8}$ in./sec/Hz. Under actual field operating conditions the noise levels were often considerably higher and as a result, research has been underway to develop suitable filters to decrease various forms of environmental noise. The use of filters to improve the sensitivity of AE/MS monitoring systems are described by Kimble (29).

During long-term AE/MS studies at a natural gas storage reservoir in central Michigan (New Haven) a special on-site monitoring facility was developed (21). The facility incorporated two systems, an eight-channel system optimized for detailed multi-channel recording on magnetic tape, and a three-channel system incorporating a three pen oscillographic recorder (Helicorder), for long-term continuous recording. A portable three-channel radio telemetry system for transmission of AE/MS data from remote transducer stations was also developed for use at the site (30).

In the Spring of 1980, the Mark I mobile microseismic monitoring facility was destroyed by fire at a mine site. During the period 1981-1983, a new facility (see Figure 8A) was designed and assembled (31). Following extensive laboratory evaluation the completed facility was successfully field tested during the summer of 1984. Since then it has been employed with great success at a number of AE/MS field sites. The Mark II mobile microseismic monitoring facility is housed in a specially designed truck. It provides a 14-channel capability and incorporates an extremely versatile bandpass and notch filter capability. As in the Mark I facility data is recorded in analog form using a multi-channel magnetic tape recorder.

On-going Studies

Effects of blasting. AE/MS techniques appear to provide a practical means of evaluating the stability of natural and man-made structures during and after blasting. In this regard, studies have been underway by the author since 1974 to evaluate the influence of underground and surface blasting on the stability of the strata overlying a shallow underground limestone mine north of Baltimore, Maryland (1). The study has involved the installation of an array of AE/MS transducers in and over the mine and the monitoring of AE/MS activity generated in the overlying strata prior to, during and following blasting. The AE/MS studies have been carried out to answer three specific questions, namely: is the mine and overlying strata generally stable?; does the ground shock due to adjacent quarry and/or underground blasting induce short- or long-term instabilities in the mine structure or the overlying strata?; and if such instabilities are present or are triggered by blasting, where are they located?

To date the effects of some 30 or more surface and underground blasts have been monitored at the Baltimore site. Although increases in local AE/MS activity were noted immediately following a number of these blasts, in all cases, the AE/MS rate dropped to the normal background level some hours later. Detailed computer analysis of a number of AE/MS parameters indicate that, although peak ground velocities as high as $5.3x10^{-3}$ m/s were observed during blasting, the underground mine and overlying strata are highly stable. Studies at this site are continuing on a routine basis.

Mine and tunnel roof stability. Unpredicted falls of rock from tunnel and mine roofs, both during development and subsequent use, continue to represent a major cause of fatalities and serious injuries. To date, no reliable, cost-effective means of predicting such roof falls has become commercially available. The objective of the proposed studies is to develop a practical, on-line, AE/MS system for continuous monitoring of mine and tunnel roof stability, location of potential areas of instability, and automatic advance warning of roof falls.

Since 1985 studies have been underway relative to the use of high frequency AE/MS signals to monitor roof stability in underground coal mines (32). During these studies, laboratory and field evaluation of a single-channel commercial AE/MS monitoring instrument, the McPhar Rock Stress Monitor (MRSM), have been underway. Under certain field conditions the MRSM was found to provide a well defined warning of such instabilities as roof falls, floor heave and pillar loading within a restricted monitoring zone.

In order to monitor larger areas (e.g. long sections of mine entry roof) studies are presently underway to develop mechanical waveguide systems which could be attached to existing roof bolts, perhaps over distances of many hundreds of feet. Since the location of unstable zones will be accomplished using linear source location techniques, only two AE/MS transducers will be required. Possibly a third will be utilized for automatic calibration.

Subsidence -- sinkhole studies. The uncontrolled development of cavities near the earth's surface due to natural or man-made causes, pose a major engineering hazard. Such cavities can result in localized instabilities such as the collapse of highway or airport runway pavements, foundation subsidence or collapse, and dam leakage. Since 1984 the Penn State Rock Mechanics Laboratory has been carrying out research, sponsored by the Federal Aviation Administration and The Pennsylvania Department of Transportation, relative to the location and stability evaluation of "sinkholes." Such features are common in karst terrain, namely terrain produced by the solution of carbonate rocks such as limestone and dolomite.

The Capital City Airport, located near Harrisburg, Pennsylvania, has experienced a large number of sinkhole occurrences throughout its existence. The feasibility of a composite monitoring program involving both AE/MS and seismic tomography techniques was evaluated in a recent research study conducted at the airport (33,34). Figure 8 illustrates the original runway condition and AE/MS monitoring equipment used in this research study. Based on the results obtained to date, the AE/MS technique, utilizing zonal and hit-sequence location, was shown to be capable of accurate source location in such an environment. Seismic tomography, utilizing signal amplitude data, proved to be sensitive to cavity prone regions. The combination of these two techniques provides an effective means of locating cavities and monitoring their stability.

Slope stability. Although AE/MS research in the field of slope stability, associated with both open pit mining and various civil engineering projects, has been relatively limited in the past, research interest in this area has increased in recent years. At present the Penn State Rock Mechanics Laboratory is carrying out a study for the Pennsylvania Department of Transportation (PennDOT) to evaluate the feasibility of using AE/MS techniques to evaluate stability of highway rock slopes. As part of this study tests are underway at two sites on I-81 south of Wilkes-Barre, Pennsylvania, and on I-80 east of Lamar,

(A) Runway sinkholes at Capital City airport, near Harrisburg, Pennsylvania.

(B) Electronics rack from Mark II mobile AE/MS monitoring facility.

FIGURE 8. Runway condition and AE/MS monitoring equipment associated with sinkhole monitoring project.

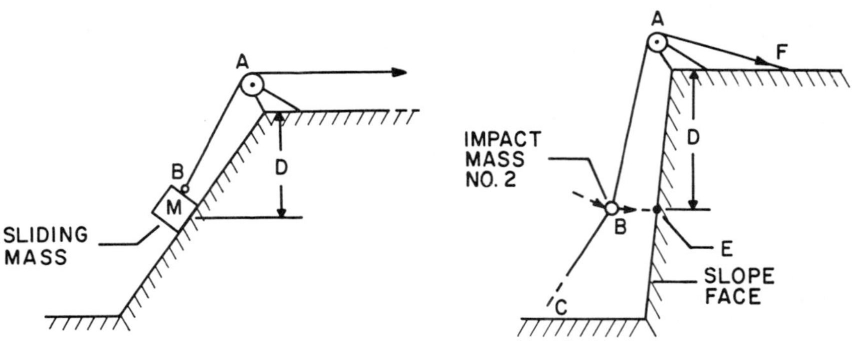

(A) Surface sliding tests.

(B) Surface impact tests.

FIGURE 9. Methods for simulating effects of surface sliding and surface impact at Lamar slope test site.

Pennsylvania. At the I-80 site an array of 10 geophones have been permanently installed in holes drilled at the top of the slope, a six geophone linear array temporarily installed in soil along the toe region, and a series of monitoring sites, for short-term installation of accelerometers and AE-transducers, established on the slope face. Similar sites for short-term measurements were also established on slope faces at two I-81 sites.

A variety of studies have been conducted at the three slope sites. AE/MS data has been monitored over a wide range of frequencies using geophones (≈10 Hz - 500 Hz, an accelerometer (0 - 2 kHz) and an AE-transducer (15 kHz - 45 kHz). Data was recorded using the new Mark II mobile AE/MS monitoring facility and a commercially available, portable, high frequency, single-channel AE/MS monitoring system. Studies include the evaluation of AE/MS background noise due to highway traffic; AE/MS propagation through fractured rock; evaluation of the characteristics of AE/MS activity generated by a number of sources, including rock-block sliding, small scale surface fracture, surface impact, and falling rocks. Source location studies using simulated sources, have also been undertaken. Figure 9 illustrates techniques used to evaluate sliding and surface impact at the Lamar site. An automated AE/MS system for monitoring rock falls has also been developed. The final report on this feasibility study will be available in late 1988.

Remote monitoring and prediction of rock drillability. At present a suitable method for on-line remote monitoring and prediction of rock drillability is not available. Recent studies, carried out by Niitsuma and Chubachi (35) at Tohoku University in Japan, indicate that the basic characteristics of AE/MS data collected during rock drilling appear to be dependent on such factors as the type of rock and its mechanical properties, and on operating factors such as drill bit type and condition, bit load and rate of advance. Such studies indicate that the AE/MS technique has the potential of providing remote on-line evaluation of rock drilling operations. For example, AE/MS data collected during preliminary drilling from surface over a proposed excavation site, or ahead of a mine or tunnel face, could provide basic data on the "drillability" of the associated rocks and changes in rock characteristics with distance or position. Furthermore, during routine drilling operations, AE/MS data could provide continuous information on bit wear and drilling efficiency.

Research relative to rock drillability was initiated at Penn State in early 1987. This involves a three phase study, namely: a fundamental laboratory study to test the concepts under well controlled conditions; a preliminary field study to provide basic data relative to the application of the concept under field conditions; and the development and testing of a prototype AE/MS field monitoring and analysis system.

Optimum AE/MS transducer array design

Source location is one of the most important features of the AE/MS technique. Since 1985 studies have been underway in regard to the analysis of source location methods and procedures for the design of optimum transducer array geometries. Three specific, although related problems, have been investigated, namely: the effect of errors in the assumed velocity model; the uniqueness of computed source location coordinates; and the effect of transducer array geometry. A recent thesis by Ge (36) and a number of published papers (37,38) have discussed these problems in detail.

AE/MS data processing. For over two decades, the Penn State Rock Mechanics Laboratory has been actively involved in the development of AE/MS analysis techniques for use in geotechnical applications. A suitable in-house data processing facility was assembled as shown in Figure 10, and suitable software has been under development for rapid and accurate processing and analysis of a variety of AE/MS field data. The first version of this software package was developed by Hardy and Mowrey in the late 1970s (20,39,40). The primary application of this software at that time was to process AE/MS data associated with an underground natural gas storage reservoir under study on behalf of the American Gas Association (AGA). It was decided in 1984 to further modify this software package in order to meet the rapidly growing applications of the AE/MS technique (41).

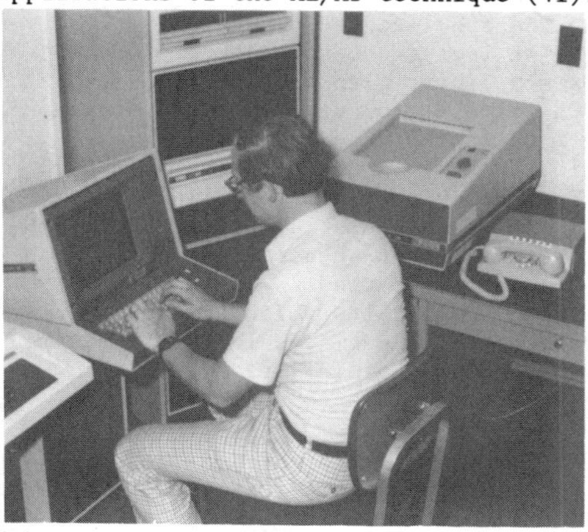

(A) Dr. G. L. Mowrey at graphics terminal.

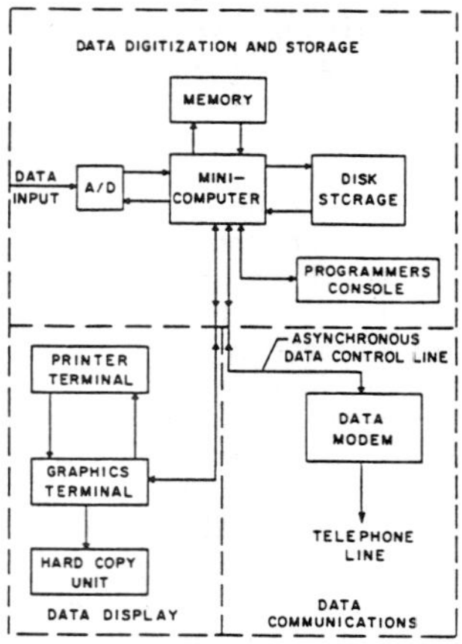

(B) Block diagram of processing facilities.

FIGURE 10. Penn State AE/MS data processing facilities.

Mechanical waveguide development. The application of AE/MS techniques in near-surface monitoring of geologic structures have to date been extremely limited due to the highly attenuative characteristics of the associated media (soil, and highly fractured and weathered rocks). In such applications the transducer/structure coupling system plays an extremely important role. One approach, aimed at resolving the problems due to high attenuation is the use of mechanical waveguides. The obvious advantage of this type of coupling system is that it increases the monitored volume of the structure, however, the major disadvantage is that a single transducer waveguide arrangement does not provide sufficient data to actually locate the source of the AE/MS activity. In a recent study by Taioli (42), a new source location concept, using dual-transducer waveguides, was tested in the laboratory. In addition, several fundamental aspects of waveguide attenuation were investigated, and the methodology necessary to study mechanical waveguide behavior and associated attenuation under confined and unconfined conditions were also developed. Continuing studies are underway which will include the development of special dual-transducer waveguide systems and associated field evaluation.

Laboratory evaluation of field stress -- Kaiser effect. A basic parameter required in geotechnical design is the state of the in-situ stress. Conventional measurement of this parameter is normally accomplished using various flat-jack or borehole techniques. If absolute stresses are necessary some type of stress relief technique is required which is costly, time consuming, and often difficult. Recently, studies have been carried out by a number of researchers to investigate the feasibility of using AE/MS techniques as a means of determining such stresses (43). Using the so-called Kaiser Effect, AE/MS evaluation of specimens, prepared from standard rock cores obtained from the desired field sites, are carried out in the laboratory. In such tests the specimens are suitably instrumented with high frequency AE-transducers, then loaded in a testing machine while the resulting AE/MS activity within the specimen, and the applied load, are recorded as a function of time. At low stresses the AE/MS rate is found to be low and often approximately linear with applied stress. At a critical stress level, however, the rate is found to suddenly increase and to continue at this new higher rate as the load is further increased. The stress level at which the rate suddenly changes is assumed to be the maximum stress level to which the material has been stressed in the past. Studies by a few researchers have included a comparison of the values of the in-situ stresses obtained by the AE/MS (Kaiser) method and by underground overcoring methods. In the majority of cases, the two methods appear to provide results which are in reasonable agreement. Since the cost of the associated laboratory tests are very inexpensive, compared to conventional field stress measurements, there is a strong economic incentive to further develop tests based on the Kaiser Effect.

At Penn State, preliminary Kaiser Effect studies were initiated in early 1982 (44). In these studies it was decided to investigate the AE/MS behavior of rock specimens for which the recent stress history was well known. Tests were carried out on specimens which had been uniaxially and triaxially preloaded under well controlled conditions. These studies verified the feasibility of the technique, based on tests on a limited number of rock types. They indicated that "triaxial" (axial stress plus confining pressure) preload stresses could be estimated on the basis of a single uniaxial test; and substantiated the fact, stated by some earlier Japanese workers, that special care must be taken in regard to specimen preparation and loading to insure that anomalous AE/MS activity is not generated at the ends of the test specimens. In the Penn State tests, maximum deviations between actual stresses and those estimated from Kaiser Tests ranged from 2.4% - 12.0% (mean 6.5%) for uniaxial preload specimens. For triaxial preload specimens maximum deviations ranged from 1.5% - 32.6% (mean 8.9%) for axial stress and from 0.3% - 21.3% (mean 12.7%) for confining pressure.

In the last few years, research has continued (45). In particular studies have been underway to develop more refined laboratory testing procedures. Detailed Kaiser Effect studies are presently underway on a variety of rock types.

DISCUSSION

It is clear, from the preceding review of recent and on-going studies, that the Penn State Rock Mechanics Laboratory has been active in a wide range of basic and applied AE/MS research in a variety of geotechnical areas. Although the applications are diverse, new techniques and equipment developed in one area often have direct application to another on-going study. Although initially the Penn state studies were more-or-less restricted to studies in competent rock, more recently a number of studies have been undertaken relative to the behavior of weak rocks and soils.

In the future, more emphasis will be given to the improvement of monitoring and analysis techniques. For example, more field research is planned using 3D-transducers; more use will be made of sophisticated digital signal processing techniques; and additional AE/MS parameters (e.g. energy, amplitude distribution, frequency spectra, etc.) will be incorporated in the analysis of field data. A recent review of the international literature (19) clearly indicates the increasing interest in the development and application of the AE/MS technique. The future offers those involved both a challenging and rewarding field of study.

REFERENCES

1. Hardy, H. R., Jr., (1981), "Applications of Acoustic Emission Techniques to Rock and Rock Structures: A State-of-the-Art Review," Acoustic Emissions in Geotechnical Engineering Practice, STP 750, V. P. Drnevich and R. E. Gray, Ed., America Society for Testing and Materials, Philadelphia, Pennsylvania, pp. 4-92.

2. Miller, R. K. and P. McIntire, (1987), Nondestructive Testing Handbook (2nd ed.) Vol. 5, Acoustic Emission Testing, ASNT, Columbus, Ohio, 603 p.

3. Obert, L., (1941), "Use of Subaudible Noise for Prediction of Rock Bursts," USBM RI 3555.

4. Obert, L., (1977), "The Microseismic Method-Discovery and Early History," Proceedings First Conference on Acoustic Emission/ Microseismic Activity in Geologic Structures and Materials, Pennsylvania State University, June 1975, Trans Tech Publications, Clausthal, Germany, pp. 11-12.

5. Kaiser, J., (1950), "Untersuchengen über das Auftreten Geräushen beim Zugversuch (An Investigation into the Occurrence of Noises in Tensile Tests or a Study of Acoustic Phenomena in Tensile Tests)." Ph.D. Thesis (in German), Technische Hochschule, Munich, Germany.

6. Mason, W., H. J. McSkimin and W. Shockley, (1948), "Ultrasonic Observation of Twinning in Tin," Physical Review, Vol. 73, No. 10, (May 1948), pp. 1,213-1,214.

7. Drouillard, T. F., (1979), Acoustic Emission: A Bibliography with Abstracts, Plenum Data Company, New York, 787 pp.

8. Hardy, H. R., Jr., (1984), "Geotechnical Applications of Acoustic Emission Techniques: Present Status and Future Goals," Proceedings of 7th International Acoustic Emission Symposium (Sendai, Japan, October 1984), pp. 632-641.

9. Hardy, H. R., Jr., (1972), "Applications of Acoustic Emission Techniques to Rock Mechanics Research," prepared for presentation at the ASTM Symposium on Acoustic Emission (Bal Harbor, Florida, December, 1971). ASTM STP-505, Acoustic Emission, pp. 41-83, Spring.

10. Koerner, R. M., W. M. McCabe, and A. E. Lord, Jr., (1981A), "Acoustic Emission Behavior and Monitoring of Soils," Acoustic Emissions in Geotechnical Engineering Practice, STP 750, V. P. Drnevich and R. E. Gray, Eds., American Society for Testing and Materials, pp. 93-141.

11. Lord, A. E., Jr., and R. M. Koerner, (1979), "Acoustic Emission in Geologic Materials," Fundamentals of Acoustic Emission, Editor - K. Ono, School of Engineering and Applied Science, University of California, Los Angeles, California, pp. 261-307.

12. Hardy, H. R., Jr., and F. W. Leighton, Editors, (1977), Proceedings First Conference on Acoustic Emission/Microseismic Activity in Geologic Structures and Materials, Pennsylvania State University, June 1975, Trans Tech Publications, Clausthal, Germany, 489 pp.

13. Hardy, H. R., Jr. and F. W. Leighton, Editors, (1980), Proceedings Second Conference on Acoustic Emission/Microseismic Activity in Geologic Structures and Materials, Pennsylvania State University, November 1978, Trans Tech Publications, Clausthal, Germany, 491 pp.

14. Hardy, H. R., Jr. and F. W. Leighton, Editors, (1984), Proceedings Third Conference on Acoustic Emission/Microseismic Activity in Geologic Structures and Materials, Pennsylvania State University, October 1981, Trans Tech Publications, Clausthal, Germany, 814 pp.

15. Hardy, H. R., Jr., Editor, (1988), Proceedings Fourth Conference on Acoustic Emission/Microseismic Activity in Geologic Structures and Materials, Pennsylvania State University, October 1985, Trans Tech Publications, Clausthal, Germany, (In Press).

16. Onoe, M., K. Yamaguchi and T. Kishi, Editors, (1982), Progress in Acoustic Emission I, Proceedings 6th International Acoustic Emission Symposium, Susono, Japan, October/November, 1982, The Japanese Society for Non-Destructive Inspection, Tokyo, 545 pp.

17. Onoe, M., K. Yamaguchi and H. Takahashi, Editors, (1984), Progress in Acoustic Emission II, Proceedings 7th International Acoustic Emission Symposium, Zao, Japan, October 1984, The Japanese Society for Non-Destructive Inspection, Tokyo, 714 p.

18. Yamaguchi, K., K. Aoki and T. Kishi (Editors), (1986), Progress in Acoustic Emission III, Proceedings 8th International Acoustic Emission Symposium, The Japanese Society for Non-Destructive Inspection, Tokyo, 778 pp.

19. Hardy, H. R., Jr., (1987), "A Review of International Research Relative to the Geotechnical Field Application of Acoustic Emission/Microseismic Techniques," Proceedings Fred Leighton Memorial Workshop on Mining Induced Seismicity, 6th International Rock Mechanics Congress, Montreal, September 1987, pp. 69-114.

20. Hardy, H. R., Jr. and G. L. Mowrey, (1981), A Microseismic Study of an Underground Natural Gas Storage Reservoir, Volume II - Field Data, Analysis and Results, American Gas Association, Inc., Arlington, Virginia, A.G.A. Cat. No. L51397, 372 p.

21. Hardy, H. R., Jr., G. L. Mowrey and E. J. Kimble, Jr., (1981), A Microseismic Study of an Underground Natural Gas Storage Reservoir, Volume I - Instrumentation and Data Analysis Techniques, and Field Site Details, American Gas Association, Inc., Arlington, Virginia, A.G.A. Cat. No. L51396, 343 p.

22. Hardy, H. R., Jr., and L. A. Beck, (1978), Microseismic Monitoring of a Longwall Coal Mine: Volume 2 - Determination of Seismic Velocity, Final Report, USBM Grant No. G0144013, Pennsylvania State University, 230 p.

23. Hardy, H. R., Jr., G. L. Mowrey and E. J. Kimble, Jr., (1978), Microseismic Monitoring of a Longwall Coal Mine: Volume 1 - Microseismic Field Studies, Final Report U.S. Bureau of Mines, Grant G0144013 (NTIS), 318 p.

24. Hardy, H. R., Jr., B. A. Anani and A. W. Khair, (1977), Microseismic Monitoring of a Longwall Coal Mine: Volume 3 - Field Study of Mine Subsidence, Final Report, U.S. Bureau of Mines, Grant G0144013, 139 p.

25. Hardy, H. R., Jr., and G. L. Mowrey, (1976), "Study of Microseismic Activity Associated with a Longwall Coal Mining Operation Using a Near Surface Array," Journal of Engineering Geology, Vol. 10, pp. 263-281.

26. Unal, E., H. R. Hardy, Jr. and Z. T. Bieniawski, (1982), "New Laboratory Instrumentation for the Evaluation of Rock Bolt Behavior," Proceedings 23rd U.S. Symposium on Rock Mechanics, Berkeley, California, August 1982, AIME, New York, pp. 985-995.

27. Unal, E. and H. R. Hardy, Jr., (1984), "Application of AE Techniques in the Evaluation of Rock Bolt Anchor Stability," Proceedings Third Conference on Acoustic Emission/Microseismic Activity in Geologic Structures and Materials, Pennsylvania State University, October 1981, Trans Tech Publications, Clausthal, Germany, pp. 173-196.

28. Kimble, E. J., Jr., (1977), "Development of Mobile Microseismic Field Equipment," Proceedings First Conference on Acoustic Emission/Microseismic Activity in Geologic Structures and Materials, Pennsylvania State University, June 1975, Trans Tech Publications, Clausthal, Germany, pp. 338-356.

29. Kimble, E. J., Jr., (1980), "Application of Filters in Improving the Sensitivity of Microseismic Recording Systems," Proceedings Second Conference on Acoustic Emission/Microseismic Activity in Geologic Structures and Materials, Pennsylvania State University, November 1978, Trans Tech Publications, Clausthal, Germany, pp. 291-304.

30. Berezniak, J. N., (1980), "Development of a Portable Radio Telemetry System for Use in Microseismic Field Studies," Proceedings Second Conference on Acoustic Emission/Microseismic Activity in Geologic Structures and Materials, Pennsylvania State University, November 1978, Trans Tech Publications, Clausthal, Germany, pp. 375-398.

31. Kimble, E. J., Jr. and H. R. Hardy, Jr., (1988), "Development of the Mark II Mobile Microseismic Monitoring Facility," Proceedings Fourth Conference on Acoustic Emission/Microseismic Activity in Geologic Structures and Materials, Pennsylvania State University, October 1985, Trans Tech Publications, Clausthal, Germany, (In Press).

32. Ersavci, N., (1988), "Recent Studies Relative to the Application of the McPhar Rock Stress Monitor," Proceedings Fourth Conference on Acoustic Emission/Microseismic Activity in Geologic Structures and Materials, Pennsylvania State University, October 1985, Trans Tech Publications, Clausthal, Germany, (In Press).

33. Hardy, H. R., Jr., R. Belesky, M. Mrugala, E. Kimble and M. Hager, (1986), A Study to Monitor Microseismic Activity to Detect Sinkholes, Federal Aviation Administration, U.S. Dept. of Transportation, DOT/FAA/PM-86/34, 184 p., National Technical Information Service (NTIS), 184 p.

34. Hardy, H. R. Jr. and R. M. Belesky, (1987), "Use of Acoustic Emission and Seismic Tomography Techniques to Study Sinkhole Development," Proceedings 7th Symposium on Ultrasonic Electronics, Kyoto, 1986, Japanese Journal of Applied Physics, Vol. 26, Supplement 26-1, pp. 3-8.

35. Niitsuma, H. and N. Chubachi, "AE Monitoring of Well-Drilling Process by Using a Downhole AE Measurement System," Proceedings of 8th International Acoustic Emission Symposium, Tokyo, Japan, October 1986, Japanese Society for Nondestructive Testing, pp. 436-445.

36. Ge, M., (1988), "Optimization of Transducer Array Geometry for Acoustic Emission/Microseismic Source Location," Ph.D. thesis, Department of Mineral Engineering, The Pennsylvania State University.

37. Ge, M. C. and H. R. Hardy, Jr., (1988), "Effect of Velocity Data Quality on Source Location and Analysis of Optimum Transducer Array Geometry," Proceedings Fourth Conference on Acoustic Emission/Microseismic Activity in Geologic Structures and Materials, Pennsylvania State University, October 1985, Trans Tech Publications, Clausthal, Germany, (In Press).

38. Ge, M. and H. R. Hardy, Jr., (1988), "The Mechanism of Array Geometry in the Control of AE/MS Source Location Accuracy," Proceedings 29th U.S. Symposium on Rock Mechanics, Minneapolis, Minnesota, June 1988 (In Press).

39. Hardy, H. R., Jr. and G. L. Mowrey, (1984), "Analysis of Microseismic Data from the New Haven Natural Gas Storage Reservoir," Proceedings Third Conference on Acoustic Emission/ Microseismic Activity in Geologic Structures and Materials, Pennsylvania State University, October 1981, Trans Tech Publications, Clausthal, Germany, pp. 365-392.

40. Mowrey, G. L., (1980), Development and Application of a Computer Analysis System to Process Microseismic Data Associated with an Underground Natural-Gas Storage Reservoir, Ph.D. Thesis, Department of Mineral Engineering, The Pennsylvania State University, 294 pp.

41. Hardy, H. R., Jr., and M. C. Ge, (1986), "A Computer Software Package for Acoustic Emission/Microseismic Field Applications," Proceedings 27th U.S. Symposium on Rock Mechanics, Tuscaloosa, Alabama, AIME, New York, pp. 134-140.

42. Taioli, F., (1987), Laboratory Evaluation of Waveguides for Acoustic Emission/Microseismic Monitoring, M.S. Thesis, Department of Mineral Engineering, The Pennsylvania State University, 144 pp.

43. Zelanko, J. C., (1983), "Determination of In-Situ Stresses Using Acoustic Emission Techniques: A Literature Review," Internal Report RML-IR/83-7, Geomechanics Section, Department of Mineral Engineering, The Pennsylvania State University.

44. Zhang, D. L., (1982), "Use of the Kaiser Effect for Estimation of the Previous State of Stress in Rock," Internal Report RML-IR/82-4, Geomechanics Section, Department of Mineral Engineering, The Pennsylvania State University.

45. Hardy, H. R., Jr., D. Zhang and J. Zelanko, (1988), "Recent Studies Relative to the Kaiser Effect in Geologic Materials," Proceedings Fourth Conference on Acoustic Emission/Microseismic Activity in Geologic Structures and Materials, Pennsylvania State University, October 1985, Trans Tech Publications, Clausthal, Germany, (In Press).

RADIOGRAPHY

NEUTRON RADIOGRAPHY TECHNIQUES AND APPLICATIONS

G. C. McClellan
Argonne National Laboratory
Idaho Falls, ID 83403

INTRODUCTION

Soon after James Chadwick confirmed the existence of neutrons in 1932, experimenters found that these uncharged particles could be used much like x-rays to produce a film image of an object through which they passed. For nearly 25 years, this new technique was only used experimentally. Then in 1956, at the 8 MW BEPO reactor at Harwell (UK), neutron radiography was used to inspect some of the neutron shielding materials and heavy metals common to the nuclear industry.[1] This was one of the first practical applications of neutron radiography. In the 1960's, neutron radiography finally became recognized as a potentially important nondestructive testing tool for industrial applications. Neutron radiography is used today for an interesting variety of nondestructive testing applications although its practitioners still form a relatively small group when compared to those using x-radiography.

In general, neutron radiography and x-radiography tend to complement one another when applied to the same object. Because materials exhibit different attenuation characteristics for neutrons than they do for x-rays, the resulting radiographs emphasize different features of the object. Neutron attenuation is governed by the nuclear scattering and absorption cross-sections of the material, which may vary even between isotopes of a given element, rather than density or any other physical or chemical property of the material. For example, lead, which is very difficult to penetrate with x-rays, is easily penetrated by a neutron beam. Hydrogenous materials such as o-rings, corrosion products, and explosive compounds have high scattering cross-sections and excellent energy moderation characteristics for neutrons and can be imaged easily using neutron radiography even when surrounded with a denser material such as metal.

NEUTRON RADIOGRAPHY TECHNIQUES

Because neutrons carry no electrical charge, they are non-ionizing and produce little effect on radiographic film. To produce a neutron radiographic image on film, it is necessary to use a conversion screen that, upon neutron capture, will emit ionizing radiation or light that can expose film or be observed with a video camera.

Direct Method

The direct exposure method of neutron radiography provides high resolution and excellent contrast and has become the standard against which other neutron radiography techniques are judged. In the direct method, photosensitive film and a gadolinium conversion screen are placed together in a light-tight cassette, preferably a vacuum cassette, and exposed to the neutron beam. The film is sensitized by the emission of 70 keV electrons from the conversion screen upon neutron capture. Light-emitting fluorescent screens such as gadolinium oxysulfide may also be used for direct neutron radiography using film or for real-time viewing with a video camera. Direct neutron radiography methods utilizing photosensitive film or other gamma-sensitive detectors cannot be used if the neutron beam contains substantial gamma radiation or if the object is highly radioactive.

Track-etch Method

Track-etch neutron radiography[2] is a direct exposure method that provides high resolution and is insensitive to both gamma radiation and light. This method utilizes cellulose nitrate film with a converter screen that emits heavy charged particles, such as protons or alpha particles, as a result of neutron capture. These emitted particles cause tiny damage tracks in the film that can be enlarged and made visible by chemical etching. The radiographic image that results from this process has very low contrast but offers extremely good resolution. Some means of enhancement is almost always necessary for viewing. Track-etch radiographs are often evaluated with sophisticated optical equipment such as a scanning microdensitometer. Cellulose nitrate having a lithium borate coating is commonly used for track-etch neutron radiography. The lithium borate coating acts as a built-in converter screen which emits alpha particles during the neutron capture process.

Real-time Method

Real-time (RT) neutron radiography has great potential for the inspection of parts and processes requiring instant feedback of information. Neutron RT systems typically utilize a scintillation or fluorescent converter screen such as gadolinium oxysulfide to produce a light signal that is fed into video equipment for viewing. With the exception of the source and converter screen, neutron and x-ray RT systems utilize similar equipment to process and display the image signal. Advantages of neutron RT systems include their ability to present a dynamic view of the process being observed and their potential portability. The major limitation of neutron RT systems is their relatively low spatial resolution. Also, real-time systems are not, as presently configured, practical for the inspection of highly radioactive components because of the gamma sensitivity of the detector system.

Indirect Method

An indirect exposure method that is insensitive to gamma radiation is used for the inspection of highly radioactive components. This method utilizes metal conversion screens that become temporarily radioactive when exposed to the neutron beam. The radiograph is made by placing the activated foil into a light-tight cassette with film after the neutron exposure is terminated. Radiation emitted by the decay process of the activated screen sensitizes the film which can then be processed. Dysprosium and indium are the most commonly used screens for indirect radiography.

It is sometimes desirable to use a screen that is sensitive to neutrons of a higher energy. Indium is a good choice because of the strong resonance peak it has for capturing 1.46 eV (epithermal) neutrons. An epithermal neutron radiograph can be made using an indium screen if a thermal neutron absorber, such as cadmium, is placed on the source side of the screen to filter out the thermal neutrons from the beam. Epithermal neutron radiographs are useful for imaging the internal features of nuclear fuel because neutrons of this energy are more likely to penetrate the fuel and provide the necessary detail in the radiograph. A common practice is to make up a detector "package" consisting of dysprosium, cadmium, and indium screens (in that order) to produce simultaneous thermal and epithermal images of an object (Figs. 1,2).

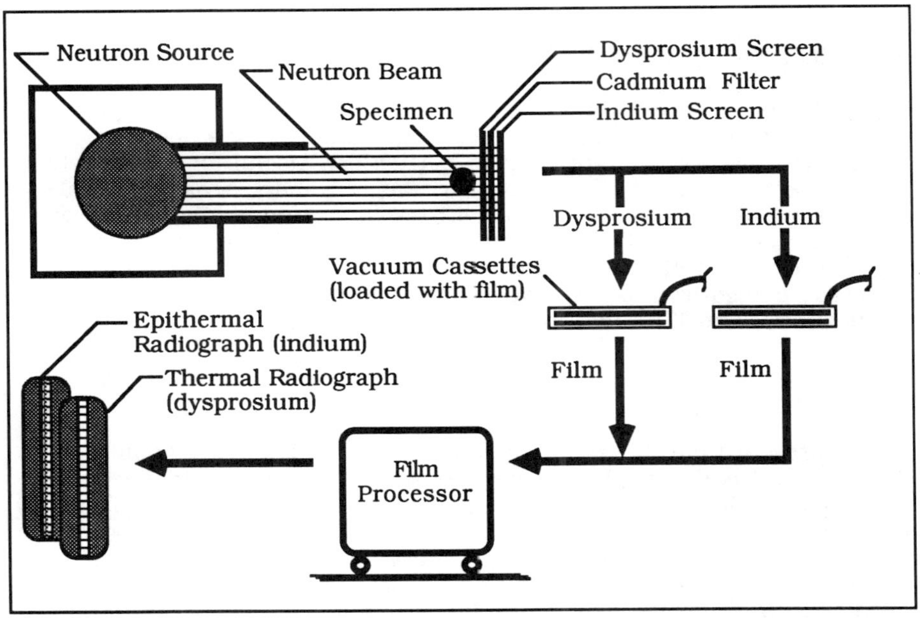

FIGURE 1. Schematic of indirect technique used for obtaining simultaneous thermal and epithermal (indium-resonance) neutron radiographs using dysprosium and indium screens.

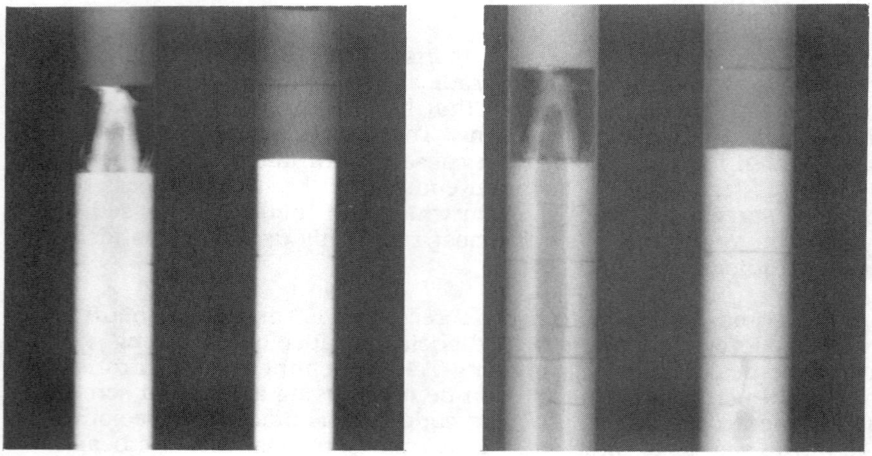

(a) (b)

FIGURE 2. Thermal (a) and epithermal (b) neutron radiographs of two highly-enriched U-PuO_2 fuel pins showing increased detail visible in fuel region on epithermal neutron radiograph. Note the fuel eruption on the left fuel pin and the difference in contrast between the enriched fuel pellets and the two depleted uranium pellets at the top of each stack.

NEUTRON RADIOGRAPHY APPLICATIONS

The differences, rather than the similarities between x-radiography and neutron radiography, make each technique valuable for a specific application. Neutron radiography is particularly useful for the detection and inspection of lighter elements encased in a more dense material such as metal, for the inspection of materials having similar densities (including isotopes of the same element), and for the inspection of highly radioactive objects. Neutron radiography applications may be generally categorized as: (1) nuclear applications, (2) commercial applications, and (3) other specialized applications.

Nuclear Applications

At the present time, there is only one neutron radiography facility in the U.S. that is designed and operated specifically for the inspection of irradiated nuclear components. This facility is the Neutron Radiography (NRAD) Facility[3] operated by Argonne National Laboratory and located at the Idaho National Engineering Laboratory (Fig. 2). The NRAD facility utilizes a 250 kW TRIGA research reactor as a neutron source and is located within a large, modern, hot-cell complex equipped with state-of-the-art remote handling and examination equipment. Two radiography stations are provided. One station directly interfaces with the main hot cell and the other station is located in a separate, clean, shielded room outside the hot cell. Both radiography stations are equipped with remote positioning devices for the specimens and detector screens. The NRAD facility has been used to inspect thousands of irradiated fuel rods and components from several US and non-US reactors. During the ten years this facility has operated, nearly 14,000 neutron radiographs have been produced.

Experimental nuclear fuels and materials are irradiated to evaluate their performance. These experimental fuels and materials are removed from the reactor periodically for nondestructive examination in a hot cell then returned to the reactor for further irradiation, sometimes to end-of-life failure. Both thermal and epithermal neutron radiographs of the individual fuel pins are made using the indirect radiography method. Dysprosium and indium screens are used. These are used to identify features such as fuel pellet separations, fuel central-void formation, fuel pellet cracking, evidence of fuel melting, and fuel column dimensional changes that may have occurred during irradiation (Fig. 3). Dimensional measurements of the internal features of fuel pins are measured from the radiographs with the aid of a computerized scanning microdensitometer. Accuracies of ±0.025 mm are achieved. Neutron radiography has proven to be one of the most valuable methods available for the nondestructive evaluation of irradiated materials.

Complete fuel assemblies containing an array of fuel pins are also neutron radiographed to evaluate features that would ordinarily be lost or disturbed during disassembly. Nuclear safety tests conducted at the Idaho National Engineering Laboratory simulate extremely abnormal operating or accident conditions. Fuel assemblies that have been subjected to these safety tests are often severely damaged as a result. Neutron tomography[4] is used at the NRAD facility for evaluating these assemblies. Neutron tomography yields computer-generated cross sections of the assembly and consists of the following basic steps: (1) epithermal neutron radiographs are made of the assembly at many radial angles, (2) these radiographs are then digitized with a scanning microdensitometer to convert the film density data to digital data, and (3) the

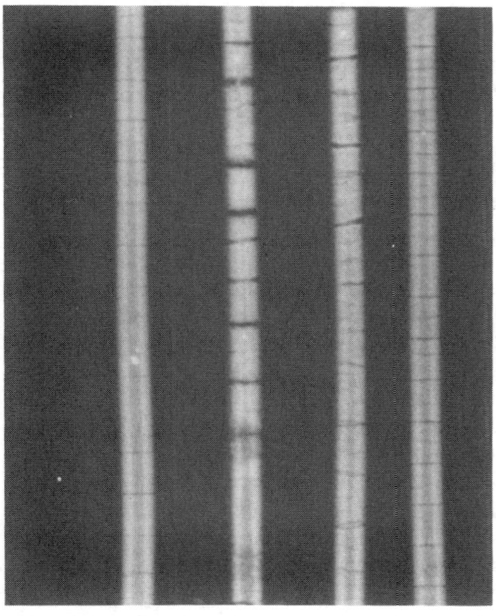

FIGURE 3. Epithermal neutron radiograph of high-burnup nuclear fuel pins showing cracked pellets, central void, and fission-product ingots (bright spots in central void).

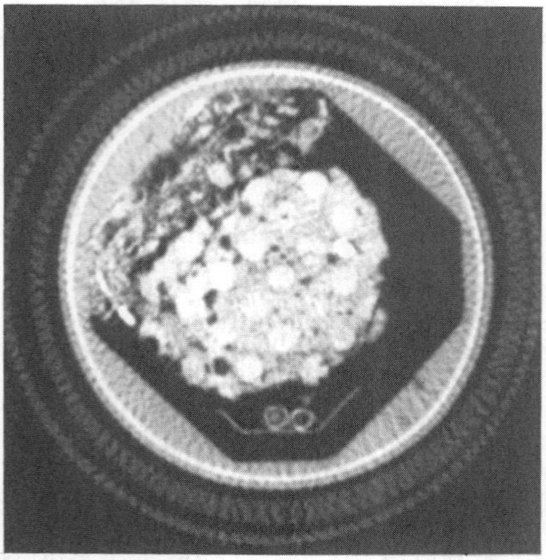

FIGURE 4. Neutron tomography cross section of severely damaged nuclear fuel assembly that has been intentionally subjected to abnormal operating conditions.

digital data set is then analyzed and transformed to produce computer-generated images of the cross sectional view of the fuel assembly. These cross sectional images can be generated at any desired axial location and provide early assessment of damage (Fig. 4) and failure mechanisms without risking displacement of the internal components by cutting. In fact, neutron tomography reduces or even eliminates the need to conduct costly and time-consuming destructive examination of the assemblies.

Commercial Applications

Commercial neutron radiography services are provided primarily for the manufacturing quality control of nonradioactive objects that are to be used in critical applications. The facilities required for neutron radiography are very expensive and are governed by complex federal regulations in the case where a reactor is used as the neutron source. These and other factors have made the installation of on-site neutron radiography services impractical and has led to the establishment of commercial neutron radiography facilities to which manufacturers can send their products for radiography. The radiographs are usually evaluated by the customer using personnel specially trained in the interpretation of neutron radiographs. The current rate of production in commercial neutron radiography facilities has risen steadily and now exceeds 18,000 radiographs per year.

There are only two commercial neutron radiography facilities currently operating in the U.S. General Electric operates a small reactor in Pleasanton, California,that is used for commercial neutron radiography; and Aerotest Operations, Inc., provides neutron radiography services at their facility in San Ramon, California. The Aerotest Operations facility is equipped with a 250 kW TRIGA research reactor and is designed for high-throughput of commercial specimens. The direct radiography technique is used wherein photosensitive film is exposed along with a gadolinium converter screen to produce a high-quality image. Explosive components, turbine blades, electronic components, and mechanical parts are received from a variety of manufacturers and radiographed in this facility.

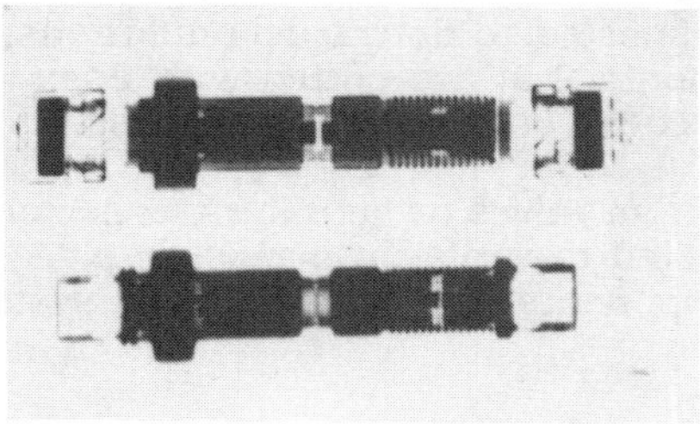

FIGURE 5. Neutron radiograph (top) and x-radiograph (bottom) of Apollo explosive bolt showing the complimentary character of the two techniques.
(Courtesy Aerotest Operations, Inc.)

Detonators, initiators, and other explosive devices used in critical aerospace applications are radiographed to verify that the explosive charge is adequate and that the internal components are positioned to guarantee proper detonation (Fig. 5). The hydrogenous materials used in these devices make them ideal candidates for inspection with neutrons. Both neutron and x-radiography must sometimes be used to completely inspect these parts. The complimentary images obtained, and the high reliability required of these devices, justifies extensive radiographic examination.

Gas-cooled aircraft turbine blades are routinely neutron radiographed to detect residual core material remaining from the investment casting process (Fig. 6). Even a small amount of this residual material left in a cooling passage could lead to failure of the turbine blade and serious damage to the engine. Since neither x-rays or neutrons can adequately image the ceramic casting material, gadolinium oxide is used to "dope" the casting material prior to casting. This provides increased radiographic contrast of the ceramic material which makes it visible in a neutron radiograph. Turbine blade inspection is currently the largest single application of neutron radiography.

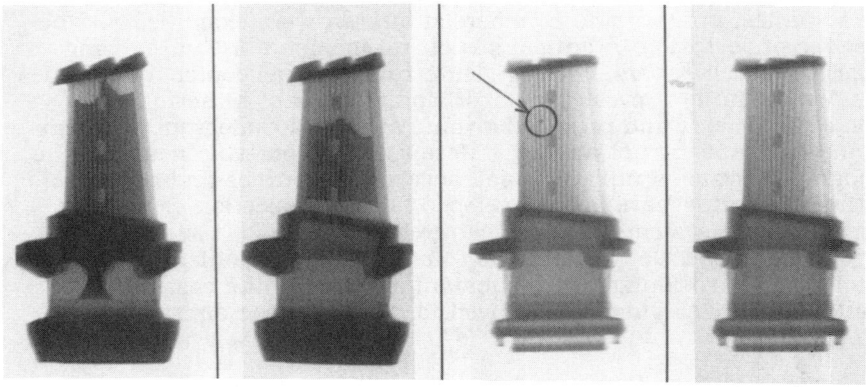

FIGURE 6. Neutron radiographs of gas-cooled turbine blades showing varying amounts of residual casting material remaining in cooling channels.
(Courtesy Aerotest Operations, Inc.)

Many of the materials and components used in modern electronic circuitry can be inspected effectively with neutron radiography. For example, ceramic capacitors are inspected to detect delaminations, and electronic filters and other components that are potted with epoxy are inspected to determine the effectiveness of the potting process.

Other critical mechanical parts such as valves or actuators, especially those containing o-rings, are inspected to verify proper assembly. These parts are usually critical components used in man-rated systems whose failure could cause loss-of-life.

Other Specialized Applications

In addition to the nuclear and commercial applications that have been described, some other interesting applications for neutron radiography have been reported.[5,6]

Neutron radiography is an ideal method for detecting corrosion products inside metal structures because of the difference in neutron attenuation between the hydrogenous corrosion products and metal. One such application is the inspection of mechanical pipe connectors used on the North Sea oil drilling platforms. These connectors are subject to serious corrosion and mechanical damage due to the ingress of sea water through faulty or damaged seals. Due to the size of the steel connectors, which have a wall thickness of 4 cm and are 75 cm in diameter, a cold neutron (energies $\leq$ 0.005 eV) beam is used to provide penetration through the steel. This method has been successful in showing the condition of the internal o-rings and mechanical joints, as well as showing corrosion products only a few microns thick that may have formed inside the structure.

There is a growing concern about the failure of aircraft structures due to hidden corrosion. In the early 1970's, it was shown that a portable californium-252 source could be used effectively to inspect aircraft structures for corrosion without costly disassembly of the aircraft.[7] In these studies, military and commercial aircraft were examined for the presence of corrosion in critical structural members including wing tanks, stabilizers, spars, landing gear struts, and helicopter rotor blades. This demonstration revealed a surprising number of hitherto undetected defects and proved the value of such a radiography system for on-line inspection of aircraft. More recently, portable neutron radiography systems utilizing small accelerator sources and real-time viewing apparatus have been developed for the inspection of military aircraft. These systems are designed to be portable for use on the flight-line. They have the potential to reduce the risk of failure by early detection of corrosion and also substantially reduce the cost of maintenance by eliminating the need to dismantle the aircraft for inspection.

The same materials properties that have made corrosion detection possible with neutron radiography also permit the quantitative evaluation of hydrogen in metals. A process was developed in the late 1960's at General Electric's Vallecitos Nuclear Power Laboratory that utilized only the neutrons scattered by hydrogen atoms in the object to produce a radiograph. Radiographs produced using this method were similar to an autoradiograph of the hydrogen distribution within the object. This NDT service was offered commercially under the name of Hysen™ and was capable of detecting and quantifying concentrations of hydrogen as low as 50 ppm. Although this service is no longer offered commercially, other laboratories have more recently used similar techniques to determine the presence and quantity of hydrogen in aluminum, titanium, zirconium, and palladium.

Some interesting civil engineering applications for neutron radiography include studies of the water permeability of concrete, analysis of water movement in partially frozen soils, and detection of microcracking in concrete structures. These studies permit engineers to evaluate the suitability of concrete structures for harsh environments.

Biological applications for neutron radiography have thus far been limited to the evaluation of surgically removed specimens. Some encouraging experiments have shown that neutron radiography may be valuable for the examination of tissue immediately adjacent to metal implants. In an effort to identify promising tumor-seeking boron compounds that can be used for the boron capture therapy treatment of inoperable brain tumors, thin histological sections have been neutron radiographed. The images were recorded on cellulose nitrate film by alpha particles emitted as a product of neutron capture in the boron.

Art historians are finding neutron radiography useful for the investigation of old paintings and other art objects. Underlying layers of paint or the contents and structure of art objects are revealed in a way not achievable using conventional techniques. Neutron activation and autoradiography of paintings is a powerful diagnostic tool when coupled with x-radiography and infrared-reflectometry. This combination of techniques has been used recently to examine paintings of the Rembrandt era.[8]

CONCLUSIONS

Neutron radiography is not used as extensively as other methods of industrial radiography, but the range of applications is constantly expanding. In some applications, such as critical aerospace components having appropriate neutron attenuation properties, neutron radiography has become an indispensable part of the quality assurance process. The use of neutron radiography has been limited to some degree by the cost and complexity of the facilities required. This has prevented the on-site inspection required in many potential applications. However, the small portable accelerator and isotopic neutron radiography sources and efficient real-time detection systems now being developed will hopefully permit more widespread use of this important technique.

REFERENCES

1. Berger, H., *Neutron Radiography*, Elsevier Publishing Co., Amsterdam, 1965.

2. Green, D. R., Tomlinson, R. L., and Gray, W. H., High Resolution Neutron Radiographic Methods for Radioactive Components, *Materials Evaluation*, pp. 1226-1231, December, 1981.

3. Richards, W. J., and McClellan, G. C., Hot Fuel Examination Facility Neutron Radiography Reactor Design, *Neutron Radiography*, Proceedings of the First World Conference, pp. 257-262, D. Reidel, Dordrecht, The Netherlands, 1981.

4. McClellan, G. C., and Tow, D. M., Neutron Tomography of Damaged Reactor Fuel Assemblies, *Neutron Radiography*, Proceedings of the Second World Conference, pp. 711-718, D. Reidel, Dordrecht, The Netherlands, 1986.

5. Barton, J. P. and von der Hardt, P., *Neutron Radiography*, Proceedings of the First World Conference, D. Reidel, Dordrecht, The Netherlands, 1981.

6. Barton, J. P., Farny, G., Person, J. L., and Röttger, H., *Neutron Radiography*, Proceedings of the Second World Conference, D. Reidel, Dordrecht, The Netherlands, 1986.

7. John, J., Californium-Based Neutron Radiography for Corrosion Detection in Aircraft, *Practical Applications of Neutron Radiography and Gaging*, ASTM-STP No. 586, 1975.

8. Fischer, C. O., Kelch, J., Laurenze, C., Leuther, W., and Slusallek, K., Autoradiography of Paintings After Neutron Activation at a Cold Neutron Guide, *Atomkern-energy Kerntechnik*, 51, No. 1, 1987.

A NEUTRON RADIOGRAPHY SYSTEM FOR FIELD USE

J. J. Antal and A. S. Marotta
U.S. Army Materials Technology Laboratory
Watertown, MA 02172

INTRODUCTION

The movement of military personnel and materiel utilizing aircraft to provide rapid deployment and the increasing concern for fuel conservation have provided an incentive to develop new lightweight structural materials. These new alloys, high-performance ceramics and composite materials have brought with them new challenges in their processing, forming and characterization. The characterization of structural materials over their full life-cycle, from the selection of raw materials to the performance of the material in the finished article in its operating environment has become important to the production and successful application of costly, high-technology lightweight materials. Along with many new sensor systems and technical instruments, neutron radiological examination has arisen as a useful characterization tool for several new materials systems.

Neutron radiography, which uses neutrons as a penetrating radiation to produce transmission images of material structures in a manner identical to better-known x-ray radiography techniques, was a fully developed technique by 1960's, when nuclear research reactors could readily provide convenient beams of thermal neutrons. The special characteristics of neutron radiogrpahy did not seem to be required to a great extent for many years afterward and it remained a "solution in need of a problem" [1]. Even the routine application of neutron radiography to critical elements of space hardware by NASA and its extensive application to fuel element analysis in the nuclear industry [2,3] did not spawn a wider interest in the technique. In recent years, however, the damage to older commercial and military aircraft by hidden long-term moisture initiated corrosion combined with the need for new lightweight structural materials within the military has had a marked affect on the development of neutron radiological techniques. In 1979 the U. S. Army demonstrated a prototype mobile and transportable neutron radiography system for helicopter examination developed from previous research by the Vought Corporation of Dallas, Texas.* A great deal of interest has been shown in this system and its construction has provoked a major renewal of interest in neutron radiography because it promises to bring neutron radiography to the workplace, a convenience provided by most nondestructive examination tools, particularly x-ray radiography [4].

---

*Presently the Vought Missiles and Advanced Programs Division of LTV Aerospace and Defense Company, Dallas, Texas.

ADVANTAGES OF NEUTRON RADIOGRAPHY

Neutron radiography is still a less readily available and more costly NDE technique than other radiological methods and thus is usually applied in special cases where other techniques fail. Neutron radiation is generally more penetrating than gamma-ray or x-ray radiation because these zero-charge particles do not interact with the sea of electrons present in materials. The primary neutron interaction is with the nuclei of atoms in materials, where both scattering and absorption processes play a major role in removing neutrons from a beam directed at an object. Neutrons have a very high sensitivity for interaction with hydrogen regardless of its chemical form within a material. Thus the new lightweight structural materials formed of organic molecules (epoxy composites, plastics) and the organic adhesives with which they are bonded are susceptible to neutron examination. The detection of moisture and moisture-initiated corrosion products is a major application of neutron radiography to structures. Other light elements which are found in lightweight materials, particularly lithium and boron, are also readily imaged by neutron radiography.

NEUTRON RADIOGRAPHY LIMITATIONS

A major reason neutron radiography has yet to be applied on as large a scale as other NDE techniques is the unavailability of a small, highly efficient source system. U. S. Army needs have demanded such a source and we review here the steps we have taken toward meeting this need.

Neutron radiographic examination has generally been applied as a supplement to x-ray radiography. By comparing the essentials of the two radiographic techniques we can readily come to an understanding of the past and present limitations of neutron radiography. Figure 1 compares the two techniques. They differ in two essential respects, the source

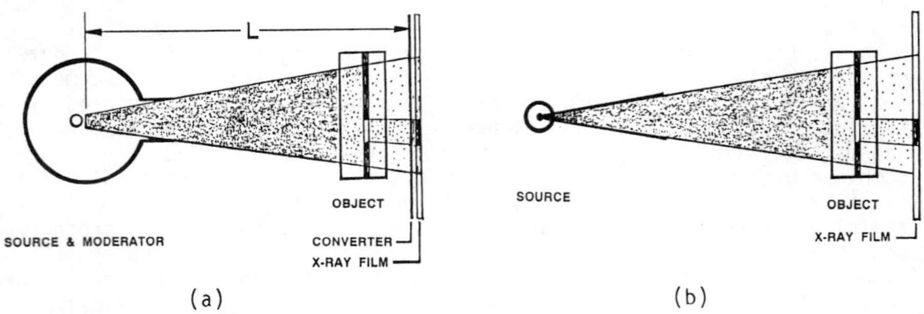

FIGURE 1. A schematic representation of the neutron (a) and x-ray (b) radiographic techniques. Neutrons are generated with kinetic energies greater than 1 MeV as a result of nuclear reactions. To be useful for general radiological purposes their energy must be reduced to an average of 0.025 eV. This is accomplished by random scattering of the neutrons in an hydrogen-rich moderator material such as polyethylene or water. The neutron beam is formed with a collimator, passes through the object being examined, and is recorded as an image on x-ray film. The neutrons which would fall on the film are intercepted by a "converter" screen, a plate containing elements such as gadolinium or lithium, which absorb the neutrons and emit electrons or gamma-rays to produce the latent image.

construction and the film recording system. The need for a moderator results in a relatively bulky source system. Neutrons, being uncharged particles, cannot be focused as is the electron beam within an x-ray source tube. This results in an effective source of extended dimensions, perhaps 100 mm in diameter, rather than the appoximately 2 mm diameter point source of x-rays. The consequence of these source characteristics is that the resolution in the final neutron image is dependent upon care in the design of the collimator and often cannot be attained without placing the source at a long distance, L in Figure 1, from the film plane. In fact a commonly applied measure of resolution in neutron radiography is the "L/D ratio", where D is the diameter of the effective source, defined by the entrance apeture of the collimator. Much inefficiency is involved in the moderation and collimation processes, and consequently low beam intensity relative to x-ray sources is at times a major limitation in neutron radiography.

NEUTRON SOURCES

Neutrons are the product of nuclear reactions and neutron sources consequently are machines which originated in nuclear physics laboratories: nuclear reactors, particle accelerators and radioactive materials. These represent a wide variety of sources in physical size, intensity, reliability of operation, and convenience of application.

Nuclear Reactors

The most common source for neutron radiography is the nuclear reactor. It is able to provide a well-collimated, high intensity beam so that exposure times and image quality can equal those of x-ray radiography. The nuclear reactor is a very large, complex and expensive facility to construct and operate. In addition, its operation is highly regulated by government agencies and access to the source is limited. The objects to be examined must be transported to and from the reactor site, and this is very often a limiting constraint. These limitations notwithstanding, commercial neutron radiography services in the United States are provided reliably by vendors using nuclear reactor sources.

Particle Accelerators

In an attempt to provide a more tractable source for neutron radiography, nuclear particle accelerators have been employed as neutron sources. These range in size from a large several-room facility such as those housing linear accelerators and Van de Graff or Dynamitron accelerators to small, compact accelerators designed for use in an analytical chemistry laboratory to provide neutron activation analysis. The large accelerators are capable of producing good intensities for neutron radiography and have the advantage over reactors of requiring only moderate governmental licensing. All accelerators have the advantage of electrical beam-on and beam-off control for easy radiographic exposure. Beam intensity varies with the size of the accelerator and may be rather low for the small, compact accelerators. Accelerators are used as high-energy x-ray sources, so it is reasonable to assume that accelerators would find a place in radiological examinations as neutron sources.

Radioactive Materials

Radioactive neutron sources fall into the "small, compact" category with the practical sources providing low intesities similar to the small accelerator sources. Neutrons are produced by radioactive sources either by direct emission or by the emitted radiation striking a target material. Neutron emission is continuous and continuously decreasing so that they must be renewed from time to time if they are to be used in a practical radiological system and exposures must be made by the mechanical interposition of shielding. The man-made isotope Californium-252 has been the most successful radioactive neutron source for radiological use to date. It has a half-life of 2.65 years which is long enough to be useful yet short enough to have a high specific activity for its direct, self-fissioning neutron emission. The advantages of all radioactive sources are their small size, their ultimately simple, reliable operation, and their low heat generation with no input power requirement. Heat generation by Californium-252 is particularly low. A summary of neutron source characteristics is given in Table 1.

SYSTEM DESIGN FOR ARMY NEEDS

Both the U. S. Navy [5] and the U. S. Air Force [6] with the IRT Corporation of La Jolla, California have explored the possibilities for neutron radiography with small radioactive source systems which could be set up at a maintenance site. The exposures required were longer than maintenance personnel thought were time- and cost-effective and the development of these systems was not pursued extensively. The most desireable neutron radiography system for the U. S. Army would provide an ability to move about and survey the airfoil and skin structure of a helicopter before dissasembly at a maintenance depot and would provide neutron radiography side-by-side with x-ray radiography in laboratory and proving ground nondestructive testing facilities. The system would also have a simple beam on-off control to meet safety and mobility regulations. To meet these requirements, one would like to combine the high intensity of reactor sources with the on-off beam convenience of accelerators with the portability, efficiency and reliability of radioactive sources. Clearly, some engineering compromises have to be made in taking even the first steps toward this goal. Based on earlier studies by the the Vought Corporation which provided a concept of a transportable neutron radiography system providing on-off beam convenience, the U. S. Army

TABLE 1. Some Average Characteristics of Neutron Sources Suitable for Radiological Systems

SOURCE	VOLUME (cm^3)	HEAT GENERATED (watts)	THERMAL NEUTRON RADIOGRAPHY FLUX ($n/cm^2/sec$)
REACTOR	2×10^8	100,000	1×10^7
VAN DE GRAAF	1×10^7	50,000	5×10^6
KAMAN A-711	1.5×10^3	300	4×10^4
Pu-Be	67	200	7×10^2
Cf-252	2	1.6	2×10^4

entered into an agreement with that firm to produce an engineering model of such a system suited to Army needs.

The Neutron Head Assembly

The neutron source employed in the system is the small, compact A-711 accelerator manufactured by Kaman Instrumentation Corporation of Colorado Springs, Colorado. It produces 14 MeV neutrons by accelerating deuterium ions at 180 KeV into a tritium gas-filled target and is designed and packaged for routine laboratory operation. The decision to use this accelerator was based on extensive experience with its reliability and operating charcteristics in both Army and Vought laboratories and its availability as an off-the-shelf item. The accelerator system consists of a power supply unit, a cooling unit, a control console and the accelerator head which might be arranged as shown in Fig. 2 in an analytical chemistry laboratory. As a radiography source, the accelerator head was enclosed in a 0.91 meter (36 inch) iron sphere containing transformer oil for neutron moderation and shielding. Additional oil shielding was introduced into the accelerator tube housing which effectively eliminated the leakage of high energy neutrons through the cable end of the tube housing. The water-cooled accelerator target is located about 40 mm (1.6 in) from the center of the sphere where it is out of direct sight of the collimator port and where it provides improved moderation through convection cooling of the surrounding oil. The use of a liquid reduces considerably concerns of radiation damage to the moderator.

Thermalized neutrons are formed into a beam by a divergent collimator which is an integral part of the spherical head. It is constructed of neutron and gamma-ray absorbing materials and provides for a minimum image size of 200 mm x 250 mm (8 in x 10 in).

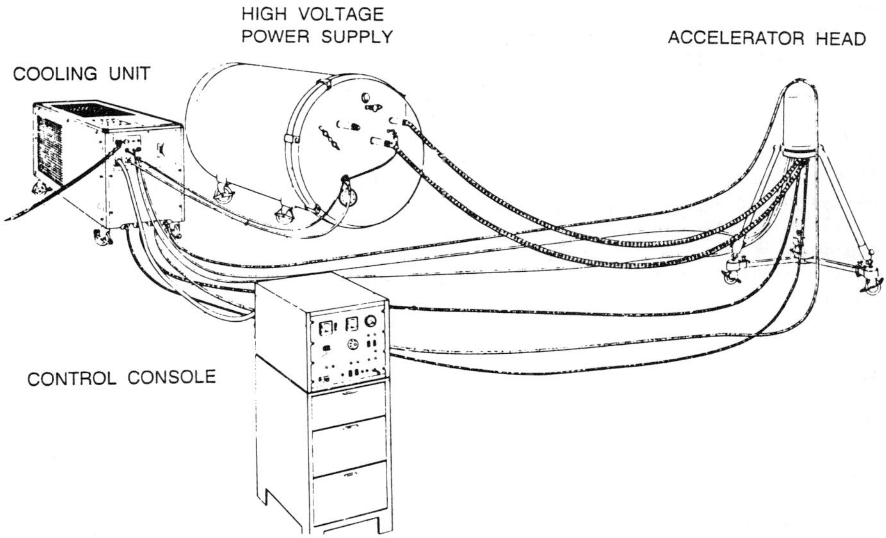

FIGURE 2. The neutron accelerator as delivered by the manufacturer for laboratory use. The accelerator head, shown on a tripod support, is 0.25 m in diameter and 0.55 m high. The power supply tank is 1.0 m in diameter and 1.8 meters long. The cooling unit size is 0.45 m x 0.55 m x 0.43 m.

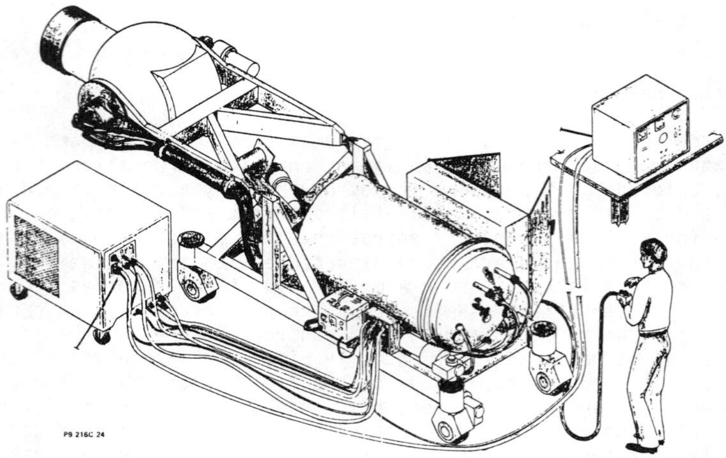

FIGURE 3. A sketch of the U. S. Army's mobile neutron radiography system. The carriage is 4.6 m long x 1.3 m wide x 1.3 m high (15 ft x 4.3 ft x 4.3 ft) and weighs 2205 kg (4860 lbs).

The Positioning Carriage

The neutron source system was completed by mounting the head on a specially-designed carriage which provides rotational, lifting, and lowering motions for the head and mobility for the complete source. A sketch of the source system is shown in Figure 3. The power supply tank, batteries, a battery charger, and oil handling utilities provide ballast to the carriage in addition to their normal functions. The orientation of the collimator and the location of the carriage is accomplished using a single operator pendant through battery-operated electrical drives and wheels. The collimator port may be positioned from ground level to a height of 2 meters (6.6 feet) and may be rotated over a 200 degree range, passing through straight-up and straight-down positions. The cooling unit was not made a part of the carriage assemblage because of a local need for the system to move through a restricted passageway. It rests on wheels on the floor nearby and its attached hoses are equipped with quick disconnect couplings. The control console is placed at a location remote from the radiation area during operation. Power for the accelerator and cooling unit is supplied by a line cord (approx. 8.5 kw).

The Complete System

The system was demonstrated to officials of the Department of Defense and its contractors and was subsequently field tested at an air force maintenance depot and an army proving gound to collect data on its performance and adaptability to field use. From the beginning the system produced useful radiographs and proved itself to be a very successful implementation of a mobile neutron radiography system. In fact it behaved more like a finished product than an engineering model. A more complete description of the neutron radiography system is given in References [7] and [8].

FIGURE 4. The mobile neutron radiography system examining the leading edge wing section of an aircraft undergoing scheduled maintenance. Personnel were not allowed in the area during an exposure period.

EXPERIENCE WITH SYSTEM TESTING

Experience with the system was gained through testing programs carried out at Defense Department locations having extensive experience with x-ray radiography and where it was thought neutron radiography might be useful to on-going operations. Our aim in this testing was to evaluate its operation further and to evaluate its acceptance as a nondestructive testing tool in a realistic workplace setting. The system was first tested in the evironment of a development laboratory in Dallas, Texas, to determine its operational characteristics. It was taken next to an U. S. Air Force aircraft maintenance center in Sacramento, California [9]. At this center it was operated jointly with the x-ray radiography staff in their daily workplace after the last shift of the day. Figure 4 shows the system in use at this site radiographing an aircraft which was scheduled for a major overhaul. At a later date, the system was transported to an Army desert test site where again the x-ray radiography staff operated the system, radiographing a variety of items common to their daily activities as an x-ray facility [10]. Figure 5 shows the system in use at the U. S. Army's Yuma Proving Grounds in Arizona. Finally it was transported to the Army Materials Technology laboratory where it now serves as a test-bed for further developments of small systems and provides radiography for materials studies on a routine basis.

The system was transported between field stations by ordinary truck with the carriage supported upon and restrained by a wooden skid. The system has sustained over 12,000 km (7700 mi) of travel in this manner with only minor damage and very little lost time attributable to the rigors of shipping. Overall, downtime has been about 1% of operating time, including the time spent in field operations. Over 700 film radiographs were taken during this testing period which we considered to be very successful in all respects.

FIGURE 5. Operational demonstration at Yuma, Arizona. Radiation protection for operators was provided by an exclusion area in this temporary situation.

SYSTEM LIMITATIONS AND IMPROVEMENTS

Although success with the system went beyond normal expectations, it was limited. Users of the system were pleased with what the system could do for them, but it always left them with a thirst for something better. Something better was primarily shorter exposure times. For radiographs to be of an acceptable quality to users in the field, exposure times of a minimum of 30-45 minutes were required. However, to keep up with the pace of x-ray radiography, exposure times under 5 minutes are desired. To reach such a goal, one can either strive to increase the source intensity by a factor of ten or increase the sensitivity of the image detector by a factor of ten. Several industrial programs are attempting to design a small accelereator with increased output, but increasing the source intensity of an accelerator without the physical size and weight of the system becoming unmanageable is proving to be very difficult.

Increasing the sensitivity of the image detection system is proving to be more rewarding. During the past few years, including during the field tests, we have been exploring combinations of film types from various manufacturers and several converter screen materials in search of useful improvements in sensitivity with some success. Most recently, we have been using an electronic imaging system of the type described by Dance and Carollo [11] which is specifically optimized for low neutron flux imaging. This electronic system is more sensitive than any of our film systems and has the added advantage of immediate viewing (no film processing time is required). The resolution of the images is limited by the line scan of the CRT screen, however. Reductions in exposure and processing times have thus been achieved with careful selection of working parameters. The data in Table 2 describes our present working environment.

TABLE 2. Neutron Imaging Systems in Use with Small Sources at the Army Materials Technology Laboratory

IMAGE RESOLUTION	TYPICAL L/D RATIO	AVE. EXPOSURE TIME	CONVERTER MATERIAL	IMAGE RECORDER
High	35:1	6 hrs	Evaporated Gadolinium Film	Kodak AA Film
Medium	25:1	1 hr	Evaporated Gadolinium Film	Kodak T-MatG Film
Low	18:1	30 min	Gadolinium Oxysulphide	DuPont NDT 75 Film
Low	15:1	3-5 min	Lithium-loaded Plastic	LTV NRTV-2 CRT Image

The data in Table 2 represent radiological work in a developmental laboratory where each radiological examination may be very different from the next, rather than a repetitive inspection for quality control of specific items or processes in a manufacturing environment. However, it shows that a trade-off is possible between the resolution in the recorded image and the exposure time. The exercise of this trade-off has been one of the most important lessons learned from the development of this small-source neutron radiographic system. We have learned that many practical problems can be effectively attacked with low-resolution radiography. This is particularly true in the aircraft maintenance area where the location of regions of corrosion and moisture entrapment in

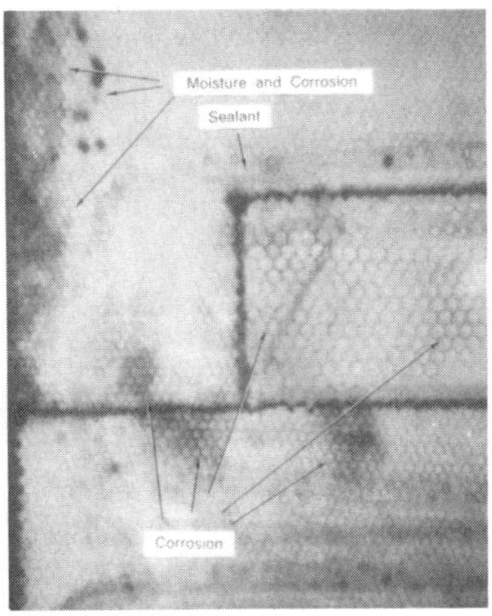

(a)

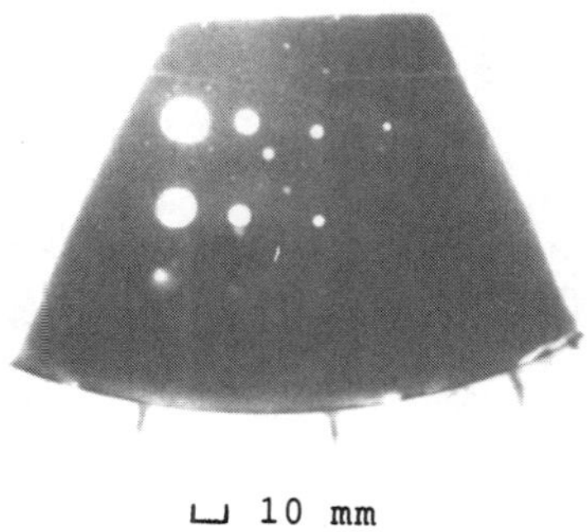

(b)

Figure 6. Typical neutron radiographs taken with the Army's mobile neutron radiography system. (a) Radiograph showing the location of moisture and corroded aluminum honeycomb in an airfoil. (b) A boron-containing, 1.3 mm-thick coating on a metal structure. The seven well-defined bright spots are holes in the coating. The remaining spots are undesireable air bubbles within the thick coating.

airfoil structures are the "defects" sought. Figure 6(a) is a reproduction of a neutron radiograph which illustrates this case. If the smallest dimension which the radiograph must reveal is the order of 0.1 mm (0.004 in), then an exposure time of about 60 minutes may be necessary (medium resolution attained) with a small source. If the smallest dimension which the radiograph must reveal is the order of 0.2 mm (0.008 in), then the exposure time may readily be reduced to below 30 minutes (low resolution attained). If this latter dimension can be seen readily (very often the case) on a 525-line raster video screen, few-minute exposure times, comparable to those usual for x-rays, can be utilized.

The maximum film density, found in areas where the incident radiation has not intercepted the object, is another parameter which can be adjusted to shorten exposure times. In our small source radiography we try to be satified with a maximum density of 2.0 density units rather than 4.0 density units radiographers generally wish to attain. Figure 6 illustrates two cases where high resolution was not required to locate the defects sought. An exposure of 30 minutes was more than adequate to show all of the detail in Fig. 6(a); the regions of corrosion and trapped moisture were well defined with even a 3-minute exposure time. Fig. 6(b) was a coating application problem where it was necessary to eliminate air bubbles of a diameter greater than 3 mm. A low resolution image taken rapidly would suffice to monitor this coating process.

One might assume that a technical person with radiographic experience would have chosen the radiographic route just described some time ago. But we found that the high state of the art which x-ray radiography has reached has worked against the acceptance of the less well developed neutron technology. X-ray radiography is standardized and equipment producing excellent image quality economically is readily available. Radiographers are trained to suspect radiographs showing a poor overall quality, including poor resolution and low density in areas where the direct beam is imaged. This is as it should be for industrial x-ray radiography where the equipment and training are readily available to meet exacting standards. To apply the same standards to neutron radiography requires that a source such as a nuclear reactor be employed. Relaxing these standards to accept near the minimum required to solve the problem at hand often allows one to find small neutron sources quite suitable for particular nondestructive evaluations.

On the other hand, we found that those who work daily with radiography readily accepted the Army's machine since its size, weight and safety requirements were not too far different from MeV x-ray units within their experience. The presence of a familiar kilovoltmeter and milliameter on a fairly simple control panel also played a role in this acceptance.

THE FUTURE FOR NEUTRON RADIOGRAPHY

During the past eight years of development and use of the Army's small accelerator source neutron radiography system, there has been a growing interest in providing industrial neutron radiology in the workplace. The reliable operation and surprisingly wide applicability of the Army's system has encouraged many to consider anew small source neutron radiological systems for in-house use. The U. S. Navy has sponsored the construction of a unit similar to that described here but configured to suit their need for examining aircraft which stand much higher off the ground than do helicopters [12]. The U. S. Air force is establishing a facility for automated neutron radiological scanning of complete aircraft prior to scheduling maintenance work [13]. New small neutron sources, such as compact cyclotrons, are being developed or proposed [14].

At the same time we are currently expanding the use of neutron radiography in our Neutron Analysis Laboratory in large part because of the convenience offered by small, easy to manage source systems. We have moved beyond traditional NDT examinations to supplement research and developmental activites with materials characterization and process monitoring using neutron radiological methods. Studies of high performance structural ceramics have been a particularly fertile area for neutron radiological analyses [15,16]. The development and successful fielding of this small accelerator source unit seems to have provoked a rethinking of the role neutron radiology can play in the nondestructive examination world.

REFERENCES

1. Ray, J. W., Neutron Radiography, A Solution in Search of Problems, Research/Development, Vol. 20, No. 7, July 1969, p. 18.

2. Kok, K. D., Neutron Radiography of Nuclear Fuels at the Battelle Research Reactor, in Practical Applications of Neutron Radiography and Gaging, ed Harold Berger, pp. 183-194, ASTM, Philadelphia, Pa, 1976.

3. Ross, A. M., Detecting Cladding Leaks in Irradiated Fuel Elements by Neutron Radiography, in Practical Applications of Neutron Radiography and Gaging, ed. Harold Berger, pp. 195-209, ASTM, Philadelphia, Pa, 1976.

4. Antal, J. J., A Renaissance in Neutron Radiography via Accelerator Neutron Sources, in Materials Characterization for Systems Performance and Reliability, eds. J. W. McCauley and V. Weiss, pp. 385-401, Plenum Press, New York, N.Y., 1986.

5. John, J., Orphan, V. J., Rundquist,D. E. and Sharp, R., Application of Neutron Radiography Techniques for Nondestructive Detection of Corrosion in Naval Aircraft & Aircraft Components, IRT Corporation Report No. INTEL-RT 6044-002, March 1974.

6. Larson, J. E., Parks, R., Baltgalvis, J. and John, J., Investigation of Neutron Radiographic Techniques for Maintenance Inspection of Air Force Aircraft, IRT Corporation Technical Report No. INTEL-RT-6081-001, March, 1975.

7. Antal, J. J., Dance, W. E., Carollo, S. F. and Moravec, J. D., Experience with an On-Off Mobile Neutron Radiography System, in Neutron Radiography, Proceedings of the Second World Conference, eds. Barton, Farney, Person and Rottger, pp.407-414, D. Reidel Publishing , Dordrecht, Holland, 1987.

8. Dance, W. E., Carollo, S. F. and Bumgardner, H., Mobile Accelerator Neutron Radiography System, Army Materials and Mechanics Research Center Rpt No. AMMRC TR 84-39, October, 1984.

9. Dance, W. E. and Carollo, S. F., Demonstration and Evaluation of Mobile Accelerator Neutron Radiography for Inspection of Aircraft Structures at SM-ALC/McClellan AFB, Vought Advanced Technology Center Rpt. No. R-92200/2CR-4, January, 1982.

10. Dance, W. E. and Carollo, S. F., AMMRC Mobile Accelerator Neutron Radiography System Operations at U. S. Army Yuma Proving Ground, LTV Aerospace and Defense Company Rpt. No. 3-41000/4R-110, April, 1984.

11. Dance, W. E. and Carollo, S. F., High Sensitivity Electronic Imaging System for Reactor or Non-Reactor Neutron Radiography, in Neutron Radiography, Proceedings of the Second World Conference, eds. Barton, Farney, Person and Rottger, pp. 415-422, D. Reidel Publishing, Dordrecht, Holland, 1987.

12. Orphan, V., Kedem, D. and Johansen, F., Mobile Neutron Radiography System for Aircraft Inspection, in Neutron Radiography, Proceedings of the Second World Conference, eds. Barton, Farney, Person and Rottger, pp. 447-454, D. Reidel Publishing, Dordrecht, Holland, 1987.

13. Froom, D. A. and Barton, J. P., Plans for Aircraft Maintenance Neutron Radiography Systems, in Neutron Radiography, Proceedings of the Second World Conference, eds. Barton, Farney, Person and Rottger, pp. 431-438, D. Reidel Publishing, Dordrecht, Holland, 1987.

14. Hawkesworth, M. R., Reactor-Quality Neutron Radiographs with a New Mobile Machine - A Superconducting Magnet Cyclotron, in Neutron Radiography, Proceedings of the Second World Conference, eds. Barton, Farney, Person and Rottger, pp. 183-190, D. Reidel Publishing, Dordrecht, Holland, 1987.

15. Antal, J. J., Marotta, A. S., Corbin, N. D. and Pujari, V. K., NDE of Structural Ceramics Using Neutron Radiography, in 16th Symposium on Nondestructive Evaluation, eds. C. D. Gray and G. A. Matzkanin, pp. 241-251, Southwest Research Institute, San Antonio, Texas, 1987.

16. Corbin, N. D., Pujari, V. K., Antal, J. J. and Marotta, A. S., A Preliminary Assessment of Neutron Radiography for Detecting Inhomogeneities in Ceramics, in Nondestructive Testing of High-Performance Ceramics, eds. A. Vary and J. Snyder, pp. 114-127, The American Ceramic Society, Westerville, Ohio, 1987.

REVIEW OF CONTEMPORARY NEUTRON RADIOGRAPHY IN EUROPE AND IN THE U. S. A.

E. Heiberg
University of Chicago
Chicago, IL 60637

Since the 1960's, neutron radiography (NR) has evolved into a mature NDT/E technology, essentially supplementing information obtainable from that of X-ray radiography (XR). Unique capabilities, however, are the abilities to radiograph radioactive material like nuclear fuel, and to differentiate between isotopes of the same element such as those found in such fuel. Other applications are the detection of inherent hydrogen in corrosion compounds and explosives, and the inspection of the proper location of O-rings in metallic casings. X-ray emulsion film is most commonly employed for the neutron imaging, but track etch offers the superior resolution. Just as for XR, electronic (real time) imaging has great potentially for NR.

INTRODUCTION

On the return from Europe a few weeks ago, I received the program of this NDT/E Conference. As you probably all are aware of, there are other speakers on neutron radiography. This should provide an excellent review of the present imaging potentialities of fast, epithermal, thermal, and cold neutrons. As many of you are not familiar with NR terminology, the references include recently published handbooks [1,2], Conference Proceedings [3-5], and Newsletters [6-8].

As a member of the ASTM E7.05 committee on neutron radiography, I participated in two different NR Conferences in Europe last May. The first one in Karlsruhe was the 10th Annual Conference of the European Neutron Radiography Working Group (NRWG), and the second one in Berlin was the first Working Group (WG 4) meeting on thermal neutron radiography. The latter was sponsored by the International Organization for Standardization (ISO). A formidable task lies ahead for ISO to agree on international NR standards equivalent to those of XR. It was a surprise to experience how little various NR working groups, including the Japanese IONOE, know about each others interpretation methods.

In this context a great help would be if the USA could accelerate the process of going metric. As this country is the world's sole non-metric industrial nation, it is constantly very confusing in international discussions to convert, for instance, 14" x 17" X-ray film to that of 35.56 cm x 43.18 cm. As you know, integral numbers are much easier to handle. A complete metric conversion should even reduce this country's trade deficit if the Europeans and the Japanese finally realize that our products are not expressed in gallons, pints, pounds, ounces, yards, feet, inches, or whatever - an impeding factor in foreign trade. Today foreigners frequently hesitate to buy American because they always are uncomfortable with a measuring system they have no feel for. A suggestion is to visit Canada and encounter how well the Canadians have adjusted to their changeover some years ago. My plea is therefore - hurry up USA, complete the metric conversion so American goods and services more readily are appreciated abroad!

Some back-tracking is now necessary as it cannot be assumed that you all are familiar with the supplementary nature of NR to that of XR, and with neutron-imaging foils like gadolinium (Gd), dysprosium (Dy), and track etch plastics [1, 9-11]. Hence we will explain some of the physics to conceive why NR is an important probe for the detection of hydrogen in various compounds like corrosion, concrete, plastics (not teflon), and so forth. You can, for example, visualize hydrogen inside steel like that of fuel in a cigarette lighter. Another approach is to superimpose an x-radiograph with a neutron image. An example is that of a toy locomotive where NR depicts the plastic parts, and XR the metallic parts. Super imposition details the whole locomotive [12].

PHYSICS OF NEUTRON RADIOGRAPHY

A unique capability is the ability to differentiate between isotopes of the same element; a differentiation that cannot be accomplished with XR as X-rays only interact with the atomic-shell electrons. Neutrons, however, don't interact at all with these electrons, but with the nucleons in the nucleus itself. These nucleonic interactions are very complex and of a quite different nature than those of photons with electrons. Because of this fact, it is feasible to differentiate between the isotopes of respective elements.

An example is natural boron, consisting of 20% boron-10 and of 80% boron-11. A 5-mil thick disc with 95% enriched boron-10 will attenuate most of the thermal neutrons. On the other hand, a 5-mil disc of 95% enriched boron-11 will transmit most of the neutrons [8]. A practical application is boron poison in nuclear fuel to determine burnup factors. Other examples of this isotopic selectiveness are hydrogen and deuterium, and of isotopes in nuclear fuel such as of uranium and plutonium.

Thermal neutrons are most copiously utilized for NR. The option to vary radiographic contrast is offered by cold and/or epithermal neutrons. Harwell in England made about 20,000 commercial radiographs last year. In the USA, Aerotest Operations in San Ramon, California is the biggest industrial NR contractor, employing a gadolinium foil with a single-coated emulsion SR-film for the imaging (Gd/SR). Their source of neutrons is a TRIGA-cored reactor [8]. A normal exposure time is 15-20 minutes with a thermal beam intensity of the order of 10^7 n/cm^2-s.

Because neutrons are electrically neutral, they cannot generate a film image as charged particles and neutral photons can. A converter to produce the ionizing radiation for the radiographic process is thus necessary when illuminating with neutrons. The most common converters are gadolinium for the direct method, and dysprosium for the transfer method. The latter is employed for radioactive materials like nuclear fuel.

TRACK ETCH

There is another imaging practice developed in Europe, namely track-etch radiography that is not commonly utilized in this country. Hence it is relevant to enlighten the technology in detail [13].

A cellulose-nitrate plastic sheet looks like a clear "window". After exposure and etching, the track-etch image appears "frosty". The "frost" patterns are caused by alpha particles from the $B^{10}(n,\alpha)$ and $Li^6(n,\alpha)$ reactions. These alphas from the boron and lithium converter produce invisible pits in the cellulose nitrate. Eventually they integrate into an invisible object-image, so initially you can't visualize anything. But when the "damage" caused by the pits in the plastic is enhanced by an etching process, you finally end up with a faint picture. Still that is not much of a radiograph, but with optical enhancement, preferably with polarized light, you easily obtain something acceptable. An overhead viewer readily demonstrates the effect.

The enhancement is achieved, as in an optical system like an overhead viewer or an enlarger, you only focus light that is transmitted through the unperturbed areas of the plastic. The light which hits the etched pits, essentially scatters out of the optics. Thus you obtain a very convincing visualization of the etched image, offering radiographic contrast compatible with that of film [14].

The development of track-etch radiography is basically the result of a cooperative effort between CEN Saclay and Kodak-Pathe' [15] in France. To the best of my knowledge no equivalent effort has been attempted in the USA. For this purpose it is emphasized that track etch has potentialities that exhaustively should be explored.

Some of the advantages of track-etch neutron-imaging are as follows:

1. Superior resolution than with any other imaging method (10μm - 100μm).

2. Integrating capability - not feasible in emulsion film radiography. With one single track-etch exposure, it is possible

to obtain the equivalent information as that of 3-4 film exposures.

3. Insensitivity to photon irradiations such as light, y- and X-rays; no cassette is thus needed.

4. The imaging plastic readily can be wrapped around an object, like a cylindrical one, if deemed necessary.

A disadvantage is that today a beam intensity of the order of 10^6 to 10^7 neutrons/cm^2-s is required for a reasonable track-etch exposure of about 20 minutes. Transportable NR units with either an accelerator or a radioactive source providing the imaging neutrons, usually offer lower beam intensites than reactors. It should be only a matter of time, however, before track-etch also advantageously is used in such units.

Kodak-Pathe's CN 80 is presently the most common track-etch plastic, replacing their red-colored CA 80-15 [15]. Apparently, for lower densities (to about D = 1,3), the CA 80-15 is more sensitive than regular silver-halide emulsion. For higher densities it appears to be the other way around. As the CN 80 is more light transparent than the previous CA 80-15, it's sensitivity characteristics are thus improved.

Examples of success are exposures of hydraulic jacks used in the European space program, depicting details that can't be observed by emulsion film. This is one of the reasons why nuclear fuel is always track-etched at Saclay just outside Paris. A classic example of depicting fine details is that of a rabbit brain. Blood vessels within the bone structure are readily visualized without applying any contrast agent. X-radiography would not show these patterns [14].

At the Jozef Stefan Institute in Ljubljana, Yugoslavia, track etch has found metallurgical applications. An example is the study of the boron distribution in steel. As you know, boron enhances the steel resilience. It was initially believed that boron patterns itself around single iron atoms, but track etch showed that this is not the case. This micro-radiography distinctively imaged the boron precipitation in characteristic hexagonal-shaped chains or bands, providing a network of boron around macro-areas of iron. The amount of boron was 0,1 wt.% in carbon steel [16].

Finally, track etch with fast neutrons knocking out protons in a hydrogen-containing plastic sheet, such as 1,6-mm thick CR-39 converter, has potentiality [17].

NEUTRON XEROGRAPHY

Neutron xerography is another imaging modality, and at the National Bureau of Standards (NBS) in Gaithersburg the technique has been studied. As you know, xerography is a low-priced reproduction method, and offers also superior edge enhancement in areas where change of object densities occurs. A famous xerograph is that of an ancient

Chinese ceremonial urn made of lead, exhibited at the Smithsonian in Washington, D.C. Because the lid was corroded to the urn itself, it was impossible to take it of and see what was inside. A neutron xerograph indicated that the urn had been damaged and then repaired. The interesting observation, however, was that of a feather-like leaflet. If any of you are experts on tomography, the challenge exists to look for a Chinese "cookie" reading [18]!

STANDARDS

To interpret a radiograph in detail, it is necessary with reference standards. Most NR is done with a spectral n-distribution like that from a reactor. Generally, it is not as easy to produce monoenergetic imaging neutrons as, for instance, in the case of X-rays. Although the present n-standards are not perfect, the most commonly used is a Beam Purity Indicator (BPI), and an Image Quality Indicator (IQI). A radiograph of a BPI indicates the spectral distribution of the imaging neutrons, including the gamma background. From the IQI-image an indication is obtained of the spatial resolution [19,20].

Cooperation exists between ASTM and NRWG about these matters, but it is still rather difficult to compare a neutron radiograph with an allegedly identical one taken at another facility. There are simply too many parameters involved that should be separated in more detail than what so far is accomplished. An example is the resolving power or resolution of an NR system. Normally, it is defined by the collimator ratio "L/D", where "L" is the length of the collimator and "D" is the input diameter [21,22].

A Canadian group under Professor Archie Harms at McMaster University in Hamilton, Ontario, has introduced a lambda (λ) resolution function that experimentally readily can be determined [9]. It is a very sensitive parameter, and at the NRWG-10 conference last May, tentative λ-values of 66,7mm^{-1} and 126mm^{-1} were reported for Gd/SR and track-etch, respectively. The higher the λ-value, the better the resolution. All NR facilities should measure their lambdas for intercomparison. Another recommendation is that a "λ-indicator" should be inherent in a future BPI and/or IQI.

ELECTRONIC IMAGING

As a final venture, I will conclude by mentioning electronic (real-time) imaging [2,23]. You will probably hear more about that from other speakers. What will be illustrated now, is a film-digitizing unit at the University of Chicago Hospitals [24].

A problem with radiography is that a trained human eye only can differentiate somewhere between 50 to 100 gray levels. In medical diagnosis it is frequently necessary to differentiate between tissues of a small contrast difference that cannot be observed visually. You can do a micro-densitometric analysis, but that is time consuming. However, a digital imaging unit can differentiate between, say 1024 gray levels, and then visualize the levels of interest. Incidentally, commercial units are today available that can differentiate 4072 gray levels.

One approach is to digitize the original radiograph like that of a pacemaker in a human chest. In medical as well as in technological diagnosis it is important for improved detection to provide radiographic images of superior quality that have high contrast, high resolution, and low noise. This is important whether the system is conventional or digital, but with digital image-processing techniques the detectability of low-contrast radiographic patterns can be improved.

A system for digital processing is two transparent "film" drums that synchronously rotate at 3 rps. The original radiograph on the output "read" drum is scanned, and the image signals are digitized by an analog-to-digital (A/D) converter. After digital image processing, the image data are converted back to the digital-to-analog (D/A) converter. Finally, on the input "write" drum, an emulsion film is exposed to a laser beam where the light output is modulated by the processed image signals. This system is a powerful tool for studies with digital radiography, because many parameters can be varied independently over a wide range. The "pixel" or aperture sizes are variable in steps of 50, 100, 200 microns-square.

The above discussion obviously indicated that low-contrast information is achieved faster by digitizing than by micro-densitometry. But please remember, digitizing a radiograph doesn't provide any other information than what is not already inherent in the film. The requested or desired facts are just furnished speedier by using "integrated densitometry".

CONCLUSION

It is a little more than 50 years ago that neutron radiography was "born" in Berlin, a few years before WW-2. When visiting Berlin for three months ago, an attempt was made to find more about this initial endeavor than what is described in, for example, Harold Berger's book [10]. Due to the war destruction of Berlin, especially in 1945, that appears to be a rather impossible task.

Today, however, NR has evolved into a mature NDT/E tool, representing informative and unique supplements to that of X-ray radiography. Unfortunately, it is much more expensive to produce N-rays than X-rays. An alleviation of the cost problem would be a transportable field unit as discussed at this conference [25].

REFERENCES

1. Neutron Radiography Handbook; complied by the Neutron Radiography Working Group (NRWG) under the auspices of the Commission of the

European Communities Joint Research Centre (EUR 7622 - 1981); edited by P. von der Hardt & H. Rottger, C.E.C. Joint Research Centre (170 pages); D. Reidel Publishing Company.

2. Nondestructive Testing Handbook, 2nd edition, ASNT, vol. 3, Radiography & Radiation Testing, section 12, p. 532-563, and section 13, p. 564-592. (Section 14, p. 593-640 concerns real-time radiography; section 16, p. 674-704 concerns radiation gaging; section 17 concerns X-ray diffraction and fluorescence), 1985.

3. Neutron Radiography; Proceedings of the 1st World Conference, San Diego, CA., 7-10 December 1981 (1073 pages); edited by John P. Barton & Peter von der Hardt; D. Reidel Publishing Company.

4. Neutron Radiography; Proceedings of the 2nd World Conference, Paris, France, 16-20 June 1986 (932 pages); edited by John P. Barton, Gerard Farny, Jean-Lois Person & Heinz Rottger; D. Reidel Publishing Company, 1987.

5. Practical Applications of Neutron Radiography and Gaging; edited by Harold Berger (319 pages): symposium sponsored by NBS & ASTM Committee on NDT; ASTM, Gaithersburg, MD, 10-11 February 1975.

6. International Neutron Radiography Newsletter; published irregularly in the British J. of NDT.

7. Neutron Radiography Newsletter 1964-1977; editor John P. Barton; distributed by The American Society for NDT (ASNT), 4153 Arlingate Plaza, Caller #28518, Columbus, OH 43228-0518.

8. Aerotest Operations, Inc., 3455 Fostoria Way, San Ramon, CA 94583. (AO's Newsletter published very irregularly).

9. Mathematics and Physics of Neutron Radiography, A.A. Harms & D.R. Wyman; D. Reidel Publishing Company, 163 pages, 1986 (see also review in Mat. Eval. 46-9, 1171, Aug. 1988).

10. Neutron Radiography: methods, capabilities and applications, Harold Berger; Elsevier Publishing Company, 146 pages, 1965.

11. Principles of Neutron Radiography, N.D. Tyufyakov & A.S. Shtan' (195 pages); published for NBS and the National Science Foundation, Washington, D.C., by the Amerind Publishing Company, Pvt. Ltd., New Delhi, 1979 (Atomizdat Publishers, Moscow, 1975).

12. Private communication: GE, Pleasanton, CA.

13. Neutron Radiography on Nitrocellulose Film; J. Markgraf & J.C. Domanus; to be published by D. Reidel Publishing Company.

14. Private communication: Ludwig Greim, GKSS Reaktorstation, D-2054k Gee sthacht.

15. Kodak-Pathe, 30 Rue des Vignerons, F-94300 Vincennes.

16. Private Communication: Joze Rant, Institut "Jozef Stefan", P.P. 199, YU-61001 Ljubljana.

17. Fast Neutron Radiography Testing for Components of Launch Vehicles by a Baby-Cyclotron, Y. Ikeda et al.: Presented at Conference on NDT/E for Manufacturing and Construction, held at the University of Illinois, Urbana, Ill., 10-12 August 1988.

18. Private communication: D.A. Garrett, NBS, Gaithersburg, MD (Dimensions/NBS, p22, March 1977).

19. Standard Method for Determining Image Quality in Thermal Neutron Radiographic Testing: ASTM E545-81 (American Society for Testing and Materials, 1916 Race Street, Philadelphia, PA 19103).

20. Standard Practices for Thermal Neutron Radiography of Materials: ASTM E748-85.

21. Standard Method for Determining the L/D ratio of Neutron Radiography Beams; ASTM E803-81.

22. Collimators for Neutron Radiography - an Overview; compiled by J.C. Domanus, Riso National Laboratory, Denmark, and edited by J.F.W. Markgraf, Commission of the European Communities, Joint Research Centre, Petten Establishment, the Netherlands; D. Reidel Publishing Company, (108 pages), 1987.

23. Position-Sensitive Detection of Thermal Neutrons; edited by Pierre Convert, ILL., and J. Bruce Forsyth, Rutherford Appleton Laboratory, Didcot, UK (430 pages); based on Proceedings of the workshop on position-sensitive detection of thermal neutrons, held at Institut Laue-Langevin (ILL), Grenoble, France; Academic Press, 1983.

24. High Quality Digital Radiographic Images: improved detection of low-contrast objects and preliminary studies; M. Ishida, P.H. Frank, K. Doi, J.L. Lehr; Radiographics, 3-2, 325-338, Spring 1983.

25. A Neutron Radiography System for Field Use, J.J. Antal and A.S. Marotta, US Army Materials Technology Laboratory, Watertown, Mass.: Presented at Conference on NDT/E for Manufacturing and Construction, held at the University of Illinois, Urbana, Ill., 10-12 August 1988.

FAST NEUTRON RADIOGRAPHY TESTING FOR COMPONENTS OF LAUNCH VEHICLES BY A BABY- CYCLOTRON

Y. Ikeda, K. Ohkubo, and G. Matsumoto
Nagoya University
Nagoya, Japan

T. Nakamura
National Space Development Agency
Tokyo, Japan

Y. Nozaki
Nissan Motors, Ltd.
Tokyo, Japan

S. Wakasa, Y. Toda, and T. Kato
Japan Steel Works, Ltd.
Hokkaido, Japan

1. INTRODUCTION

Recently, Neutron radiography (NR) has become an important means of nondestructive testing (NDT) in Japan. Especially thermal neutron radiography testing (NRT) has been used for the NDT of various explosive devices of launch vehicles, which are developed as a H-series program by the National Space Development Agency (NASDA) of Japan[1]. The NRT for launch vehicles has been carried out at the NR facility of a Baby-Cyclotron which is installed in the factory of Japan Steel Works (JSW) Ltd. In the NRT a conventional film method based on silver-halide emulsion has been exclusively employed to inspect various testing objects including components, and many valuable results have been obtained so far successfully. However, recently, as the H-series program advances, the launch vehicles to be shot up have become much larger. With larger launch vehicles, the parts used in them have also become larger and thicker.

One main disadvantage of the NRT by thermal neutrons is somewhat weak penetrability through objects because the energy is small. With the conventional thermal neutron radiography (TNR), steel objects being thicker than 40 to 50 mm are difficult to test through them because scattered neutrons obstruct real image of the object[2]. Some hydrogen-rich materials, such as organic elastomers, plastics and some explosive ones, are also difficult to be radiographed with the same reason if their thicknesses become more than 20 mm. From the facts, it has been recognized that the conventional TNR testing is no longer successful to inspect the new types of components used in the next large H-2 launch vehicles to be shot up in 1992.

Consequently a new method of NRT should be developed instead of TNR and applied to the new components of H-2 launch vehicles. In order to cope the requirement, fast neutron radiography (FNR) has been studied and has been tried for testing the new components of H-2, such as large separation bolts.

The FNR has been previously studied for biological and medical purposes[3]. The advantage of the FNR is that the fast neutrons generated with the Baby Cyclotron can penetrate through much thicker objects with some extent regardless of their composing elements, and the spatial distribution of the attenuation of fast neutrons with the objects makes the fast neutron image of them. Further the fast neutron beam is produced at the Be target of the Baby-Cyclotron without any special thermalizing system which is needed for TNR.

The image recorder used in FNR seems not to be much expensive though it is now under development. In this work, a solid state track recorder, i.e. CR39 (poly-allyl diglycol carbonate) plastic sheet[4] is employed, which is manufactured with Solar Optical Japan Co. Ltd. In the CR39 plastic sheet, fast neutrons interact with the hydrogen atoms composing the plastic and the impinged hydrogen atoms become recoil protons running forward through the plastic. After passing in the plastic, the recoil protons leave the very small areas, tracks, damaged along the flight, which are enlarged and become observable as etch-pits after appropriate etching process of the plastic. The FNR image is recorded with the distribution of the etch-pit density and finally observable after this etching process. Since usually gamma rays make no tracks, the CR39 plastic image recorder is essentially free from the fogging by them.

One main disadvantage of the CR39 sheet for FNR imaging is the low contrast of the optical density of the sheet. Since the etch-pits cannot absorb transmitting light as usual silver-halide emulsion films do, and only make shadows by scattering partially the light, then the difference of the optical densities between different etch-pit densities is rather small. To overcome this drawback, some contrast enhancement techniques have been studied[5]. With some of them, well-defined radiographs by FNR have been obtained.

2. NRT EQUIPMENT AND CR39 PROCESSING

2.1 NRT equipment of Baby-Cyclotron Facility

The Baby-Cyclotron used for the NRT is the one developed and constructed by JSW LTD., which is shown in Fig. 1. The Baby-Cyclotron produces a high energy proton beam for several purposes including neutron production. Other main purposes are the production of positron emitters for medical uses, and particle induced x-ray emission (PIXE) analyses of materials. The similar small type of cyclotron have been gradually popular nowadays.

The NR equipment installed is usually composed of a Be beam target for fast neutron generation, a stack of neutron moderators for reducing the neutron energy to thermal range and a neutron collimator for obtaining parallel beam as well as an imaging device holder. The accelerating voltage for proton beam is 16 to 32 MeV, and the beam current is nearly 50 μA. The target is made of a beryllium disk cooled by water. At a typical power condition about 4×10^{12} n/sec of fast neutrons can be

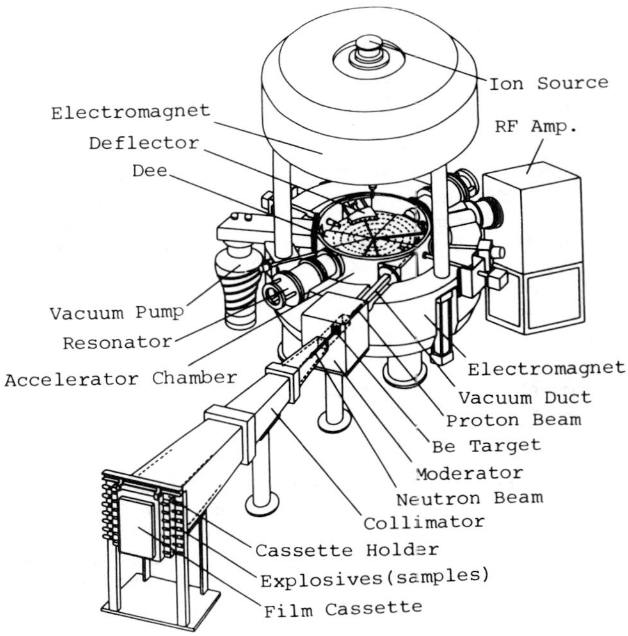

FIGURE 1. Baby-Cyclotron and its NR Facility used for NRT.

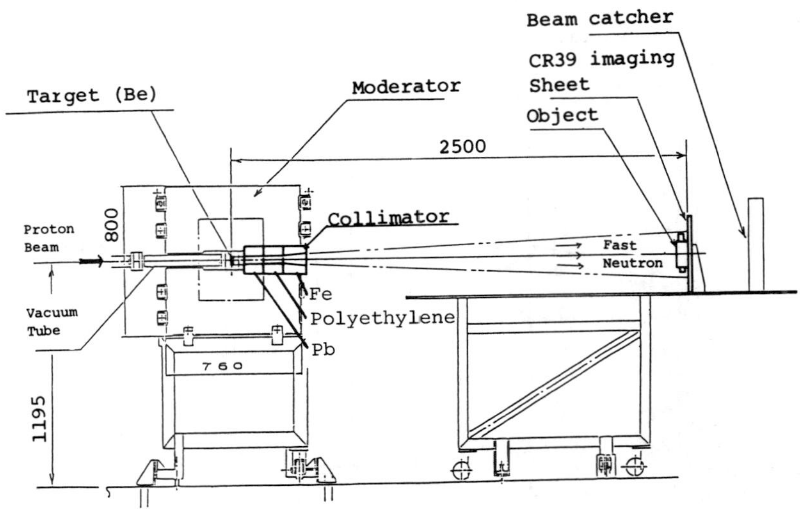

FIGURE 2. Setup of NRT by fast neutron radiography.

generated by $^{9}Be(p,n)$ reactions. The peak energy of the fast neutrons is estimated about 1 to 2 MeV[6].

However, in the FNR tests, in order to achieve the beam intensity of fast neutrons as high as possible, the part of moderator which is located in front of the target has been removed, and a new collimator system for fast neutrons is setup in front of the target, as shown in Fig. 2. The inlet diameter, D, of the new collimator hole is 20 or 40 mm. The imaging position of the FNR is changed around the distance, L, of 2 m from the target. Therefore, the geometrical collimator ratio, L/D, which is the geometrical index of the beam paralleling[7], is about 50 to 100. The fast neutron flux intensity at the imaging position is about $8x10^6$ $n/cm^2 sec$.

2.2 CR39 image recorder

Exposure and etching Figure 3 shows the principle and the basic arrangement of the FNR testing and chemical etching of CR39 sheet after exposure.

The imaging of the FNR has been carried out with CR39 plastic image recorder, in front of which objects are closely placed. The CR39 plastic sheet converts the fast neutrons into recoil protons by elastic scattering reactions with hydrogen atoms contained in the CR39 and the tracks of recoil protons are recorded in the CR39 sheet after appropriate etching procedures, as shown in Fig. 3a. In this figure, the exposed CR39 is etched out from both surfaces (front and back) with a bulk etching speed V_B, and then the thickness of it is reducing. Further along the tracks of recoil protons, the CR39 is etched out with track

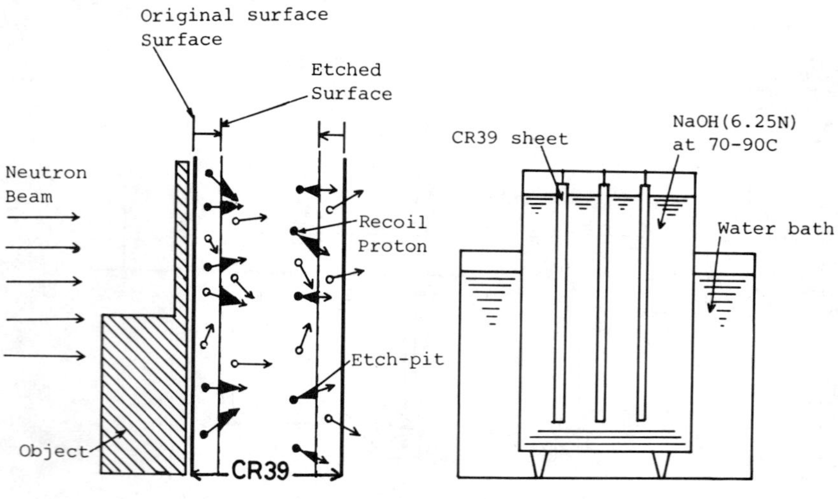

(a) Principle of FNR with CR39 (b) Etching of exposed CR39 sheets

FIGURE 3. Exposure and etching of CR39 plastic track detector etching speed V_T much faster than the other parts of the balk surface with V_B. Therefore, etch-pits are generated at the tracks on the surfaces if the angles of recoils are appropriate (closed circles), i.e. smaller than $90°-\theta_C$, where θ_C is the critical angle, $\sin^{-1}(V_B/V_T)$. Whereas if the angles of recoils are much wider than $90°-\theta_C$ (open circles), no etch-pits of the tracks are produced.(8)

The CR39 used in this work has the thickness of 1.6 mm. The etching of CR39 has been done with NaOH solution, which normality is 6.25 N. The etching temperatures of the solution are appropriately set at 90°C with accounting for etching time needed and neutron fluence exposed. It takes about 40 minutes to complete etching.

After finishing the etching of CR39, the etch-pit yield per fast neutron was measured by using a microscope. The etch-pit yields of 1.3×10^{-4} pit/n and 9×10^{-5} pit/n are obtained for the back surface and the front one, respectively. In order to obtain good quality of FNR image, more than 3×10^6 pits/cm^2 of etch-pit density is needed, which means that the appropriate exposure time of this facility is nearly 1 hr.

Image enhancement Usually the image on CR39 sheet has rather low contrast optically, because the etch-pits don't absorb any lights as the silver grains in conventional films do and only scatter it to make the shadows of the images. Therefore in order to make them visible much better, some image enhancement technique must be applied. For the important purpose, a new image transfer method has been devised and applied in this study. In the image transfer method, a photography camera takes the images on the CR39 sheet by using a high contrast optical film with high resolution, such as Minicopy film of Fuji Film Co.. The light for taking photography is polarized and exposed to the imaged CR39 sheet from both oblique side to avoid the direct light from the sources entering into the camera, so that only the scattered light at the surface of the CR39 having etch-pits can enter the camera and transfer the image to optical film. Since the optical film used for this purpose, Minicopy, has very high contrast property, even rather low contrast image on CR39 sheet has been strongly enhanced. Figure 4 shows a result of the image contrast enhancement. In this figure, the optical densities of both original CR39 sheet and transferred optical film (Minicopy) are plotted against the logarithm of the exposed fast neutron fluence. The optical densities have been measured with a microphotometer. As can be seen, under the exposure less than 3×10^6 n/cm^2, the contrast (slope of the line) for the optical film is about three times larger than that of the original CR39 sheet. The optical density of Minicopy shows rather rapid satulation as the neutron fluence increases in this figure. This is due to the satulation characteristics of the film itself. If the shutter time of the camera are properly adjusted for such higher region of pit density, much larger slope of the optical density curve of the Minicopy can be obtained for such higher fluences.

In this work, the FNR photographs shown later are obtained with the method.

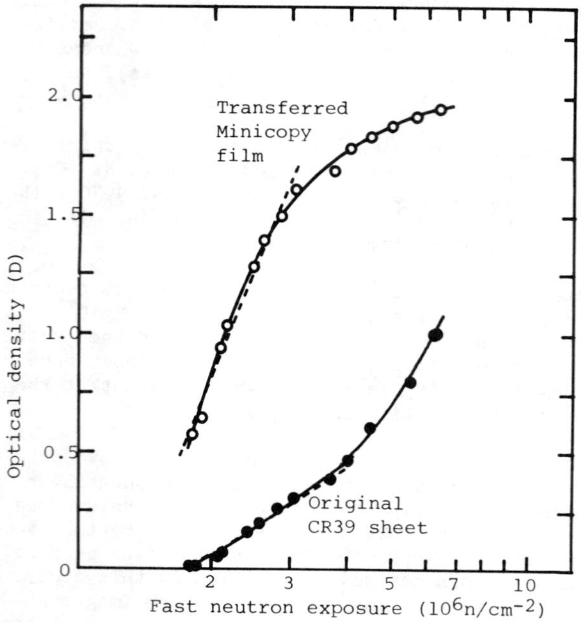

FIGURE 4. Typical optical density of original exposured CR39 sheet and those of Minicopy films used for image contrast enhancement. (Dotted lines are given for comparison of contrasts of both films.)

2.3 The aims of NRT for explosive devices of launch vehicles[1]

Figure 5 shows the illustration of H-2 type launch vehicles, which is a third stage space rocket with boosters. It is equipped with many explosive devices to separate the components. Since their reliability cannot be confirmed by any repeating testing, the NRT at the time of manufacture process is very important.

The items of the NRT for the explosive devices are as follows,

1. To inspect the explosive, its existence and no impurity.
2. To assure the existence of '0' rings.
3. To check the perfection of explosives, their no gaps, no cracks and no dips.

To satisfy these items it is necessary for the FNR to have high contrast and high resolution for explosives and '0' rings in thick steel containers.

Previous experiment of FNR has shown[5] that a cubic carbohydrate (1 cm in thickness), as a matter for simulation of explosive, can be clearly imaged through a steel plate of 90 mm in thickness. Besides, if the L/D of collimator is sufficiently high, smaller than 0.5 mm gap distances could be detected.

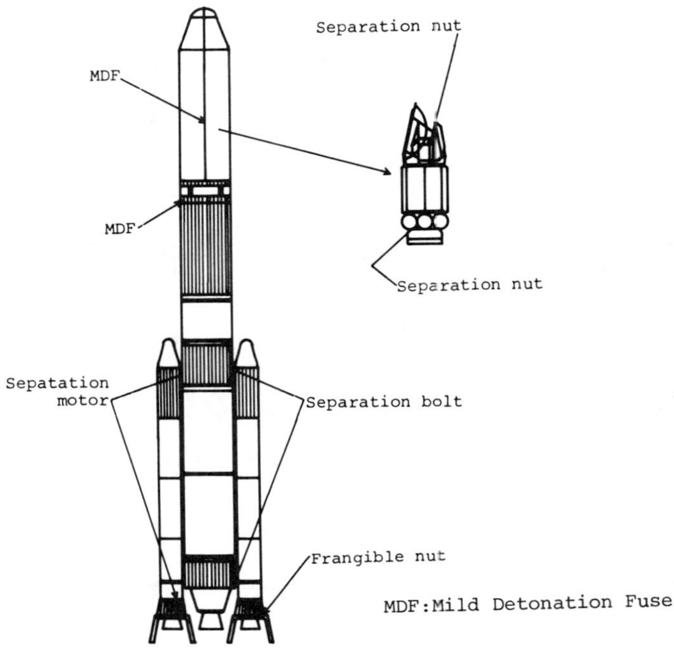

FIGURE 5. The illustration of H-2 launch vehicle.

3. RESULTS OF FNR IMAGING

3.1 FNR of simulation models

Various objects such as explosive devices and their simulators are imaged with the FNR. Figure 6 shows the illustration of two large objects which are prepared as simulators of the large explosive separation bolt employed in H-2 launch vehicles. As shown in the figure, both are made of steel in substantially large thickness. The largest diameters are 100 mm. They include various 'O' rings and vacancies as well as some other materials, such as water in the A type, and lithium hydro-oxide, boron carbide, small beads of polyvinyl alcohol and lead in the B type.

The result of FNR for them are shown in Fig. 7 (a) and (b). In both FNRs, 'O' rings inserted in the objects are clearly discernible and some other including materials are also discriminated by the contrast differences. It is especially noticed that lithium hydro-oxide and boron carbide show no longer strong absorption of neutrons. On the other hand, water and organics can yield substantially strong contrast even to the thick steel. Some cavities over water (7a) and over organics (7b) can be also observable. The filler of lead (7b) yields only slightly higher attenuation of neutrons than the air cavity.

The measured film's optical densities for various parts of the simulators are given in Table 1. In spite of the low contrast of the images on the original CR39, the transferred optical films yield substantially higher

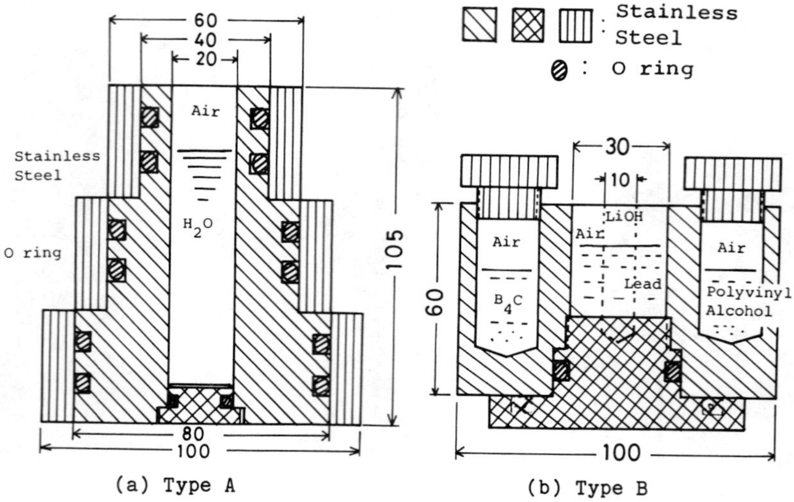

FIGURE 6. The illustration of simulation models for large explosive devices of H-2, Type A and Type B.

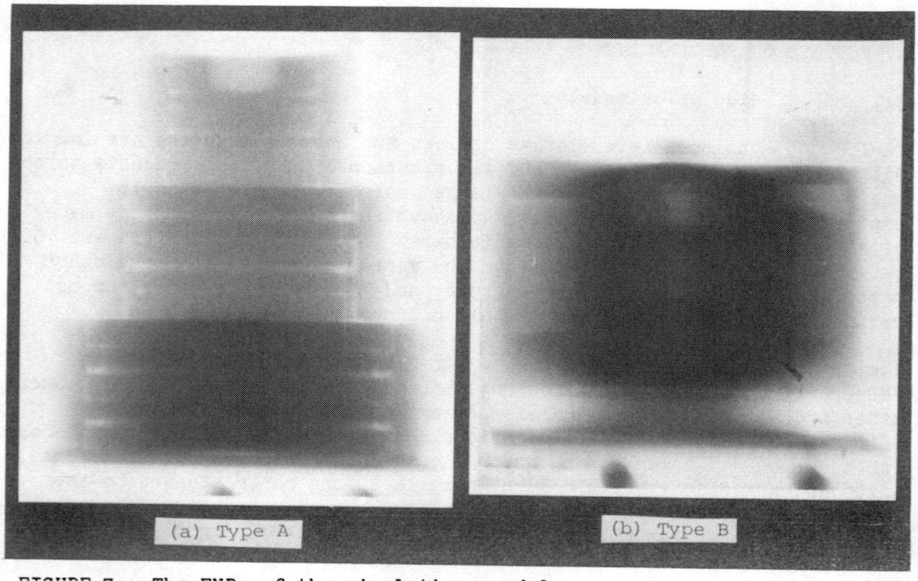

FIGURE 7. The FNRs of the simulations models

TABLE 1 Typical optical density measured for the image of Type B in the CR39 and transferred optical film (Minicopy)

Location Materials (thickness, mm)	Optical density CR39 (original)	 Optical film (transferred)
1. Center		
Fe(100)	0.37	0.16
Fe(50) / LiOH(10) / Air(40)	0.47	0.70
Fe(50) / LiOH(10) / Air(10) / Pb(30)	0.42	0.51
Fe(80) / LiOH(10) / Air(10)	0.40	0.32
2. Left Fe(50) / Air(30)		0.50
Fe(50) / Pb(30)		0.45
Fe(80)		0.31
Fe(30)		0.60
3. Right Fe(30) / Air(20)	0.46	0.68
Fe(30) / $(C_2H_4O)_n$(20)	0.40	0.37
4. Background (no object area)	0.62	1.40

contrast for the images. The optical film, Minicopy, yields the optical densities of the lightest background area (no objects) and the thickest part of steel as 1.40 and 0.16, respectively. However, the corresponding optical densities on the original image of the CR39 sheet are only 0.62 and 0.37. Therefore, the contrast difference between the maximum and the minimum densities of the optical film is about five times larger than that of the CR39 in this case. This is a quite big improvement.

3.2 FNR test of large explosive devices of H-2 launch vehicles

The large explosive devices of H-2 launch vehicles have been imaged by the FNR. Figure 8 shows the picture of the typical explosive devices, i.e. the center bigger one (a) is a separation bolt, and the two smaller ones by the both sides are a separation nut (b) and a separation motor (c). The main bodies of these parts are made of steel. The diameter of the largest separation bolt is 90 mm. The inner configuration of it should be briefly described. It is composed of some parts, such as the pressure cartridge at the head containing explosions attached with fuse plugs, which is followed with a large and thick main body (cylinder), the wall thickness being nearly 15 mm, and the inner diameter of it being 63 mm. In the inside of this main body, there is a movable piston assembly, which are composed of two steel pistons and one lead coupling. Between the piston and the cylinder, several 'O' rings are used to prevent the exploded pressure leaking when the explosive is fired.

FIGURE 8. The photograph of the outward of explosive devices, of H-2 (a) separation bolt, (b) separation nut and (c) separation motor

The resulting FNRs show the inner structures of the separation bolt and the separation nuts, fairly clearly. The small screws for fuse plugs on the cartridge can be observable. All four 'O' rings and the lead coupling are well imaged, as well as the outer and the inner fringes of the main cylinder, and the inner piston being clearly imaged. The width of 'O' rings is 7 mm, then in this object, the thickness of 14 mm of the elastomer has been well imaged through 76 mm thickness of steel.

For the results of smaller separation nuts, fairly details of the composition such as screws and 'O' rings can be successfully observed. The resolution of the FNRs for such smaller objects are sufficiently high, and they almost seem to satisfy the main items of NRT mentioned previously for such smaller explosive devices.

In this case, the devices could not be treated as containing any explosives. Therefore there is no explosives in the inside. The possibility for the detection of narrow gaps and small cracks existed in explosives is still uncertain for such large devices. The problem should be cleared near future. However, the results obtained so far give us much great hope for the NRT of such large components of H-2 launch vehicle with the FNR.

REFERENCES

1) Yamawaki K., Fukushima Y., Nakamura, T. and Sekita, J., "Neutron Radiography Facility For Explosive Devices of H Launch Vehicles and Its Application" in 'Neutron radiography' (Proc. 2nd World Conf. NR) eds, Barton J. P., Farny, G., Person, J. L. and Rottger, H., D. Reidel Pub. Co, Dordrecht, Netherlands, pp.313-320, 1987.

2) Ikeda, Y., Ohkubo, K., Suzuki, H., Tomatsu, Y. and Matsumoto, G., "An Evaluation of Neutron Radiography for Non-Destructive Testing of Defects" in 'Neutron radiography' (Proc. 2nd World Conf. NR), pp.305-312, 1987.

3) Dühmke, E. and Greim, L.:, "Fast neutron radiography in biology and medicine" in 'Neutron Radiography' (Proc. 1st World Conf. NR) eds. Barton, J. P. and Von der Hardt, P., D. Reidel Pub. Cd., Dordrecht, Netherlands, pp.537-580 1983. Also see the next paper,'Dühmke, E., Greim, L. and Schmitz, H. W., "Fast neutron radiography (FNR) with track etch technique" in 'Neutron Radiography' (Proc. 2nd World Conf. NR), pp.639-646 1978.'

4) Somogyi, G., "Status of development in the field of CR39 track detectors" in 'Solid State Nuclear Track Detectors' (Proc. 11th Int. Conf.) eds Fowler, P. H. and Clapham, V. M., Pergamon Press, Oxford, pp.101-113 1982. And also see the next paper,'Cartwright, B. G., Shirk, E. K. and Price, P. B., Nucl. Instr. Methods, 153, pp.457-460 1987.'

5) Ikeda, Y. et al., 'Fast neutron radiography tests at the YAYOI-reactor, University of Tokyo' Nucl. Instr. Methods, forthcoming.

6) Lone, M. A., Ferguson, A. J. and Robertson, B. C., "Characteristics of neutrons from Be targets bombarded with protons, deuterons and alpha particles" Nucl. Instr. Methods 189, pp.515-523 1981.

7) Berger, H., 'Practical Application of Neutron Radiography and Gaging', ASTEM, Special Technical Pub. 586. See 'Neutron Sources for Radiography and Gaging' by Cutforth, D. C. 1975.

8) Fleischer, R. L., Price, P. B. and Walker, R. M., 'Nuclear Tracks in Solid', University of California Press, Berkeley, p.50 1975.

IMPROVEMENT OF RECONSTRUCTED IMAGE QUALITY OF NEUTRON COMPUTED TOMOGRAPHY

Y. Ikeda, M. Koike, S. Honda, Y. Tomatsu, and G. Matsumoto
Nagoya University
Nagoya, Japan

ABSTRACT

Neutron computed tomography has been studied. Endeavor has been given to obtain high image quality of CT reconstruction. Film method is comparatively preferred to dynamic neutron TV one. Some models for nuclear fuels have been reconstructed. Dispersion of 300 μm Eu-particles in TiO_2 pellets, which simulate PuO_2/UO_2 nuclear fuel, have been well reconstructed.

1. INTRODUCTION

Neutron Computed Tomography (NCT) has been studied for the inspections of nuclear fuel elements, such as rod bundles and fuel pellets, as well as some purposes of widening the application of neutron radiography (NR) technique. One of the goals of this investigation is to obtain the high quality of reconstructed image, i.e. to attain the high resolving power, low noise level and artifact free of CT image is aimed. In order to attain the goal, various conditions for NR imaging, such as the characteristics of NR facilities, and NR imaging systems has been studied. Several computational processes for NCT including the methods of data inputting and computational algorithms have been studied.

The NR facilities employed in this work are the ones installed in Kyoto University Reactor (KUR) and Rikkyo University Reactor (RUR). Both NR facilities have moderate neutron flux intensity, tolerable neutron collimator ratios, and sufficiently high neutron/gamma ratios.

Fundamentally, two NR methods for neutrons CT have been studied, one being high resolving film method[1,2] and another being high speed electronic imaging method by using neutron TV systems[3].

The computational processing for the former film method has been carried out with the large scale computer, FACOM 382, of Nagoya University, whereas for the latter electronic imaging method, an image processing computer, TOSPIX-2, has been employed. For the simulation tests of nuclear fuel elements, and their bundles, various modeling samples have been prepared and investigated. By the film method, very fine particles (less than 300 μm in dia.) of neutron absorber (Eu_2O_3) can be clearly reconstructed.

2. PRINCIPLE OF NEUTRON CT

2.1 Parameters of neutron CT

The principle of neutron CT is fundamentally similar to X-ray CT. However, for neutron CT, the object is treated as the material composed of neutron cross section. The purpose of neutron CT is to reconstruct the distribution of neutron cross section in the object. Except for this point, both CT have the same mathematical frames. Then the computational algorithms developed for X-ray CT are quite applicable to neutron CT.

The principle of CT was firstly given by J. Radon[4], and has been developed and reported by many researchers. Therefore, the explanation about it should be very limited. For only brief explanation, Fig. 1 is given. In this case, the object is represented by a function f(x,y), which is the distribution of neutron cross section for the object. The x-y coordinate system is fixed in the object itself. By exposing a neutron beam to the object, a projection of it is obtained. Providing the Y axis of the X-Y system is taken as the same direction as the neutron beam, the projection is expressed with a function $g(X,\theta)$, which is the integral neutron cross section along the Y axis as following,

$$g(X,\theta) = \int_{-\infty}^{\infty} f(x,y : X,Y)dY , \tag{1}$$

where θ is the rotation angle made by the two coordinate systems. For CT, many projection data are required, i.e., the angle θ has to be changed with rotating the two systems. However, in neutron CT, usually the source of neutrons is a reactor, which is stationary. Then in neutron CT the X-Y system is fixed in the place and the object, i.e. the x-y system, is only rotated with a small and constant increment $\Delta\theta$. The numbers of the projection, NT, is given with $\pi/\Delta\theta$ because π rotation of object is enough to execute CT.

The projection $g(X,\theta)$ is recorded in NR image as the pattern of neutron attenuation, which is expressed as following,

$$\Phi(X,\theta) = \Phi_0 \exp\{-\int_{-\infty}^{\infty} f(x,y:X,Y)dY\} , \tag{2}$$

$$= \Phi_0 \exp\{-g(X,\theta)\}, \tag{3}$$

where Φ_0 and $\Phi(X,\theta)$ are the neutron flux before and after transmission of the object. The modified neutron flux $\Phi(X,\theta)$ is the projection NR image and recorded in radiographic films as the pattern of optical density, D, or with electronic devices as electric charge distribution.

The projection NRs are read out and scanned with several methods. In this work, mostly a very fine drum scanner is used for reading optical density as the inputting device for the computer. The sampling pitch, SP, of the device is changeable between 25 and 1000 μm. Then if the length of projection data is LP, the numbers of data sampling per projection, NS, is given by the relation, NS = LP/SP.

Since the sampling data are optical density, they are converted to the integral cross section $g(X,\theta)$ before they are used in reconstructuring

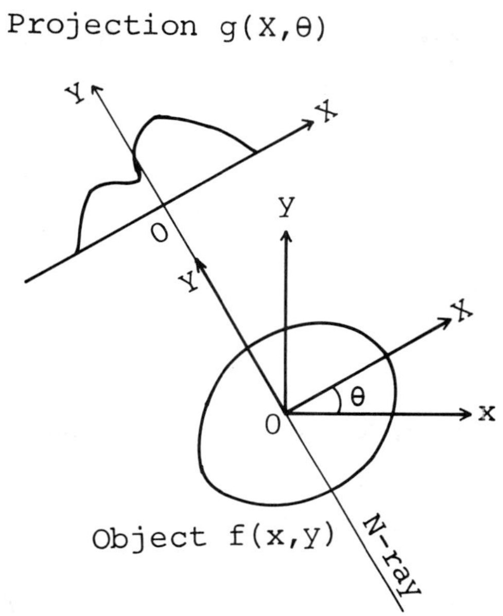

FIGURE 1. Function g(X,θ) as a projection of the object function f(x,y) at angle θ, where X-Y axes are taken as the Y being along the neutron ray, and the x-y axes are fixed in the object.

TABLE 1. Notation of parameters used in neutron CT

Parameter	Notation
1 Numbers of Projection	NT
2 Length of 1-dimensional projection	LP(mm)
3 Sampling pitch	SP(μm)
4 Numbers of Sampling	NS
5 Numbers of reconstructing pixels (1-dimension)	NP

TABLE 2. Characteristics of NR facilities

	KUR	RUR
Thermal neutron flux ($n/cm^2/s$)	$1x10^6$	$3x10^6$
Collimator ratio (L/D)	100	100
Cd ratio	400	5
Neutron/gamma ratio ($n/cm^2/mR$)	$1x10^6$	$6x10^5$
Imaging area (cm in dia.)	16	20

process according to previously obtained D-g curves.

Reconstructuring is performed with some CT algorithms. For comparison, Fourier Transformation (FT)[5], Filtered Back Projection (FBP)[6] and Convolution Integral (CI)[7] algorithms are referenced, coded and employed.

The mesh pitch of reconstructure coordinate system, RP, are appropriately selected to give much better quality of CT image. The numbers of pixel for reconstruction CT image, NP, is given as, NP =(LP/RP) x (LP/RP).

The parameters used here are summarized in the table 1.

3. AUTOMATIC IMAGING OF NEUTRON RADIOGRAPHS FOR CT

3.1 NR facilities

The most important factors of NR facility for neutron CT are the flux intensity and the collimator ratio. Since the neutron CT needs lots of projection, the exposing time per one projection should be small, which requires high flux intensity of neutrons. The distance between the object and imaging device is somewhat larger for neutron CT than usual neutron radiography testing, which causes the blurring of the image larger when the low collimating of neutron is employed. The level of neutron collimating is given by the collimator ratio, L/D of NR facility, where L is the length and D is the inlet diameter of the collimator used in NR facility. Since the geometrical unsharpness of NR is given as the product $(L/D)^{-1} \cdot L_s$, where L_s is the distance between object and image screen, the larger L/D ratio is very important for obtaining fine projection data for CT.

Two reactor facilities for neutron radiography have been used in the present CT work. Table 2 shows the characteristics of both NR facilities. The NR facility of KUR has moderate neutron flux intensity, high L/D ratio being 100, high neutron/gamma ratio being $1x10^6 n/cm^2/mR$ and very high Cd ratio being 400. Therefore, good image quality for NR can be proved with this NR facility.

Comparing with the KUR facility, the RUR facility has rather lower neutron/gamma ratio and Cd ratio, though the neutron flux intensity is higher than KUR. Therefore, the thermal content of NR films obtained at RUR is somewhat lower. However, suitable choice of radiography films can result in sufficiently tolerable images.

3.2 NR imaging with an automatic camera and the process for obtaining projection data

In order to obtain many projection data for variously different angles, imaging of NRs has been devised automatically. For this purpose an automatic camera has been employed, which configuration is shown in Fig. 2. Detail description about the camera is given the reference [2]. For the supplement, the camera is using a roll film, which is mostly IX-100, Fuji Film Co., the length of one roll and the width of the film being 30 m and 70 mm, respectively. The type IX-100 film has moderate high speed and also moderate resolving power, but sufficiently high contrast and dynamic range. By using one roll of the film, usually about 50 projection are imaged. The exposure time of the film at KUR or RUR is about 2 minutes per one projection, which means it takes more than 3 hours to obtain a set of 100 projections. But since the camera has been completely operated by an automatic sequence controller, the human labor work is very small.

A godolinium metal foil (25 μm thickness) is used as an image screen because it has high resolving power and insensitivity to gamma-rays. To attain close contact between the metal foil and the film, the screen is pressed to the film with two magnetically operated plungers. The walls of the camera are coated with boron containing paint to shield neutrons and the lead shieldings are used in- and out-side of the camera.

The exposed films are developed with a industrial X-ray film developer. The highest optical density of the film has been chosen to be about 3.0.

After developing the roll film has been cut to give many projection films and the projection NRs are lined in the drum of the drum scanner. The data-sampling aperture and pitch of this unit are variable from 25 μm to 1 mm. However, since choosing too small aperture results too noisy data, then the aperture size and sampling pitch of 50 μm have been employed. The optical density (D) being from 0-4.0 is possible to measure, but usually the maximum D are set about 2.5-3.0 because of good contrast resolution. After sampling the data, the optical density is converted to integral neutron cross section (Σ). Sometimes a procedure for redistributing the histogram of Σ is applied to those projection data in order to increase contrast resolution.

4. TESTING AND COMPARING OF CT ALGORITHMS

There are several computational algorithms for neutron CT as mentioned above. In order to compare the characteristics of them, both computer simulation and reconstruction from experimental data have been performed for several test objects. Figure 3 shows the results of the three algorithms for reconstruction of one of the test objects. As can be seen, FT algorithm shows much noisy image in this case, but FBP and CI yield less noisy images in the center. The filter function used here is the Shepp & Logan function[7].

Figure 4 shows the change of the amount of noise or artifact according to the numbers of projection, NT. In this test, the model (given in Figure 3) to be reconstructed is an aluminum disc of 50 mm in dia., which contains four trains of rods made of four different elements. This testing model represents the low contrast inclusions in aluminum matrix in order to study the contrast resolution between different algorithms, then the inclusions, such as (1) copper, (2) lead, (3) lead-tin alloy and (4) duralumin are used for rod arrangements. In order to evaluate the amount of the noisy artifacts in the CT reconstructed image, the standard deviations of the output levels for the whole aluminum matrix have been calculated, and the amounts of artifact have been defined as the ratio, $A=\sigma_n/\bar{g}_{CT}$, where A is the amount of artifact, σ_n is the standard deviation and $\bar{g}_{CT}$ is the average output for CT reconstructed image of aluminum matrix. As shown in fig. 4, the amounts of artifact for FBP and CI method are remarkably decreased as the numbers of projection, NT, increase. The amounts of artifact have reached at a constant levels fairly independent to the NT themselves when they become larger than 100

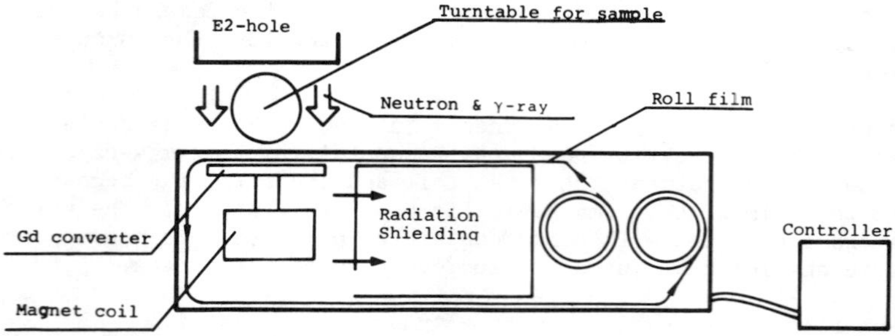

FIGURE 2. Configuration of the automatic neutron CT camera.

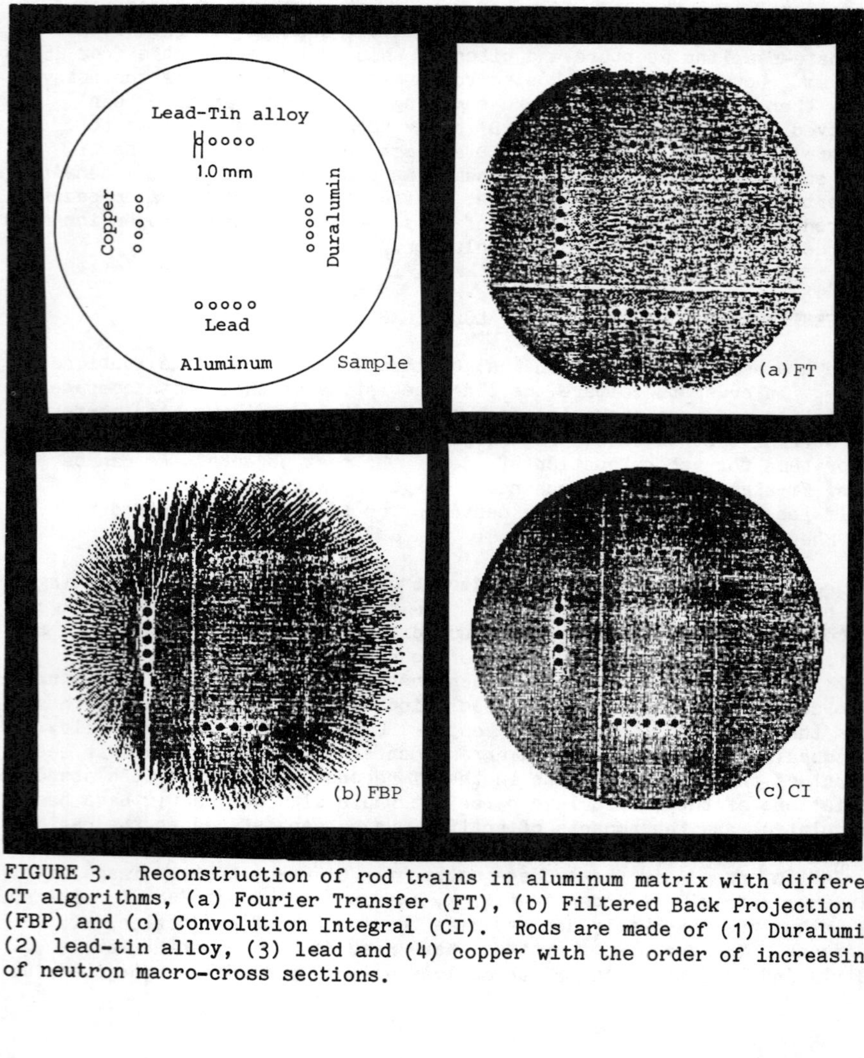

FIGURE 3. Reconstruction of rod trains in aluminum matrix with different CT algorithms, (a) Fourier Transfer (FT), (b) Filtered Back Projection (FBP) and (c) Convolution Integral (CI). Rods are made of (1) Duralumin, (2) lead-tin alloy, (3) lead and (4) copper with the order of increasing of neutron macro-cross sections.

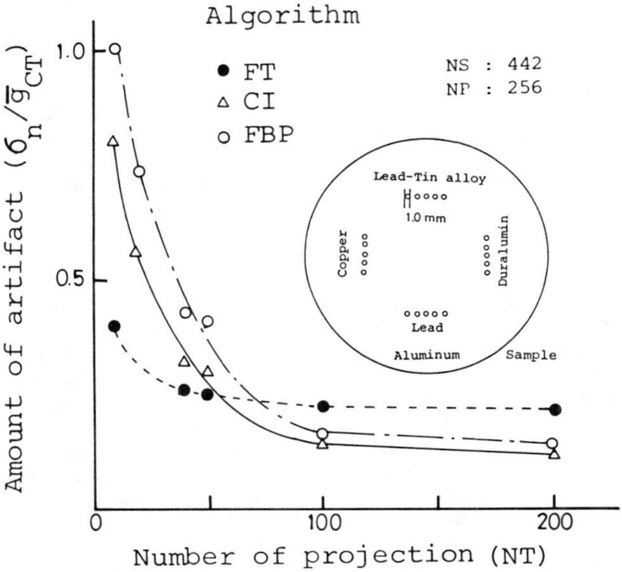

FIGURE 4. Variation of artifacts in reconstructed CT image as a function of number of the projection, NT.

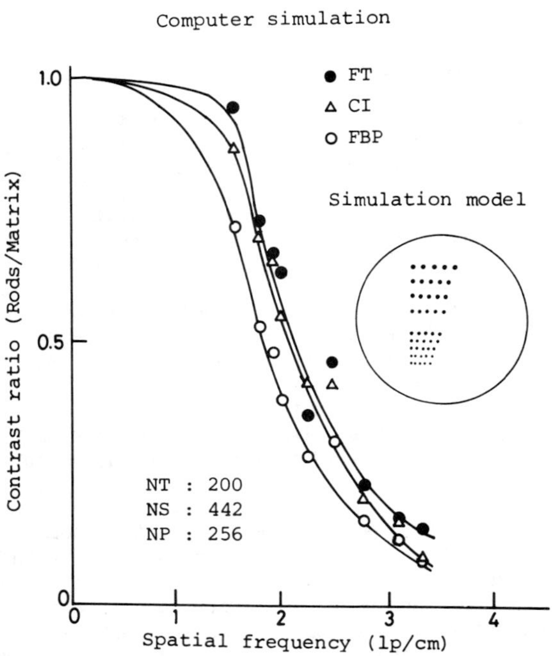

FIGURE 5. Variation of the contrast of rod arrangements in reconstructed CT image as a function of spatial frequencies.

for FBP and CI, and much faster for FT. However, as mentioned previously, the last constant level of FT is substantially higher than those of two others.

Figure 5 shows another results of the testing, in which the contrasts of reconstructed CT images are compared. As space frequency increase, the contrasts between the rod arrangements and the aluminum matrix quickly decrease. In this case, CI and FT show only slightly higher contrast than FBP. Although the CI algorithm needs the longest computing time, for the sake of good image quality and easier handling of computer programming, the method has been mostly used in this work.

For above results brief discussion should be given. For the amount of artifact or noisiness, the FT method seems to be inferior to other two method. This reason may be clear. The FT is using one-dimensional Fourier Transformtion of projection data with a polar coordinate system, but when reconstructuring and carring out inverse Fourier Transformtion of two-dimension, one must use a Cartesian coordinate system, and the nodes are different between them. Therefore, interpolation must be needed for the process. In this work, two-order interpolation is used. Probably, the incompleteness of the interpolations is the main source of the artifact. Second, the FBP is slightly inferior to others for the contrast transfer with spatial frequency. The reason of this is somewhat speculative, but in this case some noise filter function is used with Shepp and Logan filter function. Probably this process seems to be affect the image unsharpness. However, more examination will be needed for the problem.

5. RESULTS OF NEUTRON CT

5.1 Neutron CT for the simulators of nuclear fuels

One important application of neutron CT is in the field of nuclear technology, especially, the inspection of nuclear fuels, e.g., fuel bundles, fuel rods and fuel pellets. Both irradiated and non-irradiated fuels can be the objectives for neutron CT. However, it is very difficult to execute directly neutron radiography of post-irradiation materials in Japan. Therefore, in this work, some simulators for them have been prepared and reconstructed.

Figure 6 shows the illustration of the simulator for a fuel bundle of fast reactor used in the neutron CT. In this case stainless steel rods of 5 mm in diameter simulate the fuel rods in the bundle. The rods are arranged with regular hexagonal. However, to test the ability of detection for some defects, it has one blank and three failure rods in the arrangement. Each failure rod has a hole being 1 mm, 2 mm and 3 mm dia. in the inside. The outer casing is made of aluminum with the space filled with normal air.

Figure 7 shows the reconstructed CT images with three different method, but using similar CI algorithm: (a) the result with film method, (b) with film method improved with an image restoration technique, and (c) with neutron TV. As can be seen, the aimed characteristics of the simulator have been well reconstructed essentially, i.e. the rod arrangement is perfectly reconstructed, especially (a) and (b), and the blank and failure rods are well reconstructed. Especially in (b) even the smallest hole of 1 mm dia. in the rod can be observed with no difficulty.

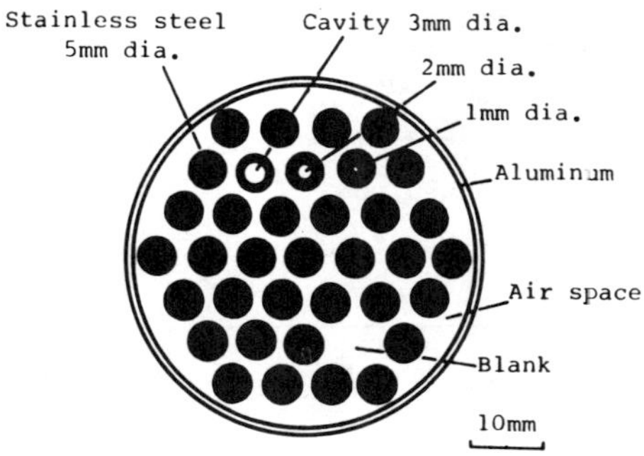

FIGURE 6. Illustration of simulation model for reactor fuel bundle.

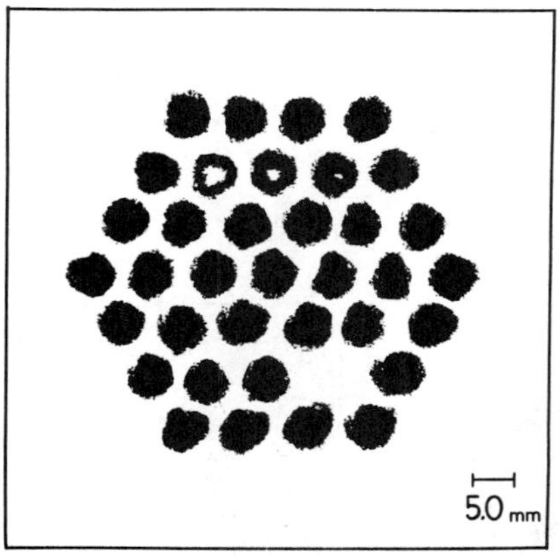

(a) Film method without image restoration

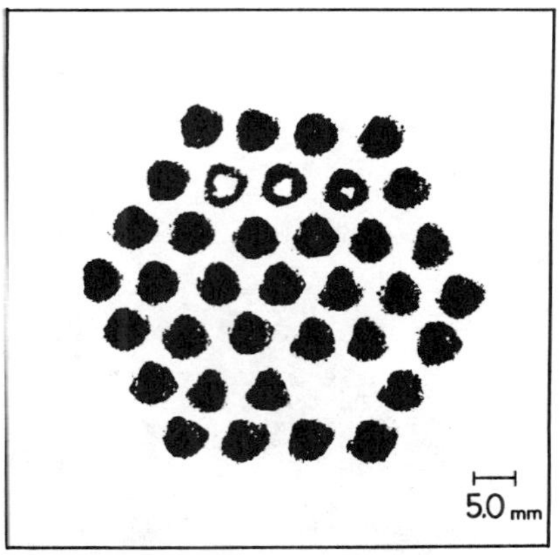

(b) Film method with image restoration.

Condition of film method
NT:100
NS:500 (SP 100μm)
NP:512 x 512

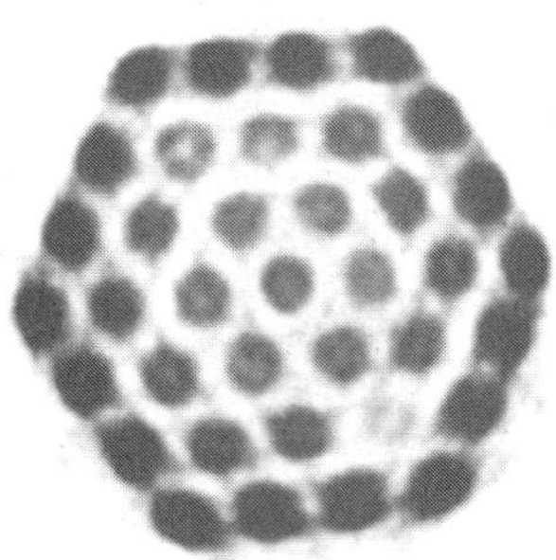

(c) Neutron TV method

FIGURE 7. Reconstructed image of the simulation model of fuel bundle, (a) Simple film method, (b) Improvement with image restoration (Wiener filter being used) and (c) Neutron TV method.

TABLE 3. Samples of TiO_2 pellets including various sizes and densities of Eu-particles for the simulation of Pu-particles in nuclear fuels

Sample	Eu-particle Diameter (μm)	Eu_2O_3 Density (mole %)	Neutron Cross section (cm^{-1})
A	250 - 300	100	70.8
B	125 - 150	50	35.5
C	250 - 300	50	35.5
D	250 - 300	20	14.3
E	44 - 64	100	70.8
F	125 - 150	20	14.3

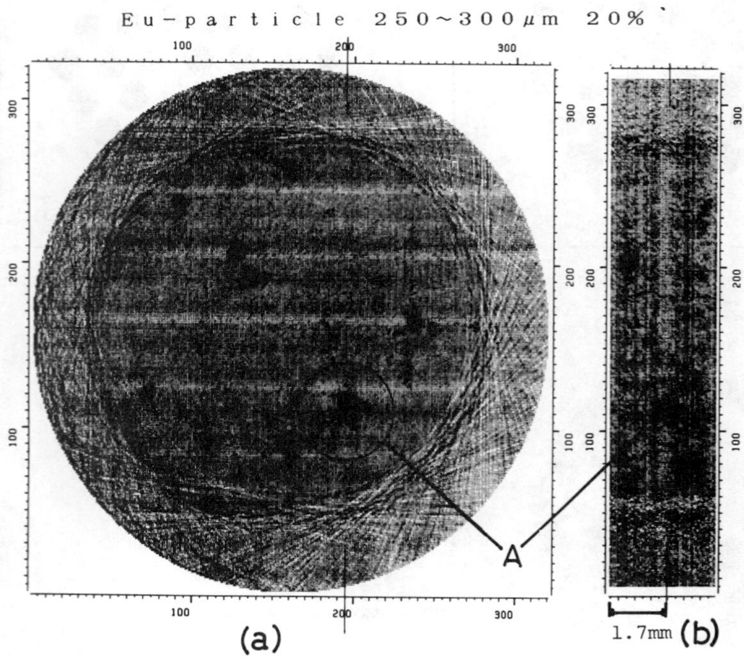

FIGURE 8. Reconstructed image of Eu-particles (20%) in TiO_2 pellet with neutron CT, (a) sectional CT image, (b) result of 3-dimensional reconstruction of the large Eu-particle, A (circled).

5.2 Reconstruction of hot spot particles in fuel pellets

In order to obtain high breeding gain, plutonium oxide, PuO_2, is used as the mixture with UO_2 in the fuel pellets of fast reactor, or advanced thermal reactor with higher conversion ratio. In those pellets, usually Pu-particles are inevitably produced, in which the density of plutonium is much higher than other areas. The Pu-particles give rise to hot spots in fuel rods, because heat generation rate by Pu is much larger, which causes severe detorioration in the performance of the mixed-oxide fuel. Therefore, the Pu-particle must be detected and depressed in fuel manufacturing. According to the specifications, the pellet having a Pu-particle, which diameter is larger than 50 μm, is not allowable for a fast breeder reactor, and a Pu-particle larger than 300 μm is not permissible for an advanced thermal reactor.

In order to inspect such Pu-particles in fuel pellets, some simulation pellets have been prepared, in which, europium oxide (Eu_2O_3) particles of various sizes and atomic percent are dispersed in TiO_2 matrix. The table 3 shows the prepared pellets with Eu-particles, their average diameters, densities and involving neutron macro-cross sections. These pellets are neutron-radiographed with using IX-100 films and reconstructed with CI algorithm.

Figure 8(a) shows the reconstructed sectional image of neutron CT for the pellet 'D', the Eu-particle sizes being 250 to 300 μm and density is 20 mole %. After the 2-dimensional reconstruction, the largest particle (marked with a circle) in the section has been reconstructed in the vertical 3-dimensional direction. Figure 8(b) shows the result of this reconstruction of the particle. As can be seen, the particle seems to be simple one. However, another example of the similar 3-dimensional scanning shows that the one particle at a section is not truly one, but two smaller particles are overlapping. From the reconstructed images of the pellet 'D', the Eu-particle sizes are measured. It is shown that the reconstructed particles are somewhat larger than the preparation. The reason of the enlargement seems to be dependent to the contrast of the original projection data of neutron radiographs. For smaller particles, it is still difficult to reconstruct by the neutron CT. A new roll type of film with much higher resolution and stronger contrast is expecting to be available.

This work has been performed at Research Reactor Institute, Kyoto University. and newly started at Institute for Atomic Energy, Rikkyo University.

REFERENCES

(1) Matsumoto, G. and Kurata, S., 'The neutron computer tomography' in Proc. First World Conf. NR 'Neutron Radiography' eds Barton, J. P., and Van der Hardt, P., pp.899-906, D. Reidel Pub. Dordecht, 1983.

(2) Matsumoto, G., Sendo, T., Honda, S., Ohkubo, K. and Ikeda Y., 'Improvement of reconstructed image quality in neutron CT' in Proc. Second World Conf. NR 'Neutron Radiography' eds Barton, J. P., Farny, G., Person, J. L. and Rottger, H., pp.745-752, D. Reidel Pub. Dordrecht, 1987.

(3) Ikeda, Y., Sakai, H. and Matsumoto, G., 'Neutron computed tomography with a high speed image processor 'Materials Evaluation', forthcoming.

(4) Radon, J., On the determination of Functions from their Integrals along Certain Manifolds, Ber. Verh. Sachs Akad. Wiss. 69, pp.262-277 (1917).

(5) Mersereav, R. M. and Oppenheim, A. V. 'Digital Reconstruction of Multidimensional Signals from Their Projections' Proc. the IEEE, 62, 1319(1974).

(6) Budinger, T. F. and Gullberg, G. T. 'Three-dimensional reconstruction in nuclear medicine emission imaging' IEEE Trans. Nucl. Sci., NS-21, 2(1974).

(7) Shepp, L. A. and Logan, B. F. 'The Fourier reconstruction of a head section' IEEE Trans. Nucl. Sci., NS-21, 21(1974).

ELECTROMAGNETIC METHODS

EMAT EXAMINATION FOR CRACKS IN RAILROAD WHEEL TREADS

R. E. Schramm, P. J. Shull, and A. V. Clark, Jr.
National Bureau of Standards
Bolder, CO 80303

D. V. Mitrakovic
University of Belgrade
Belgrade, Yugoslavia

INTRODUCTION

Cracks in railroad wheels may result from high stresses due to dynamic or static loads and residual stresses generated by such events as heating during braking. These flaws generally originate in the tread surface or flange and can lead to catastrophic wheel failure resulting in considerable equipment damage and possible derailment. This threat to personnel safety and the potential costs in time and money are inducements in the search for an effective, automated method for nondestructive examination that will identify damaged wheels needing replacement.

The current method of examination is visual observation. In the early 1970's, an ultrasonic method was introduced (1-4). This basically involves an in-rail system of piezoelectric transducers generating Rayleigh waves, electronics for signal generation and processing, and a method for tagging suspect wheels. The goal is to examine each wheel of a train as it rolls by a checkpoint in a railyard. There are currently two of these systems in operation in this country. A companion portable hand-held system subsequently verifies any flaw indications. The Fraunhofer Institute (IzfP) in Saarbrücken, Federal Republic of Germany, has been developing another ultrasonic system along this line (5) but using electromagnetic-acoustic transducers (EMATs). IzfP has placed a prototype into operation.

Our current research in this area (6, 7) is the development of an ultrasonic system using EMATs designed to work on American-style wheels and rails. The objective is to check every wheel on a train, also in a roll-by mode. Like the earlier works, our system will utilize Rayleigh waves that travel around the wheel tread. We are using EMATs for two main reasons: 1. No ultrasonic path in the transducer itself and low sensitivity to phase-transformed signals mean a simplified, low-noise signal that is relatively easy to interpret with high reliability, even in an automated system; 2. Their non-contact nature eliminates the need for ultrasonic couplants, i.e., no water sprays or liquid-filled boots. Two extensive reviews of EMAT designs and applications have recently been published (8,9).

EQUIPMENT

In the electronic system, a function generator provides the rf signal for the gated MOSFET power amplifier which drives the transmitter. The burst consists of 5-10 cycles at 500 kHz with a pulse repetition frequency (PRF) of 60 Hz. The receiver preamplifier is a very-low-noise design. Impedance matching is very important for both transmitter and receiver to ensure maximum efficiency.

Our EMAT device is designed for insertion into a recess cut out of the rail. Therefore, one goal in the transducer design is to keep the configuration as compact as possible to minimize both the amount of required machining in the rail and the loss of weight-bearing surface. Toward this end, we use a single permanent magnet (Nd-Fe-B for maximum field strength) 32 mm X 26 mm X 52 mm with the magnetization direction along the 32 mm direction and oriented normal to the tread. For pitch-catch operation, there are separate transmitter and receiver coils. These are meanderlines with a periodicity of 6 mm to generate a Rayleigh wave with this wavelength at 500 kHz (velocity = 3 km/s). The two coils are stacked with a slight shift to prevent their shielding each other, and then laid on top of the magnet to contact the wheel tread (Fig. 1). To prevent signal complications from ultrasound production in the magnet, we place a thin foil of copper or aluminum between the coils and the magnet to minimize eddy current generation in the Nd-Fe-B. The original coil design contained five cycles since this length was likely to correspond to the expected length of contact with the curved wheel.

The receiver coil is AWG 36 enamel-coated wire that has been wound through the pattern six times on acetate-based adhesive tape. This multiplicity adds to the sensitivity since the repetitions are series-connected. The flexible tape allows the coil to conform to the wheel's curvature.

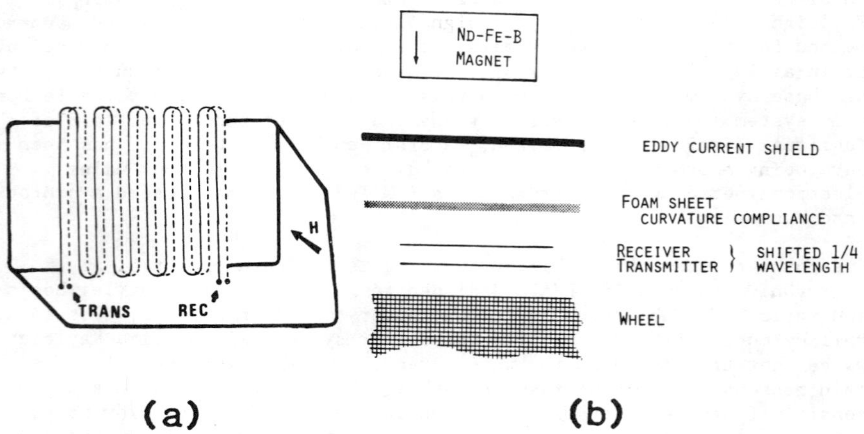

FIGURE 1. Rayleigh-wave EMAT construction.

(a) Two meanderline coils, shifted to prevent shielding, placed atop a magnet with field H.

(b) Exploded sideview showing sequence of transducer parts.

The transmitter coil has taken two forms. The first is a printed circuit on polymer film. The conductor is about 1 mm wide and 0.025 mm thick. In this form, the impedance is very low (about 1.3 Ω, including a current limiting resistor). Our power amplifier can deliver about 140 A of rf current into this coil. The second form is identical to the wire-wound receiver, and the much higher impedance limits the drive current to only about 30 A.

The greater the current flow through the transmitter, the greater is the eddy current induced in the specimen and the greater is the ultrasound generated. While the printed circuit coil permits the maximum current, the wirewound coil passes the current through the system multiple times and the current density (A/mm^2) induced in the specimen is roughly the same, i.e., the size of our ultrasonic signal is nearly identical. Using the lower current coil will increase the longevity of the power amplifier, and will allow an increase in the pulse repetition frequency (PRF) should this prove desirable later. At present, the limit on the PRF is due to two factors in the power amplifier: ohmic heating of the MOSFET output and the time required to recharge the power supply filtering capacitors.

While EMATs are noncontact devices, their efficiency decreases rapidly with liftoff. To assure the greatest coupling with the curved wheel tread, we have been inserting a few millimeters of compliant material between the magnet and the coil pair. This material compresses under the weight of the wheel and the flexible coil backing assumes the tread shape.

By its design, this transducer should be bidirectional and generate identical sound pulses from both ends of the coil. Their travel direction is normal to the legs of the meanderline. Experimental measurements on a flat steel plate with a saw-cut notch confirmed that this was the case. Translating and rotating the transducer on this same plate indicated that the ultrasonic energy radiates in a beam spreading at about 8-9^0 (using the 3-dB down points). Locating the sensor midway in the tread means that the signal must travel approximately 70 cm before it irradiates the entire width of the tread.

EXPERIMENTAL RESULTS

For the initial wheel measurements, the transducer was mounted in a small aluminum box and moved manually around a stationary wheel. On a used but unflawed wheel, this arrangement produces a signal which is still detectable after 14 round trips or more than 36 m of travel.

A small cut only about 1 mm deep generated a distinct reflection. The critical flaw depth, however, is about 6 mm so our interest is in distinguishing between cracks of greater depth (no longer safe) and those more shallow (safe for continued use). Accordingly, we sawcut an initial circular (half-penny) flaw into the center of the tread along a wheel radius. This flaw depth into the wheel was 5.6 mm at the maximum; the length was oriented along the tread width and 38 mm long at the surface.

In any field measurements, there will be many physical variables that may change the transducer-wheel coupling; among them will be variations in the tread profile due to different degrees and types of wear. Another factor is the taper of the wheel tread, i.e., the tread and the top surface of the rail are not parallel. The use of compliant material under the coils to allow them to conform to each wheel will minimize any separation. Figure 2 gives an initial indication of what the allowable limits may be. In this case the echo amplitude from the 5.6 mm slot yielded the signal-to-noise ratio (S/N) as a function of the liftoff from the surface. An S/N of 2.5, while rather small, ought to be workable. This lower limit on S/N corresponds to a liftoff of 1.5 mm which should be quite possible to obtain in most circumstances. These preliminary data show that the test system will be useful even before future refinements.

To determine how far a wheel can move and still return a useful signal, we first built a wooden form to simulate the track surface and then examined the signal strength as the transducer moved past the wheel contact point. Again using a 3-dB down point as the outer limits, a usable signal should be available during at least 2 cm of train travel. This is the space window in which the wheel must be located to both transmit and receive our test signal. While it would be desirable to introduce multiple pulses into the wheel in order to improve statistics for a higher signal-to-noise ratio, this may not be feasible since the Rayleigh wave travels multiple times around the circumference. To avoid processing confusion, it will be necessary to delay any new pulse until the prior one has sufficiently decayed. Since each round trip takes nearly a full millisecond, this means a delay interval of at least 10 ms. To get even two test pulses into the wheel (one at the beginning and end of the 2 cm window) means the train cannot travel faster than 2 cm/10 ms or about 7 km/h. This would be unacceptably slow for a field system so we shall have to settle for a single shot. A likely scenario is to use a trigger device to start the pulse as the wheel enters the window. Recording just the data arriving before the return of second round trip signal (see below) takes nearly 2 ms so the maximum train speed could be about 2 cm/2 ms, or 36 km/h. This is much better and likely improvement in the window length could increase the allowable speed.

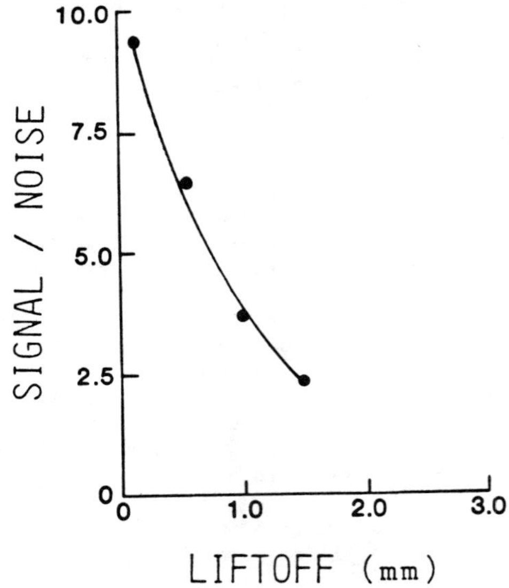

FIGURE 2. Signal-to-noise ratio as a function of separation between EMAT and wheel.

As noted above, the transducers are bidirectional. Figure 3 indicates schematically the signal paths to a flaw and the resulting scope trace. Signals A and C are echos that have traveled about 30% and 70% of the way around the wheel circumference to a flaw and then back. B and D are first and second round trip signals. Frequently some splitting occurs in these signals and this likely has at least two sources. Each round trip signal is a combination of counter-rotating signals. The path length is quite long (about 2.6 m) and some phase incoherence between these two becomes likely over this distance before they again combine on arriving back at the transducer. Also, the Rayleigh wave has some tendency to spread out to the wheel flange where it has a somewhat longer path than along the tread. The amount of ultrasonic energy in a wheel depends on the closeness of EMAT coupling and the wheel condition, e.g., grain size and residual stress state. Consequently, we are presently using the amplitude of the first round trip signal as a normalizing factor; the amplitude ratios A/B and C/B are our current flaw measurements.

To determine how the flaw position relative to the transducers affects the ultrasonic data, we measured the amplitudes of signals A, B, and C at 5 cm intervals around the entire circumference of the wheel. To obtain an idea of the sizing capability of this approach, the data were repeated for two flaw sizes, both circular sawcuts in the center of the tread; one was 5.6 mm deep and 38 mm long at the surface, and the other was the same cut enlarged to 12.1 mm deep and 53 mm long.

Data for the small flaw are in Fig. 4 and show three distinct regions. The first has a very large negative slope up to about 70 cm. With the beam divergence noted above, the ultrasonic energy needs this distance to spread out to the full tread width of 10 cm. Inside this distance, the flaw intercepts a very large fraction of this concentrated energy and generates a very large reflection. Once the beam has spread out to full width, in the second region, the flaw and round-trip signals remain fairly constant. In the third region, past about 225 cm, the flaw intercepts a large fraction of the counter-rotating ultrasonic beam; from this side, the flaw blocks a large portion of the round-trip signal and so increases the signal ratio.

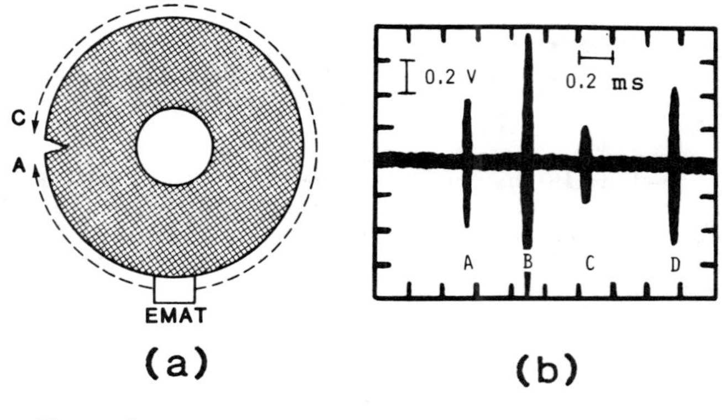

FIGURE 3. Flaw echoes.

(a) Schematic of bidirectional Rayleigh-wave travel to a flaw. A is the short path and C is the long path.

(b) Typical oscilloscope signal. A and C are short and long path flaw echoes. B and D are first and second round trip signals.

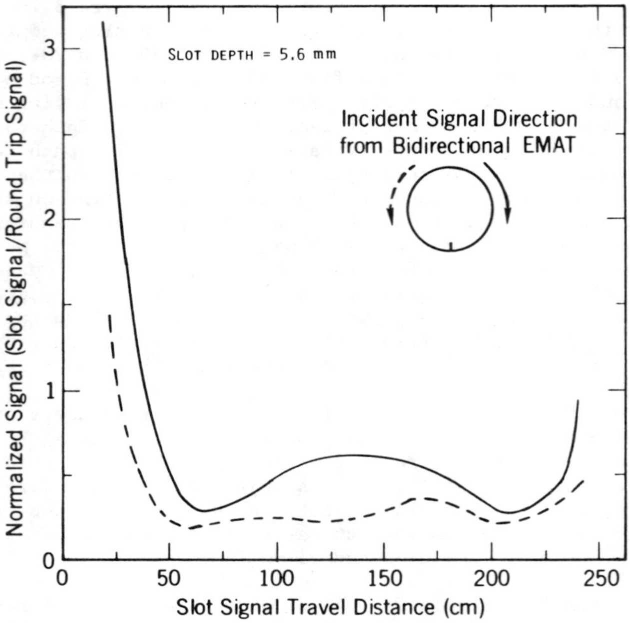

FIGURE 4. Signal amplitude as a function of flaw position around the wheel circumference. The simulated flaw was a saw cut in the center of the tread at 0 cm (262 cm). The two curves indicate data from signals traveling in opposite directions.

The data from the larger flaw show the same general trend (7) although the center region had somewhat higher values. This encourages our expectations that we will be able to achieve meaningful size discrimination.

The signals traveling in the opposite directions should generate the same data. While this appears basically true (the solid and dotted lines in Fig. 4 are close to each other), there does seem to be a slight systematic shift. Possibly there is some difference in the smoothness of the two sides of the sawcut.

Signal B's arrival time remains constant, of course, while A and C move closer to B as the EMAT-flaw distance increases until they all coincide when the flaw is exactly opposite the transducer. Thus, there are two zones on the circumference where meaningful measurements are not possible. Both of these are about 24 cm long; one is centered at the transducer and the other is exactly opposite. The first is due to recovery of the receiver amplifier following the transmitter pulse; the second is due to the merger of the round-trip and flaw signals when their acoustic path lengths become identical. As a result, approximately 80% of the tread can be inspected with each pass.

FUTURE WORK

We have recently constructed a short (4 m) section of track in which to mount the transducer (Fig. 5). Rolling an actual wheel set over the device will much more closely simulate field conditions and permit a reasonable debugging of several parts of the system: mechanical mounting, compliant coil backings, geometric variations in the size and position of magnet and coils, triggering (if necessary), and signal collection and storage. A new meanderline with 8 cycles (60% longer) will take advantage of the full length of our largest magnet. Testing will show if this will increase the space window for injecting the ultrasonic pulse and so increase the allowable train speed. With multiple wheels now available, it will be possible to generate a variety of artificial flaw sizes in the pursuit of a workable sizing algorithm. Since the wheels have a few centimeters of side-to-side play in normal operation, this rolling test will also show us the importance of the wheel position with respect to the transducer.

Automatic signal processing will be necessary before moving the system to a railyard to look for actual cracks in the field.

ACKNOWLEDGMENTS

We are grateful for discussions with H. J. Salzburger of IzfP as well as John Cowan and Richard Conway of The Fax Corporation. This work was sponsored by the U.S. Department of Transportation, Federal Railroad Administration, Washington, D.C. We also wish to thank the Transportation Test Center, American Association of Railroads, Pueblo Colorado, and, in particular, Britto Rajkumar and Dominic DiBritto for providing test wheels.

FIGURE 5. Current test configuration.

(a) Standard wheel set rolled along a short length of track.

(b) EMATs mounted in rail. A protective film covers the coils and they are represented schematically in this photo.

REFERENCES

1. Inspect Wheels "In Motion," Progressive Railroading, Jan., 1974.

2. Spotting the Defects in Railroad Wheels, Business Week, p. 44, Sept. 28, 1974.

3. Cracked-Wheel Detector "Could Save Millions," Railway Systems Control, May, 1975.

4. U.S. Patent 3,978,712: Sept. 7, 1976, John Vincent Cowan, Gerald De G. Cowan, and John Gerald Cowan.

5. Salzburger, H. J., and Repplinger, W., Ultrasonics International 83, conference organizers Z. Novak and S. L. Bailey (Butterworth & Co. Ltd., Kent, Great Britain), pp. 8-12, 1983.

6. Clark, A. V., Schramm, R. E., Fukuoka, H., and Mitrakovic, D.V., Ultrasonic Characterization of Residual Stress and Flaws in Cast Steel Railroad Wheels, Proceedings, IEEE 1987 Ultrasonics Symposium, Vol. 2, B. R. McAvoy, editor (Institute of Electrical and Electronic Engineers, New York), pp. 1079-1082, 1987.

7. Schramm, R. E., Clark, Jr., A. V., Mitraković, D. V., and Shull, P. J., Flaw Detection in Railroad Wheels Using Rayleigh-Wave EMATs, Review of Progress in Quantitative Nondestructive Evaluation, Vol. 7B, D. O. Thompson and D. E. Chimenti, editors (Plenum Press, New York), pp. 1661-1668, 1988.

8. Maxfield, B. W., Kuramoto, A., and Hulbert, J. K., Evaluating EMAT Designs for Selected Applications, Materials Evaluation, vol. 45, no. 10, pp.1166-1183, 1987.

9. Alers, G. A., and Burns, L. R., EMAT Designs for Special Applications, Materials Evaluation, vol. 45, no. 10, pp. 1184-1194, 1987.

PREDICTIONS OF MAGNETOSTATIC LEAKAGE FIELDS AROUND P.W.S. ROPES BY 3-D DYNAMIC RELAXATION METHOD

K. Hanasaki, K. Tsukada, and Y. Fujinaka
Kyoto University
Kyoto, Japan

INTRODUCTION

Parallel Wire Strand (P.W.S.) ropes are used as major elements in suspension roofings and suspension bridges. Long-term exposure to weather conditions deteriorate them by corrosion internally rather than on the surface. There has been a great need to evaluate this deterioration to decide whether or not they require replacement.

Disturbance of the magnetic field around a P.W.S. rope caused by its internal defects cannot be measured by an ordinary flux density meter, because the output signals are so small that they may be badly affected by mechanical vibrations of the sensor and/or the rope, electromagnetic waves in the air and sometimes terrestrial magnetism. In such a case, differential output signals from a couple of sensors, set closely together, give us the important and useful information about defects in the rope.

The authors are developing a new magnetic testing machine for P.W.S. ropes, which consists of a leakage flux detector assembly with Hall-effect sensors and two exciting solenoid coils set co-axially with the rope to be inspected. In order to design the optimized dimensions of the sensors, the three-dimensional distribution of leakage fields around the defective P.W.S. ropes must be predicted. Ida and Lord presented the 3-D finite element method for predictions of leakage field[1]. The authors have also developed a new computer code, based on the dynamic relaxation method by which the three-dimensional magnetic flux density of each element of the discretized volume can be calculated. In this paper, the numerical modeling is described and the predictions by the code are compared with the results obtained from the experiments of inspecting the P.W.S. rope with artificial defects.

NUMERICAL METHOD

The dynamic relaxation method is a kind of explicit type finite difference method in which, at the beginning of the computation, all time-dependent variables at each node, side and element of the discretized volume are given from the initial conditions, and those at each subsequent discretized time are calculated step by step from the previous ones based on the governing equations and the boundary conditions. In this method, it is easy to take any peculiar conditions into account at any time because the governing equations or conditions can be alternated dynamically. Moreover, as unknown values at the next time step are calculated explicitly from the present values, the computer code can be relatively simplified. However, as is well-known about such an iterative computation, we should take care of the stabilities and the convergence, otherwise the accuracy of the result may not be guaranteed. This means that the increment time of the calculation is significant for the guarantee as well as for saving machine time of the computation. In this study a variable time step method[2)] was applied, that is, the step time is alternated dynamically at each step of a calculation as mentioned below.

MAGNETIC FIELD EQUATIONS

In many studies on the leakage flux inspection method, the region of interest usually does not include current sources. For a source free region the magnetic field strength vector, H, can be derived from the magnetic scalar potential, ϕ as follows

$$H = -\mathrm{grad}\ \phi \tag{1}$$

The flux density vector, B, is given by

$$B = \mu H \tag{2}$$

where μ is the magnetic permeability of the material. In the Maxwell equations, since the magnetic charge is conceived not to be divided into the N-charge and the S-charge, that is, the conception of the magnetic charge density is not there, the divergence of the magnetic scalar potential is always equal to zero. However,in this work, in order to apply the dynamic relaxation method to the magnetic field problem, the concept of magnetic charge density which is proportional to the rate of change of the magnetic scalar potential, similar to that of the electric charge density, has been introduced here. Therefore, the relation between the magnetic flux density and the magnetic scalar potential is given by an artificial equation,

$$\mathrm{div}\ B = -\alpha \frac{\partial \phi}{\partial t} \tag{3}$$

where α is a relaxation constant. The divergence on the left side represents increase or decrease of the magnetic charge density in the unit time, and on the right side is the rate of change of the scalar potential.

NUMERICAL MODELING OF MAGNETIC FIELD AROUND P.W.S. ROPE

Discretization

Since the cross sectional shape of the P.W.S. rope is as peculiar as that of a honeycomb as shown in Fig. 1(a), the ordinary rectangular coordinate system is not convenient for this purpose. Hence, a new hexagonal cylindrical element (Fig. 1(b)) had been developed for this work, which saved computer machine time, core memory and geometric handling in the computer code.

Fig. 2 shows a general view and three kinds of coordinate systems for the numerical model; one is for the spatial length (x, y and z), one is for the naming of the volume elements (i, j and k) and the other is for the direction of the magnetic flux (x_1, x_2, x_3 and z).

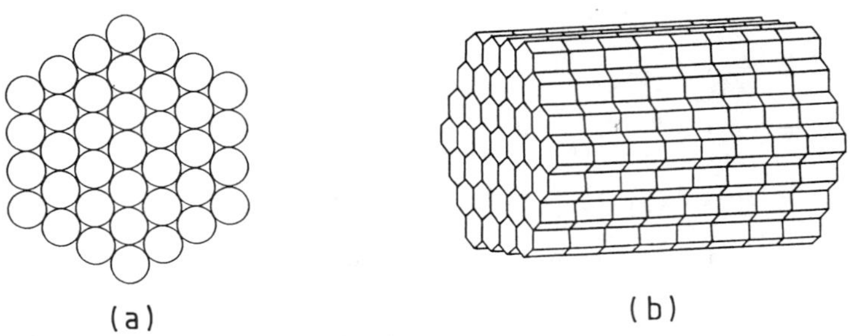

FIGURE 1. Cross-section of a P.W.S. rope (a) and hexagonal cylindrical elements for calculation (b).

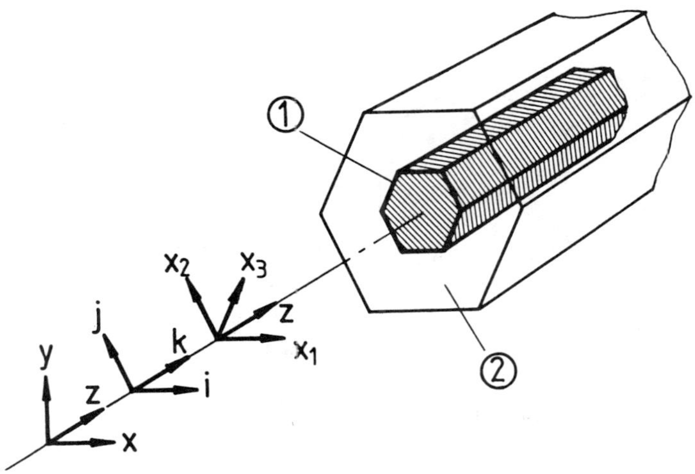

FIGURE 2. General view and three kinds of coordinate systems for the numerical model.
1---Steel 2---Air

Formulation

The magnetic potential is assumed to be uniform within a discrete element in the material, and the flux density at some time t, which flows perpendicularly to each boundary face of the adjacent elements, is calculated by the following formula as shown in Fig. 3.

$$
\begin{aligned}
B^t_{x_1,i,j,k} &= \mu \bullet H^t_{x_1,i,j,k} \\
B^t_{x_2,i,j,k} &= \mu \bullet H^t_{x_2,i,j,k} \\
B^t_{x_3,i,j,k} &= \mu \bullet H^t_{x_3,i,j,k} \\
B^t_{z,i,j,k} &= \mu \bullet H^t_{z,i,j,k}
\end{aligned} \qquad (4)
$$

where,

$$
\begin{aligned}
H^t_{x_1,i,j,k} &= -(\phi^t_{i+1,j,k} - \phi^t_{i,j,k}) / \Delta x_1 \\
H^t_{x_2,i,j,k} &= -(\phi^t_{i,j+1,k} - \phi^t_{i,j,k}) / \Delta x_2 \\
H^t_{x_3,i,j,k} &= -(\phi^t_{i+1,j+1,k} - \phi^t_{i,j,k}) / \Delta x_3 \\
H^t_{z,i,j,k} &= -(\phi^t_{i,j,k+1} - \phi^t_{i,j,k}) / \Delta z
\end{aligned} \qquad (5)
$$

where $\Delta x_i (i=1,2,3)$, and Δz are the distances between the centers of the adjacent elements in each direction, respectively.

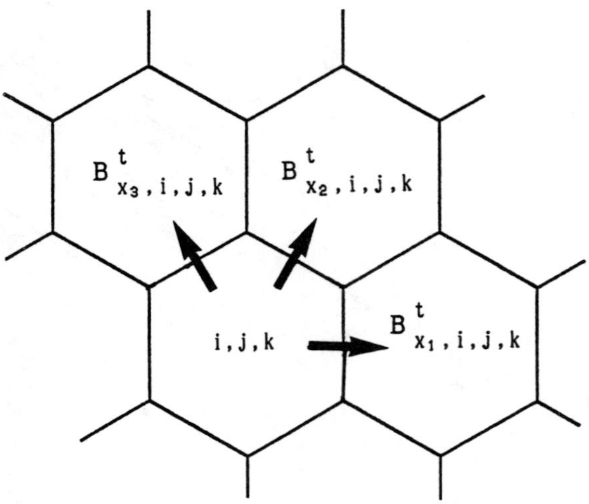

FIGURE 3. Flux densities on the x_1-x_2-x_3 plane which flow out of the element i,j,k (z component is perpendicular to the paper surface).

The rate of change of magnetic charge of a volume element i,j,k at time t is calculated from the total flux flowing into or out through the eight faces of the element as shown in Fig. 4, and also is considered to be proportional to the rate of change of magnetic potential, which is given by

$$\begin{aligned}\dot{\phi}^{t}_{i,j,k} = -\frac{1}{\alpha}\{ &(B^{t}_{x_1,i,j,k} - B^{t}_{x_1,i-1,j,k})\Delta S_{x_1} \\ &+(B^{t}_{x_2,i,j,k} - B^{t}_{x_2,i,j-1,k})\Delta S_{x_2} \\ &+(B^{t}_{x_3,i,j,k} - B^{t}_{x_3,i-1,j-1,k})\Delta S_{x_3} \\ &+(B^{t}_{z,i,j,k} - B^{t}_{z,i,j,k-1})\Delta S_{z}\}\end{aligned} \tag{6}$$

where $\Delta S_{xi}(i=1,2,3)$ and ΔS_z are area of the faces of the hexagonal cylindrical element, respectively.

Then, change of the magnetic potential for a small time step Δt is approximated by the following relation.

$$\Delta\phi^{t}_{i,j,k} = \dot{\phi}^{t}_{i,j,k} \bullet \Delta t^{t} \tag{7}$$

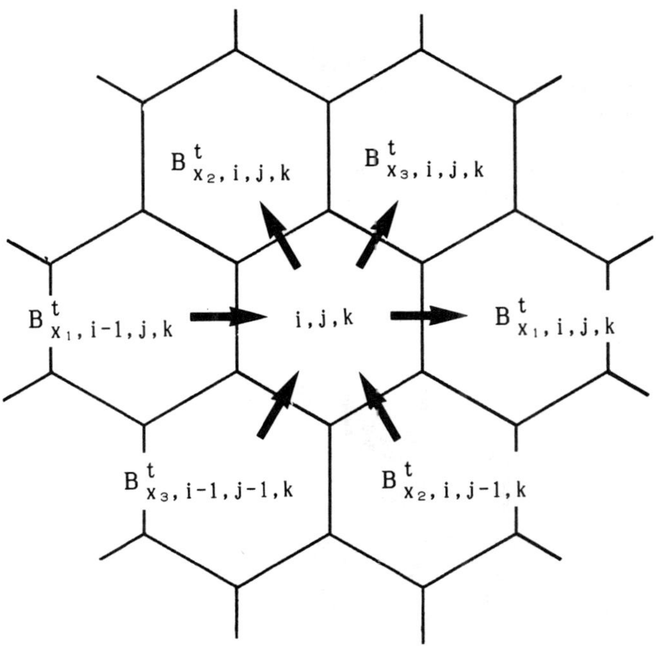

FIGURE 4. Flux densities on the x_1-x_2-x_3 plane which contribute to the potential change of the element i,j,k (two z components, which are perpendicular to the paper surface, also contribute).

Hence, as mentioned above, stability conditions for the time step Δt used in this calculation must be considered. In this work a method has been applied such that the time step Δt is not constant but variable through the calculation, that is, it is determined by

$$\Delta t^{t} = \frac{\Delta \phi_{limit}}{\left| \dot{\phi}^{t}_{i,j,k} \right|_{max}} \tag{8}$$

where $\Delta \phi_{limit}$ is a constant value by which change of the potential during the step time is limited, and the denominator of the right side is the maximum value of the change rate all over the elements. Then, the magnetic potential at a new time $t+\Delta t$ given by

$$\phi^{t+\Delta t}_{i,j,k} = \phi^{t}_{i,j,k} + \Delta \phi^{t}_{i,j,k} \tag{9}$$

Having calculated the new magnetic potential in each element, we may now repeat the entire process.

Boundary Conditions

Potential. In an element such that the potential is fixed by the given condition, the potential must be replaced compulsorily with the value of the condition just after it is calculated by Equation 9 every time.

Flux density. Where the flux density is defined by the given conditions, it must be determined so that the conditions may be satisfied after the calculation of Equations 5.

For instance, at the outer element where the flux is not allowed to flow out or in through the surface, the flux density must be set to zero as shown in Fig. 5, that is

$$B^{t}_{x_1,i,j,k} = 0 \tag{10}$$

$$B^{t}_{x_2,i,j-1,k} = 0$$

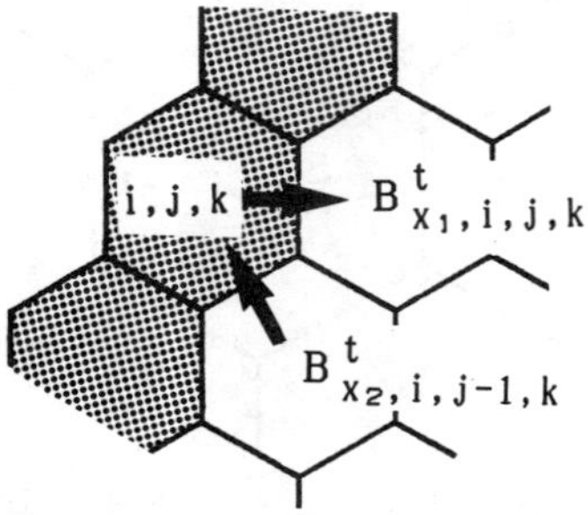

FIGURE 5. Flux densities which are concerned in the boundary condition at an impermeable zone.

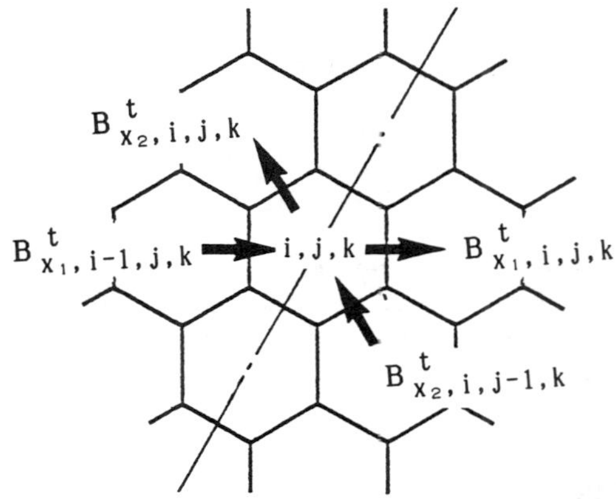

FIGURE 6. Flux densities which are concerned in the boundary condition at a mirror image zone.

In case of a mirror image condition, that is, at a symmetry plane, the following relations exist among the flux densities as shown in Fig. 6.

$$B^t_{x_1,i-1,j,k} = B^t_{x_2,i,j-1,k}$$
$$B^t_{x_2,i,j,k} = B^t_{x_1,i,j,k} \tag{11}$$

Initial Conditions

The initial values of potential and flux density are usually set to zero except those defined by the boundary conditions. In case the initial values are known by some means, for instance, by equivalent dipole approximations or by experiments, those values may be used for the purpose of saving computation machine time.

General Assumption for the Simulation

In this work, qualitative simulations were carried out with the following assumptions applied:

1) The potential in the element of the N-pole side for the boundary condition is +1 and the S-pole side is -1 respectively, for it is sufficient to express the calculation results in values relative to a standard.
2) No flux flows out or in through the surfaces of all elements on the outer boundary except ones of N- or S-pole.
3) μ is 1 in the air, 160 in the direction x_i(i=1,2,3) and 7000 in the direction of z in the steel, according to the result of the measurement for the apparent magnetic permeabilities.
4) The magnetic force is not so strong as the steel reaches magnetical

saturation.

5) α can be set to 1 all over the region of computation for the same reason as in 1).

Checking the Code by Calculation of Typical Problems

To verify the code based on this method, calculations were carried out for some typical problems such that the flux distribution on the z-plane is clear and understood intuitively. The results, as shown in Fig. 7, are considered to be reasonable.

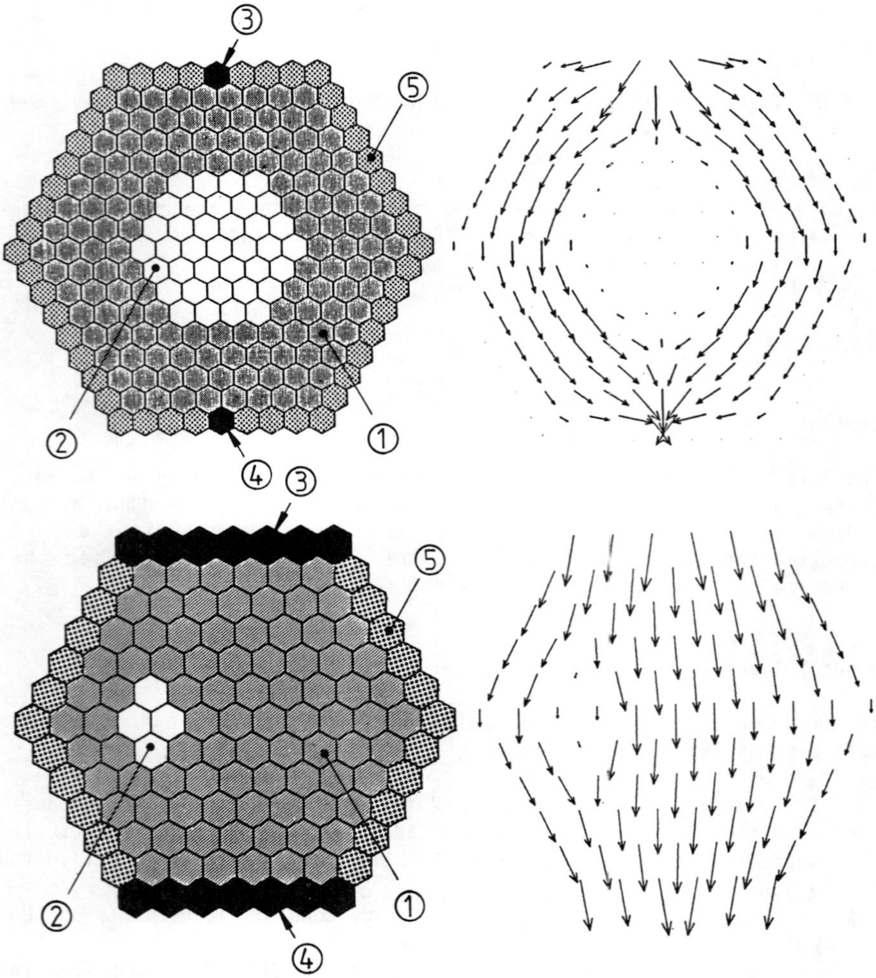

FIGURE 7. Flux density profiles in a ferromagnetic material having a non-magnetic region inside obtained by the method described here.
1...Steel 2...Air 3...N-pole 4...S-pole 5...Impermeable region

RESULTS OF APPLICATION TO A DEFECTED P.W.S. ROPE

The method described above was applied to the prediction of the leakage flux around a P.W.S. 37 rope which was constructed of 37 wires of 5 mm diameter, and had a defect corresponding to three missing wires over a length of 30 mm at a depth of 10 mm from the surface. The results were compared with the data obtained from the experimental setup used to measure flux density profile.

Experiment

The block diagram of the experimental setup is shown in Fig. 8. The P.W.S. rope was magnetized by two exciting coils which were set up 800 mm away from each other. Each coil had 40 turns and was supplied with a direct current of 150 A, thus the magnetromotive force was 12000 AT. The leakage flux density was measured by three Hall element sensors, one of which faced the normal direction of the rope surface, one faced the rope axis direction and the other one was perpendicular to those directions. Measurements were taken by scanning the surface of the rope at various lift-offs with the sensor assembly mounted with those Hall elements. One of the results are shown in Fig. 9.

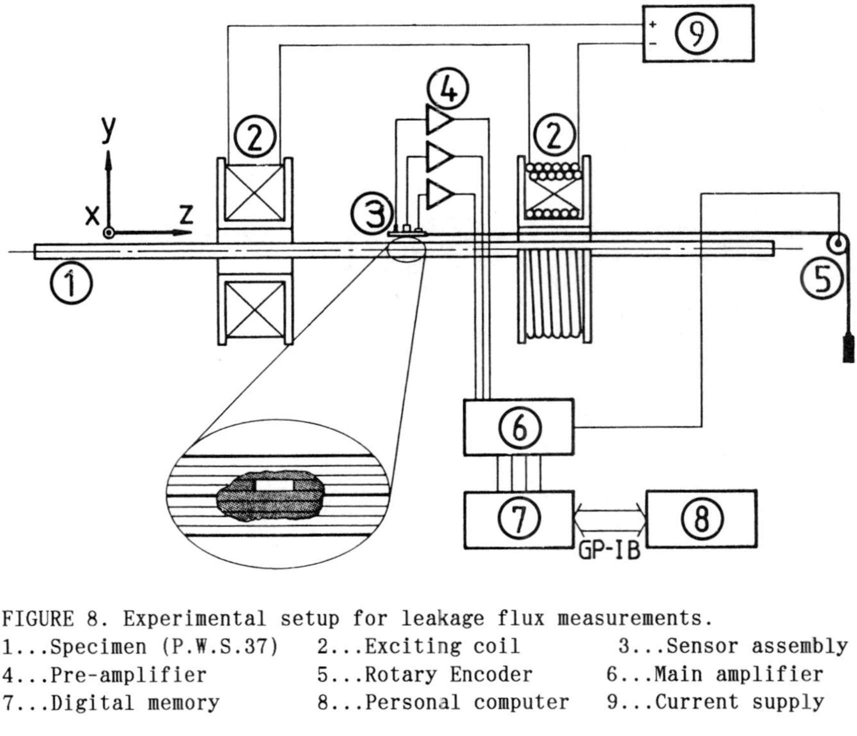

FIGURE 8. Experimental setup for leakage flux measurements.
1...Specimen (P.W.S.37) 2...Exciting coil 3...Sensor assembly
4...Pre-amplifier 5...Rotary Encoder 6...Main amplifier
7...Digital memory 8...Personal computer 9...Current supply

The short lines in Fig. 9 present the profiles of the leakage flux densities on the plane at a lift-off of 5 mm from each surface of the rope, and also the profiles on vertical planes to the top and bottom surfaces which are closest and farthest to the defect, respectively.

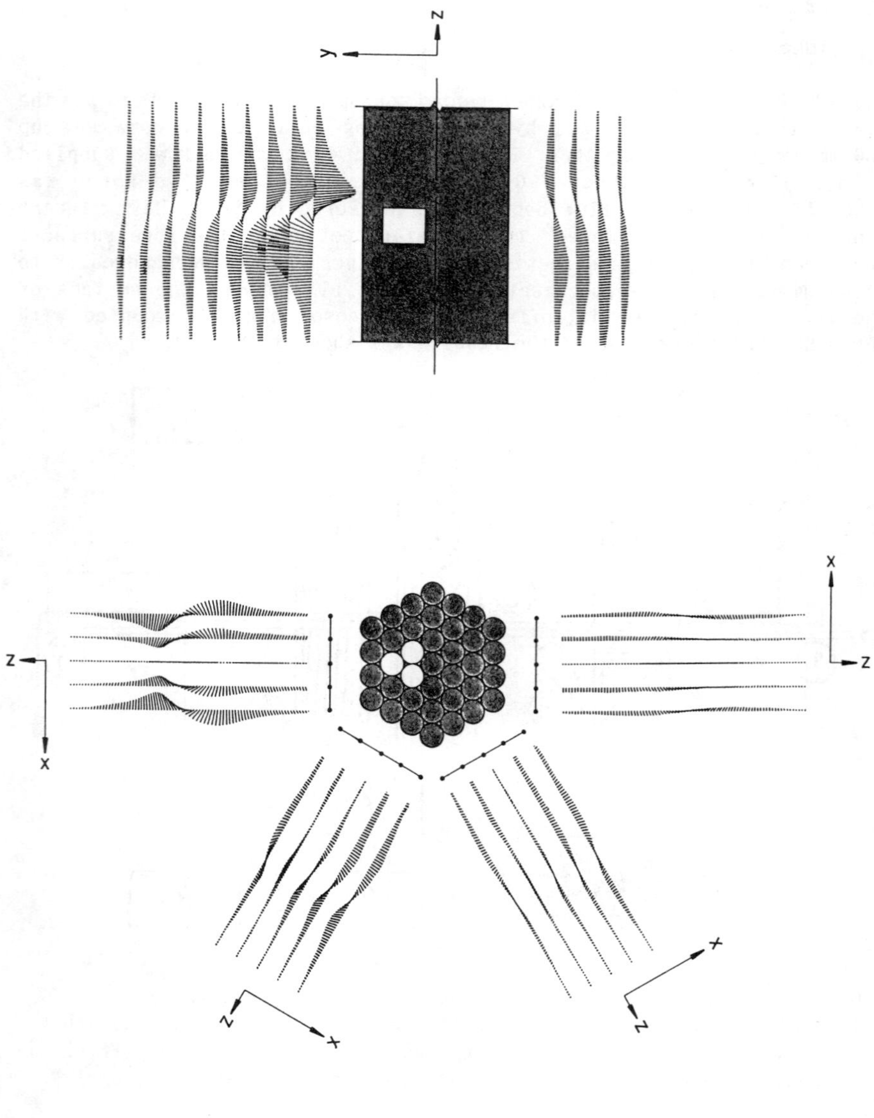

FIGURE 9. Experimental leakage flux density profiles.

Calculation

Fig. 10 is the results of the calculation for predicting leakage flux around the P.W.S.37 rope, and shows the profiles in a similar manner to the experimental results in Fig. 9. The numerical model used for this calculation is shown in Fig. 11, and $\Delta\phi_{limit}$ in Equation 8 was set to the value between 0.001 and 0.01

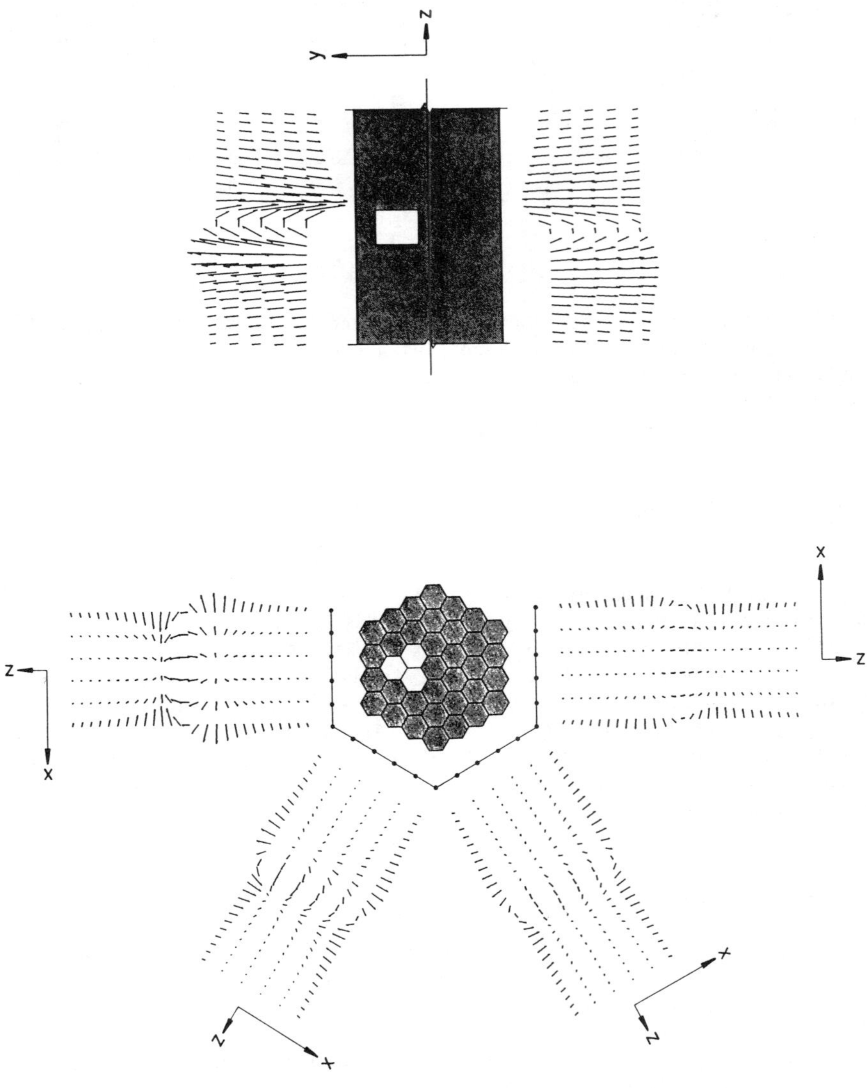

FIGURE 10. Calculated leakage flux density profiles.

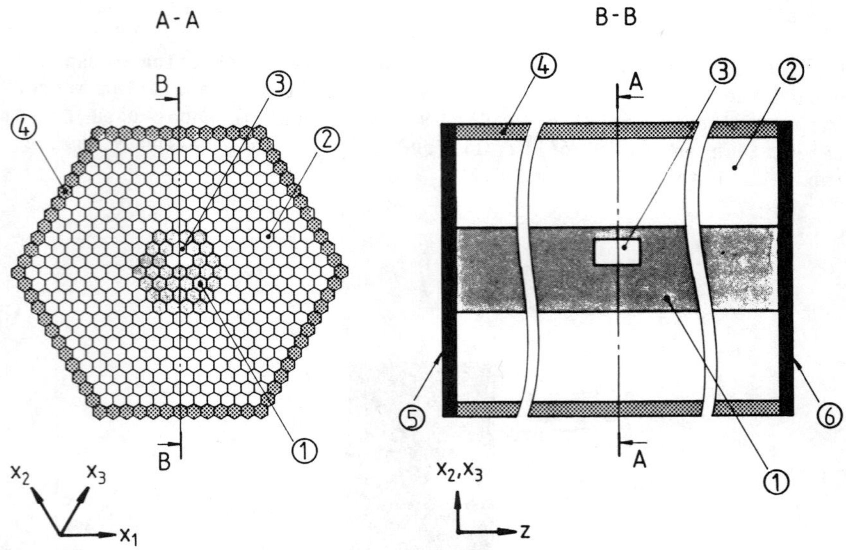

FIGURE 11. Set up of the numerical model and boundary conditions of the calculation for predicting leakage flux around the P.W.S.37 rope. The number of discretized elements in z-direction was 50.

1...Steel (P.W.S.37)	2...air	3...air (defect)
4...Impermeable region	5...N-pole	6...S-pole

DISCUSSION AND CONCLUSION

Fig. 9 and Fig. 10 agree qualitatively. However the reduction rate of the leakage flux with distance from the defect position in the later is a little smaller than that in the former. The reasons are considered to be due mainly to the finite domain to be calculated, that is, the leakage fluxes are confined within limited space and do not flow out to the open air as in the real phenomenon. The accuracy of the calculation can be improved by domain enlargement and element refinement. This method can be a powerful tool for defect characterization when an accurate non-linear relation of B-H curve is taken into account.

REFERENCE

1. Ida, N., Lord, W., 3-D Finite Element Predictions of Magnetostatic Leakage Field, IEEE Transaction on Magnetics, vol. mag-19, no. 5, pp. 2260-2264, September 1983

2. Hatta, N., Kokado, J., Hanasaki, K., Numerical Analysis of Cooling Characteristics for Water Bar, Transaction of Iron and Steel Institute of Japan, vol. 23, pp. 555-564, 1983

STRESS MEASUREMENT VIA THE ACOUSTOELASTIC EFFECT AND WATER-COUPLED ULTRASONIC WAVES

A. M. Koshti
Rockwell International
Downey, CA 90240

D. M. Egle
University of Oklahoma
Norman, OK 73019

INTRODUCTION

In fabrication processes, such as drawing, welding, rolling, etc., stresses are locked into the material. These residual stresses may be relieved to a certain extent by heat treatment processes wherever such processes are possible and practical. However, the residual stresses often remain in the components. Unless the stresses are measured or estimated, components need to be designed conservatively, leading to disadvantanges such as bulkiness and waste of material. Underestimation of residual stresses may lead to the failure of components, thus there is a practical need for measuring residual stresses.

There are a number of methods for measuring residual stresses. These can be categorized as destructive (material removal strain gage methods), semidestructive (hole drilling), or nondestructive (x-ray diffraction, Barkhausen noise analysis, acoustoelasticity). X-ray diffraction methods are restricted to surface layers and Barkhausen noise analysis can be applied only to ferromagnetic materials. Thus, nondestructive methods, such as ultrasonic (acoustoelasticity), for measuring residual stresses within the bulk of a component are urgently required.

One of the more recently developed ultrasonic methods of measuring stresses is based on the acoustoelastic effect. Acoustoelasticity refers to the perturbation of ultrasonic velocity due to presence of stress. This effect can be easily observed by applying different magnitudes of stress to a component and measuring velocity of an ultrasonic wave.

If the stress field is not isotropic (equal principal stresses), the effect of stress is to remove the isotropy of the ultrasonic velocity in an otherwise isotropic material. One of the most useful acoustoelastic methods is based on the birefringence of two shear (transverse) waves traveling in the same direction along one of the principal stresses but polarized in the directions of the other two principal stresses. Acoustical birefringence refers to the difference of velocities of the two shear waves. However, in order to determine the stress from acoustical anisotropy, the velocities in the absence of stress must be calculated or preferably measured.

Apart from the dependence on stress, the ultrasonic velocity is an equally strong function of the texture of the material. The texture (or preferred grain orientation) results from fabrication processes such as rolling and extrusion. In other words, the texture also has an effect of removing isotropy of ultrasonic velocity. In order to successfully measure residual stresses in real materials, the texture anisotropy needs to be measured separately and subtracted from the total anisotropy to determine the net acoustic anisotropy due to residual stress.

SIGNIFICANCE OF THIS RESEARCH

One of the drawbacks of the contact transducers, which are typically used for acoustoelastic measurements, has been poor reproducibility of results. To avoid contact, noncontact electromagnetic transducers (EMATs) have been used [1]. An alternative method is to use water immersion techniques in which both specimen and transducer are immersed in water. There is a considerable space between the transducer and the specimen occupied by the water, thus avoiding contact. Since it is very easy to move the transducer over the specimen in water, the technique is mostly used to scan the specimens. In the present research, a water immersion technique suitable for measuring through the thickness average of in-plane in thin plates is developed.

Acoustic birefringence measurements have advantages over longitudinal wave velocity measurements. One of the advantages is that one may determine the principal acoustic directions in a plane-stressed plate. The difference in the shear wave velocities with displacements in two principal directions has been shown to be proportional to the difference of principal stresses. In past research, contact transducers were used for the technique. In this research, water-coupled ultrasonic transducers are used for the measurements. This combines the advantages typical of immersion techniques with those of the birefringent technique. The material in this paper was a part of the first author's thesis [2].

PRINCIPLES

Different waves such as longitudinal, transverse and Rayleigh waves have been used in acoustoelastic stress measurement. Shear waves can be generated by using shear transducers, but they can also be generated by using longitudinal transducers as shown in Figure 1. A longitudinal wave generated by transducer D1 is incident at an oblique angle on a platelike specimen through water. The incident ray (R1) is reflected (R2) in the water. Ray (R1) also generates two refracted rays, one longitudinal and the other shear, in the aluminum plate. The angle of incidence is chosen to be greater than the first critical angle which is $\cong$ 13.56° for water-aluminum interface. At 13.56°, the refracted longitudinal wave just disappears from the aluminum. The second critical angle corresponds to just vanishing of shear wave in the aluminum and it is 29.2°. The angle that has been chosen is about 18.5°. Therefore, the longitudinal wave is absent but the shear wave (R3) is present.

Ray R3 is reflected as a shear wave (R4) at the bottom of the specimen. Ray R3 also generates a refracted longitudinal wave in water. Shear wave R4 is reflected as a shear wave in the aluminum plate and refracted as a longitudinal wave (R5) in water. Waves R2 and R5 are parallel to each other and they impinge at normal incidence on the receiving transducer D2. A pulse generated at transducer D1 will be divided in many rays traveling in the same direction as rays R2 and R5. Rays R2, R5, and other rays parallel to R2 from multiple reflections in the plate may return to the receiving transducer. A pulse returned by R2 will have a different path length (say P1) from the one returned by R5 (say P2).

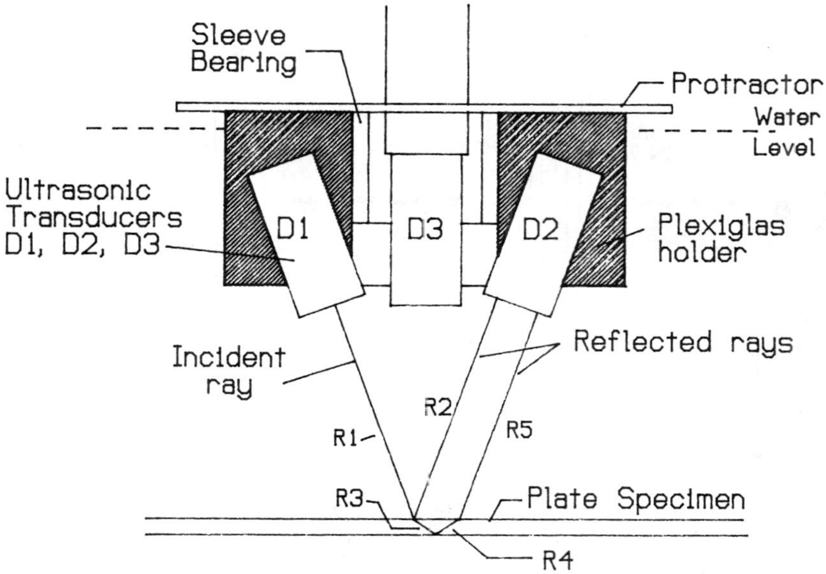

FIGURE 1. Transducer fixture.

The time taken by a pulse traveling along path P1 (say T1) will be different from the time taken to travel path P2 (say T2). The path lengths are given by,

$$P1 = R1 + R2$$
$$P2 = R1 + R3 + R4 + R5 \quad . \qquad (1)$$

The difference in travel time (T2 - T1) can be measured on the oscilloscope. If the transmitted pulse is short in duration, the two pulses which travel paths P1 and P2 will be easily separable. Figure 2 shows a typical oscilloscope trace. The first signal is from the top surface echo and the second signal which is inverted with respect to the first one, is from the bottom surface echo. The time delay (T2 - T1) between the two echoes can be measured accurately by many ways. The time delay can be used to find the velocity of sound in the specimen. This velocity then can be related to the state of stress in the plate [2].

TRANSDUCER FIXTURE

Two longitudinal transducers (D1 and D2) are rigidly held in a plexiglas block as shown in Figure 1. The angle between the axes of the two transducers is about 37°. There is also a facility for fixing the third longitudinal transducer (D3) on the bisector (centerline) of the above two axes. There is a sleeve bearing integral with the top part of the plexiglas block. The inner race of the bearing can be mounted on the bottom part of an angle fixture. The angle fixture is fixed to a scanning shaft of an ultrasonic immersion scanning bridge. The scanning shaft can translate in three directions, i.e., along its length (vertical) and two movements in mutually perpendicular directions in a horizontal plane.

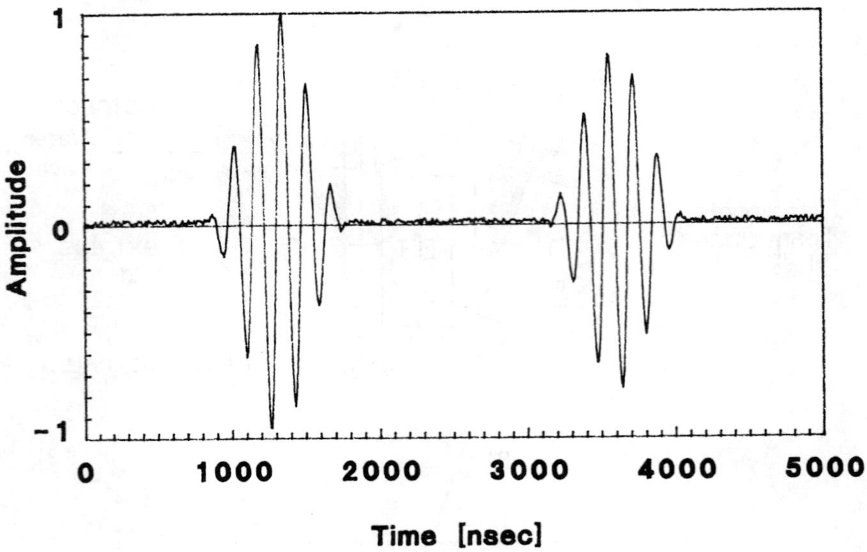

FIGURE 2. Typical oscilloscope trace.

The plane formed by the centerlines of transducers D1 and D2 is referred to as the principal plane. The principal plane also contains the bisector of the above two centerlines. The bisector can be referred to as the principal centerline. In short, the different movements of the fixture do the following functions:

1. The three translatory movements of the scanning shaft can locate the transducer fixture at any desired location within the immersion tank.

2. The double yoke joint sets the principal centerline at any acute solid angle from the shaft centerline to achieve the best alignment, i.e., when the principal centerline is normal to the specimen.

3. The sleeve bearing allows rotating the principal plane (or the fixture) about the principal centerline to any angle between 0° to 360° with respect to the specimen geometry.

Vertical movement of the fixture causes diminishing of one of the echoes and increasing of the other. Therefore, the amplitude of the echoes can be equalized by locating the fixture at a correct distance from the specimen. By adjusting the yoke joint, the principal centerline can be set normal to the specimen, yielding best received signal from the reflected echoes in that particular vertical position. The alignment is considered complete when the echoes are exactly identical in shape and size (but inverted to each other) and have maximum possible amplitude.

VELOCITY MEASUREMENT

Figure 3 shows the geometry of the wave paths. The quantity h is the thickness of the plate. The line CD is perpendicular to the path R2. If t_1, t_2 are the times for a pulse to travel the paths

$L_1 = AD; \quad L_2 = AB + BC$ (2)

then the difference in arrival times for the wave reflected from the top surface, R2, and the wave which passes through the plate once, R5, is

$T = t_2 - t_1$ (3)

Using the geometry shown in Figure 3, this time difference may be shown to be

$T = [2h/(V_s \cos \phi) - (2h/V_w)\sin \phi_1 \tan o]$ (4)

where

V_s = speed of shear waves in the plate

V_w = speed of ultrasonic waves in the water

Using Snell's law,

$(\sin \phi_1)/V_w = (\sin \phi)/V_s$

Eq. (4) may be simplified to

$T = (2h/V_s)\cos \phi \quad ; \quad V_s = (2h/T)\cos \phi$ (5)

SMALL CHANGES IN SPEED

If small changes are assumed in the variables, then Eq. (5) may be used to calculate the relative change in wave speed

$dV/V = -\tan \phi \, d\phi + dh/h - dT/T$ (6)

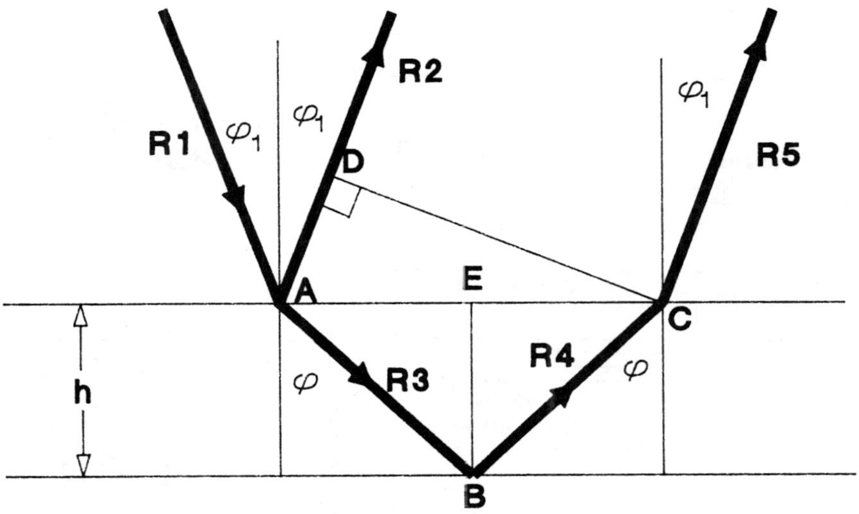

FIGURE 3. Geometry of the wave paths.

Using Snell's law, onw may show that

$$d\phi = \tan \phi \ (dV/V) \tag{7}$$

Thus

$$(1 + \tan^2\phi)(dV/V) = dh/h - dT/T \tag{8}$$

The quantity dh/h is the average normal strain ε_z in the direction perpendicular to the plate. For thin plates, ε_z is constant through the thickness; hence Eq. (8) may be written

$$dV/V = \cos^2\phi[\varepsilon_z - dT/T] \tag{9}$$

In Eq. (9), the thickness, h, is absent. Fractional change in velocity can be directly found by measuring fractional change in travel time, if ϕ and ε_z are known. The fractional changes in the velocity can be used to compute stresses directly. Thus, the requirement to measure the thickness is eliminated. Thickness is usually measured by mechanical instruments such as Vernier calipers or micrometers. Fukuoka et al. [3] discuss the difficulty of making accurate thickness measurements. Usually the order of error in thickness measurement is so large that the resulting error in the stress is too great for the acoustoelastic analysis to be of practical value. Thus, avoiding thickness measurement greatly increases the accuracy and applicability of this technique.

INSTRUMENTATION

There are different methods by which time delay between any two successive echoes can be measured. In the present approach, crosscorrelation of the two echoes is used to determine the delay between the two received echoes. This method was successfully used by Egle [4]. A schematic diagram of the measuring system is shown in Figure 4. It utilizes a conventional ultrasonic pulser-preamplifier, the transducer fixture, and an ultrasonic scanning bridge. The transient recorder is used to digitize

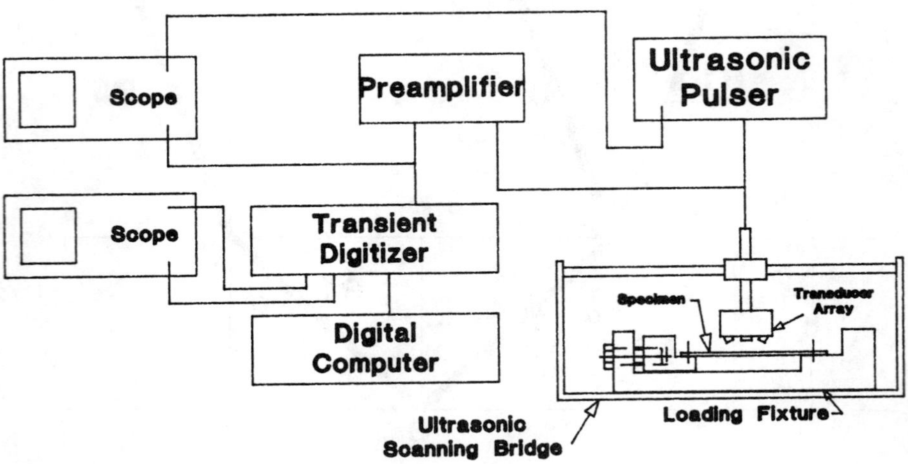

FIGURE 4. Instrumentation for acoustoelastic measurements.

the two echoes, one from the top surface and the other from the bottom surface of the specimen. The signal is averaged over 100 repetitions to improve the resolution and linearity of the transient recorder and to reduce the noise in the signal. The sample interval of the digitizer was 10 nanoseconds which was the smallest interval available on that equipment. An interpolation scheme [4] is used to determine the relative delay to about 0.1 nsec.

An example of the captured waveform is shown in Figure 2. It was obtained by using plane 10 MHz transducers in the transducer fixture and a 3/16-inch thick aluminum plate specimen.

PROCEDURE

The experiments involve loading a dogbone-shaped aluminum plate subject to a uniform tensile load and the axis of the specimen. Both the applied strain and the delay time between the two echoes were measured for several applied loads. The specimen was a 3/16-inch thick aluminum plate. The long axis of the specimen coincides with the rolling direction. The fixture used for loading the specimen is shown schematically in Figure 4. The specimen can be pulled lengthwise by turning the bolt at the end of the fixture. To measure strain in the specimen, a rosette strain gage was bonded on the specimen.

EXPERIMENT #1

Experiment #1 was performed to observe the variation in the delay time with different angular positions of the transducer fixture. This angle, θ, is defined as the angle between transducer fixture plane of polarization and the principal tensile stress σ_1. Three levels of applied stress were used. These levels, in terms of strain, were 0, 620, and 1135 μs.

Figure 5 shows the plot of measured delay time angle θ. It should be remembered that $\theta = 0°$ is in the direction of principal stress and also in rolling direction. The measurement repeatability error was a maximum of ± 0.05 for any two consecutive measurements in exactly the same conditions. The signal was averaged 100 times for each time measurement.

Temperature at the time of the experiment was 20°C in the immersion tank. The angle was changed from 0° to 180° in steps of 10°.

The following observations can be made from Figure 5:

1. All plots are symmetric about 0° and 90°. Since the specimen material axis (rolling direction) is along the 0° axis, the graph for zero stress is symmetric about 0° or 90°. Since the applied stress is also along the rolling direction, the plots for stressed conditions are also symmetric in the same way.

2. The delay time plot for zero stress is typical of the anisotropy caused by rolling of the plate.

3. The plots of delay times for the applied stress are not exactly sinusoidal but are a little flat from 50° and 130°.

4. All the plots intersect at about 50° and 130°.

Information about the effects of the stress on the delay time can be

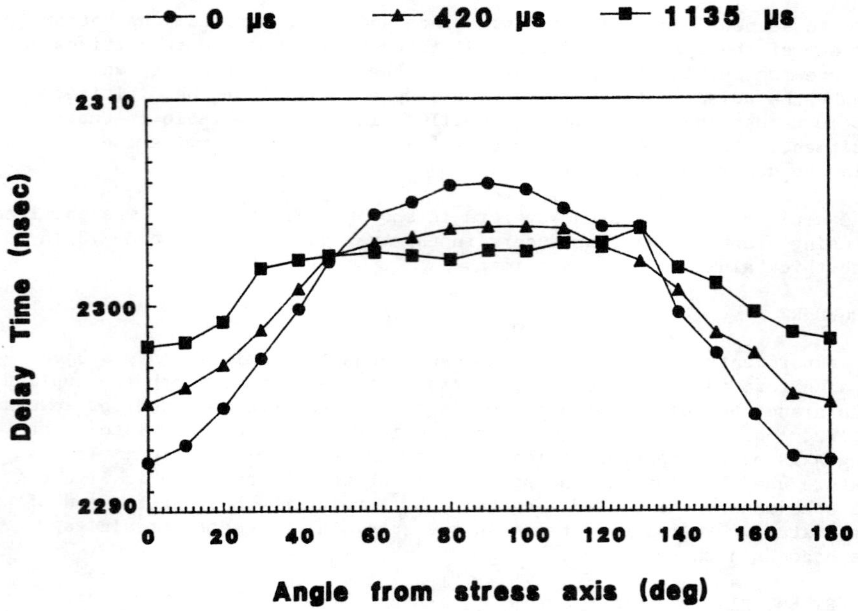

FIGURE 5. Delay time as a function of orientation.

obtained by subtracting the zero stress readings from the corresponding stress level 1 and 2 readings. These computations are shown in Figure 6.

The following observations can be made from Figure 6:

1. At 0° and 180°, the velocities are less than that at 90°.

2. The curves are approximately sinusoidal of nature cos 2θ.

3. The variation in delay time increases with applied stress.

4. The delay time at 0° is greater (about 1.5 times) than that at 90° for both stress states.

5. At about 50° and 130°, there is no change in time.

EXPERIMENT #2

This experiment was performed on the same specimen at 21.5°C. Different tensile loads were applied and delay times at 0° (T_0) and at 90° (T_{90}) were measured. These data are plotted in Figure 7 along with a least squares straight line fitted through the data point.

The delay times T_0 and T_{90} vary linearly with the strain. The absolute slope of the line for T_0 is higher than that for T_{90}. Also for T_0, the points are closer to the line than for T_{90}. Based on the observed linearity of transit times T_0 and T_{90} with stress, an acoustoelastic constant can be defined as follows

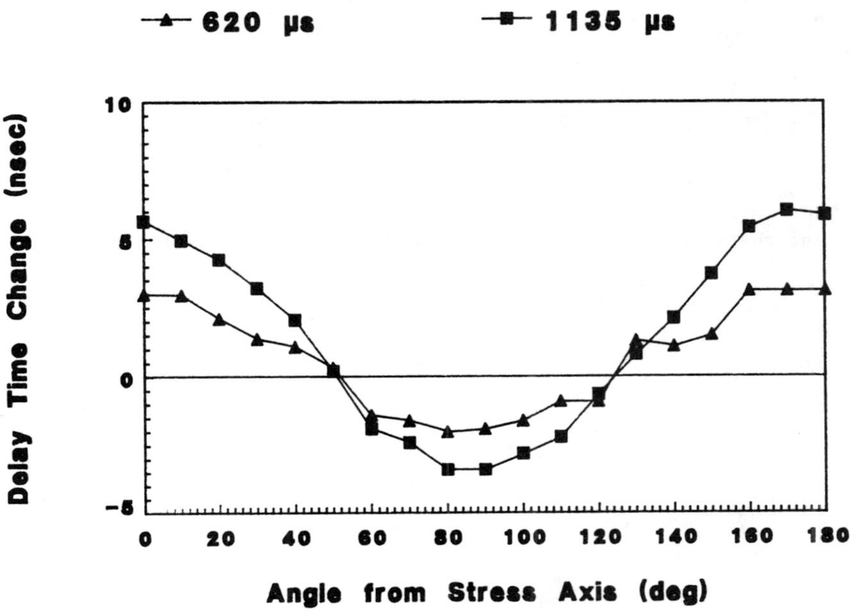

FIGURE 6. Change in delay time with applied strain.

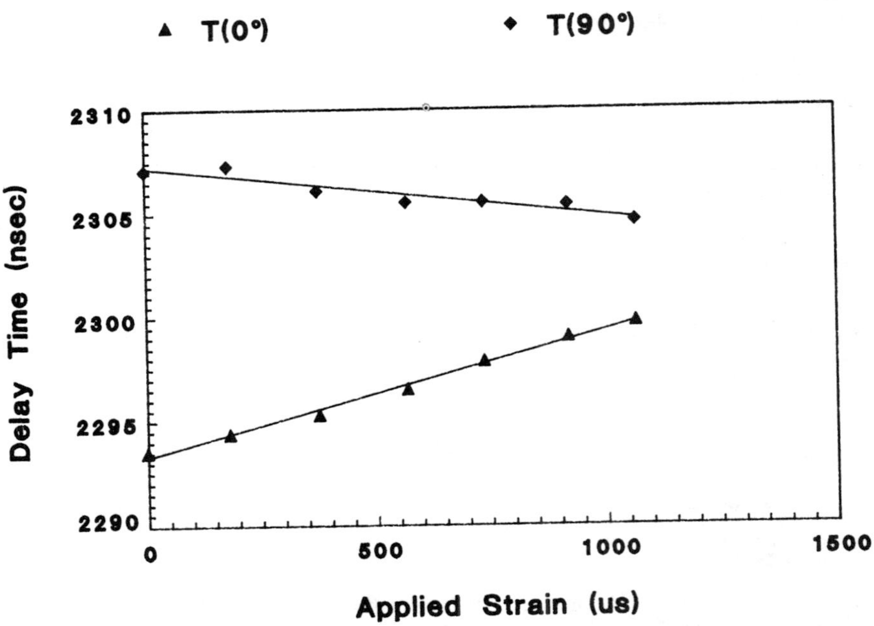

FIGURE 7. Delay time as a function of strain.

$$C_1 = \frac{V_0^2 - V_{90}^2}{\sigma_1 - \sigma_2} \tag{10}$$

where V_0 and V_{90} are the speeds of shear waves in the plate in the directions $\theta = 0°, 90°$, and are the velocities measured by using T_0 and T_{90} in Eq. (5). The terms σ_1 and σ_2 are principal stresses in directions 0° and 90°, respectively.

Assuming the stress-induced changes are small, C_1 may be expressed as

$$C_1 = \frac{2V^2}{\sigma_1 - \sigma_2} \frac{(V_0 - V_{90})}{V} \quad , \tag{11}$$

But

$$V_0 = V + dV_0 \text{ and } V_{90} = V + dV_{90} \tag{12}$$

$$C_1 = \frac{-2V^2 \cos^2\phi}{\sigma_1 - \sigma_2} [dT_{0,90}/T] \tag{13}$$

where

$$dT_{0,90} = T_0 - T_{90}.$$

In practice, materials are rarely 100% isotropic. They are nearly isotropic. The anisotropy of the material changes the delay time even under no stress conditions. We denote this quantity as,

$$\frac{dT°_{0,90}}{T°} = \frac{T°_0 - T°_{90}}{T°} \tag{14}$$

The material anisotropy can be subtracted to isolate the stress effects and C_1 can be written as

$$C_1 = \frac{-2V^2 \cos^2\phi}{\sigma_1 - \sigma_2} \left[\frac{dT_{0,90}}{T} - \frac{dT°_{0,90}}{T°}\right] \quad . \tag{15}$$

Using Eq. (15) and the data in Figure 7, the constant C_1 may be calculated.

The properties of the aluminum specimen are

$\rho = 2.737$ gm/cm^3,

$\lambda = 5.70 \times 10^6$ bar,

$\mu = 2.76 \times 10^6$ bar,

where λ and μ are Lamé constants and ρ is density. The angle ϕ was measured to be 18.25 ± 0.25. The experimental values of C_1 are tabulated in Table 1.

TABLE 1. Experimental Values of Acoustoelastic Constant

Microstrain	σ/μ	C_1^*
180	0.48×10^{-3}	-0.2286
374	1.0	-0.4740
567	1.5	-0.5161
734	2.0	-0.5224
917	2.5	-0.5029
1064	2.8	-0.5250

* Units of $(\text{cm/sec})^2/(\text{dyne/cm}^2)$.

The second column of the table is the applied stress normalized by the shear modulus μ. For the lowest stress, the constant differs significantly from those calculated for higher stress levels. For low stresses the change in velocity is also very small so that measurement accuracy (repeatability) is of the order of change in time. Thus, constants evaluated for the lowest stress may be neglected. The average value from the remaining data set is

$$C_1 = -0.508 \pm 7\% \quad (\text{cm/sec})^2 \cdot (\text{dyne/cm}^2)$$

Thus, for measuring residual stresses, constant C_1 is useful.

Equation (15) can be used in practical situations if C_1 is known. The parameter $dT_{0,90}$ can be measured at a location where stresses are to be measured. However, $dT°_{0,90}$ may not be measured at the same location but at a suitable location where the anisotropy can be approximately same but the residual stress is small. For a flat rolled plate, $dT°_{0,90}$ can be found by measuring at a location which is on the same rolling line but is at the edge of the specimen. This type of measurement was done by Fukuoka et al. [3].

COMMENTS

From the above experiments, one can conclude that the technique can be used for measuring the average (through the thickness) longitudinal residual stresses in aluminum plates. Once constant C_1 is determined, this technique can be used to find the difference of principal residual stresses in platelike specimens. The direction of maximum stress for isotropic materials is the one in which maximum change in time is observed. The material axis is assumed to be along the principal-stress direction for applicability of Eq. (15). Thompson et al. [6] developed a theory for SH waves in which the material axis can be at any angle with respect to the principal stress. The approach may possibly be useful for this technique also.

REFERENCES

1. Blessing, G.V., Hsu, N.N., and Proctor, T.M., "Ultrasonic-Shear-Wave Measurement of Known Residual Stress in Aluminum", *Experimental Mechanics*, 1984, p. 218.

2. Koshti, A.M., "A New Acoustoleastic Technique for Stress Measurement", Master's thesis, University of Oklahoma, 1985.

3. Fukuoka, H., Toda, H., and Naka, H., "Nondestructive Residual Stress Measurement in a Wide-Flanged Rolled Beam by Acoustoelasticity", *Experimental Mechanics*, 1983, p. 120.

4. Egle, D.M., "Using the Acoustoelastic Effect to Measure Stress in Plates", UCRL-52914, Lawrence Livermore Laboratory, 1980.

5. Smith, R.T., Stern, R., and Stephens, R.W.B., "Third-Order Elastic Moduli of Polycrystalline Metals from Ultrasonic Velocity Measurements", *Journal of the Acoustical Society of America*, Vol. 40, No. 5, 1966, p. 1002.

6. Thompson, R.B., Lee, S.S., and Smith, J.F., "Angular Dependence of SH Velocity in Stressed, Orthorhombic Plates", Paper presented at IEEE Ultrasonics Symposium, Nov. 1984.

PROBING A CONVEX INTERFACE WITH AN ULTRASONIC BEAM

H. C. Choi and J. G. Harris
University of Illinois
Urbana, IL 61801

1. INTRODUCTION

The purpose of this paper is to examine how a well-collimated ultrasonic beam is scattered from a convex circular fluid-solid interface. The incident beam is emitted from a transducer in the fluid. The radius of curvature of the interface is assumed to be so large that the incident beam never grazes the interface and therefore only reflection, transmission and the excitation of a leaky Rayleigh wave need be considered. We are primarily concerned with studying the combined effects of diffraction spreading of the incident beam and of the curvature of the interface.

The need to transmit an ultrasonic beam across a curved interface arises frequently in nondestructive testing. For example, Elsley, et al. [1] discuss the detection of cracks lying below the surface of a borehole in an aircraft engine disc. In such work the frequency of operation is typically on the order of 5 MHz. If we assume that the radius a of the transducer is 0.5 cm and that the transducer is immersed in water, then $k_o a$ equals 1.05×10^2, where k_o is the wave number in water. Moreover, a transducer radiates a well-collimated beam out approximately to a distance σ/π, where σ equals $k_o a^2/2$. For the values considered above σ/π equals 8.3 cm. Therefore, in many nondestructive testing situations, the curved interface is struck, not by a plane or spherical wave, but by a beam; and therefore it is important to understand how the diffraction spreading of the beam affects the wavefields scattered from the curved interface.

Here we model the incident beam as a two-dimensional, time-harmonic wavefield whose initial particle displacement is either Gaussian or uniform over the aperture. The Gaussian beam remains Gaussian in profile but spreads slightly as it propagates outward. The uniform beam, in contrast to the Gaussian beam, is composed in part of edge-diffracted waves. As a result it develops a complicated profile that changes and spreads as it propagates outward. We expect that the propagation properties of a real beam lie somewhere between these two models. To calculate the scattered geometrical wavefields and the leaky Rayleigh wave we decompose each beam into a sum of plane waves, calculate the wavefield scattered from a local region of the curved interface when struck by a plane wave and then compose the scattered beam fields by summing over the plane wave responses. These integrals are then evaluated asymptotically.

The calculations leading to the scattered beam fields are quite involved.

We discuss these calculations in detail in Choi and Harris [2]. In this paper we merely formulate the integral representations of the scattered beam fields and then give numerical results.

There is an extensive literature on scattering from penetrable and impenetrable cylinders, on the excitation, at plane interfaces, of leaky surface waves by an acoustic beam and on the asymptotic evaluation of integrals. This we shall, in a limited way, review as we develop the topics in the paper. However, two papers having goals similar to ours are that by Ruan and Felsen [3], and that by Randall and Stanke [4]. Ruan and Felsen calculate the reflection and transmission of an electromagnetic Gaussian beam at a convex cylindrical boundary separating two dielectrics using the complex ray method. Randall and Stanke develop a numerical model that describes the pulse-echo response of an ultrasonic transducer insonfying a cylindrically layered medium from the inside. Their beam model has an initial profile that is Gaussian across the aperture but that drops abruptly to zero at the aperture's edge.

2. INTEGRAL REPRESENTATIONS OF THE SCATTERED WAVEFIELDS

2.1 The Geometry

We consider a circular fluid-solid interface in a two-dimensional space. The geometry for the convex interface is shown in Fig. 1. The interface has a radius R_o; the fluid has a density ρ_o and wave speed c_o; and the solid has a density ρ, a compressional wave speed c_L and a shear wave speed c_T. A two-dimensional ultrasonic beam emitted, in the fluid, from a transducer whose half width is a and whose center is located at (r_o, θ_o) strikes a circular interface at the angle α, which is defined as the angle between the beam axis and the normal to the interface (Fig. 1). The observation point is located at (r, θ). The interface lies close enough to the transducer that it is struck by a well-collimated beam. In this paper we are concerned only with the single interaction at the interface and therefore we shall not consider the possibility that other waves, that have been excited by subsequent reflections or refractions, strike the interface. Moreover various surface waves are excited at the interface. These are discussed at length by Überall [5]. Here we consider only the excitation of the leaky Rayleigh wave, which can be thought of as the lowest whispering gallery mode. Stated differently, we are concerned with scattering when only a very localized part of the interface is insonified. Therefore we assume a to be much less than R_o.

2.2 Gaussian and Uniform Beams

We express the wavefields in the fluid in terms of a displacement potential rather than a velocity potential because we use displacement potentials to describe the wavefields in the solid.

Harris [6] has shown how two-dimensional, time-harmonic Gaussian and uniform beams can be constructed. The displacement potential for the incident Gaussian beam is described by

$$\Phi = \frac{iaA}{2\sqrt{\pi}} \int_S \Phi_G^*(\nu)\ exp[ik_o r cos(\pi/2+\theta-\alpha-\nu)]\ d\nu \qquad (2.1)$$

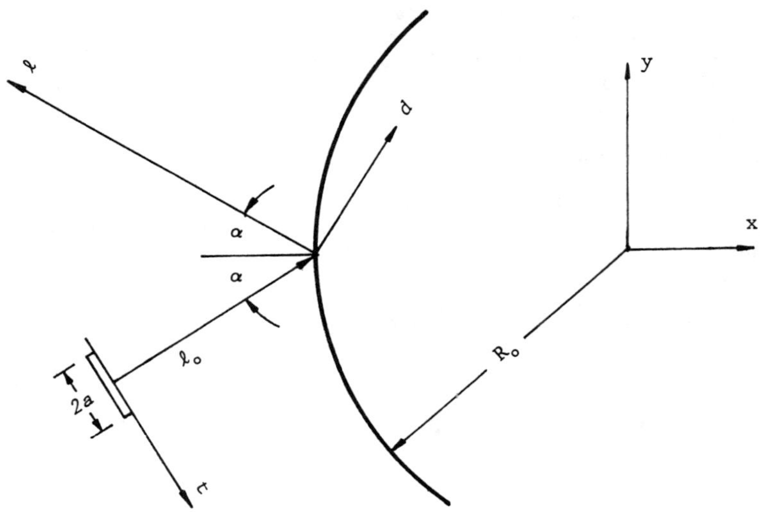

FIGURE 1. The geometry describing the scattering of a bounded beam from a circular interface.

where the weighting function $\Phi_G^*(\nu)$ is given by

$$\Phi_G^*(\nu) = exp[ik_o r_o sin(\theta_o - \alpha - \nu) - (k_o a\ cos\nu/2)^2]. \tag{2.2}$$

Here $k_o = \omega/c_o$, where ω is the angular frequency (the time dependent term $exp(-i\omega t)$ is suppressed). The beam profile at $\ell_o = 0$ is given by

$$\partial\Phi/\partial\ell_o = A\ exp(-t^2/a^2) \tag{2.3}$$

where the constant A is the magnitude of the beam at the center of the transducer and (t, ℓ_o) are the incident beam coordinates shown in Fig. 1. Their origin is located at (r_o, θ_o). Note that we also use ℓ_o to indicate the distance from (r_o, θ_o) to the interface. The contour S is the Sommerfeld contour and is the same contour, but shifted by $\pi/2$ to the right, as is the contour S' in Fig. 2. For a uniform beam we use a slightly different construction for the potential from that used in [6]. The displacement potential is written as a sum of line sources over the aperture, where each line source is constructed as the Gaussian beam, Eq. (2.1), was. It is described by

$$\Phi = -\frac{iA}{2\pi}\int_{-a}^{a}\int_S \Phi_U^*(\nu,t)\ exp[ik_o r cos(\pi/2+\theta-\alpha-\nu)]\ d\nu \tag{2.4}$$

where

$$\Phi_U^*(\nu,t) = exp[ik_o r_o sin(\theta_o-\alpha-\nu) - ik_o t\ cos\nu]. \tag{2.5}$$

The beam profile at $\ell_o = 0$ is given by

$$\partial\Phi/\partial\ell_o = A\ [\ H(t+a) - H(t-a)] \tag{2.6}$$

where H(t) is the Heaviside function. The double integral representation for the incident uniform beam makes it possible to calculate accurately the leaky Rayleigh wave and the reflected wavefield when the angle of incidence α is near the Rayleigh angle η_R.

2.3 Integral Representations of the Scattered Wavefields

Note that Eqs. (2.1) and (2.4) represent weighted sums of plane waves. Thus, to calculate the wavefields scattered from the circular fluid-solid interface when struck by an incident beam, we calculate the wavefields scattered by an incident plane wave and then sum them over the variable that distinguishes each plane wave in the incident beam. In carrying out our work we were most influenced by the methods and results of Ludwig [7] and Doolittle, et al. [8]. Here we briefly indicate how the calculations are done.

Ludwig [6] shows that an incident plane wave propagating in the x-direction can be written as

$$\Phi^i = exp(ik_o r cos\theta)$$

$$= \sum_{m=-\infty}^{\infty} \int_{-\infty}^{\infty} J_\varphi(k_o r)\ exp[i\varphi(\pi/2 - \theta + 2\pi m)]\ d\varphi \qquad (2.7)$$

where $J_\varphi(k_o r)$ is a Bessel function of the first kind of order φ. Because we are interested only in a single interaction with the interface, it is enough to consider the case $m = 0$ [6]. Let Φ_R be the displacement potential for the reflected wave in the fluid, and Φ_L and Φ_T be the displacement potentials in the solid. Then the displacements are given by

$$\underset{\sim}{u} = \nabla\ \Phi^i + \nabla\ \Phi_R \qquad (2.8)$$

in the fluid and

$$\underset{\sim}{u} = \nabla\ \Phi_L + \nabla\times(\Phi_T \hat{e}_3) \qquad (2.9)$$

in the solid, where $\hat{e}_3$ is the unit vector in the z-direction. These potentials satisfy the reduced wave equations with the wave numbers k_o, $k_L = \omega/c_L$ and $k_T = \omega/c_T$. The boundary conditions at the interface are that the normal displacement and traction are continuous, that the tangential traction is zero and that the scattered wavefields propagate away from the region of insonification. Solving this boundary value problem with the incident plane wave given by Eq (2.7), we obtain somewhat lengthy expressions for the reflected and transmitted waves which contain Hankel functions whose arguments are the appropriate wave number times R_o. For large arguments and observation points in the 'illuminated' region, we can use Debye's asymptotic expansion for the Hankel function. The approximate expressions for the scattered waves can thus be expressed in the form

$$\Phi_I = B \int_{S'} G_I(\eta) g_I(\eta)\ exp[ik_o \sigma f_I(\eta)]\ d\eta \qquad (2.10)$$

where the contour S' is shown in Fig. 2 and the phase $f_I(\eta)$ is given by

$$f_I(\eta) = \kappa_I \bar{R}_o cos\eta_I - \bar{R}_o cos\eta - \kappa_I \bar{r} cos\gamma_I + (\eta_I - \eta - \gamma_I + \pi - \theta)\bar{R}_o sin\eta. \qquad (2.11)$$

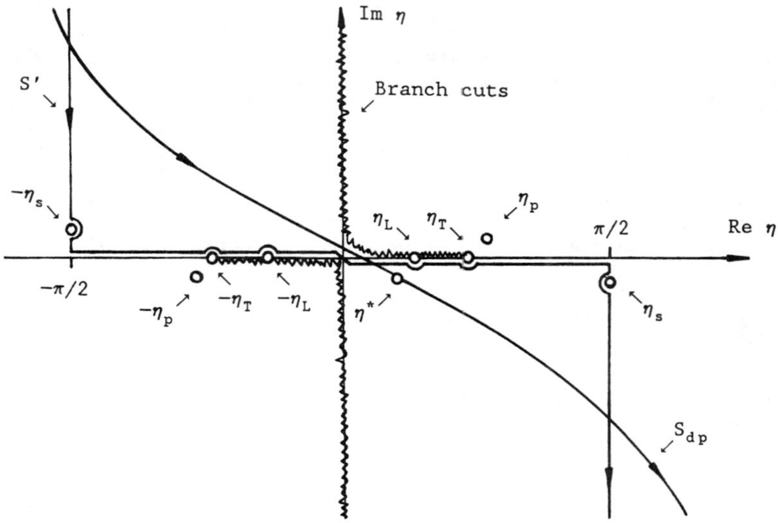

FIGURE 2. The complex η plane is sketched to show the poles, the branch cuts, the integration contour S' and a typical path of steepest descent S_{dp}.

Hereafter the subscript I assumes the value R to indicate the reflected wave, L to indicate the transmitted compressional wave or T to indicate the transmitted shear wave. The geometrical factor $g_I(\eta)$ has the form

$$g_I(\eta) = \bar{R}_o[\cos\eta\cos\eta_I/(\bar{r}\cos\gamma_I)]^{1/2} \tag{2.12}$$

and $G_R(\eta)$, $G_L(\eta)$ and $G_T(\eta)$ are the reflection coefficient, the transmission coefficient for the compressional wave and the transmission coefficient for the shear wave, respectively. A detailed discussion of these coefficients can be found in Brekhovskikh [9]. The other terms are

$$\eta_I = \sin^{-1}(\sin\eta/\kappa_I) \tag{2.13}$$

$$\gamma_I = \sin^{-1}[\bar{R}_o\sin\eta/(\kappa_I\bar{r})] \tag{2.14}$$

where $\kappa_I = k_I/k_o$, $\bar{R}_o = R_o/\sigma$ and $\bar{r} = r/\sigma$. Note that $k_R = -k_o$ so that $\kappa_R = -1$. The parameter $\sigma = k_o a^2/2$. Further, the constant $B = \pm\sqrt{k_o\sigma/2\pi}\,\exp(\pi i/4)$ where the upper sign is used for the reflected and transmitted compressional wave and the lower sign for the transmitted shear wave.

The parameter σ (more correctly σ/π as we indicated in the introduction) measures the distance over which a beam remains collimated. It is known variously as the Fresnel or Rayleigh number. Harris [10] has shown that an asymptotic expansion in inverse powers of $(k_o\sigma)$ can be used to describe a collimated beam. Consequently, when we asymptotically approximate the integrals representing the scattered wavefields this will be the large parameter. Note that when $k_o\sigma$ is large $k_o a$ is large implying that the aperture is many wavelengths across.

The reflection and transmission coefficients have four poles (two

Rayleigh wave poles $\pm\eta_p$, two Stoneley wave poles $\pm\eta_s$) and four branch points as shown in Fig. 2. The branch cuts have been chosen so that the imaginary part of each radical is always non-negative. A Rayleigh pole gives a leaky Rayleigh wave, a Stoneley pole a Stoneley wave and the branch points lateral waves as discussed in Pott and Harris [11], and Harris and Pott [12]. It is important to note that the Rayleigh poles and the outermost branch points associated with the shear wave in Fig. 2, for some materials such as metals, can be quite close together. In this paper we will not consider the effect of this branch point because Harris and Pott [12], in their study of the plane interface problem, have shown that the branch point has only a weak effect upon the reflected wavefield.

Using the weighting function $\Phi_G^*(\eta)$, Eq. (2.2), the scattered wavefields corresponding to an incident Gaussian beam are constructed from Eq. (2.1) and Eq. (2.10). They are described by

$$\Phi_I/aA = C \int_S \int_{S'} G_I(\eta) g_I(\eta)\ exp[ik_o\sigma F_I(\eta,\nu)]\ d\eta\ d\nu. \tag{2.15}$$

The phase function is given by

$$F_I(\eta,\nu) = f_I(\eta) + (\nu+\alpha-\pi/2)\bar{R}_o sin\eta + \bar{r}_o sin(\theta_o-\nu-\alpha) + icos^2\nu/2. \tag{2.16}$$

The phase $f_I(\eta)$ is given by Eq. (2.11). The constant $C = iB/2\sqrt{\pi}$. The scattered wavefields corresponding to an incident uniform beam are constructed as triple integrals using Eqs. (2.4) and (2.10). They are described by

$$\Phi_I/aA = D \int_{-1}^{1} \int_S \int_{S'} G_I(\eta) g_I(\eta)\ exp[ik_o\sigma F_I(\eta,\nu,t)]\ d\eta\ d\nu\ dt \tag{2.17}$$

where the constant $D = -\ iB/2\pi$ and the phase function is given by

$$F_I(\eta,\nu,t) = f_I(\eta) + (\nu+\alpha-\pi/2)\bar{R}_o sin\eta + \bar{r}_o sin(\theta_o-\nu-\alpha) - (2t/k_o a)cos\nu. \tag{2.18}$$

3. EVALUATION OF THE SCATTERED WAVEFIELDS

The coordinates $(d,\ \ell)$ shown in Fig. 1 describe the scattered beams. Their origin lies at the interface. Each ℓ-axis makes the angle $\alpha_I = sin^{-1}(sin\alpha/|\kappa_I|)$ with the normal; that is, each ℓ-axis corresponds to a geometrically reflected or refracted ray.

As we indicated in the introduction, the asymptotic evaluation of these integrals, Eqs. (2.15) and (2.17), is quite involved. The details are given in [2]. In this paper we have contented ourselves with reporting various numerical results which will make clear the relevant physical phenomena.

3.1 Gaussian Beam

To illustrate the scattered wavefields consider a convex water-metal interface. Assume that a = 3 mm, the frequency is 4 MHz and $\ell_o = \ell$ = 5 cm. The wave speed in water $c_o = 1.5\times10^3$ m/s. In this case σ = 7.5 cm and $k_o\sigma = 1263$. Further, assume that Poisson's ratio $\nu = 0.3$, $k_o/k_T = 2$ and $\rho_o/\rho = 0.125$. The Rayleigh angle $\eta_R = 32.7°$ and the leaky Rayleigh wave pole $\eta_p = 32.6° + i0.401°$.

First, consider the case in which α is much less than any critical angle. Then the excitation of a lateral or a leaky Rayleigh wave will not be strong enough to affect the specularly scattered beam. In Fig. 3 the magnitude of the reflected beam $|\Phi_R/aA|$ is plotted as a function of the normalized transverse distance d/a for R_o = 7.5 cm and 30 cm when α = 10°. The magnitude of the transmitted beams $|\Phi_{L,T}/aA|$ is similarly plotted in Fig. 4 for R_o = 15 cm and α = 10°. In Figs. 3 and 4 we have also plotted the magnitude of the beam reflected from and transmitted across a plane interface, respectively. Also plotted in Fig. 3 is a reference beam that represents a beam that has been specularly reflected from a perfectly rigid plane interface.

Because the reflection and transmission coefficients are smooth functions when their arguments are not close to any critical angles, the magnitudes of the scattered potentials, $|\Phi_I/aA|$, depend largely on the geometry of the interface. Therefore the beams scattered from a convex interface undergo spreading that is in addition to their diffraction spreading. Fig. 3 shows this curvature effect on the reflected beam. Moreover, the maxima of the scattered beams are shifted slightly from the beams' central axes, which are determined using geometrical acoustics. Ruan and Felsen [3] have also noted this phenomenon. These shifts are especially noticeable in the transmitted beams as shown in Fig. 4.

When the angle of incidence α is near the Rayleigh angle η_R, the leaky Rayleigh wave is strongly excited. This leaky Rayleigh wave interacts with the specularly reflected beam. Moreover this interaction is frequency dependent, as Ngoc and Mayer [13] have shown in their study of

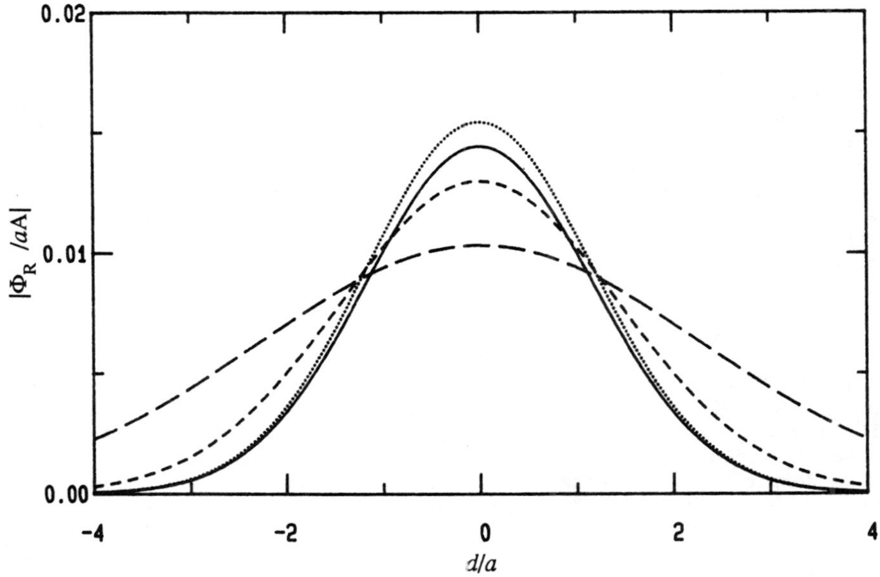

FIGURE 3. The specularly reflected Gaussian beam is plotted; the solid line is used for the plane interface, the dashed line for the convex interface whose radius R_o = 30 cm, the long-dashed line for R_o = 7.5 cm and the dotted line for the reference beam.

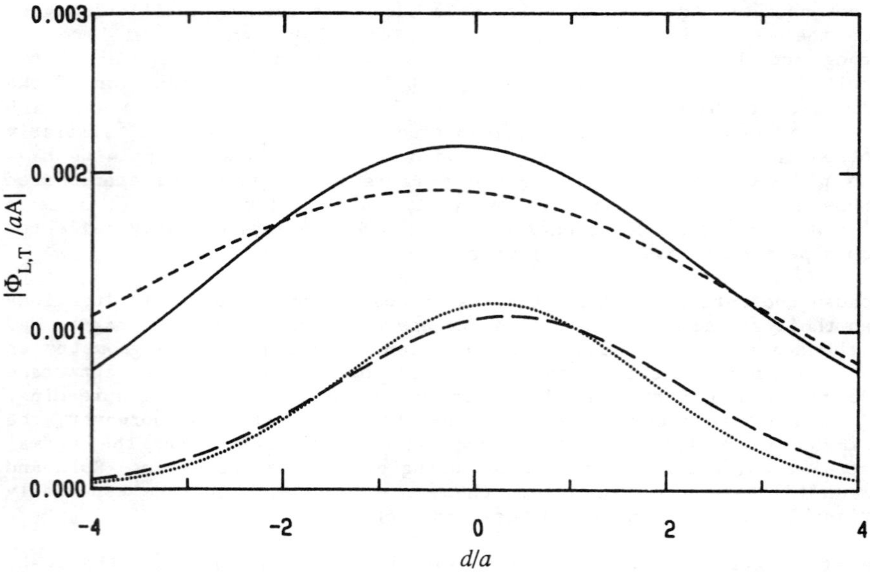

FIGURE 4. The transmitted beams for the incident Gaussian beam are plotted; the solid line indicates the compressional beam and the dotted line the shear beam transmitted across a plane interface; the dashed line indicates the compressional beam and the long-dashed line the shear beam transmitted across a convex interface.

the plane interface case. As the frequency increases the leaky Rayleigh wave assumes a more dominant role. In Fig. 5 we have illustrated this interaction at two frequencies, 4 MHz and 10 MHz, and for a plane and a convex interface. The angle α = 32.7°, the Rayleigh angle, and R_o = 7.5 cm for the convex interface. Note that at 10 MHz σ = 18.8 cm and $k_o\sigma$ = 7896. All the remaining parameters have the same values as they had in Fig. 3. In addition to the effect of frequency, note that the curvature of the interface causes the null to be shifted and the scattered beam to undergo additional spreading.

3.2 Uniform Beam

The incident uniform beam is composed of two separate wavefields; namely, a plane wave emitted from the face of the transducer and diffracted waves emitted from its edges. The Gaussian beam by contrast consists of a single wavefield. Therefore the scattered wavefields for the uniform beam have a complexity that those for the Gaussian beam do not. To illustrate this increased complexity the magnitude of the reflected beam, $|\Phi_R/aA|$, for the incident uniform beam is plotted in Fig. 6 using the same values for the parameters as were used in Fig. 3. In addition, in Fig. 7 the magnitude of the transmitted beam, $|\Phi_{L,T}/aA|$ is plotted using the same values for the parameters as were used in Fig. 4. These figures show that the curvature of the convex interface plays the same role as it does in the case of the incident Gaussian beam. Because of the large difference in wave speeds between the fluid and the solid, the transmitted beams

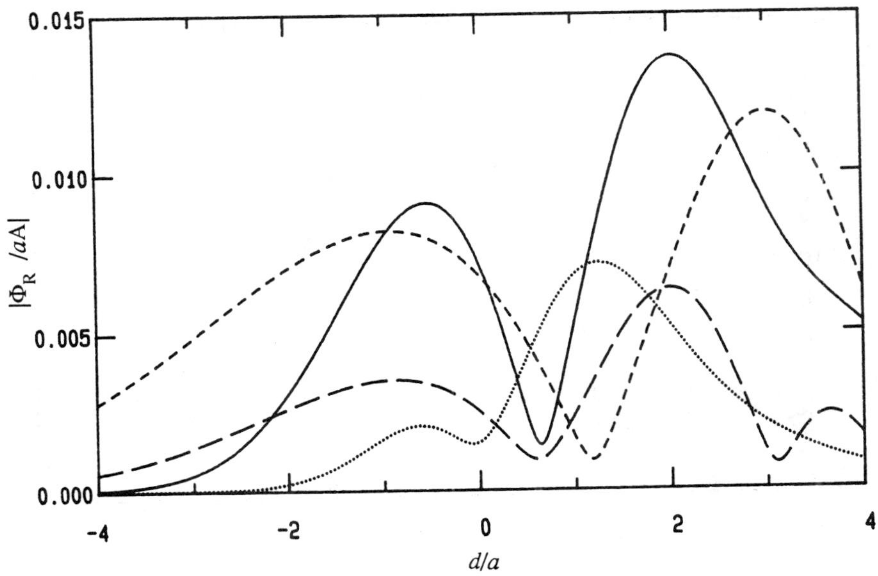

FIGURE 5. The frequency dependence of the nonspecularly reflected Gaussian beam is illustrated; the solid line indicates scattering at 4 MHz and the dotted line scattering at 10 MHz from a plane interface; the dashed line indicates scattering at 4 MHz and the long-dashed line scattering at 10 MHz from a convex interface.

spread much more rapidly than does the reflected beam. This is why sidelobes are present in Fig. 6 and absent in Fig 7. If we consider only the central lobe then the scattered wavefields for the incident uniform beam and those for the incident Gaussian beam are somewhat similar.

Next consider the scattered uniform beam in the fluid when the incident angle α is near the Rayleigh angle η_R. As in the Gaussian beam case the leaky Rayleigh wave is now excited along with the reflected beam. The magnitude of the nonspecularly reflected beam in the fluid, $|\Phi_R/aA|$, is plotted in Fig. 8 using values for the parameters identical to those used in Fig. 5. We again note that as the frequency is increased the leaky Rayleigh wave plays a more dominant role and that the curvature of the interface causes the profile of the scattered wavefield to change. Further, there is more structure here than in the Gaussian beam case.

4. CONCLUSIONS

The preceding graphs have shown how a convex interface can cause an ultrasonic beam to undergo spreading that is in addition to the diffraction spreading it undergoes because of the finite size of the aperture. For nondestructive testing it may be preferable to use a transducer that emits a Gaussian beam because the absence of edge-diffracted waves means that there are no sidelobes present that might confuse a measurement of a defect.

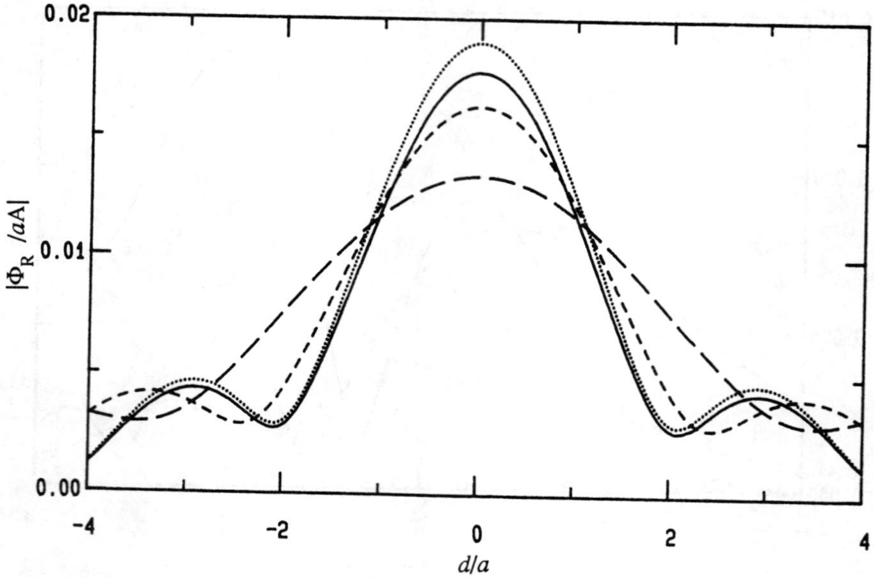

FIGURE 6. The specularly reflected uniform beam is plotted; the conventions used here are identical to those used in Fig. 3.

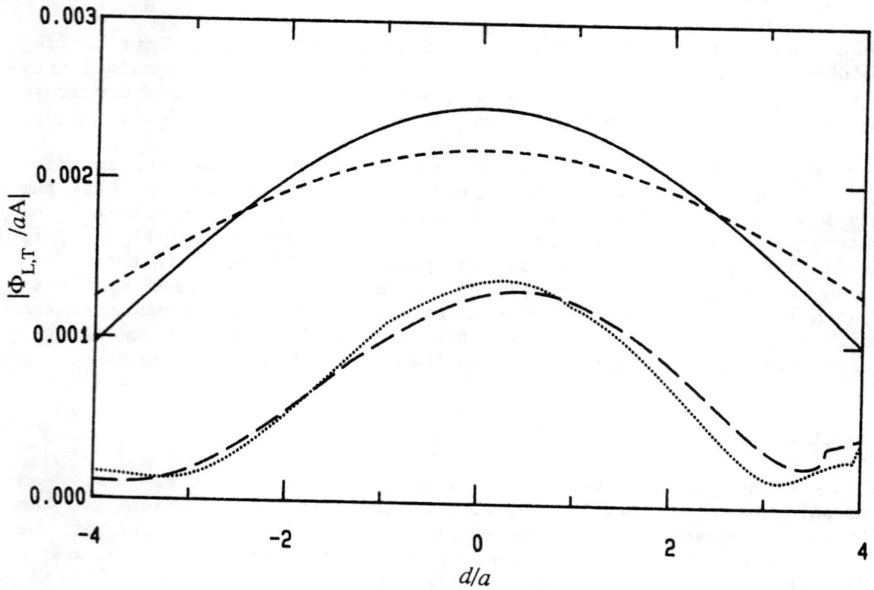

FIGURE 7. The transmitted beams for the incident uniform beam are plotted; the conventions used here are identical to those used in Fig. 4.

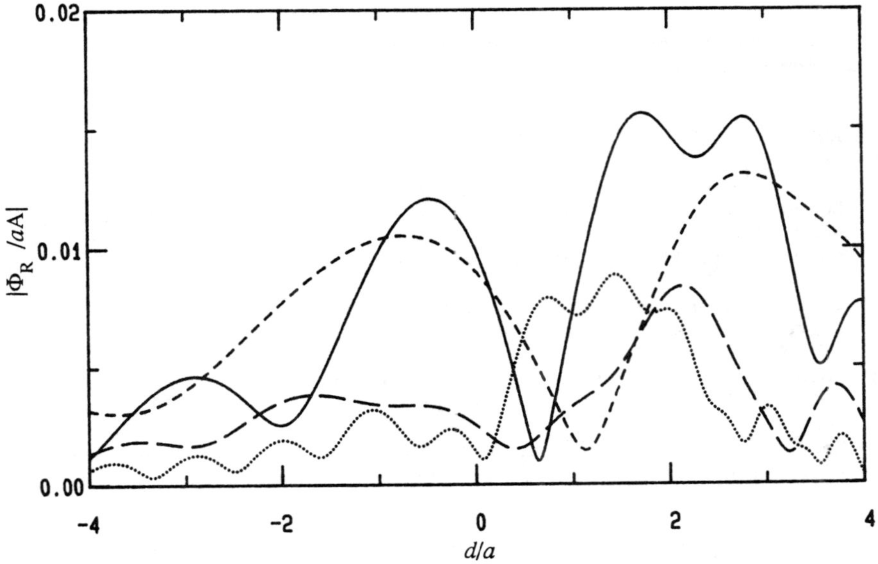

FIGURE 8. The nonspecularly reflected uniform beam is plotted; the conventions used here are identical to those used in Fig. 5.

Specular reflection does not always take place because for angles of incidence near the Rayleigh angle a leaky Rayleigh wave is excited. The pronounced null that occurs in the reflected wavefield can be used to measure the Rayleigh wave speed. However, as our graphs indicate, one must take account of both the curvature of the interface and the profile of the incident beam.

In many nondestructive testing situations, the compressional transmitted beam, excited by a normally incident beam, is the one used. If we assume that no crack or other defect could readily be detected unless its maximum length parallel to the tangent of the interface, call this d^*, is equal to one-half the beam width, then we would find that

$$d^* = 2\{[\ell_o+\ell k_o/k_L+\ell_o \ell(k_o/k_L-1)/R_o]^2+\sigma^2[1+\ell(k_o/k_L-1)/R_o]^2]\}^{1/2}/\sqrt{2k_o\sigma} \quad (4.1)$$

for the Gaussian beam and

$$d^* = \pi[\ell_o+\ell k_o/k_L+\ell_o \ell(k_o/k_L-1)/R_o]/\sqrt{2k_o\sigma} \quad (4.2)$$

for the uniform beam. These are approximate formulas that involve some arbitrary but quite reasonable definitions of beam width and, for the uniform beam, the assumption that the crack lies in the farfield of the transducer. Nevertheless, they clearly indicate how the beam spreading limits the resolution of a crack-detection system.

ACKNOWLEDGMENT

This material is based upon work supported by the National Science Foundation under Grant No. MSM-8513928.

REFERENCES

[1] Elsley, R.K., Addison, R.C. and Graham, L.J., Detection of Flaws below Curved Surfaces, in *Review of Progress in Quantitative Nondestructive Evaluation*, Vol. 2A, ed. D.O. Thompson and D.E. Chimenti, pp 113-128, Plenum, New York, 1983.

[2] Choi, H.C. and Harris, J.G., Scattering of an Ultrasonic Beam from a Curved Interface, forthcoming.

[3] Ruan, Y.E. and Felsen, L.B., Reflection and Transmission of Beams at a Curved Interface, *J. Opt. Soc. Am.*, Vol. A3, pp 566-579, 1986.

[4] Randall, C.J. and Stanke, F.E., Mathematical Model for Internal Ultrasonic Inspection of Cylindrically Layered Structures, *J. Acoust. Soc. Am.*, Vol. 83, pp 1295-1305, 1988.

[5] Überall, H., Surface Waves in Acoustics, in *Physical Acoustics, Vol. 10,* ed. W.P. Mason and R.N. Thurston, pp 1-60, Academic, New York, 1973.

[6] Harris, J.G., Edge Diffraction of a Compressional Beam, *J. Acoust. Soc. Am.*, Vol. 82, pp 635-646, 1987.

[7] Ludwig, D., Uniform Asymptotic Expansion of the Field Scattered by a Convex Object at High Frequencies, *Comm. Pure Appl. Math.*, Vol. 20, pp 103-138, 1967.

[8] Doolittle, R.D., Überall, H. and Ugincius, P., Sound Scattering by Elastic Cylinders, *J. Acoust. Soc. Am.*, Vol. 43, pp 1-14, 1968.

[9] Brekhovskikh, L.M., *Waves in Layered Media*, 2nd ed., Academic, New York, pp 43-47, 1980.

[10] Harris, J.G., The Wavefield Radiated into an Elastic Halfspace by a Transducer of Large Aperture, *J. Appl. Mech.*, forthcoming, 1988.

[11] Pott, J. and Harris, J.G., Scattering of an Acoustic Gaussian Beam from a Fluid-solid Interface, *J. Acoust. Soc. Am.*, Vol. 76, pp 1829-1838, 1984.

[12] Harris, J.G. and Pott, J., Further Studies of the Scattering of a Gaussian Beam from a Fluid-solid Interface, *J. Acoust. Soc. Am.*, Vol. 78, pp 1072-1080, 1985.

[13] Ngoc, T.D.K. and Mayer, W.G., Numerical Integration Method for Reflected Beam Profiles Near Rayleigh Angle, *J. Acoust. Soc. Am.*, Vol. 67, pp 1149-1152, 1980.

ULTRASONIC EVALUATION OF ALUMINUM TO ALUMINUM ADHESIVE BOND DURABILITY

E. J. Chern, B. B. Djordjevic, J. S. Ahearn, D. K. McNamara, and A. Desai
Martin Marietta Laboratories
Baltimore, MD 21227

ABSTRACT

The nondestructive evaluation (NDE) of adhesive bond integrity requires an understanding of the physical and chemical adhesive/cohesive bonding mechanisms as well as the properties of the adhesive itself and therefore the NDE method must be sensitive to these dominant bond parameters. The quality of adhesive bonds can be viewed from two aspects, strength and durability. The durability of aluminum adhesive bonds has been shown to depend on aluminum oxide morphology and on interfacial chemical reactions. Because adhesive bond durability is closely related to the micromorphology of the adherend surface, it is necessary to operate at high ultrasonic test frequencies so that interaction at the adhesive/adherend interface can be increased. In this paper, after reviewing the factors affecting the bonds in aluminum-to-aluminum (Al/Al) adhesive bond quality,we report 50-MHz acoustical microimaging results that qualitatively demonstrate the applicability of using high-frequency stress-waves for durability evaluation of Al/Al adhesive bonds.

INTRODUCTION

Adhesives offer unprecedented advantages, e.g., light weight, uniform stress distribution, and sealing effect, over mechanical fasteners in material joining technology. However, the quality of adhesive bonds are affected by many variables, e.g., adhesive type, surface treatment, curing condition, and assembly process. Thus, it is very important to assess adhesive bond quality by destructive and nondestructive methods to establish adhesive joint structural integrity criteria. Whereas destructive techniques provide indirect statistical data about the quality of a specific bonding process, NDE techniques promise direct assessment of the structural bond quality.

Adhesive bond quality can be viewed from two aspects, strength and durability. Many destructive testing methods such as tensile pull, lap shear, T-peel, fatigue strength, and creep rupture are used to determine adhesive bond strength of the witness coupons. However, since each technique measures a slightly different aspect of bond quality, the most appropriate bond strength measurement is not well defined. Conventional ultrasonic pulse echo and other techniques [1-3] have also been widely used to nondestructively measure parameters and correlate with one of the previously mentioned destructive techniques. However, they have had limited success.

Durability, the other important aspect of adhesive bond quality, most commonly is destructively determined using a combined stress/environmental exposure test of the test coupons, such as the wedge test. Durability is affected by three factors, namely surface morphology, surface chemical treatment, and contamination. The purpose of this study is to determine the feasibility of using a high-frequency acoustical microimaging technique for Al/Al bond durability assessment due to contamination.

In the next section, after reviewing the Al/Al adhesive bonding mechanism and the factors that affect Al/Al bond durability, we report the results of our work whereby we qualitatively demonstrate the applicability of using high-frequency ultrasound for the durability evaluation.

BACKGROUND

The durability of Al/Al adhesive bonds depends on physical interlocking and on chemical interfacial interaction mechanisms [4]. Physical interactions are controlled by the surface roughness of the adherend. Figure 1 shows a typical Al/Al bonding structure. The controlling factors in the durability of Al adherends bonded with an epoxy adhesive have been attributed to various conditions, including surface morphology, adherend surface chemical treatment, and contamination from bonding processes. The debonding mechanisms of aluminum adhesive bond durability are summarized in Table 1.

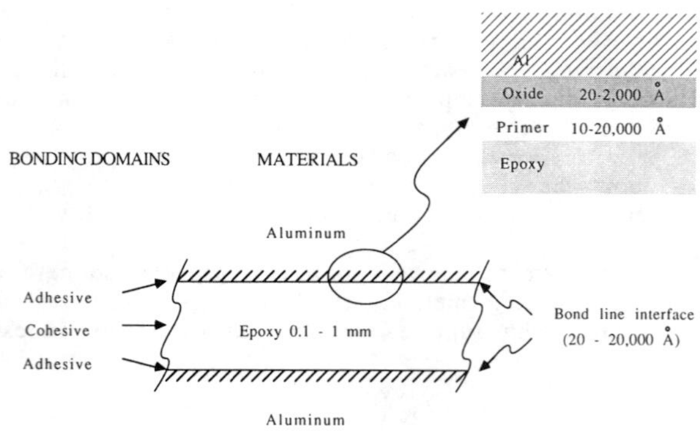

FIGURE 1. Typical aluminum-to-aluminum adhesive bonding structure.

Bonding Parameter		Debonding Mechanisms
Smooth surface Dispersion force	-------->	Water attack of dispersive bonds
Smooth surface Stable interface	-------->	Water attack of chemical bonds
Rough surface Unstable oxide	-------->	Water attack of oxide (ALOOH formation)
Rough surface Stable oxide	-------->	Water attack of oxide or plasticization of adhesive
Rough surface Stable oxide Stable interface	-------->	Plasticization of adhesive

TABLE 1. Debonding mechanisms of aluminum adhesive bond durability. Increasing durability from top to bottom of table.

Figure 2 [5,6] shows the effect of surface morphology on the durability of the Al/Al adhesive bonds. Three surface roughnesses are used: a) smooth, resulting from the commercial deoxidizer Smut-Go #1, b) intermediate, resulting from the Forest Products Laboratory (FPL) process, and c) rough, resulting from the phosphoric acid anodization (PAA) process. According to the wedge test results, the PAA adherend performed significantly better than either FPL or Smut-Go #1, and FPL performed better than Smut-Go #1. These results strongly indicate that increasing oxide microroughness improves bond durability. However, the question remains concerning the effect of the surface chemistry on durability.

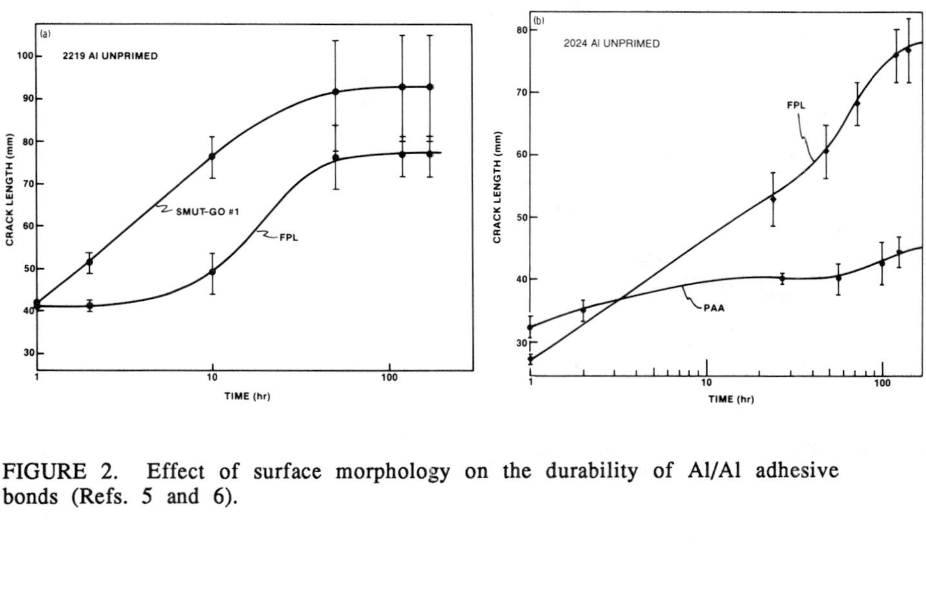

FIGURE 2. Effect of surface morphology on the durability of Al/Al adhesive bonds (Refs. 5 and 6).

Figure 3 [7] shows the wedge test results for four FPL-treated adherends, three of which subsequently underwent further surface chemical treatment with either methyl phosphonic acid (MP), phosphoric acid (PA), or nitrilotris methylene phosphonic acid (NTMP). Several significant comments should be made regarding the data. First, adsorption of a monolayer of MP degrades durability. This suggests that passivation of the Al oxide surface by MP adsorption can decrease durability, presumably by decreasing the contribution of chemical bonding to crack propagation resistance. Second, adsorption of a monolayer of PA has little effect on bond durability. This implies that the difference in behavior between FPL and PAA adherends is primarily caused by differences in oxide microroughness and not surface chemistry. Finally, the adsorption of a monolayer of the chemically active molecule, NTMP, markedly increases bond durability to the point where it rivals that of PAA adherends. Thus, by forming strong chemical bonds across the Al/epoxy interface, durability of an intermediate-rough oxide can be improved to nearly that of a rough oxide such as PAA.

The other important factor affecting the Al/Al bond durability is contamination introduced during the surface treatment and bonding processes. The most common contamination from processes is silicone contamination from handling and acid contamination from surface preparation. Silicone, a common component of mold release agents, is often introduced during bond lay-up, and acid is often introduced through inadequate rinsing during the "cleaning" process. Such contamination is inherent to actual bonding practices and frequently causes the most common quality control problems. However, it is subtle and hard to detect. Figure 4 [8] shows tensile pull results for Al/Al samples after soaking in rinse

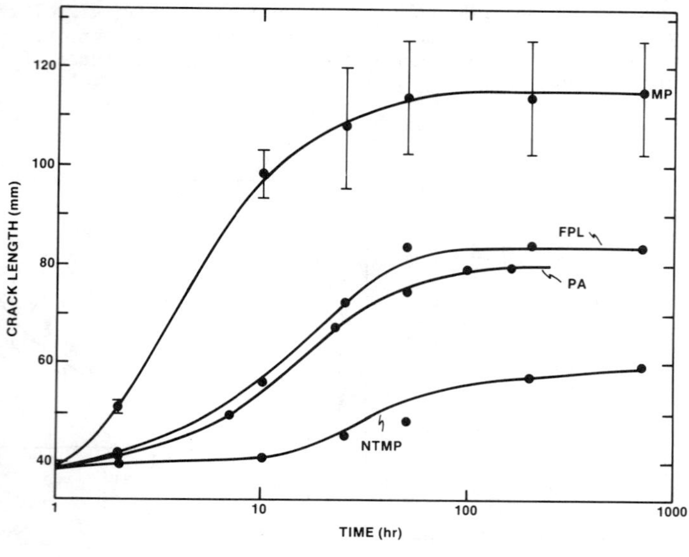

FIGURE 3. Wedge test results for FPL-treated adherends with methyl phosphonic acid (MP), phosphoric acid (PA), and nitrilotris methylene phosphonic acid (NTMP) treatments (Ref. 7).

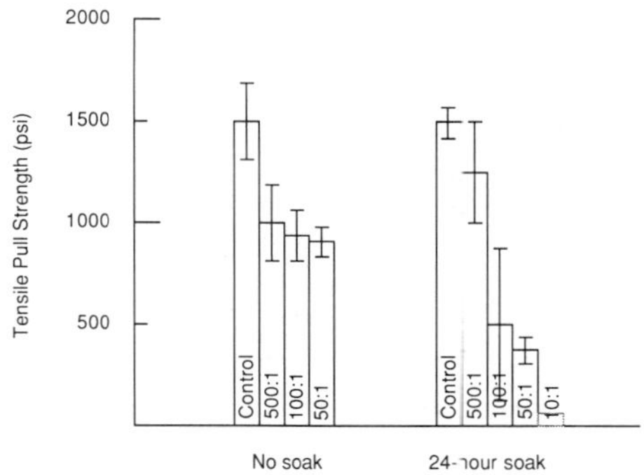

FIGURE 4. Bond tensile strength for different ratios of contamination in the rinse of the aluminum bonding surfaces (Ref. 8).

solutions of various degrees of contamination. These types of contamination are very critical to the bond durability and can cause dramatic failure. The objective of this paper is to investigate NDE techniques to detect this adverse factor introduced during the surface preparation processes.

EXPERIMENTAL METHODS

As described in the previous section, the durability of adhesive bonds was shown to depend on oxide morphology and on interface chemical reactions. Because adhesive bond durability is closely related to the microstructure of the adherend surface, we need to operate at high-frequency ultrasound to increase the interactions with the adhesive bondlines on a microscopic scale.

Two 6 in. x 6 in. x 1/8 in. 2219 bare T-87 panels were bonded with American Cyanamid FM-300-M adhesive. Prior to bonding, the panels were degreased in Turco 4215 solution, rinsed with distilled/deionized water, and deoxidized in Smut-Go #1 at room temperature for 15 minutes. Half of each panel was then immersed in a contaminated bath as follows: the Sample #1 panel was dipped into a solution of one part Smut-Go #1 to one part water and Sample #2 was dipped into a solution of one part Smut-Go #1 to ten parts water. After air drying, the dipped panels were bonded with the Smut-Go #1 cleaned panels using the FM-300-M adhesive.

Tensile strength tests performed on similarly prepared samples show that the bonds between panels soaked in contaminated solution were weaker than those panels that did not undergo the extra soaking (Fig. 4 [8]). Tensile strength for the two specimens are 50 psi and 20 psi for 1:10 and 1:1 contaminations, respectively.

The two Al/Al adhesive bond samples were scanned using a high-frequency ultrasonic imaging technique. The acoustical microimaging system (Fig. 5) consists of a mechanical scanning assembly, high-frequency pulser/receiver, and computerized data acquisition system. The stepless time-gate delay was set at the second Al/epoxy interface (top surface of the

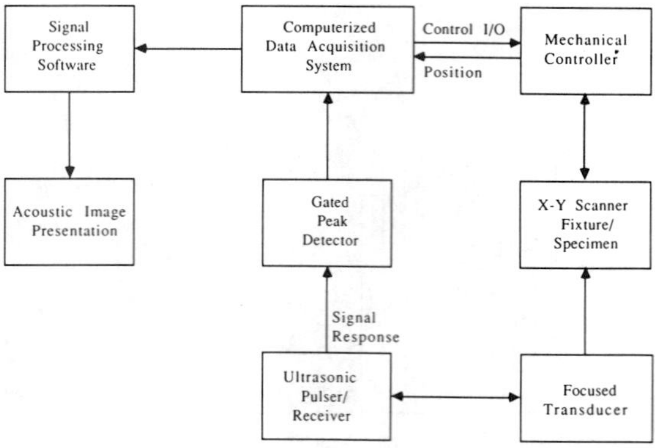

FIGURE 5. Block diagram of the acoustical microimaging system.

second Al panel). Acoustical microimaging scans were performed at a resolution of 0.005 in. and covered a total area of 2 in. x 1 in. Transducers operating at 10, 20, 30, 50, and 100-MHz were used for the experiments.

RESULTS AND DISCUSSION

As mentioned in Section II, Smut-Go #1 treatment provides the least durable Al/Al adhesive bond system tested in a controlled manner and therefore it has been chosen for our study. Figure 6 is a typical oscilloscope trace of amplitude versus time from a sampling point. Figure 7 is a 50-MHz

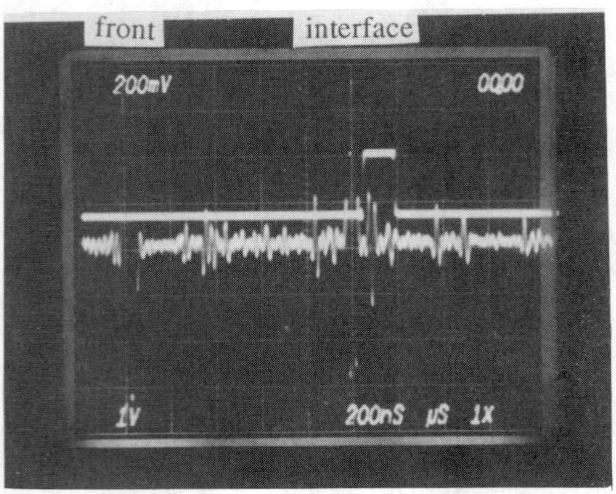

FIGURE 6. Typical oscilloscope trace showing amplitude versus time from a sampling point.

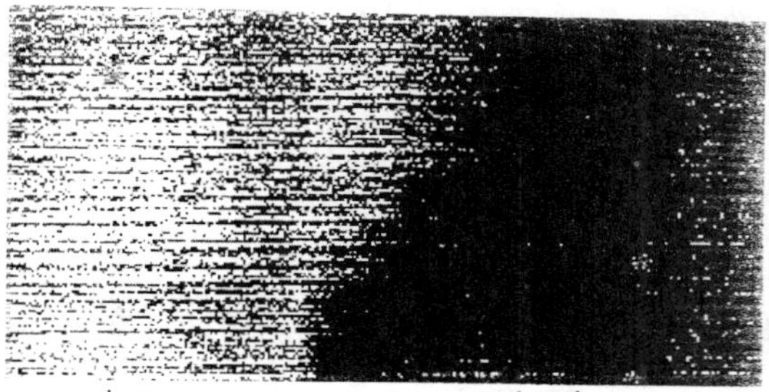

clean contaminated 2" x 1"
(1 Smut Go#1 : 1 Water)

Frequency: 50 MHz

Focus: interface

Gate: interface

Figure 7. A 50-MHz acoustical microimage of Sample #1 with the sound beam focused on the second Al/epoxy interface The clean area is on the left and the contaminated area is on the right.

acoustical microimage of Sample #1 with the sound beam focused on the second Al/epoxy interface. The contaminated area is on the right half and the clean area is on the left half. Figure 8 is a 50-MHz acoustical microimage of Sample #1 focused on the back surface of the second Al adherend. Similar microimages were also obtained for Sample #2, i.e., one part Smut-Go #1 : ten parts water.

clean contaminated 2" x 1"
(1 Smut Go#1 : 1 Water)

Frequency: 50 MHz

Focus: back surface

Gate: interface

FIGURE 8. A 50-MHz acoustical microimage of Sample #1 focused on the back surface of the second Al adherend. The clean area is on the left and the contaminated area is on the right.

The reflected amplitude difference of the two sides observed at 50-MHz (Figs. 5 and 6) shows that acoustical microimages can indicate a difference between the contaminated and clean sides of Sample #1. The amplitude difference between the two is ~5%. Similar acoustic results were observed on Al/Al Sample #2 at 50-MHz. The contamination of Sample #2 is an order of magnitude smaller than that of Sample #1 and the reflected acoustic amplitude difference between the contaminated and the clean side was about 2-3%. However, some results have shown inconclusive differences or differences less than 2-3%. These results may be due to uneven contamination resulting from the preparation.

There was little difference between the clean and contaminated sides of acoustic images at 10, 20, and 30-MHz operating frequencies for Samples #1 and #2. Due to the high attenuation nature of the adhesive at 100-MHz, the reflected acoustic amplitude is too small for meaningful analysis under the detection limits of this experimental setup. Adhesive bond durability is closely related to the adherend oxide microstructure, bond contamination, load, and environment. We have demonstrated the feasibility of using a high-frequency acoustical microimaging technique to investigate Al/Al adhesive bond contamination and thus the durability. Future efforts will be directed toward systematic evaluation and analysis using a range of durability samples.

REFERENCES

1. Segal E. and J.L. Rose, "Nondestructive Testing Techniques for Adhesive Bond Joints," Research Techniques in Nondestructive Testing, Vol. IV., ed. R. S. Sharpe, Academic Press, New York (1980), p. 275.

2. Rokhlin, S.I., "Adhesive Joint Characterization by Ultrasonic Surface and Interface Waves," in Proc. International Symposium on Adhesive Joints: Formation, Characterization, and Testing, Plenum Press, New York (1984), p. 307.

3. Teller, C.M., K.J. Diercks, Y. Bar-Cohen, and N.N. Shah, "Nondestructive Evaluation of Adhesive Bonds," in Proc. of the Sixteenth Symposium on NDE (1987), p. 139.

4. Ahearn, J.S. and G.D. Davis, "Adhesive Bond Durability of Aluminum," in Proc. Second International Conference in Adhesion, York, UK (1987) p. 29/1.

5. Ahearn, J.S., G.D. Davis, T.S. Sun, and J.D. Venables, "Correlation of Surface Chemistry and Mechanical Properties of Aluminum/Polymer Bonds," Adhesion Aspects of Polymer Coatings, K.L. Mittal, ed., Plenum Press, New York (1983), p. 281.

6. Hardwick, D.A., J.S. Ahearn, and J.D. Venables, "Environmental Durability of Aluminum Adhesive Joints Protected with Hydration Inhibitors," Journal of Materials Science, Vol. 19 (1984), p. 223.

7. Davis, G.D., J.S. Ahearn, L.J. Matienzo, and J.D. Venables, "Use of Hydration Inhibitors to Improve Bond Durability of Aluminum Adhesive Joints," Journal of Materials Science, Vol. 20 (1985), p. 975.

8. McNamara, D.K. and A. Desai, "External Tank Primer Adhesion," Martin Marietta Laboratories Technical Report MML TR 87-69 (1987).

ULTRASONIC DIFFUSE FIELD MEASUREMENTS OF GRAIN SIZE

R. L. Weaver
University of Illinois
Urbana, IL 61801

The incoherent transport of diffuse ultrasound was studied in polycrystalline steel specimens as a function of heat treatment. Observed diffusion constants were found to vary with frequency and grain size in a manner consistent with theoretical expectations. Strong correlations with Charpy toughness values were also observed. It is suggested that ultrasonic diffusivities may provide quantitative nondestructive evaluation of metallic microstructures with utility comparable to that of the ultrasonic attenuation method and the ultrasonic backscattering method.

I. INTRODUCTION

At times long after an ultrasonic source has acted in a multiply-scattering solid body, long that is compared to the typical time between the randomizing scatterings of an ultrasonic ray, the ultrasonic field has lost most of its spatial phase coherence. The resulting total wave field is essentially incalculable by the theorist. The wave field in this regime is termed diffuse, and the analysis appropriate thereto diffuse field analysis. While there exists a substantial body of theoretical and experimental literature concerned with the average, or non-diffuse part of the wave field [*e.g.* 1-3], at these late times this average wave has been too attenuated to be of any significance. The average, or "coherent" field is small compared to the incoherent diffuse field. Methods for laboratory spatial averaging do exist[*e.g.* 4] whereby the coherent field in a polycrystal can be enhanced relative to the incoherent field, but they are of considerable complexity and only marginally effective.

If the scattering is high, as it is if one is employing frequencies capable of probing the shorter microstructural length scales, the time period over which the coherent field retains substantial amplitude and may therefore be easily extracted is very short. The coherent field may even fail to propagate a single thickness of the sample and thus cannot be measured. In such circumstances it is the incoherent field which must be queried if any material information is to be recovered.

The incoherent field consists of rays which have been significantly scattered, by random scatterers, one or more times. There exists a small body of literature concerned with that part of the wave field which has scattered through a significant angle once; the interest there is usually in those rays which backscatter and can therefore be detected with the source transducer[5]. Backscatter analysis has received increasing interest because it promises to extract microstructural information in a high frequency regime inaccessible to the more conventional coherent wave attenuation methods, and to extract this information as a function of depth. There are, however, some difficulties associated with the method.

Any incoherent field, including one dominated by the singly backscattered, is inescapably stochastic. As such it fluctuates, in frequency space and time, in a large and characteristic fashion. Averaging of the field will not reduce the relative fluctuations. Such an averaging would be of value only in order to reduce the contribution from electronic noise. The noise-like fluctuations of the actual ultrasonic signal, which are in fact not truly noise, are a consequence of the stochastic nature of the scattering. These fluctuations can be reduced by averaging the square of the signal, a quantity related to ultrasonic energy density, over time or frequency or space. This in turn requires substantial regions of time, frequency or space over which the signal is statistically stationary. Such is rarely available in the backscattering configuration.

The backscattering configuration is further complicated by the lack of a quantitative theory connecting microstructure to rms amplitudes as perceived by finite sized transducers. This is partly due to the complexity of the off-axis near field of a transducer and partly due to the inescapable inclusion of both singly and multiply-scattered rays in the measured wave field.

It is suggested here that the multiply scattered and therefore fully developed diffuse field may provide alternative possibilities for the probing of material microstructure. As will be seen in subsequent sections, such fields can be approximately stationary over large domains in time and frequency and space and may therefore be sufficiently averagable that their stochastic fluctuations may be rendered unproblematic. In addition, certain simplifications will obtain in the theory of these late time fields which render their interpretation less difficult and the connection with microstructure more direct.

II. THEORY

A wave field in a multiply scattering medium, when analyzed over length scales greater than the mean free ray paths between successive scatterings, *i.e.*, on length scales sufficiently greater than the inverse of the attenuation, and on time scales sufficiently greater than mean free ray travel times, can be expected to evolve in accordance with a heat, or diffusion, equation. These critical length and time scales depend on the strength of the scattering, but are typically, in ultrasonic applications in metallic polycrystals, of order centimeters and microseconds. The heat-equation like evolution of ultrasonic energy density can be established by appeal to purturbative multiple scattering formalisms[6], and also, more physically, by arguing that typical rays execute incoherent random walks upon multiple scattering. In either case, one then anticipates that the ultrasonic spectral energy density $E(\mathbf{x},t,\omega)$ at position $\mathbf{x}$, time t, and frequency ω evolves according to

$$D \nabla^2 E - \sigma E + P = E_{,t} \tag{1}$$

$D(\omega) > 0$ (assumed position independent in a statistically homogeneous medium) is the diffusivity of the ultrasound. It is the key microstructure dependent parameter. $\sigma(\omega) > 0$ is the linear internal friction (in units of nepers of power per unit time) and P is the energy production rate density. The t subscript indicates a partial time derivative.

The present contention is that the study of ultrasonic energy density evolution, on sufficiently smoothed length and time scales, will allow, upon a fit to the predictions of equation (1), extraction of the energy diffusivity D and the material internal friction σ, each potentially a significant NDE parameter.

Internal friction, though often studied in isolated specimens at low frequencies, and in single crystals or amorphous materials at very high frequency, is perhaps impossible to study at ultrasonic frequencies in structures composed of heterogeneous materials except by the above means. Inasmuch as internal friction is dependent upon chemical state and deformation history, it may, in some materials, be worth measuring nondestructively.

The diffusivity D, in contrast, is very clearly of nondestructive interest. The diffusivity will certainly scale roughly with some kind of ultrasonic mean free path; it is therefore an inverse measure of the scattering strength and thus of microstructure. The well known high sensitivity of the scattering strength to ultrasonic frequency leads one to anticipate diffusivities with strong frequency dependencies. These will in turn reveal information on microstructural length scales, that is, grain sizes. In light of the well known correlations between grain size and fracture toughness it appears that D could be a useful nondestructive parameter, at least as useful as the conventional ultrasonic attenuation, and perhaps more easily measured.

One anticipates diffusivities equal to wave speed times mean free path, divide by three.

$$D = c\, \ell \,/\, 3 \tag{2}$$

This expression is based on the random walk model. If the mean free path ℓ is identified with the distance over which coherent intensity diminishes by a factor of e

$$\ell = 1 \,/\, 2\alpha \tag{3}$$

where α is the usual coherent wave amplitude attenuation, then

$$D = c \,/\, 6\alpha \tag{4}$$

is found to be a simple inverse measure of attenuation. Equation (4) is a very simple model and it does capture much of the qualitative physics, but it is in need of substantial quantitative modification.

The argument must be modified to account for different wave modes, the longitudinal and the transverse, each potentially capable of carrying diffuse energy, and each with its own wave speed and attenuation. One postulates

$$D = f_T\ (c_T/6\alpha_T) \ + \ f_L\ (c_L/6\alpha_L) \qquad (5)$$

where f_T and $f_L = 1\text{-}f_T$ are the mixing weights for the transverse and longitudinal contributions to the diffusivity.

In the Rayleigh, low frequency, long wavelength, limit the conjecture (5) was found to be correct in a detailed calculation [6] based on an assumption of cubic crystallites if the mixing weights are taken to be

$$f_T/f_L = 2\ (c_L/c_T)^3 \qquad (6)$$

which is between about 10.0 and 16.0 for most metals and corresponds to the equilibrium partition of energy in a fully developed diffuse wave field [7].

Inasmuch as attenuations scale with the fourth power of frequency in the Rayleigh regime and the second power on the high frequency stochastic asymptote (and become frequency independent only in the very high frequency geometrical optics limit), the conjecture (5) predicts a frequency dependence of D with a power between -2 and -4 over most of the relevant frequency range.

In contrast, the detailed calculation [6] showed an inverse log frequency dependence when the frequency is on the stochastic asymptote, very different from equation (5)'s predicted inverse second power dependence. The error is apparently in the identification of ℓ with $1/2\alpha$. ℓ should be taken to be the mean free path against ray scattering through a significant angle. $1/2\alpha$ is merely the mean free path against any scattering and is therefore only a lower bound on ℓ. At high frequencies where scattering is increasingly biased towards the forward direction, the difference can be significant.

While the theoretical calculation was confined to the special case of an untextured aggregate of cubic crystallites, we can with some confidence assert that the formulas (5) and (6) are good estimates for D in the Rayleigh limit and serve as underestimates at higher frequencies. We furthermore suggest that D should scale inversely with the fourth power of frequency at low frequency and inversely with the logarithm of frequency on the stochastic asymptote in all materials with a well defined microstructural length scale which can serve to nondimensionalize frequency. Furthermore, in the very high frequency geometrical optics limit one expects D to become frequency independent.

Laboratory diffusivities have been studied, it appears, in only two published instances. Weaver [8] investigated the incoherent transport of sub MHz ultrasound in an aluminum sample with an artificially induced centimeter scale heterogeneity. The scattering, though not related to polycrystallinity, nevertheless led to a diffuse field energy density evolution in accordance with a heat equation like that of equation (1). Guo, Holler and Goebbels [9] studied ultrasound in the 5 to 12 MHz range in polycrystalline half spaces. Spatial averaging was employed to reduce the fluctuations, and diffusivities were recovered as functions of frequency. They recovered an inverse first power dependence for their observed diffusivities. While one imagines that D could vary roughly in this manner in the transition region between the expected Rayleigh inverse fourth power and the expected stochastic inverse logarithmic regimes, the ubiquity of Guo *et al*'s inverse first power is striking and deserves further investigation. In the present article we seek to repeat measurements such as that of Guo *et al*. We shall, however, study D as a function of both frequency and heat treatment, and also attempt to extract D without spatial averaging. We shall also modify the geometry. Each of these changes is envisioned as progress towards an eventual nondestructive implementation of measurements of D.

III. THE EXPERIMENTS

Three bars of AISI 1045 steel were prepared with different heat treatments. Each had dimensions of 2.5 x 3.0 x 20 cm. The normalized specimen was given a heat treatment consisting of two hours at 850°C followed by an air cooling in accordance with standard prescriptions. The "annealed" specimen experienced one and one half hours at 850°C followed by an oven cooling. The "overheated" specimen experienced two hours at 1000°C and an air cooling. Charpy fracture toughness values taken at 100 degrees C were found to correlate as expected with the heat treatments. Normalized: 21.0 foot pounds, Annealed: 16.4 foot-pounds, Overheated: 11.0 foot pounds, corresponding to the expected decrease in toughness following abusive heat treatments.

The ultrasonic tests were carried out with a pair of 15 MHz 1/4" diameter broadband piezoelectric transducers. The contact points on the specimens were polished for maximum coupling efficiency. A light oil was used for coupling. One transducer was placed on one of the small ends of the bar, the other on a lateral face. The distance between source and receiver was varied between tests. The source transducer was driven by a conventional broad band ultrasonic pulse generator with negligible rise time. The received signal was preamplified by 50 dB, high-pass filtered at about 6 MHz in order to remove the strong but unimportant weakly scattered low frequencies and then digitized at 50MSamples/sec for 200 microseconds. The waveform was then passed to a microcomputer for analysis. The resulting Nyquist frequency of 25MHz was well above the highest significant frequency present in the diffuse wave signal.

Due to both the presence of internal friction dissipation, and to the volumetric distribution, or dilution, of the diffuse field energy, diffuse field signals tend to be weaker than those of conventional ultrasonics. Electronic noise can therefore be problematic. In order to improve signal to noise ratios, the repetitive signals were averaged, as much as 200 times in some cases, by microcomputer, without disturbing the physical configuration. According to a simple calculation, this procedure should have reduced noise power levels by $\log_e(200) = 5.3$ nepers of power = 23 dB and increased the signal to noise ratio by the same amount. These improvements were in fact observed.

This need for extensive repetition averaging of the signals (<u>not</u> the energies), while not prohibitive and without the contact-reproducibility problems often associated with spatial averaging, is perhaps the major current practical difficulty with the proposed technique. It is conjectured, however, that improvements in transducers, preamplifiers and contact efficiency, and also more rapid waveform processing, will lead to amelioration.

Upon construction of the repetition averaged waveform, a time-spectral analysis was effected by FFT of each of 20 time windows of 10 microsecond duration. The resulting spectra were absolute value squared and smoothed by summing neighboring spectral values into "bins" of width 1.25 MHz each. By this process, and for each frequency bin, a temporal record was constructed expressing spectral energy density as a function of time at 20 different times with 10 microsecond spacings. According to standard arguments which assume Gaussian statistics, stationarity across the 10 microsecond time windows, and weak variations with respect to frequency across the 1.25 MHz binwidths, the resulting spectral energy density should retain a fractional level of rms fluctuation only of order $(\Delta f\ \Delta t)^{-\frac{1}{2}}$, or to one part in $\sqrt{12.5}$, or 28%. By averaging over the time windows of 10 microseconds and frequency bins of 1.25 MHz we have potentially obviated the need for the problematic spatial averaging commonly employed in backscattering studies, and also by Guo *et al*.

A typical result is shown in Fig 1. For the case of the overheated specimen, with a source-receiver separation of 5 cm, the recovered natural logarithms of the power spectral densities are displayed for four different frequency bins at each of the 20 time windows. The topmost record is that of the bin centered on 8 MHz; the next records are those of the bins at 10.5 , 14.25, and 16.75 MHz respectively. The source acted at a point in time late within the second time window, at about 20 microseconds. Thus time windows number 1 and 2 indicate the level of electronic noise remaining after the repetition-averaging.

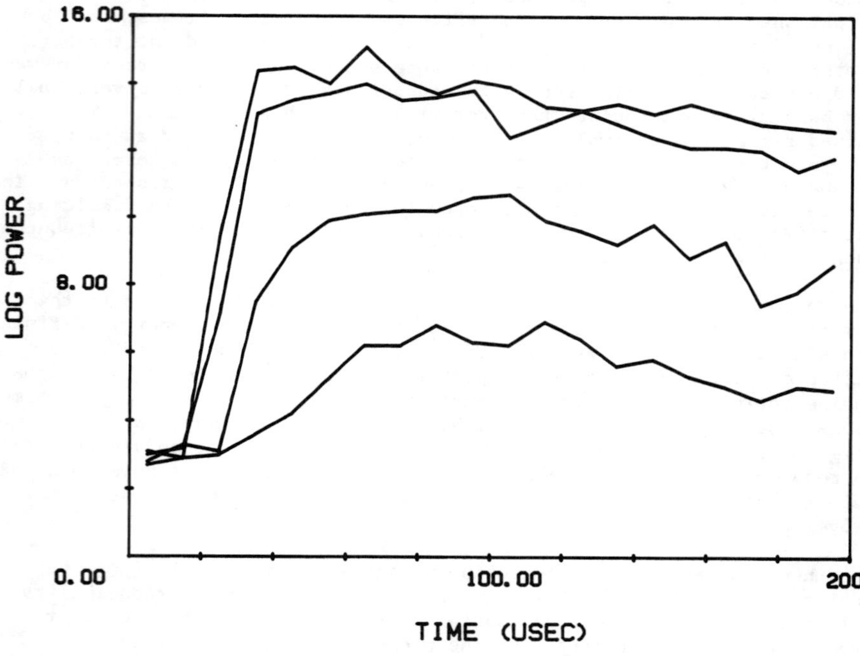

Figure 1. The power spectral densities are shown as a function of time for a source-receiver separation of 2.5 cm in the overheated steel specimen. The uppermost record is that of the bin centered on 8 MHz; the succeeding records are those of the bins centered on 10.5, 14.25 and 16.75 MHz respectively.

Attention is first directed to the apparent level of stochastic fluctuation. The predicted rms variation of plus or minus 28% (or 0.28 nepers of power) is clearly seen in the electronic noise. The ultrasonic signals, though, occasionally fluctuate to a greater degree. This is attributable to the very obvious non constancy of the power spectral densities across a bin of width 1.25 MHz. (Levels between adjacent bins differ by as much as a factor of e^4.) If this is in fact the case, more sophisticated frequency averaging, using an average with significant weighting variations across the bin, could be expected to decrease the fluctuation level to something of order 28%.

Attention is also directed to the very different character of the temporal behaviors at the different frequencies. The lower frequencies show a sharp arrival followed by a steady decay. The higher frequencies require as much as 80 microseconds to reach maximum strength. The late time decays are attributable to dissipative processes (internal friction) and also to continued diffusion of the ultrasonic energy away from the region near the source transducer and towards the far end of the specimen. It is clear that, at this 5 cm distance from the source, the lowest frequencies show an arrival on a time scale comparable to that appropriate for pure propagation (5cm/10microseconds) and the higher frequencies show a much slower diffusive arrival.

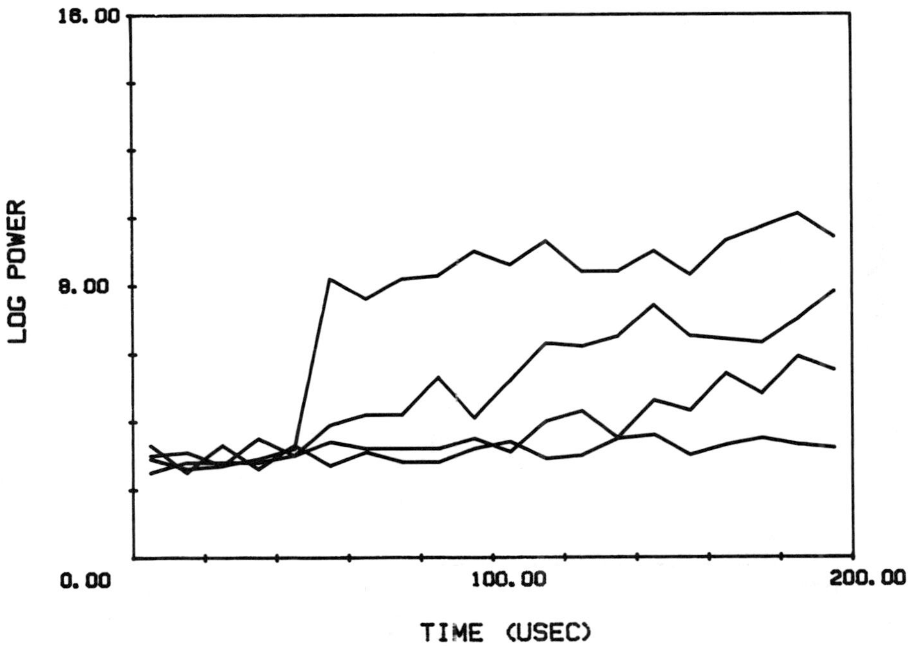

Figure 2. Power Spectral Densities for the case of a source-receiver separation of 17.5 cm in the overheated specimen. The records shown are for the bins centered on 8.0, 9.25, 10.5, and 11.75 MHz. All higher frequency bins remain below the level of the electronic noise throughout the observation period.

Fig 2 shows a similar set of data obtained from a receiver position 2.5 cm from the end of the bar, and far from the source. A prompt arrival is again observed, but only at the lowest frequency. Again its speed is comparable to that expected for propagation (c = 17.5 cm/ 30 microseconds.) Note that its amplitude is much diminished (by 7 nepers of power) from that associated with the prompt arrival in Fig 1, corresponding to the attenuation of the coherent wave front. The prompt arrival of the lowest frequencies in Fig 2 is then followed by a steady increase in power levels as energy diffuses towards this end of the bar from points closer to the source. Thus the diffusive character of the transport is evident in this specimen even for the relatively weakly scattering lower frequencies.

Diffusion constants were extracted from data such as this by fitting these curves to solutions of the model PDE, equation (1). It was assumed that the transport in the directions perpendicular to the axis of the bar was sufficiently rapid that the bar could be modelled as one dimensional. The energy deposition term P in equation (1) is ideally modelled as impulsive and concentrated at the source position; in practice it was necessary to treat it as only exponentially well confined to the vicinity of the source, that is, the prompt arrivals were accounted for. Upon imposition of insulating boundary conditions at each end of the bar, the solution to equation (1) is straightforward.

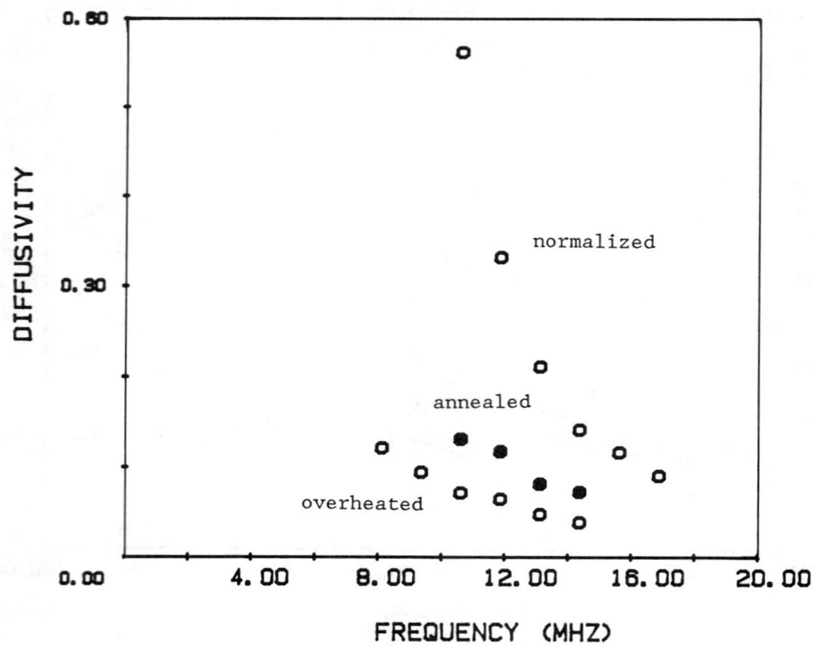

Figure 3. The Diffusivities (in square centimeters per microsecond) as extracted from power spectral density profiles like those of Figures 1 and 2. The higher values are taken from the normalized specimen, the intermediate values from the annealed specimen and the smallest diffusivities, corresponding to the greatest resistance to transport, are taken from the large grained, overheated specimen.

The solution depends nonlinearly on the parameters D and σ and therefore standard linear regression methods for best fit are not indicated. Because non-linear least squares methods pose a challenge in terms of stability, these early efforts at D extraction proceeded by hand. It is hoped that future best fits will be arrived at by computer.

Diffusivities were extracted by comparing energy levels associated with the same frequency and time at different receiver locations. Thus the effects of internal friction σ were eliminated from the analysis. For sufficiently late times, the prompt arrival is irrelevant and the energy deposition term P could be treated as concentrated at the source transducer. Exact numerical results for the case of zero internal friction given in terms of dimensionless time ($t^* = Dt/L^2$ where L is the bar length) were tabulated. D was then taken from

$$D = L^2 \; (t^*/t) \qquad (7)$$

where t was the observation time and t^* the dimensionless time at which the numerical results showed the same relative energy levels between the two receiver locations as was observed. Because several different choices of t and observer location were available, this method allowed for several quasi-independent estimates for D at each frequency.

The method is admittedly imperfect. The possibility for unconscious bias on the part of the person doing the analysis cannot be eliminated. The method is tedious as well. The method also requires power spectral density profiles from different receiver positions, with the implication that contact efficiencies are assumed to be reproduced as the receiver is moved. In the short term absence, however, of a reliable and stable nonlinear least squares fitting procedure designed for the extraction of diffusivities and internal frictions from a single curve such as that of Figs 1 or 2, this method appears to be adequate. It is reassuring that the different estimates for D based on different observation times were generally in agreement with each other.

Fig 3 shows the diffusion constants recovered by the above method. Not all frequency bins are displayed for all specimens. The missing points are for cases where the data did not allow reliable or consistent estimates of D. The expected decrease of diffusivity with both increasing frequency and decreasing grain size is apparent. The frequency dependence in the small-grained normalized sample fits well to an inverse fourth power, in accordance with the expectation for the Rayleigh regime. The other, larger grained, specimens show a frequency dependence closer to an inverse second power and in accord with expectations for smaller grain sizes. The correlation with fracture toughness is apparent at any of the frequencies.

IV. SUMMARY

While the possibility for nondestructive materials characterization by means of ultrasonic diffusivity has been demonstrated, the practical utility of the approach remains unproven as yet. There is a clear quantitative connection between ultrasonic diffusivities and microstructure, just as there is between ultrasonic attenuation and microstructure. Whether diffusivity will prove to be more easily measured, though, is not yet so clear. The weaker signals and associated signal to noise ratio difficulties must be overcome, as must also the problematic prescription for the extraction of the best fit diffusivities. The method necessarily yields only an average of material properties over a volume of at least a few mean free paths and is therefore probably less applicable than attenuation and backscatter methods when fine spatial resolution is required. Nor is this method appropriate in materials with very high internal friction such as, perhaps, polymer based composites, where the ultrasonic field may be expected to dissipate before it can be measured reliably. On the other hand and unlike other methods, it promises to deliver two parameters related to material character, microstructural scattering strength, and internal friction, and to do this with few demands on geometry. The approach seems to lack the tendency to systematic error found in attenuation studies and to show a better quantitative theoretical connection with microstructure than does the backscatter technique.

This work was supported by the U.S. Army Research Office under Contract DAAL 03-87-K0006 with the University of Illinois Advanced Construction Technology Center.

REFERENCES

1. E.P. Papadakis "*Scattering in Polycrystalline Media*" in Methods of Experimental Physics (P. Edmonds, editor) Academic Press N.Y. (1981)

2. F.E. Stanke and G. Kino "*A Unified Theory for Elastic Wave Propagation in Polycrystalline Materials*" J. Acoust. Soc. Am. 75, 665-681 (1984)

3. A. Vary "*Correlation Among Ultrasonic Propagation Factors and Fracture Toughness Properties of Metallic Materials*" Mat. Eval. 36, 55-64 (1978)

4. P.B. Nagy, D.V. Rypien and L. Adler "*Spatial Averaging in Porosity Assessment by Ultrasonic Attenuation Spectroscopy*" 683-684, in Nondestructive Characterization of Materials II (Bussiere, Monchalin, Ruud, and Green, editors) Plenum, N.Y. (1987)

5. J. Saniie and N.M. Bilgutay "*Quantitative Grain Size Evaluation Using Ultrasonic Backscatter Echoes*" J. Acoust. Soc. Am. 84, 400-408 (1986)

6. R.L. Weaver "*Diffusivity of Ultrasound in Polycrystals*" submitted to J. Mechs. and Phys. Solids (1989)

7. R.L. Weaver "*On Diffuse Waves in Solid Media*" J. Acoust. Soc Am. 71, 1608-1609 (1983)

8. R.L. Weaver "*Anderson Localization of Ultrasound*" submitted to Wave Motion (1988)

9. C.B. Guo, D. Holler, and K. Goebbels "*Scattering of Ultrasonic Waves in Anisotropic Polycrystalline Metals*" Acustica 58, 112-120 (1985)

APPLICATION OF HOLOGRAPHIC INTERFEROMETRY AS A NDT METHOD IN COMPOSITE PLATES

N. Rubayi, S. H. Liew, and M. A. Wright
Southern Illinois University at Carbondale
Carbondale, IL 62901

INTRODUCTION

In recent years, composite materials are being increasingly used in engineering applications in which high structural performance is required. As the general level of technology advances, special requirements are imposed on the materials. High strength-to-weight ratios and/or high stiffness-to-weight ratios are sought for applications in aircraft, automobiles, and space structures. To insure a safe and efficient application and performance of the composite structures, the composite must be free from flaws which could develop during the fabrication process or even later when the structure is put into service. A reliable and accurate inspection technique therefore must be utilized for detecting flaws and delaminations. Holographic interferometry is one of the major techniques which could be employed as nondestructive testing. Holography as an optical method is an ideal tool to provide a full field of information of the test object.

Figure 1 shows the general layout of holographic components used in producing holograms for nondestructive testing of composite plates. The hologram is created by the interference of two coherent light beams at a photographic plate. One beam, the reference beam, comes directly from the laser to the film plate, while the other, the object beam, is scattered from the surface of the test object. Later, when the processed photographic plate (the hologram) is illuminated by the reference beam alone, the hologram creates light waves which are identical to the waves scattered from the original object. A hologram stores a precise three-dimensional image of the object being inspected, and holographic interferometry can measure changes in the surface of the object to a fraction of a micrometer. Holography can detect flaws in a structure, providing the flaws cause the surface of a loaded structure to distort unexpectedly as discussed in references 1, 5, 6, and 7, respectively.

The primary objective of this study was to employ holographic interferometry in the investigation of the various stressing methods which could be utilized to detect flaws and delaminations in graphite/epoxy plates. These nondestructive techniques were: thermal, vacuum, and vibration stressing. The secondary objective was to study the effects of the shape of the flaws, their size, and their locations in the composite plate, as well as the effect of edge clamping upon the detection and the resulting fringe pattern. The influence of the ply orientations and stacking sequence of the laminates upon the detection of the flaw was also investigated.

EXPERIMENTAL PROCEDURE

1. The Holographic Process

The holographic components used in this study are shown in Figure 2. Real-time holography was used in both the thermal and vacuum stressing methods in conjunction with the fringe control technique. A hologram was made of the flawed composite plate at the steady state before stressing. That is, the photographic plate was chemically developed in a special tank on the holographic bench in-situ. The composite plate was then stressed by either heating or by vacuum pressure, the flaw was viewed in real-time through the fixed developed photographic plate, and the virtual image was photographed by a 35 mm. F-3 Nikon camera. Whereas, in using the method of vibration stressing, both the real-time and time-average holography were utilized.

2. Manufacturing the Flawed Composite Plates

Several graphite/epoxy composite plates with flaws were manufactured. Some were made with a unidirectional ply sequence $[0°]_S$ and some with cross ply $[0°/90°/90°/0°]_S$, while others were made with multiple directions $[0°/45°/90°/0°]_S$. All plates were made of 8 plies of prepreg tape. The plates were rectangular in shape with the dimensions 12.7 cm. x 17.8 cm. and of an average thickness of 0.178 cm. The flaws were induced into the composite plates by inserting them at different ply depths between the laminates during the manufacturing process. For instance, at a location of one ply depth, the shim was between laminates 1 and 2.

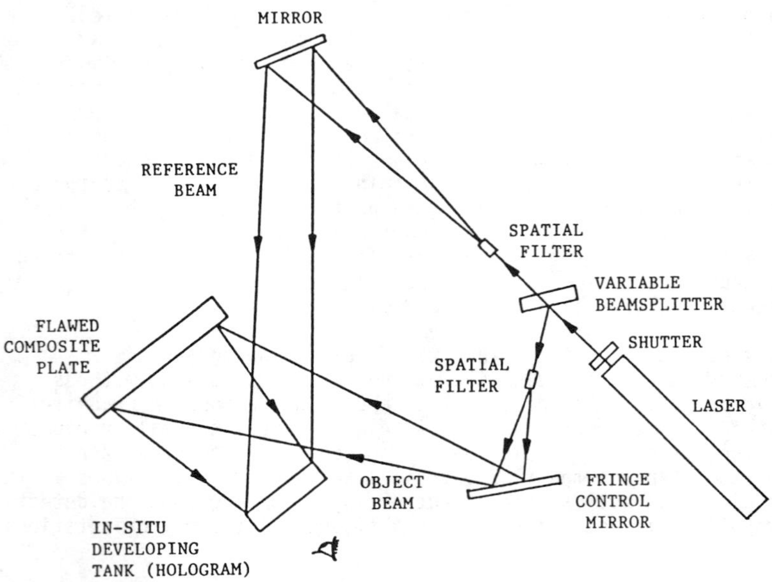

FIGURE 1. Schematic diagram of the holographic layout.

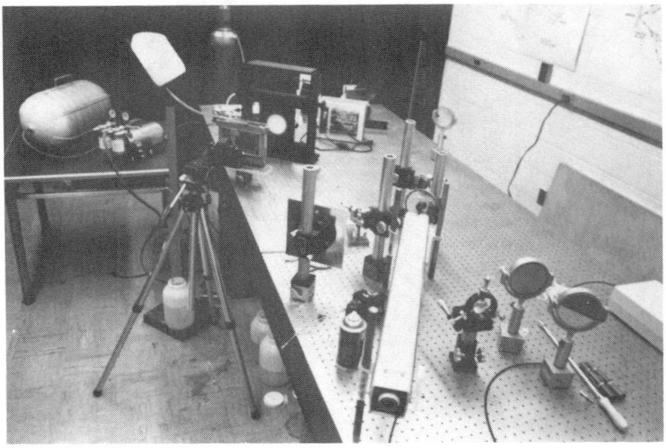

FIGURE 2. General view of the holographic components.

The flaws were made from aluminum and teflon shims of different shapes and sizes as shown in Figure 3. Three shapes were investigated: equilateral triangular, rectangular, and circular shapes. The aluminum shims were made by cementing the edges of two thin foils together so that a void was created between them. The total thickness of the aluminum shim was 0.004 in., while the teflon shim was 0.005 in. The composite plate was cured in the hot press where the temperature was increased by 1°C./min. to a maximum of 175°C. at a constant pressure of 9000 psi. The temperature and the pressure were kept constant for 1½ hours, after which the heater was switched off. The plate was allowed to cool down to room temperature, while the pressure was kept constant at 9000 psi. The entire curing cycle required about 5-6 hours.

All manufactured composite plates were sprayed on one surface with a flat white paint prior to testing so as to make the testing surface opaque, increasing the scattering and reflection of the object light waves.

3. The Thermal Stressing

Thermal stressing is a practical method for detecting flaws in composite structures. Sensitivity to delamination is due to a combination of locally reduced stiffness and a change in thermal conductivity through the flaw. To utilize this method, the flawed composite plate was held in the universal loading device on the holographic bench either by clamping two edges or four edges. A cement-on thermocouple was cemented to the back of the plate at a location far from the position of the flaw in order to monitor the temperature. The temperature was recorded on the digital readout monitor. A hologram was made at an ambient temperature. That is, the photographic plate was developed in-situ and viewed in real-time.

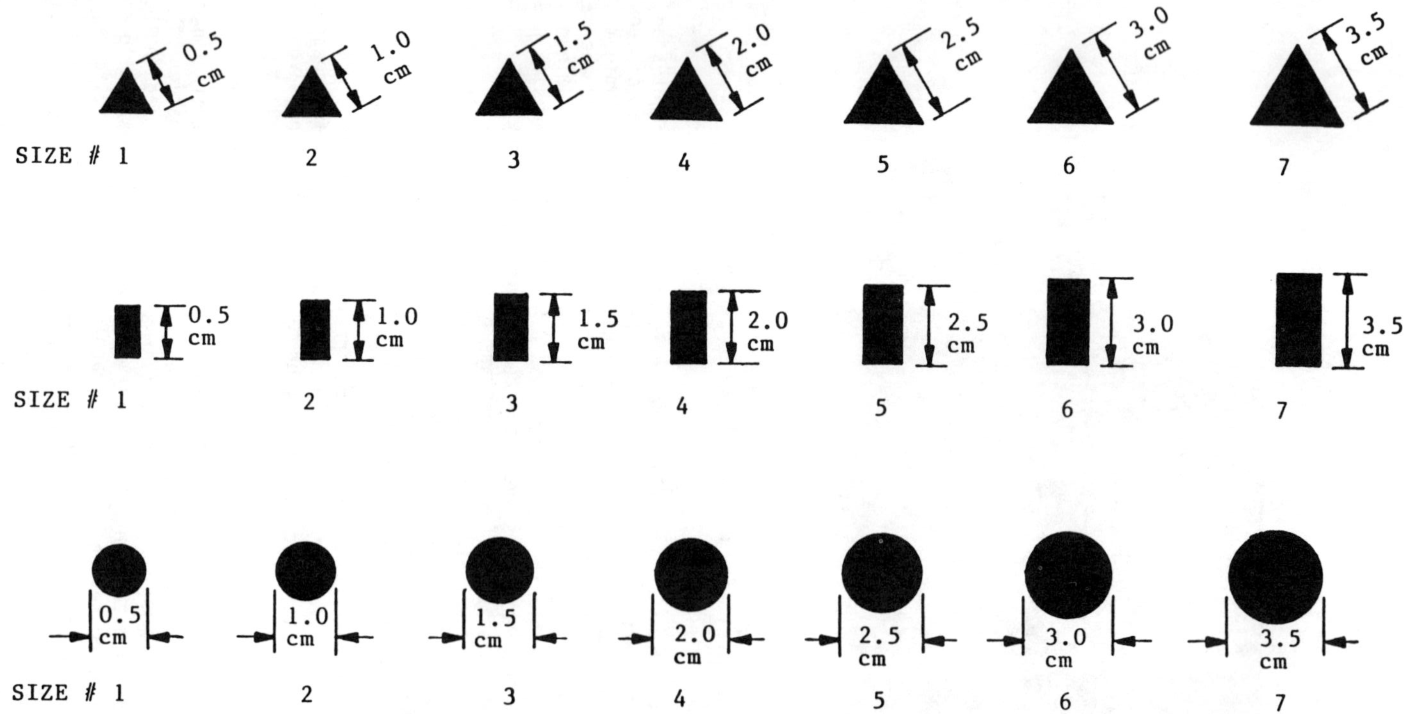

FIGURE 3. Various shapes and sizes of teflon and aluminum shims.

An electric element was placed behind the composite plate to raise its temperature by several degrees. Immediate real-time viewing of the hologram revealed a rapid motion of the resulting fringes. The flawed laminates, however, showed distortions in the fringes.

Fringe control was employed in conjunction with the real-time holography. The fringe control, as discussed in references 2, 3, and 4, respectively, is an optical technique which reduces the fringe frequency, providing the broadest spacing of the resulting fringes on the test plate, thus allowing the flaw to be more visible. To record the resulting fringes at various temperatures when the flaw was visible, a 35 mm. camera was utilized. Figures 5 and 6 demonstrate the resulting fringe patterns and the detection of flaws and delamination at various temperatures.

4. The Vacuum Stressing

A vacuum chamber was designed and attached rigidly to the loading frame on the holographic table as shown in Figure 2. The flawed composite plate was held against the chamber by an aluminum fixture frame using a double faced tape as a gasket to equalize the slight clamping pressure on the plate. The edges of the plate were sealed with Dow Corning lubricant and sealer to avoid loss of vacuum. Real-time holography was employed; that is, at zero vacuum, the first exposure of the holographic plate was made and the photographic plate was developed in-situ. The vacuum pump generated a vacuum of about 7.0 inch of Hg. When this amount of vacuum was reached, the pump was shut off, thereby trapping the vacuum in a reservoir tank by the closing of its inlet valve. The amount of vacuum in the chamber was then controlled by opening and closing the inlet valve of the tank. Generally, a small amount of vacuum was needed to load the plate and to observe the flaw.

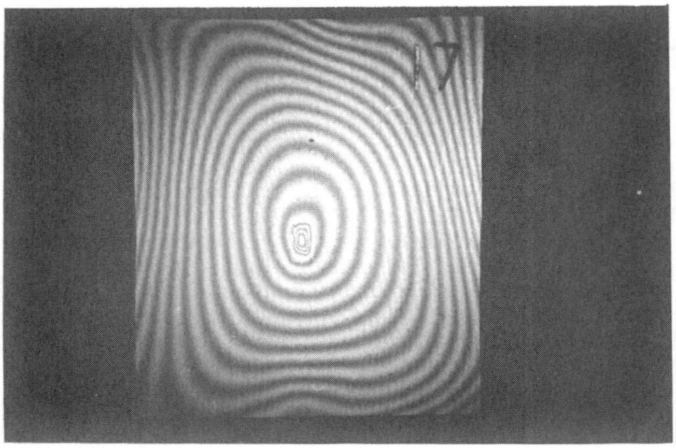

FIGURE 4. Real-time hologram of a cross ply plate containing size # 3 aluminum rectangular shim of 0.004 inch thick at one ply deep in thermal stressing at 3° F. above room temperature (R.T.=76°F.).

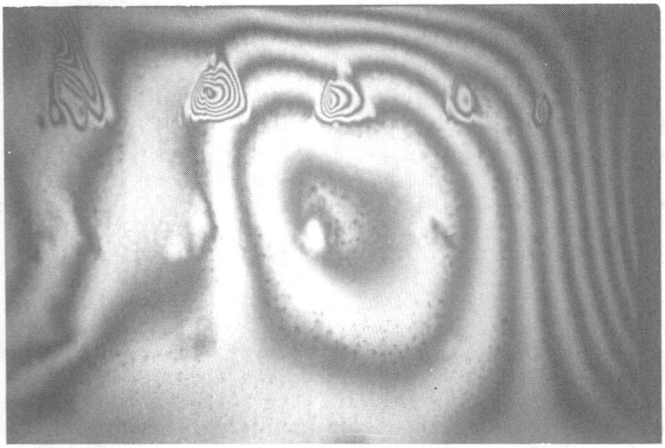

FIGURE 5. Real-time hologram of a multidirectional plate with various sizes of teflon triangular shims of 0.005 in. thick at 1.0 to 4.0 plies in thermal stressing at 2°F. above room temperature (R.T.=78°).

The amount of vacuum was in the range of 4 mm. to 8 mm. of Hg. The fringes were viewed at different magnitudes of vacuum: when a flaw was detected, a photograph was made of the hologram. As a representation, Figure 7 and 8 illustrate the resulting fringes and the detection of the flaw in composite plates.

5. The Vibration Stressing

A piezoelectric/electromagnetic vibration shaker of a maximum output of 60KHz. made by Wilcoxon Research Inc. was utilized to set the composite plates in sinusodial vibration. A schematic diagram of the vibration set up is shown in Figure 4. The composite plate was clamped either at one side or at four edges in the loading frame mounted on the holographic bench. The shaker was attached to the composite plate and a good coupling was ensured. A hologram was made at zero frequency and then the frequencies were increased slowly and viewed in real-time until a resonant occurred. When a flaw was detected, the resonant frequency was kept constant and a time-average hologram was made for the purpose of photographing and documenting the flaw. Typical results are shown in Figures 9 and 10, which depict the presence of the flaw at the center of the plates at their resonant frequencies.

DISCUSSION OF RESULTS

In this study, the majority of the composite plates being investigated contained flaws of either size # 1 or # 3. These sizes were considered to be in the range of small and medium size categories. The experimental results of investigating these plates are summarized in Tables 1, 2, and

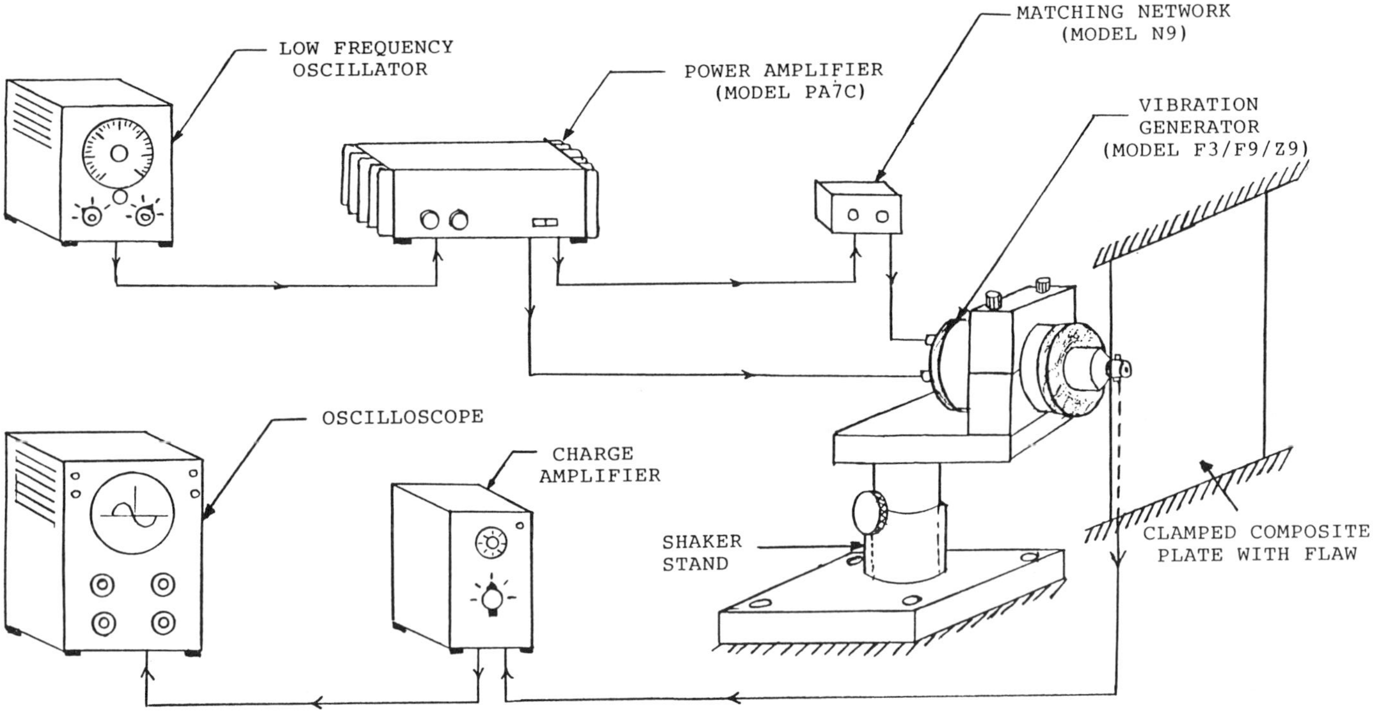

FIGURE 6. Schematic diagram of the equipment employed in vibration testing.

TABLE 1. Results of composite plates with size # 3 aluminum shims inserted between laminates for various stressing techniques.

Stressing Method	Plate Type	Maximum Ply Depth Detected for Shims of Various Shapes ▮	▲	●
Thermal	Unidirectional	2	2	2
	Cross ply	3	3	2
	Multidirectional	3	3	3
Vacuum	Unidirectional	1	1	1
	Cross ply	1	1	1
	Multidirectional	1	2	1
Vibration	Unidirectional	1**	-	1**
	Cross ply	1	1	2**
	Multidirectional	1	1	2**

Key:

- Flaw not detected at any ply depth

** Flaw detected for both one clamped or four clamped edges

Unidirectional $(0°/0°/0°/0°)_s$

Cross ply $(0°/90°/90°/0°)_s$

Multidirectional $(0°/45°/90°/0°)_s$

Thickness of a Laminate = 1 Ply depth of 0.22 mm.

TABLE 2. Results of composite plates with size # 3 teflon shims inserted between laminates for various stressing techniques.

Stressing Method	Plate Plate	Maximum Ply Depth Detected for Shims of Various Shapes		
		■	▲	●
Thermal	Unidirectional	3	1	2
	Cross ply	3	4	4
	Multidirectional	4	3	3
Vacuum	Unidirectional	1	1	1
	Cross ply	1	2	2
	Multidirectional	1	1	1
Vibration	Unidirectional	1	-	2*
	Cross ply	2*	2**	2**
	Multidirectional	2**	1	2**

Key:

- Flaw not detected at any ply depth

* Flaw detected for plates with one edge clamped only

** Flaw detected for both one clamped or four clamped edges

Unidirectional $(0°/0°/0°/0°)_s$

Cross ply $(0°/90°/90°/0°)_s$

Multidirectional $(0°/45°/90°/0°)_s$

Thickness of a Laminate = 1 Ply Depth of 0.22 mm.

TABLE 3. Results of composite plates with size # 1 aluminum shims inserted between laminates for various stressing techniques.

Stressing Method	Plate Type	Maximum Ply Depth Detected for Shims of Various Shapes		
		▮	▲	●
Thermal	Unidirectional	2	3	2
	Cross ply	2	3	2
	Multidirectional	3	3	2
Vacuum	Unidirectional	1	-	-
	Cross ply	-	-	-
	Multidirectional	1	-	-
Vibration	Unidirectional	1^{**}	-	1^{**}
	Cross ply	-	-	1^{**}
	Multidirectional	1^{**}	-	1^{**}

Key:

- Flaw not detected at any ply depth

** Flaw detected for both one clamped or four clamped edges

Unidirectional $(0°/0°/0°/0°)_s$

Cross ply $(0°/90°/90°/0°)_s$

Multidirectional $(0°/45°/90°/0°)_s$

Thickness of a Laminate = 1 Ply Depth of 0.22 mm.

TABLE 4. Results of thermal stressing method for various sizes of teflon shims detected for cross ply composite plates.

Maximum Ply Depth	Flaw Sizes Detected for Triangular Shapes	Flaw Sizes Detected for Circular Shapes
1	1,2,3,4,5,6,7	1,2,3,4,5,6,7
2	-,2,3,4,5,6,7	-,2,3,4,5,6,7
3	-,2,3,4,5,6,7	-,-,3,4,5,6,7
4	-,2,3,4,5,6,7	-,-,3,4,5,6,7
5	-,-,-,4,5,6,7	-,-,-,-,-,6,7
6	-,-,-,-,5,6,7	-,-,-,-,-,6,7

1 Ply Depth = 0.20 mm.

TABLE 5. Results of thermal stressing method for various sizes of teflon shims detected for multidirectional composite plates.

Maximum Ply Depth	Flaw Sizes Detected for Triangular Shapes
1	2,3,4,5,6,7
2	-,3,4,5,6,7
3	-,3,4,5,6,7
4	-,-,-,5,6,7
5	-,-,-,5,6,7
6	-,-,-,5,6,7

1 Ply Depth = 0.20 mm.

3, respectively. However, some cross ply and multidirectional plates were also made to contain various sizes of shims, including sizes # 1 to # 7, which were placed at different ply depths. The results of these plates are shown by Tables 4 and 5 and also by the representative Figure 6.

In examining Table 1, it is clear that the thermal stressing method was the most effective. This method gave the maximum ply depth detected as three plies for both cross ply and multidirectional plates. The effect of the three different shapes of size # 3 aluminum shims upon the detection of the flaw was investigated. In using thermal stressing, the triangular and rectangular shaped shims were the easiest to detect as compared with the circular shape. However, in vibration stressing the circular shape was the easiest to observe. In comparing the results of plates containing either size # 1 or # 3 for aluminum shims, it was evident from Tables 1 and 3 that in using thermal stressing the size of the triangular shape shim had little effect upon the flaw detection. For both sizes the maximum flaw depth was detected at the third ply, whereas in vacuum stressing the triangular shape of size # 1 was not detected at all for all types of plates. Also, in the case of vibration stressing both sizes 1 and 3 were not detected for the triangular shape.

Table 2 illustrates the effect of the teflon flaw shape and the ply orientation and stacking sequence upon the flaw detection. It was demonstrated that the thermal stressing resulted in obtaining a maximum depth of detection of four plies, whereas the vacuum stressing gave the least depth at which the flaw can be detected. Both Tables 4 and 5 represent the effect of the size and shape of teflon flaw upon the detection for both cross ply and multidirectional plates. It is interesting to note that in both types of plates as the size of the shim increases, the depth

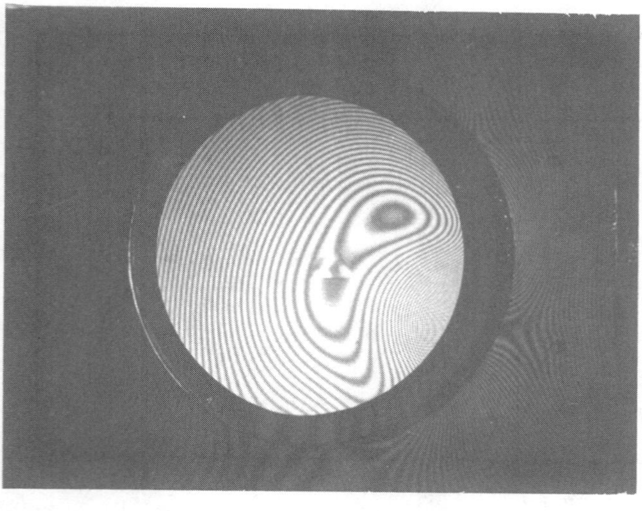

FIGURE 7. Real-time hologram with a fringe control of a cross ply plate containing a triangular teflon shim of size # 3 and of 0.005 in. thick at one ply deep in a vacuum stressing of 7.00 mm. Hg. at room temperature.

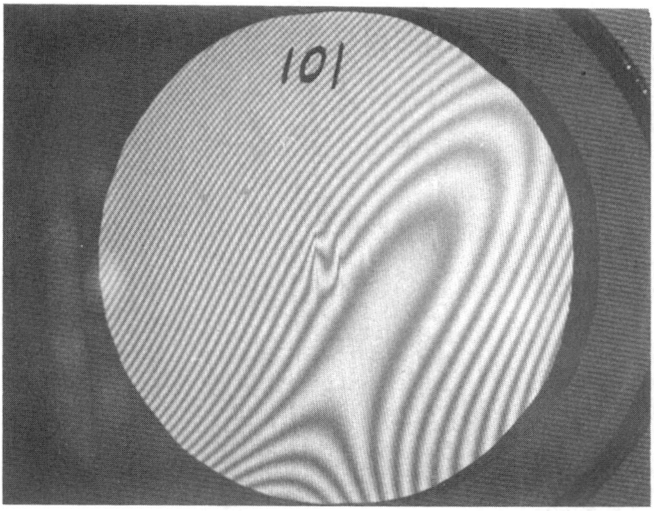

FIGURE 8. Real-time hologram with a fringe control of a multidirectional plate containing a circular size # 3 teflon shim of 0.005 in. thick at one ply deep in a vacuum stressing of 6.00 mm. Hg. and at room temperature.

of detection will also increase. In examining the results of Table 4 for the effect of the shape of the shim upon the flaw detection, it is clear that the triangular shape was the easiest to detect. It can be detected at a deeper ply depth than the circular shape. Furthermore, the results in Tables 4 and 5 of the triangular teflon shims revealed that in cross ply plates, the smaller size of shims can be detected easier at a ply depth near the surface of the plate than those in the multidirectional plates.

It is interesting to note that in utilizing the thermal stressing method, the amount of the temperature gradient above the ambient temperature which was required to detect the flaws was influenced by several factors. These factors included the thermal coefficient of expansion of the material of the shim, the ply orientation, the stacking sequence and the depth location of the shim in the plate.

In employing the thermal stressing, it was found that the effect of clamping the edges of the plate upon the flaw detection depends largely upon the location of the flaw. If the flaw is at the center of the plate, then clamping two edges or four edges will have almost the same effect on the flaw detection. However, by clamping the four edges, a more even and symmetrical fringe pattern is produced. When the flaw is located near an edge of the plate, then clamping the four edges is recommended. In employing the vacuum stressing method, the edges of the plate should be clamped all around and sealed. As for the vibration stressing, the plate could be clamped either at one edge as a cantilever or at four edges. However, clamping only one edge will produce excessive vibrational nodes and antinodes at the unclamped region of the plate.

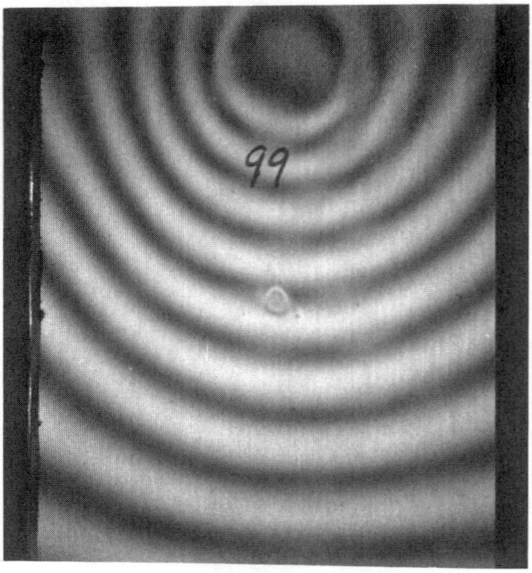

FIGURE 9. Real-time hologram of a multidirectional plate containing a triangular teflon shim of size # 3 and of 0.005 in thick at one ply deep. The plate is clamped at four edges and vibrating at resonant frequency of 32.5 KHz.

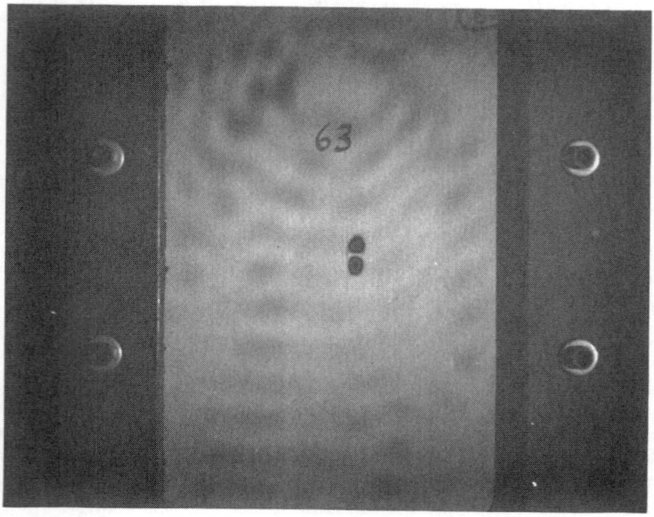

FIGURE 10. Time-average hologram of a multidirectional plate containing a rectangular aluminum shim of size # 3 and of 0.004 in. thick at one ply deep. The plate is clamped at four edges and vibrating at a resonant frequency of 33.0 KHz.

Among the three stressing methods investigated, it was found that the thermal stressing was the least laborious and the quickest method to employ in detecting flaws in composite plates. In general, the triangular shape flaw was the easiest to detect.

For the various sizes of shims investigated in this study, the thermal stressing revealed that the triangular size #5 was the smallest size that can be detected at 1.32 mm. from the surface of the plate.

In conclusion, this study has demonstrated that holographic interferometry as a NDT is a valuable and effective tool to be employed for the detection of delamination and flaws at or near the surface of composite plates.

REFERENCES

1. Erf, R. K., Holographic Nondestructive Testing, Academic Press, New York, 1974.

2. Champagne, E., and Kersch, L., Control of Holographic Interferometric Fringe Patterns (A), Journal of the Optical Society of America. 59, 1535., 1969.

3. Kersch, L., Advanced Concepts of Holographic Nondestructive Testing, Materials Evaluation, 29, pp. 125-140, 1971.

4. Champagne, E., Quantitative Data Reduction with use of Fringe Control Technique in Conjunction with Holographic Interferometry, Proceedings of the Symposium on Engineering Applications of Holography, SPIE, Redondo Beach, California, 1972.

5. Metherell, A. F., Spinak, S., and Pisa, E. J., Subfringe Interferometric Holography for Linearly Recording Small Displacements (A), Journal of the Optical Society of America, 59, 1534, 1969.

6. Schulz, W. J., Mangiapane, J. A., and Stargardter, H., Development of Borsic-Aluminum Composite Fan Blade for Supersonic Turbofan Engines, ASME Publication 71-GT-90, 1971.

7. Rubayi, N., Holographic Investigations, Final Annual Report, Materials Technology Center, Southern Illinois University at Carbondale, 1987.

ACKNOWLEDGMENT

The authors are indebted to the Materials Technology Center of Southern Illinois University at Carbondale for sponsoring this research project.

HOLOGRAPHIC NONDESTRUCTIVE TESTING

J. Gryzagoridis
University of Cape Town
Rondebosch, South Africa

ABSTRACT

While holography is mostly used to recreate beautiful 3-dimensional scenes, engineering applications of holography make use of a powerful technique, called holographic interferometry.

This technique makes use of the ability to record two slightly different scenes and display the small differences between them.

The image of an object stored in holographic film is made to interfere with another image of the same object, or with the object itself, creating sets of contour lines. These contour lines, commonly referred to as interference fringes, are a measure of the amount of dimensional differences of the object at two different states.

The most powerful feature of the Holographic N.D.T. is that information is obtained over the entire or large portion of the surface of a component rather than at just a point.

Since internal anomalies change the way a part responds to external stress, the irregularities of the contour lines are usually obvious even to the untrained viewer.

It is possible therefore to locate the position, orientation and densities of the flaws and irregularities without contact being made. This paper presents briefly examples of holographic N.D.T. work that was carried out at the Mechanical Engineering Department of the University of Cape Town. The work which involved small projects was in the form of feasibility or exploratory study aimed at gaining knowledge and expertise. Typically, both real-time and double exposure holographic N.D.T. was employed studying deformations of a cantilever beam, locating surface cracks, internal flaws in the forms of cracks and debonds. Other examples allow the visualization of the interior of a pipe and a pressure vessel aiming at monitoring erosion, corrosion and artificially created flaws. Components tested were made from metallic materials and also fibre reinforced composites.

The knowledge gained and the different techniques devised form the basis or the platform for the development of an automatic, instant, self contained holographic camera which can be easily transported to locations in industrial plants, workshops and production lines for in-situ N.D.T. applications.

INTRODUCTION

Interest in non-destructive evaluations in industries ranging from manufacturing to transport, mines and power generation is not new - in the assessment and confirmations of quality, longevity and economy. The technologies of Non-Destructive Testing (N.D.T.) have been available for some time, but has not been used by industry to the best advantage. A wide spread misconception about N.D.T. is that it is only applicable at the very end of production process. This misunderstanding has led to the concentration of N.D.T. use as a quality control procedure. With the correct approach N.D.T. could and should be applied to the design and manufacture stages as well as during operation and cease only on retirement of the component. With the continued introduction of a host of well tried and new techniques, non-destructive engineering has become recognized as a distinct discipline in its own right; it provides the necessary tools for reinforcing the efforts of designers, inspectors and operators. Development of these tools has been spectacular in recent years and an important ingredient in this has been the practical application of lasers : in laser-based materials process control and in holographic interferometry used for flaw detection. It is the second of these to which the present work addressed and the main objective is in converting the experience gained in the laboratory on holographic flaw detection towards a practical, reliable tool which can be used in the industrial workplace.

Holographic interferometry is a recent experimental method that can be used in non-destructive testing (Erf [1], Vest[2]). The ability to recreate 3-dimensional scenes makes it possible to apply interferometry to 3-dimensional objects. The image of an object stored in holographic film is made to interfere with the real object creating sets of contour lines or a fringe pattern.

Inspection of the fringe pattern arising from strain perturbation caused by the application of thermal or mechanical loads, gives a considerable amount of information on the type and magnitude of surface movements. Several methods of analysis of the fringe pattern have been proposed with a view to quantify surface strains. An excellent review of these methods has been presented by Briers[3].

Holographic interferometry is extremely sensitive, so sensitive in fact, that it often reveals peculiarities in the objects' inner structure such as debonds or delaminations. Surface flaws such as cracks and micro-cracks can also be easily detected by the same technique.

If a flaw is present in the real object , either on the surface or beneath, it disturbs the contour lines pattern which is developed as a result of the image interference. The disturbance observed is in the form of abrupt changes or discontinuities on the contour lines often easily detectable to an untrained observer. Crack detection, (Friesen[4]) and the location of debonds in composite structures (Buhman[5]) are fields where holographic interferometry has been found to be very useful.

There are two kinds of holographic interferometry which are used most often:

Real-time Holography which allows instantaneous vision of the changes occurring by looking at the holographic image of the unstressed object superimposed on the stressed object

Double-Exposure Holography requiring the exposure of the same hologram twice with stresses introduced between exposures. The difference is recorded permanently in the hologram

Because of the stringent stability requirements of the holographic set-up, the technique has not found wide practical applications in industrial environments and the best results have been obtained in laboratories.

The major difficulty so far has been the long exposure (sometimes over 1 minute) required in order to make a hologram using low powered continuous-wave gas lasers. High powered lasers up to now have been very bulky and extremely expensive. With the recent advent of the small pulsed ruby lasers (relatively inexpensive) the exposure time is cut down to say nano-seconds and therefore the stringent stability requirements are relieved.

EXPERIMENT PROCEDURE

The two key components of the system as shown in the schematic arrangement (figure 1) are the laser, which provides the coherent light necessary to form the hologram and a very high resolution photographic material on which the intricate interferometric pattern is encoded.

The laser beam is divided through suitable beam splitters and beam expanders in two parts; the object beam and the reference beam. This basic set-up is accepted by investigators as suited to many kinds of holographic work and is also often used for holographic interferometric N.D.T. work.

Two kinds of holographic interferometry were attempted during the execution of this work. At first some samples were tested for flaws using the double exposure holographic interferometry which required the exposure of the same hologram twice, with strains introduced on the samples between exposures. The difference is recorded permanently on the hologram. Double exposure holographic interferometry is much easier than real

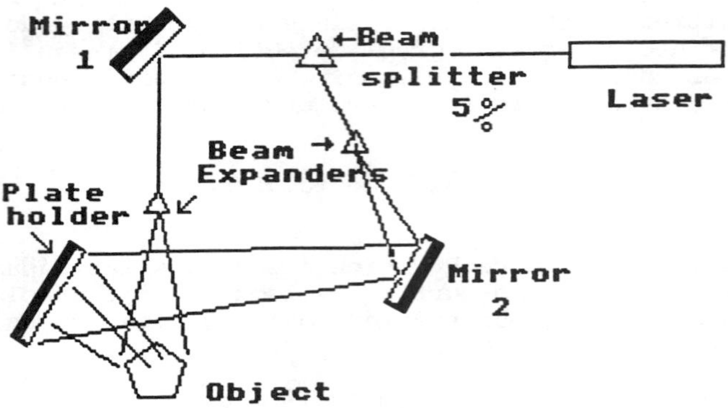

FIGURE 1.

Holographic Set-Up

time holographic interferometry, because the two waves are always reconstructed in exact register and distortion of the emulsion from the processing affects both images equally.(Hariharan[6])

Although from the onset of the investigation this approach was successful, it immediately became apparent that the method is somewhat blind. In other words, although the flaws were being detected, one was not sure whether a more dramatic manifestation of the flaw could be obtained as a result of more or less amount of stressing.

Confident of the fact that flaws were being detected, testing was continued employing real time holographic interferometry. This allowed instantaneous observation of the contour lines created by the interference of the stressed object with its unstressed holographic virtual image. In other words while looking through the hologram one can stress the object and enjoy the immediate creation of contour lines. As a rule, small amounts of stressing of the object creates wide contour lines of interference. The contour lines will become narrower and more dense and eventually impossible to discern with increased amount of object stressing.

Real Time Holographic Interferometry is far more demanding, in terms of time and precision required as compared to Double Exposure; however,it is the way to watch, without interruption, every little structural disturbance that may occur in an object. Disturbances caused by the touch of a finger or the flow of warm air from a hair dryer for example, produce effects which are observed as they occur. Hariharan states that precise repositioning of the hologram after processing is necessary to avoid the introduction of spurious fringes and in addition with photographic emulsions it is necessary to take precautions to avoid local deformation of the emulsion due to non-uniform drying. For this purpose it is recommended to expose and process the photographic plate in situ in a liquid gate (Van Deelen[7]).

In the present investigations because of relocating knife edges on the holographic plate holder, relocation was not a problem and spurious fringes only resulted from film processing. Subsequent to a private communication with Phillips[8] it was found that on using only the developer based on pyrogalol, rinsing in water and allowing it to drip dry after relocation, only one very wide residual fringe at most was visible.

All real time testing of the samples was observed through the use of a high resolution video system which enabled the production of a VHS recording. Samples of the real time tests were also captured in still photographs and form part of this paper; these are shown in Appendice A and B together with brief descriptions.

SUMMARY OF RESULTS

1) The tests have demonstrated that surface defects such as cracks (Appendies A,B) and micro-cracks (Appendix B) are detectable.

2) Voids, debonds and delaminations were also easily detected. (Appendix B).

3) Erosion and Corrosion can be monitored. (Appendix A)

The work reported has given futher credence to Holographic Interferometric N.D.T. in component flaw detection but is only laboratory-based.

Serious work has taken place (Ennos[9], Newport[10], Appollo[11]) in an attempt to transfer holographic techniques from the research laboratory to the users in Industry, and this has now become a reality with the availability of high power, double pulse ruby lasers.

APPENDIX A

Deformations Revealed

Typically, both real-time and double exposure holograms can be made when studying deformation of a part. The example is a cantilever beam where the amount of its deflection can be measured directly since the hologram of the beam appears overlaid with fringes corresponding to displacements of 0.15 microns. (FIG. 1)

Invisible Defects

Subsurface or interior irregularities constitute defects if they affect functionality. Since these irregularities show up in the way the material behaves under stress, they can often be detected visually with holographic interferometry.

There are several good methods for making subsurface defects visible under holographic scrutiny. One of the easiest is to raise or lower the part's temperature and observe the thermal expansions. The example chosen was a composite of a brittle material (glass) sandwiched between ductile material (plastic) The specimen was bent until the glass was cracked within it, while the surfaces appeared completely intact. Stressing it thermally, the position of the subsurface defect became instantly recognizable. (FIG.2)

Locating Cracks

There are several ways that cracks can be made visible. Because near the crack the deformation must be different, the fringe pattern will exhibit some discontinuity more often closely spaced at the crack. Temperature or mechanical stressing gives easily detectable results. (FIG.3)

The examples shown consist of a plate with a circular hole and flat bars, all of them having cracks as shown or indicated by the pattern.

Erosion and Corrosion

Holography allows the visualisation of the interior of a pipe, pressure vessel, etc. Depending on the particular situation, a comparison is made of the test object over a period of time or instantaneous comparison with an uncorroded sample. Under hydraulic or pneumatic pressure, defects are immediatly recognized as localised fringe patterns or as distorted fringes.

The examples shown are a corroded pipe, a corroded segment of a pressure vessel and a new pipe with three minute flaws artificially introduced in the internal wall. There is no difficulty in recognising the defects underneath the surfaces. (FIGS. 4, 5 and 6)

APPENDIX B

This Appendix contains a sample of the photographic record of the fringe patterns which were the result of thermally stressing the various specimens. The thermal stressing was effected simply by blowing hot air from a hair dryer on the specimens.

All photographs were obtained while the samples were being scrutinized using real time holographic interferometry. Short interpretations of the exhibiting patterns follow:

Photograph No. 1 shows the different specimens of fibre glass reinforced composites that were inspected for flaws using real time holographic interferometry.

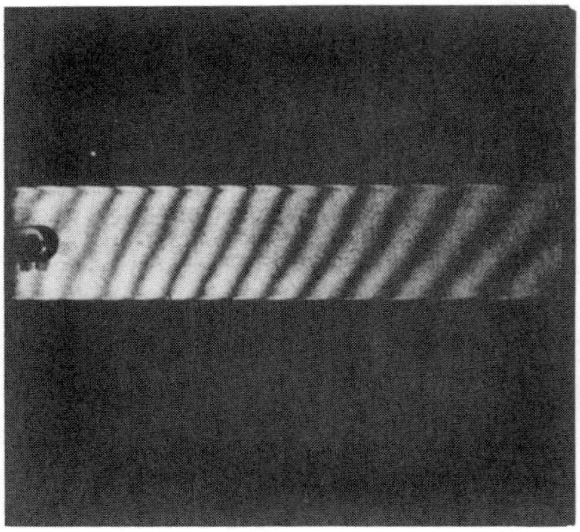

fig.1.

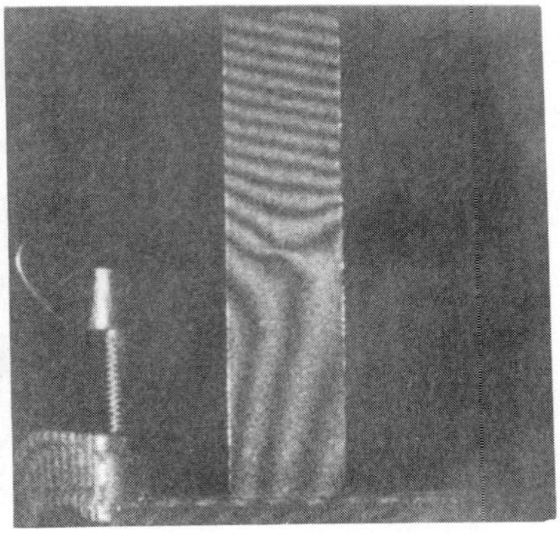

fig.2.

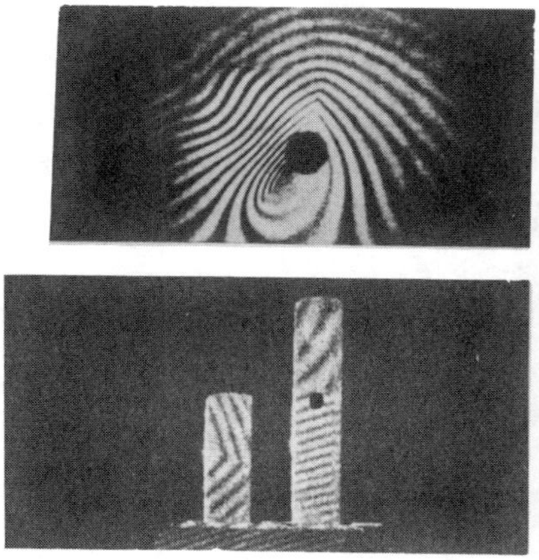

fig.3.

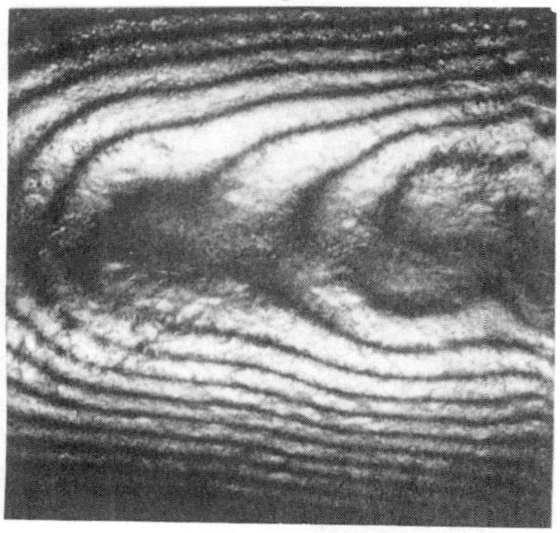

fig.4.

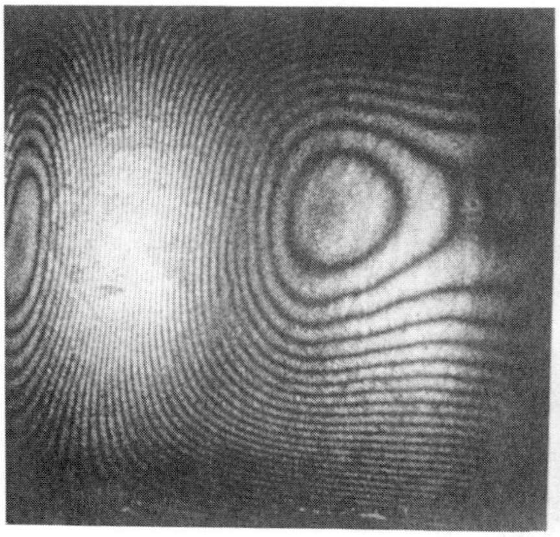

fig.5.

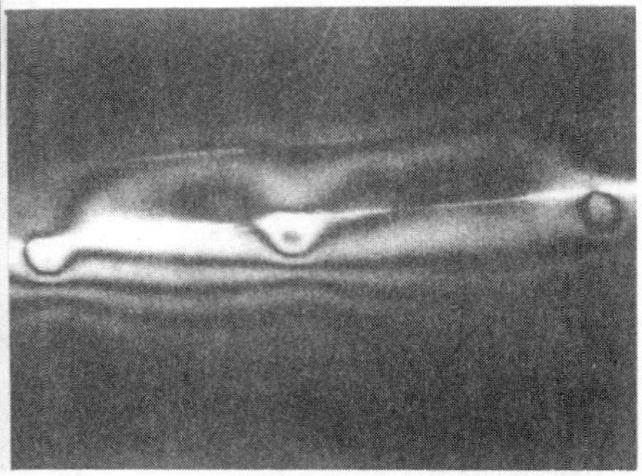

fig.6.

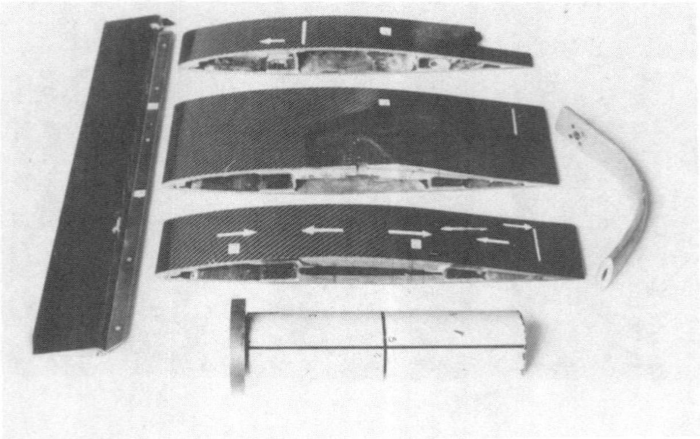

Photograph No.1

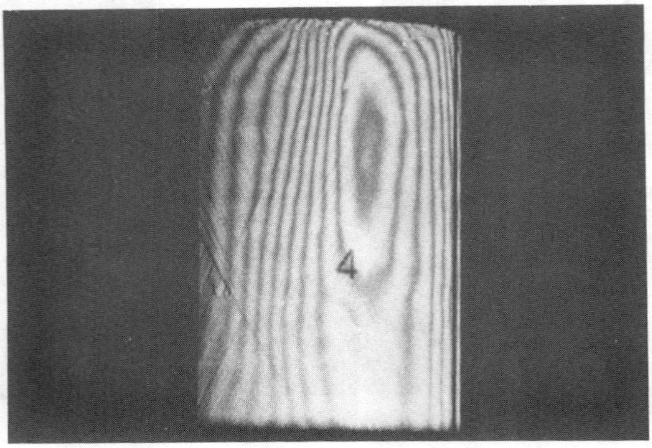

Photograph No.2

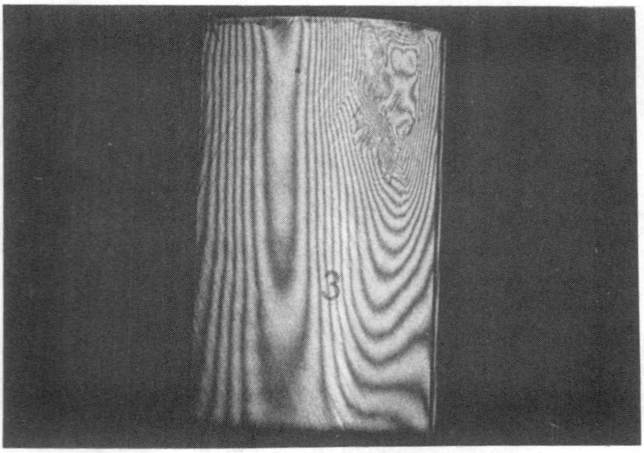

Photograph No.3

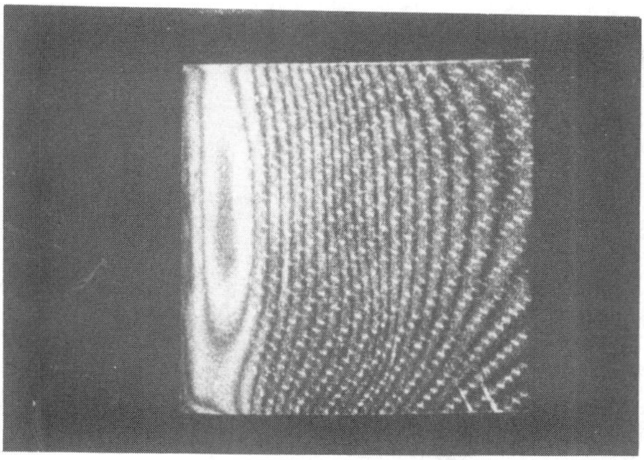

Photograph No.4

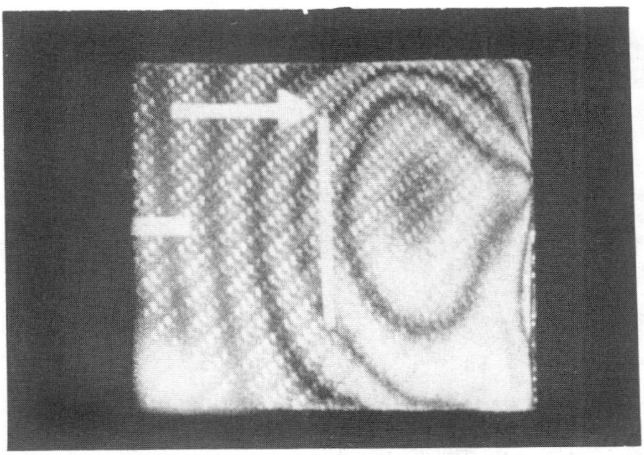

Photograph No.5

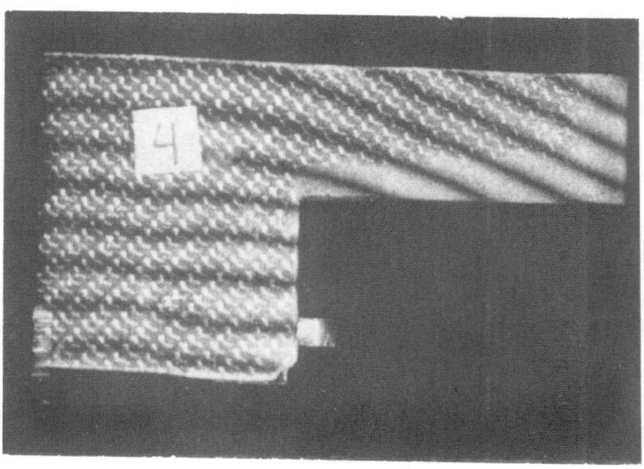

Photograph No.6

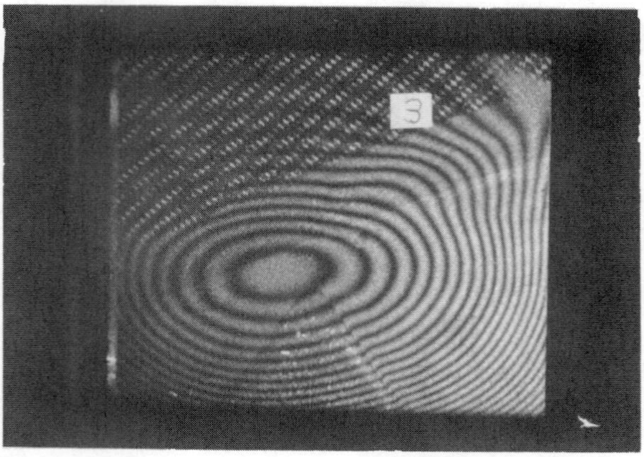

Photograph No.7

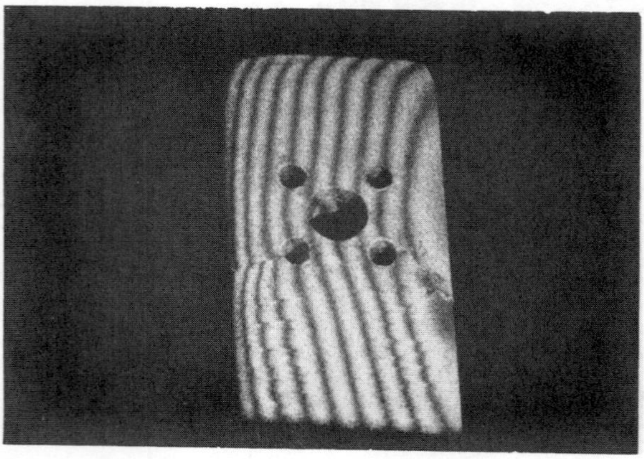

Photograph No.8

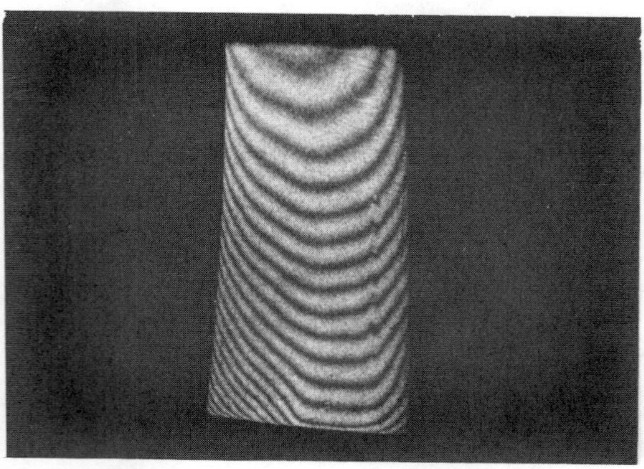

Photograph No.9

Photographs Nos. 2 and 3 exhibit samples of the fringe patterns indicating two different types of flaw on a fibre wound glass reinforced thin tube.

Photograph No. 2 shows a pattern which is the result of an internal debond about the size of a quarter of the dark centre fringe. Photograph No. 3 shows the pattern resulting from a small internal debond located almost at the lip of the cylinder on the left, and a dramatic fringe pattern on the right resulting from surface delaminations.

Photographs Nos. 4 and 5 depict the pattern that resulted by blowing hot air on the surface of the leading and trailing edges of a wing section. The leading edge contained a void and the trailing edge contained a debond to the right of the vertical white line.

Photograph No. 6 depicts the fringe pattern from a sound specimen where neither a debond or a delamination existed and serves as a comparison to photographs 4 and 5.

Photograph No. 7 depicts the characteristic pattern of a small debond centred underneath the smaller circle of the fringe pattern as well as the discontinuity of the pattern seen at about "5 o'clock", indicating the presence and magnitude of a crack.

Photograph No. 8 depicts the patterns that are created when cracks are present. In photograph 8 the directions of the large cracks emanating from the two bottom holes and extending toward the edges are clearly visible but in addition the wavy-spiky pattern of the fringes in the bottom half of the photo demonstrate the presence of surface microcracks.

Finally photograph No. 9 shows the typical discontinuity in the fringe pattern due to the presence of yet another crack.

REFERENCES

1. Erf. R.K. Holographic Non-Destructive Testing. New York: Academic Press (1974)

2. Vest, C.M. Holographic Interferometry. New York: John Wiley (1979)

3. Briers, J.D. "The interpretation of holo-graphic interferograms" Optic and Quantum Electronics, 8, (1976) 469-501.

4. Friesen, A.A. and Vest, C.M. "Detection of microfracture by holographic interferometry, Appl. Opt., 8, (1969) 1253 -1254

5. Buhman, K.P., Stelling, H.A. and Winkler, T. Nondestructive testing methods for composites Kunstoffe, 64 (1974) 750 - 760.

6. Hariharan P. Optical Holography. Cambridge University Press (1984).

7. Van Deelen, W. Nisenson. Mirror blank testing by real time holographic interfer-ometry. Applied Optics, 8, (1969) 951-5.

8. N.J. Phillips. Dept. of Physics Loughborough Univ. Private Communication (1986).

9. Ennos, A. "Holographic non-destructive testing with CW lasers away from the laboratory" SPIE Vol.349, Proc. Holo-graphic Non-destructive Testing (1982).

10. Design and Testing With Holography. Newport Corporation Technical Note.

11. Pulsed Laser Interferometry Appollo Lasers Inc. Technical Note.

AN INVERSE APPROACH TO DEFORMATION PROBLEMS FOR THE NONDESTRUCTIVE EVALUATION OF RESIDUAL STRESSES

A. Maniatty and N. Zabaras
University of Minnesota
Minneapolis, MN 55455

ABSTRACT

Most deformation processes are investigated through experimental testing or numerical simulation. With experiments, which ideally represent the actual phenomena during a process, one can only approximately measure various quantities, for example, displacements, strains, velocities, etc., at discrete points of the specimen. Furthermore, the measurements of the actual traction or displacement boundary conditions is a very difficult task, especially in cases where the specimen is in contact with another body. Therefore, one usually prefers to perform measurements in the interior of the specimen, which are easier to obtain and in general, more accurate. With computational methods, on the other hand, it is possible to calculate, among other things, the residual stresses and strains in the deformed body, but the calculations depend heavily on an accurate knowledge of the traction or displacement boundary conditions during the process. This paper will discuss an inverse formulation for deformation processes where computational techniques are used in conjunction with experimental data to predict the actual boundary conditions during the process which allows for the calculation of the residual stresses and strains at the end of the process. This work brings together the best features of both analysis and testing. Ideally, utilizing known test data and approaching the problem from an inverse point of view can result in the actual determination of boundary and internal conditions, as well as several material properties.

INTRODUCTION

In boundary value problems of solid mechanics, one deals with solving for internal displacements, strains, and stresses given sufficient boundary traction and/or displacement conditions (Fung, 1965). Such direct problems have been well studied, and the existence and uniqueness of their solution has been established. The inverse problem arises when insufficient boundary conditions are prescribed, but additional information on the solution, such as internal displacements or strains, is given. Then the problem is to calculate the displacement, strain, and stress fields all over the body, including the boundary using the experimentally available data which usually contains small but unspecified error. These inverse problems are a subset of the so-called ill-posed problems which have solutions that generally do not satisfy the conditions of existence, uniqueness, and stability with small changes in the internal data (Tikhonov and Arsenin, 1977; Hadamard, 1923; Payne, 1974; and Romanov, 1987).

In recent years, the main emphasis of inverse problems has been in transient heat conduction problems (Zabaras and Ruan, 1988), where one is interested in calculating boundary temperatures and fluxes from experimentally specified internal temperature data.

The numerical solution of these inverse problems includes stabilizing algorithms in conjunction with direct methodologies for which many different numerical methods, such as finite elements, boundary elements, or finite differences, can be employed. These inverse algorithms usually include regularization techniques, least squares techniques, dynamic programming, filters, and other optimization methods. For an extensive literature review, the book by Beck et. al. (1985) is recommended.

A finite element method analysis and implementation of two-dimensional inverse linear elastic problems has already been performed by Maniatty (1988). The developed algorithm uses an iterative sensitivity analysis, and in order to calculate the boundary traction vector, it minimizes the error between the calculated internal data and the experimentally available measurements. For stabilizing the minimization technique, regularization has been used at each iteration in order to enforce stability and smoothness on the solution of the unknown boundary traction vector. The developed method has been shown to be a very effective tool for the solution of such inverse problems. Also, it can be easily adapted to existing finite element commercial codes. The techniques developed in the above mentioned reference have been recently applied to elastic contact problems. More specific, the contact tractions between the roll and the sheet during a steady rolling process have been calculated using strain measurements at points internal to the roll (Maniatty, 1988).

Inverse problems have also been discussed in the field of thermoelasticity by Grysa, et. al. (1981), structural dynamics by Trujillo (1978), vibration problems by Gladwell (1986), and nondestructive testing of plastic strains by Mura (1985) and Mura et. al. (1986). In this last work (Mura et. al., 1986), the boundary element method is used to inversely evaluate the plastic strains in the bulk of the specimen caused by way of static loading from measurements of surface displacements after the application of the loading which are relative to the state before the load was applied. An estimate of the plastic strains is calculated for simple problems without taking into account the material constitutive representation and the deformation history of the specimen.

The present paper will start by defining a general inverse, non-linear material deformation problem. More specific, the problem of calculating the traction and displacement boundary conditions from internal strain or displacement history data of a specimen undergoing small elastoviscoplastic deformation will be analyzed. Some simple examples will be given. The potential of these techniques to NDTE and residual stress calculation will be discussed.

DEFINITION OF THE PROBLEM

Consider a three-dimensional body B with boundary ∂B and let $\sigma_{ij}(\underline{x})$, $\varepsilon_{ij}(\underline{x})$, and $u_i(\underline{x})$, i=1,3, j=1,3, denote the stress, strain, and displacement fields respectively where $\underline{x}$ defines the location of a point in the body. The body B is assumed to be homogeneous and isotropic with its elastic strains calculated from Hooke's law, while the non-elastic strain rates are specified from a given elastoviscoplastic constitutive law.

Following the familiar notation of Fung (1965), the governing equilibrium equations, neglecting body forces, and the traction and displacement boundary conditions are as follows

$$\sigma_{ij,j} = \rho \ddot{u}_i \qquad \text{on } B \tag{1}$$

$$\sigma_{ij} n_j = \hat{t}_i \qquad \text{on } \partial B_{i1} \tag{2}$$

$$u_i = \tilde{u}_i \qquad \text{on } \partial B_{i2} \tag{3}$$

where ρ is the density, $\ddot{u}_i$ is the acceleration, $n(\underline{x})$ is the unit normal at $\underline{x} \in \partial B$, $\partial B \equiv \partial B_{i1} \cup \partial B_{i2}$, and $\hat{t}_i$ and $\tilde{u}_i$ are the boundary tractions and displacements. In addition, the symbol (,) indicates a spatial derivative, and ($\dot{}$) indicates a time derivative. All variables are functions of location $\underline{x}$ and time t.

In addition to the above equations, assuming small strains and rotations, one can write following familiar tensor notation

$$\varepsilon_{ij} = \frac{1}{2}(u_{i,j} + u_{j,i}) \tag{4}$$

$$\varepsilon_{ij} = \varepsilon_{ij}^{e} + \varepsilon_{ij}^{n} \tag{5}$$

$$\varepsilon_{ij}^{e} = \lambda \sigma_{kk} \delta_{ij} + 2\mu\sigma_{ij} \tag{6}$$

$$\dot{\varepsilon}_{ij}^{n} = f(\sigma_{ij}, q_{kl}) \tag{7}$$

$$\dot{q}_{kl} = g_{kl}(\sigma_{ij}, q_{rs}) \tag{8}$$

where q_{kl} are state variables usually representing micromechanical parameters such as hardness, dislocation density, etc. Equation (4) defines the strains from the displacements, and equation (5) decomposes the total strains into the elastic (indicated with superscript e) and non-elastic (indicated with superscript n) ones. Equation (6) expresses Hooke's law with λ and μ being the Lamé constants, and equations (7) and (8) define the general form of elastoviscoplastic constitutive models with state variables, for example that proposed by Hart et. al. (1976) and by Hart (1976), for the non-elastic strain rate, $\dot{\varepsilon}_{ij}^{n}$, where the non-elastic strains are assumed to be incompressible ($\varepsilon_{ii}^{n} = 0$). The non-elastic strain rates are generally assumed to be functions only of the current values of the stresses σ_{ij} and state variables q_{kl} which follow a given evolution law given in equation (8). For a direct deformation problem, the following boundary value problem can be defined. Given $\hat{t}_i$ and $\tilde{u}_i$, solve equations (1) - (8) for the stress, strain, and displacement fields. It is obvious that the residual strains/stresses at the end of the deformation will be part of the solution. The solution of such problems using the finite element method has very well established and the reader is referred to excellent texts by Zienkiewicz (1979) and Bathe (1982).

In this paper, the goal is to calculate the boundary tractions $\hat{t}_i$ and displacements $\hat{u}_i$ given approximate information on displacements or strains at some points $\underline{x}$ internal to the body B ($\underline{x} \in B$, $\underline{x} \notin \partial B$). That is, given the following

$$u_i(\underline{x}_k, t) \cong \hat{u}_{ik}(t) \qquad \underline{x}_k \in B, \qquad \underline{x}_k \notin \partial B, \qquad k = 1,M \tag{9}$$

or

$$\varepsilon_{ij}(\underline{x}_k, t) \cong \hat{\varepsilon}_{ijk}(t) \qquad \underline{x}_k \in B, \qquad \underline{x}_k \notin \partial B, \qquad \underline{x}_k \in \partial B* \text{ for all } k,\ k = 1,M \tag{10}$$

where M is the number of internal measurement points (sensor locations), solve for the boundary tractions and displacements. In the problem of prescribed internal strains, defined by equation (10), the measurement points are restricted such that $\underline{x}_k \in \partial B*$ where $\partial B*$ is a closed curve internal to B, and in addition, certain restrictions are imposed on the displacement field to avoid rigid body motion (Maniatty, 1988). Obviously, certain additional restrictions should be imposed on the number of sensor locations and the percent error in the given data such that a physically meaningful solution can be obtained.

SOLUTION PROCEDURE FOR THE INVERSE PROBLEM

Here, an inverse problem defined by equations (1) and (4) - (9) will be considered, i.e., one with prescribed internal displacements. A finite element approach is followed here. The region B is discretized into finite elements in such a way that sensors are located at finite element nodes. The number of sensors is generally much less that the number of finite element nodes. A weak Galerkin formulation of equations (1) - (3) is involved which finally gives the following system of equations (Owen and Hinton,1986)

$$[M]\{\ddot{u}\} + [K]\{u\} = \{F\} \tag{11}$$

where $\{u\}$ is the nodal displacement vector, $\{\ddot{u}\}$ is the nodal acceleration vector, [M] is the mass matrix, [K] is the stiffness matrix, and $\{F\}$ is the vector of nodal forces which, in addition to terms associated with the boundry tractions, contains the pseudoloads created by the viscoplastic effects, see Owen and Hinton (1986) for a development of this system. Following a time stepping procedure, with t_n being the time at step n, n = 1,2,...,N, and Δt the value of the constant time step, an implicit time marching scheme for equation (11) takes the form

$$[M]\frac{\{u_{n+1} - 2u_n + u_{n-1}\}}{(\Delta t)^2} + [K]\{(1-\theta)u_n + \theta u_{n+1})\} = (1-\theta)\{F_n\} + \theta\{F_{n+1}\} \tag{12}$$

with $.5 \leq \theta \leq 1.$ for stability and where the displacement, velocity, and acceleration fields are considered known at $t \leq t_n$, and so $\{u_{n+1}\} \equiv \{u(t_{n+1})\}$ is assumed to be the basic unknown of the problem. The load vector $\{F_{n+1}\}$ is a function of the unknown traction vector $\{\tau_{n+1}\} \equiv \{\tau(t_{n+1})\}$, therefore, equation (12) cannot be used as it is to solve for $\{u_{n+1}\}$. Note that part of $\{u_{n+1}\}$ is approximately known through equation (9) for the inverse problem with given internal displacements at discrete points. Let $\{u^*_{n+1}\}$ be the part of $\{u_{n+1}\}$ at the finite element nodes which coincide with the sensor locations

$$\{u^*_{n+1}\} = [Q]\{u_{n+1}\} \tag{13}$$

where [Q] is a proper rectangular transformation matrix. The goal is to select the traction $\{\tau_{n+1}\}$ in such a way that $\{u^*_{n+1}\}$ is closest to $\{\hat{u}_{n+1}\} \equiv \{\hat{u}(t_{n+1})\}$, where $\{\hat{u}_{n+1}\}$ is the vector of experimentally determined displacements. If an iterative scheme is developed with the following equation

$$\{u^r_{n+1}\} = \{u^{r-1}_{n+1}\} + [X]\{\tau^r_{n+1} - \tau^{r-1}_{n+1}\} \tag{14}$$

where r is the iteration number and [X] is the sensitivity coefficient matrix defined as

$$[X] \equiv \left[\frac{\partial\{u_{n+1}\}}{\partial\{\tau_{n+1}\}}\right] \qquad \text{evaluated at } \{\tau\} = \{\tau^{r-1}_{n-1}\} \tag{15}$$

then a first goal can be to minimize the following function of $\{\tau^r_{n+1}\}$

$$E = \{u^{*r}_{n+1} - \hat{u}_{n+1}\}^T\{u^{*r}_{n+1} - \hat{u}_{n+1}\} \tag{16}$$

Substituting equations (13) and (14) into (16) and taking $\partial E/\partial\{\tau^r_{n+1}\} = 0$, a new value for $\{\tau^r_{n+1}\}$ can be found. Then this updated traction can be used to solve the direct problem for $\{u^{*r}_{n+1}\}$, and the process continues until convergence is acheived at each time step. As it is known from previous inverse problems this procedure is unstable, so an alternative technique must be developed.

Following previous work by Beck (1970), Zabaras and Liu (1988), and by Zabaras and Ruan (1988), the solution can be stabilized by minimizing the error not only at t_{n+1} but on the time interval $t_{n+1} \leq t \leq t_{n+p}$, where $p \geq 2$ in each time step. That is, in each time step, not only the information at the current time is used in the solution, but also future time information is used. Since it is known that the tractions must have a certain smoothness over short time periods, it can be temporarily assumed at each iteration that the traction varies linearly for the next few time steps, i.e.,

$$\{\tau^r_{n+i}\} = \frac{t_{n+i} - t_n}{\Delta t}\{\tau^r_{n+1} - \tau_n\} + \{\tau_n\} \qquad i = 1,p \tag{17}$$

Therefore, since the traction is known at time t_n, by assuming temporarily a traction at t_{n+1}, all the tractions for the times t_{n+i}, i=1,p, can be found and used in the minimization. Including this and carrying out the minimization with respect to the only independent variable $\{\tau^r_{n+1}\}$, the following final equation is obtained (for details, see Maniatty, 1988)

$$\sum_{i=1}^{p}\left(\left[\frac{\partial\{u^*_{n+i}\}}{\partial\{\tau_{n+1}\}}\right]^T\{\hat{u}_{n+i} - u^{r-1}_{n+i}\}\right)$$

$$= \sum_{i=1}^{p}\left(\left[\frac{\partial\{u^*_{n+i}\}}{\partial\{\tau_{n+1}\}}\right]^T\left[\frac{\partial\{u^*_{n+i}\}}{\partial\{\tau_{n+1}\}}\right] + [R]\right)\{\tau^r_{n+1} - \tau^{r-1}_{n+1}\} \tag{18}$$

where [R] is a positive definite regularization matrix which controls the relative magnitude of the tractions between iterations. Details on the numerical calculation of the sensitivity coefficients are given by Maniatty (1988). Using equation (18), one can update the traction to a better value for the next iteration. This scheme proves to be stable.

In summary, the algorithm proceeds as follows:

The goal is to find $\{\tau_{n+1}\}$. Everything is known at $t \leq t_n$.

1. Assume a traction at the iteration r = 0, $\{\tau_{n+1}^{o}\}$, to be zero or a better estimate if available.
2. Let r = r+1. Calculate the tractions $\{\tau_{n+i}^{r-1}\}$, i=1,p, based on the assumed traction $\{\tau_{n+1}^{r-1}\}$ and the known traction $\{\tau_n\}$ using equation (17).
3. Solve the direct problem (equation (12)) using these tractions to get $\{u_{n+i}^{r-1}\}$, i=1,p.
4. Compute the updated traction $\{\tau_{n+1}^{r}\}$ from equation (18).
5. Check for convergence on the tractions. If the root mean square norm between the tractions at the r and r-1 iterations is less than a given tolerance, then convergence has been reached.
6. If convergence has been attained, then begin the next time step, otherwise use the updated traction as the new assumed traction and continue with step 2.

EXAMPLES

Two examples will be presented to demonstrate the method. The first example is a two-dimensional static elastic problem, and the second example is a one-dimensional elastoviscoplastic problem with dynamic effects. Although both problems are geometrically simple, they demonstrate well the ill-posedness of the problem and the validity of the solution scheme.

The first example is for a linear elastic two-dimensional bar with a notch in it undergoing a constant axial load as shown in figure 1. Since the problem is symmetric about the y-axis, only the portion to the right of the y-axis was modeled. The geometry was discretized into 38 four-noded linear isoparametric elements with the elements being concentrated near the notch since a stress singularity occurs there. The elastic modulus of the material was taken as 30,000, Poisson's ratio was taken as .3, and the thickness was taken as .1 in consistent units.

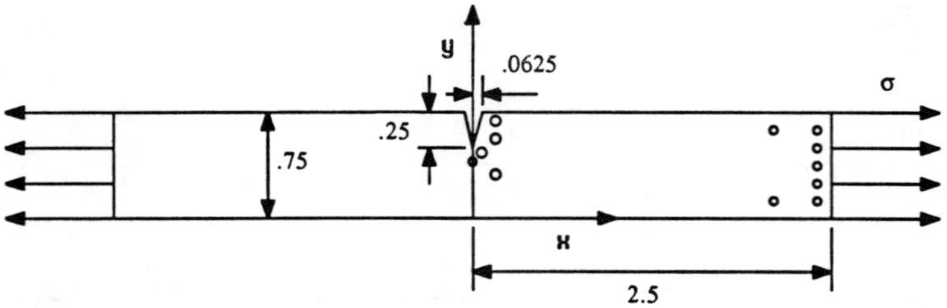

FIGURE 1. A two-dimensional bar with a notch under a uniform axial load.
o -- indicates a sensor

First, in order to generate data to represent internal displacement measurements, the direct problem was solved using the finite element method for a constant uniform traction loading of 1000 on either end as shown in figure 1. The calculated displacements at 12 discrete points, which represent sensor locations, were then multiplied by (1+.02w), where w is a random number between -1. and 1., in order to simulate random experimental error. The "sensors" which are located nearest the notch are approximately a distance of .05 from the boundary, and the "sensors" nearest the end where the load is applied are a distance of .1 from the boundary. The other "sensors" are slightly more removed from the boundary and are therefore less important, but are used so that there are at least as many known internal displacements as there are unknown boundary tractions, a requirement for the solution algorithm.

This simulated experimental data with random error was then used in the inverse formulation to solve for the traction condition in the region of the notch and on the end where the actual loading occurs. With just using a simple least squares minimization on the difference between the "experimental" displacements and the calculated displacements in the iterative scheme described, the solution diverges rapidly. By including a simple diagonal regularization matrix [R], the solution converges to a reasonably good solution. The traction condition in the notch should be zero since no loading was applied there, and in fact, the inverse solution gives a traction condition close to zero which is plotted in figure 2. The traction condition on the end where the loading was applied was predicted also reasonably well except at the corners which generally causes difficulty for these types of problems. The results for the end traction condition are plotted in figure 3 where the tractions at the corners have been obtained by extrapolating from the tractions away from the corners.

For this example, the regularization matrix was taken to be a diagonal matrix with a constant term along the diagonal. By varying the values along the diagonal, or adding terms to the off-diagonals, it is possible to introduce weighting or bias if desired. The solution is relatively insensitive to the magnitude of the term used on the diagonal as the regularization matrix controls more the rate of convergence than the actual solution because it is used to control only the difference of the tractions between iterations. It is important only to make the term large enough to force the solution to converge and to be stable, but small enough so as not to drive the solution to be flat or zero. Generally, acceptable values for this can range over several orders of magnitude making this a robust solution technique.

The second example is a one-dimensional elastoviscoplastic problem with a dynamic loading. The geometry for the problem is shown in figure 4. The material follows a very simple elastoviscoplastic model where

$$\dot{\varepsilon}^n = \gamma(\sigma - \sigma_Y - H'\varepsilon^n) \qquad \text{for } |\sigma| \geq |\sigma_Y + H'\varepsilon^n| \tag{19a}$$

$$\dot{\varepsilon}^n = 0 \qquad \text{for } |\sigma| < |\sigma_Y + H'\varepsilon^n| \tag{19b}$$

where γ is a fluidity parameter taken as .001, σ_Y is the uniaxial yield stress which is taken as 10, and H' is a strain hardening parameter which is taken as 5000. In addition, the mass of the bar is taken as 5 and the elastic modulus as 10000.

Again, a direct problem was first solved to generate simulated experimental data. Five linear elements were used. The loading condition was taken as one full sine cycle with an amplitude of 20, and the node closest to the boundary where the traction is being applied (at x = .875) was taken as the "sensor" location. A plot of the displacement at this node and the traction condition on the same graph is shown in figure 5. The inertia effect can be seen well here as the displacement lags the load, and continued oscillation occurs after the

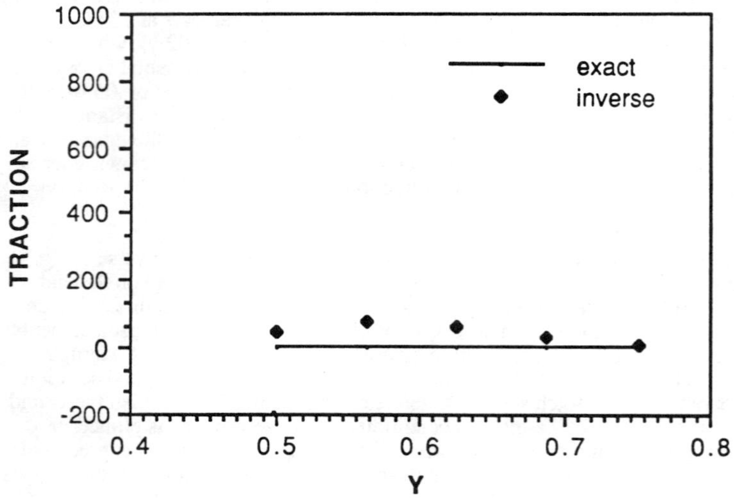

FIGURE 2. Zero traction condition on the notch as predicted with the inverse algorithm using regularization.

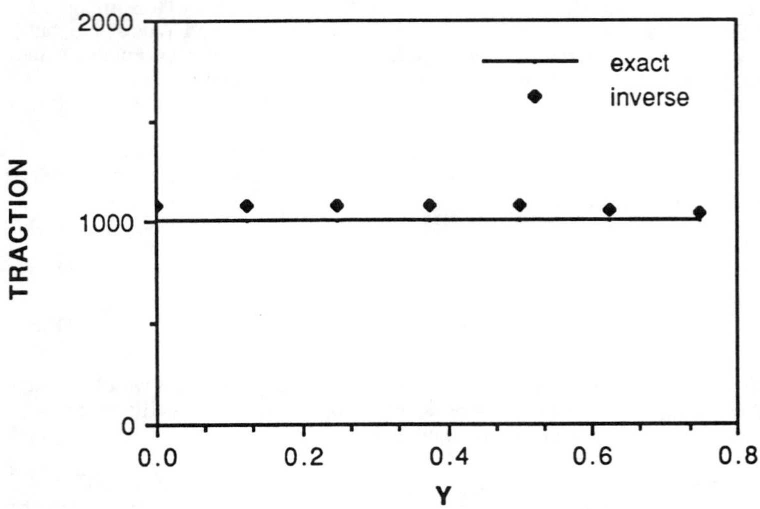

FIGURE 3. Stress condition on the end of the bar as predicted with the inverse algorithm using regularization.

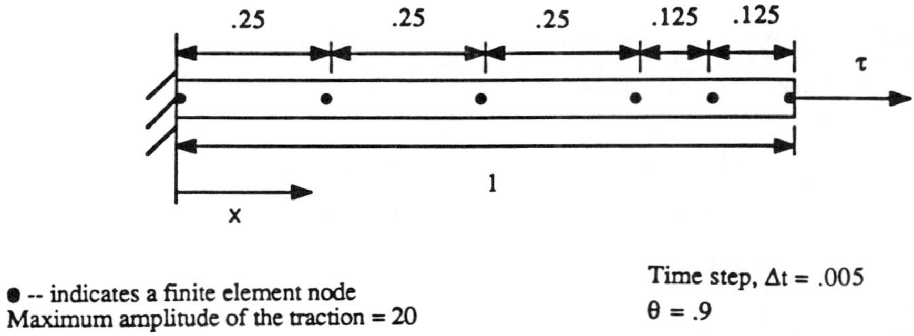

FIGURE 4. Elastoviscoplastic one-dimensional bar fixed at one end with a time varying traction on the other end. Sensor at x = .875.

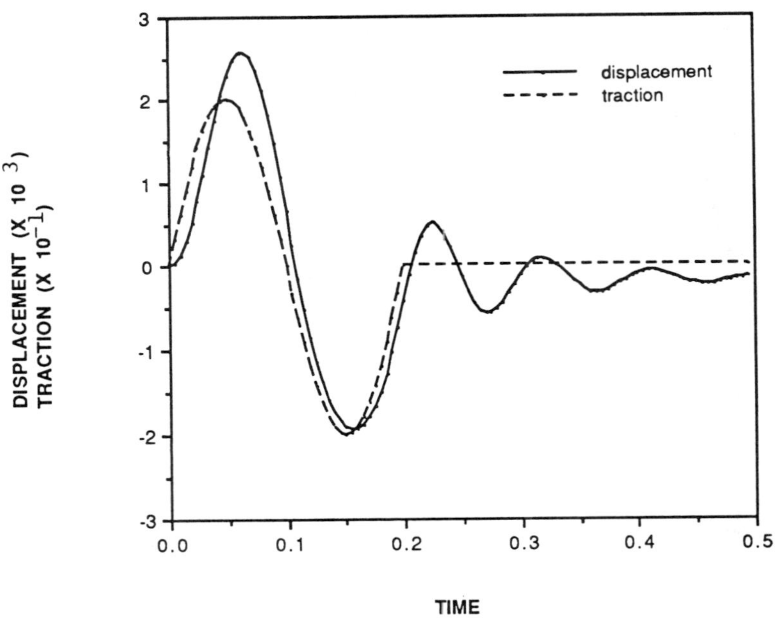

FIGURE 5. Applied traction at x = 1. and the resulting displacement at the sensor, x = .875, for the elastoviscoplastic bar.

load is removed. Again, to simulate random error, the displacements were multiplied by (1+.02w), where w is the same as before.

The simulated experimental data was then used to solve inversely for the traction condition. Using again just a least squares minimization (p = 1) in each time step, an unstable solution was obtained as can be seen in figure 6. If one future time step is used, i.e., p = 2 in equation (18), without regularization a very accurate solution is obtained as is shown in figure 7. Regularization was found not to be necessary in this case. One future time step was found to be all that was necessary to stabilize the problem in this case. Adding more future time steps introduces too much bias into the solution and causes the solution to diverge. The number of future time steps to use depends on the value of the sensor location and the time increment with respect to the variation of the traction condition in that time. With smaller time increments, more future time steps are necessary to stabilize the solution. In addition, since the displacements at internal points lag the traction on the boundary, it is necessary to use more future information to gain a valid solution for such cases. Adding addition information from other "sensors" more removed from the boundary does not improve the solution noticeably. From the computed boundary condition during the loading, one computes the residual strains a long time after the loading is removed which are plotted in figure 8. Obviously the residual stresses are zero.

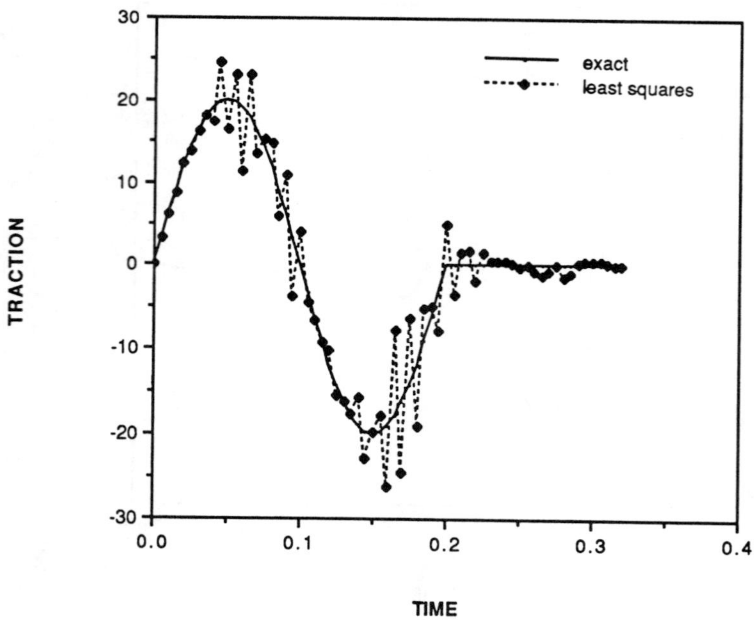

FIGURE 6. Predicted traction using a least squares inverse technique (p = 1) with 2% random error in the given displacements.

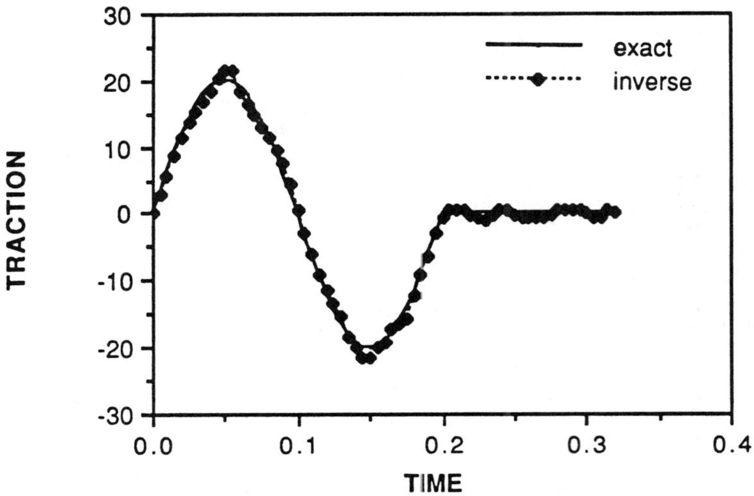

FIGURE 7. Predicted traction using an inverse technique with one future time step, i.e., p = 2, with 2% error in the given displacements.

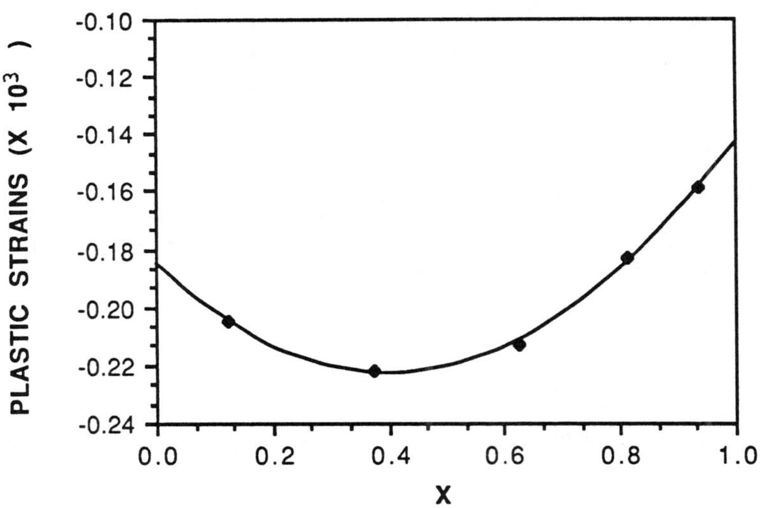

FIGURE 8. Inversely predicted plastic strains a long time after the application of the load, for the same situation as shown in figures 5 - 7.

DISCUSSION

The method reported here for the solution of one-dimensional inverse elastoviscoplastic problems can in principle be extended to two- and three-dimensional problems. In addition, the proposed techniques can be applied to inverse deformation problems somewhat different to those reported here of high significance to NDTE. For example, this paper reports a method for the calculation of the residual deformation when measurements during the loading process are available, but another inverse problem which may arise, is to calculate the residual stresses in a specimen due to some unknown loading process. So except for the final shape of the specimen, and possible information on the microstructure, plastic strains, defects, and flaws on the surface, no other measurements of any type are available. It is quite obvious that any additional, to the original deformation, non-linear deformation will depend not only on the applied loads, but also on the initial residual stresses (Washizu, 1982). In other words, measurements of displacements or strains at various internal positions of the specimen will only depend on the initial residual stress distribution, assuming that the applied boundary tractions and displacements are known.. Then a minimization procedure similar to that reported in this paper can be applied. The error between the calculated and measured displacements or strains would now be minimized with respect to the discrete nodal values of the residual stress field on the bulk of the specimen. This problem is similar to the so-called parameter estimation problem (Beck and Arnold, 1977). Note that since the boundary conditions during the loading process are known, one can minimize the above mentioned error at many time steps including the whole new deformation process. If the boundary tractions are not known, then one can employ ideas similar to those reported here in order to calculate both the boundary traction history during the additional deformation and the initial residual stresses due to the previous deformation process.

Finally, it should be emphasized that the non-linear deformation applied to the already plastically deformed specimen must be such that the new plastic deformation does not destroy the specimen. For example a sinusoidal load or boundary displacement with a zero average value like the type of loading reported here will create only a small amount of plastic deformation. Note also that an elastic deformation will probably not be sufficiently affected by the initial residual stresses, and so it cannot be used in such a method for the calculation of residual stresses.

CONCLUSIONS

The work reported here is a continuation of the research efforts by Zabaras and co-workers at the University of Minnesota to develop algorithms for the solution of inverse ill-posed problems related to the correlation of experimental and analytical techniques as applied to thermomechanical processes. Work is in progress for the extension of the techniques presented here to two- and three-dimensions as well as to cases where real experimental data are available.

ACKNOWLEDGEMENTS

The work reported here has been supported by the Alcoa Foundation. The computing has been supported by the Academic Computing Services and Systems of the University of Minnesota and by the Minnesota Supercomputer Institute.

REFERENCES

Bathe, J. (1982). *Finite Element Procedures in Engineering Analysis*, Prentice-Hall.

Beck, J.V., Blackwell, B., and St. Clair, C.R., Jr. (1985). *Inverse Heat Conduction, Ill-Posed Problems*, A Willey-Interscience Publication, New York.

Beck, J.V. (1970). "Non-Linear Estimation Applied to the Non-Linear Heat Conduction Problem." *Int. J. Heat Mass Transfer*, 13, 703-726.

Beck, J.V. and Arnold, K.J. (1977). *Parameter Estimation in Engineering and Science*, John Wiley, New York.

Fung, Y.C. (1965). *Foundations of Solid Mechanics*, Prentice-Hall.

Gladwell, G.M.L. (1986). *Inverse Problems in Vibration*, Martinus Nijhoff Publishers.

Grysa, K., Cialkowski, M.J., and Kaminski, H. (1981). "An Inverse Temperature Field Problem of the theory of Thermal Stresses." *Nucl. Engrg. Design*, 64, 169-184.

Hadamard, J. (1923). *Lectures on Cauchy's Problems in Linear Partial Diffential Equations*, Yale University Press, New Haven, CT.

Hart, E.W., Li, C.Y., and Yamada, H. (1976). "Phenomenological Theory: A Guide to Constitutive Relations and Fundamental Deformation Properties." *Constitutive Equations in Plasticity*, M.I.T. Press, pp. 149-197.

Hart, E.W. (1976). "Constitutive Relations for the Non-Elastic Deformation of Metals." *J. of Engrg. Mats. and Tech., Trans. ASME*, Series H,98, No. 3, 193-202.

Maniatty, A. (1988). Master's Thesis, Department of Mechanical Engineering, University of Minnesota, Minneapolis, MN 55455.

Mura, T. (1985). "A New NDT: Evaluation of Plastic Strains in Bulk from Displacements on Surfaces." *Mechanics Research Communications*,12 (5), 243-248.

Mura, T., Cox, B., and Gao, Z. (1986). "Computer-Aided Nondestructive Measurements of Plastic Strains from Surface Displacements." *Comput. Mech. '86*, 1.

Owen, D.R.J. and Hinton, E. (1986). *Finite Elements in Plasticity: Theory and Practice*, Pineridge Press Limited, Swansea, U.K.

Payne, L.E. (1974). *Improperly Posed Problems in Partial Differential Equations*, SIAM Regional Conference, New Mexico.

Romanov, V.G. (1987). *Inverse Problems of Mathematical Physics*, VNU Science Press, Ultrecht, The Netherlands.

Tikhonov, A.N. and Arsenin, V.Y. (1977). *Solution of Ill-Posed Problems*, V.H. Winston and Sons, Washington, D.C.

Trujillo, D.M. (1978). "Application of Dynamic Programming to the General Inverse Problem." *Int. J. Numer. Meth. Eng.*, 12, 613-624.

Washizu, K. (1982). *Variational Methods in Elasticity and Plasticity*, Third Edition, Pergamon Press.

Zabaras, N. and Liu, J. (1988). "An Analysis of Two-Dimensional Linear Inverse Heat Transfer Problems Using an Integral Method." *Numer. Heat Transfer*, 13, 527-533.

Zabaras, N. and Ruan, Y. (1988). "A Deforming FEM Analysis of Inverse Stefan Problems." *Int. J. Numer. Meth. Eng.*, forthcoming.

Zienkiewicz, O.C. (1979). *The Finite Element Method*, McGraw-Hill, New York.

VERIFICATION OF NUMERICAL CONTROL PROGRAMS - PROBLEMS AND APPROACHES

C. H. Chang
University of California, Los Angeles
Los Angeles, CA 90024

ABSTRACT Current research on NC program verification (NCPV) concentrates on the graphic representation and display of the in-process workpiece and on the cutter path defined by the NC program. Most of these NCPV systems rely on human judgement to verify an NC program and their capability to verify the dimensions of an finished workpiece is very limited. These systems cannot guarantee that an NC program can be used to fabricate the designed part to the specified dimensions and tolerances. This paper analyzes the factors that cause the rejection of NC programs, including the NC program syntax, the dynamic response of the NC control system, NC machine tool geometric errors, and the machining process plan and specifications, thereby redefining the NC program verification problem. Dimension verification taking tolerance into consideration is discussed, and the effect of NC programming systems and application environment on the design of NCPV systems is studied. The approach to the design of NCPV systems is also discussed.

1. INTRODUCTION

Since they first came on the scene more than thirty years ago, NC machines have been recognized by industry as capable of attaining high accuracy, productivity and flexibility in the fabrication of mechanical parts. The operation of an NC machine is controlled by its control unit (the NC controller) according to the input information, i.e., the NC program. The dimensions of a part made on an NC machine are essentially determined by the NC program. The productivity and flexibility of production attainable by an NC machine depend upon the availability of the required NC programs which should be verified and error free. Therefore, the verification of NC programs is crucial to the effective utilization of NC machines and flexible manufacturing systems.

The importance of NC program verification (NCPV) has been recognized since the first NC machine was developed in 1952 [1]. The reason for verification of manually written NC programs is obvious: human programming errors occur frequently. The necessity for verifying NC programs generated from a computer-assisted NC programming system, such as APT, COMPACT II, and EXAPT, has also been realized, for the part conceived in the programmer's mind may not be the same as the designed one, and the cutter path defined in a program can be different from what he desires. Programming errors may also exist in the definition of an NC machining operation by means of these systems. The utilization of CAD/CAM systems to generate NC programs allows NC programmers to define a cutter path based on the CAD model of a designed part and to visualize the part, the fixtures used and the cutter path on the screen, and hence makes it possible to detect significant programming errors.

Initial research efforts focussed on NC program verification by means of computer graphics systems or plotters for 2-D and 2½-D machining [2-7]. More recent work is aimed at the development of NCPV systems for 3-D and multiaxis machining by utilizing solid modeling techniques to simulate the cutter path and to display in-process machined workpiece on computer graphics systems [8-12]. With the exception of references [11] and [12], this research is essentially concerned with the techniques for graphic representation and display of machined parts and the programmed cutter path. The proposed systems still rely on human judgement to verify the cutter path and machining process, and the dimension verification capabilities are very limited. The NC programmers or the operators of these NCPV systems have to decide whether the displayed cutter path is acceptable. Sungurtekin and Voelcker [11] have discussed automatic verification of NC programs by displaying the in-process (update) workpiece, i.e., the result obtained by subtracting the volume swept by the cutting tool during its motion from the volume occupied by the workpiece (stock material). The verification is basically an interactive process and requires considerable human judgement. The same concept was adopted by Wang and Wang[12] who introduced tolerance into their system by shifting the envelope surface generated by the tool motion. A multiaxis machining program can be verified on their system with the aid of a routine that examines the interference between the volume swept by the tool and that occupied by the part and fixtures.

All NCPV systems consider the programmed cutter positions which are the theoretical tool positions in the machine coordinate systems without regards to tolerances (with the exception of reference [12]). Since an NC program is usually written simply on the basis of the part geometry, and the tool movement is directed by the NC controller according to the NC program, the machine tool geometric errors and the errors due to dynamic response of the machine-controller system cannot be compensated for during machining. Similarly, since most NCPV systems consider and display the cutter geometry positions defined in a part machining program, the effect of machine tool and control system errors on the deviation of the cutter path from the defined position cannot be revealed. Problems that are not taken into consideration by existing NCPV systems include also syntax errors in an NC program, tool wear and deflection during machining and the specified machining process and parameters, such as feed, feedrate, and depth of cut. Therefore, the existing NCPV systems cannot guarantee that a verified NC program can be accepted by the NC machine and can be used to fabricate the designed part to the specified dimensions and tolerances, especially for machining high precision part at high removal rate. Consequently, in most cases we have to resort to test-run an NC program on the NC machine in order to prove that the NC program is satisfactory [1,10].

2. NC PROGRAM VERIFICATION REDEFINED

As can be seen from the previous discussion, many factors contribute to errors in cutter positions. Most current NCPV systems consider only one of these, namely the definition of tool geometric positions in a theoretical coordinate system. Factors other than geometric definition of tool positions, the designed part and the fixtures should also be considered in an NCPV system.

The profile of a part is usually composed of a variety of geometric entities, such as lines, circles, planes, cylinders, torus, and others. Most NC machines can have only two types of contouring motions, namely linear interpolation and circular interpolation. A linear interpolation cutter motion is determined by the rates at which electric pulses are sent to the driving motors in the X, Y and Z axes of an NC machine. Since the ratio of these rates are kept constant during the motion, the tool motion is linear. Thus, the maximum deviation of the cutter path from the designed profile is found at the end position of the tool motion.

A circular interpolation motion can be realized in the three coordinate planes of a CNC machine. It consists of a number of linear motion vectors which approximate the desired circle. The deviation of the cutter positions from the designed circular profile during circular interpolation motion is determined by the design and parameter setting of the NC controller. Normally, it is less than the allowable limit. However, circular interpolation motion is affected also by the specified feedrate and the radius of the circle. This will be explained in Section 4.2.2.

Cutter paths other than linear, including also circular paths, can be approximated by a series of linear motion vectors based on the linear interpolation concept (Fig. 1). These motion vectors are determined by the part profile and the specified tolerance. Any change in tolerance results in variation of the number, directions and starting and end points of these motion vectors.

With a computer-aided NC programming system, such as APT, or a CAD/CAM system, a two-dimensional non-linear part profile is approximated by a

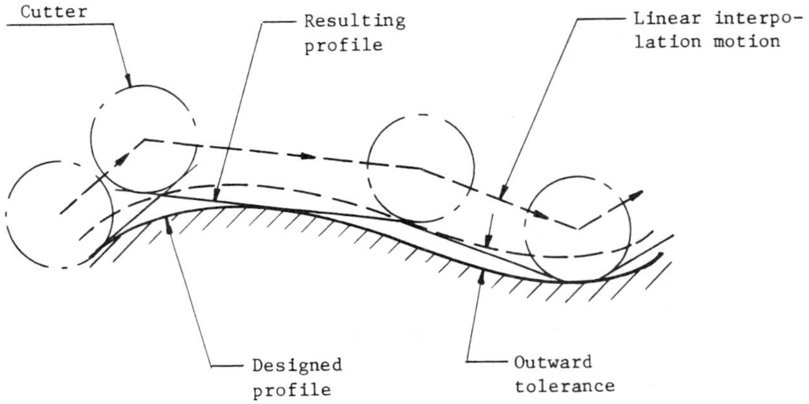

Figure 1. The approximation of a part profile by linear interpolation motion.

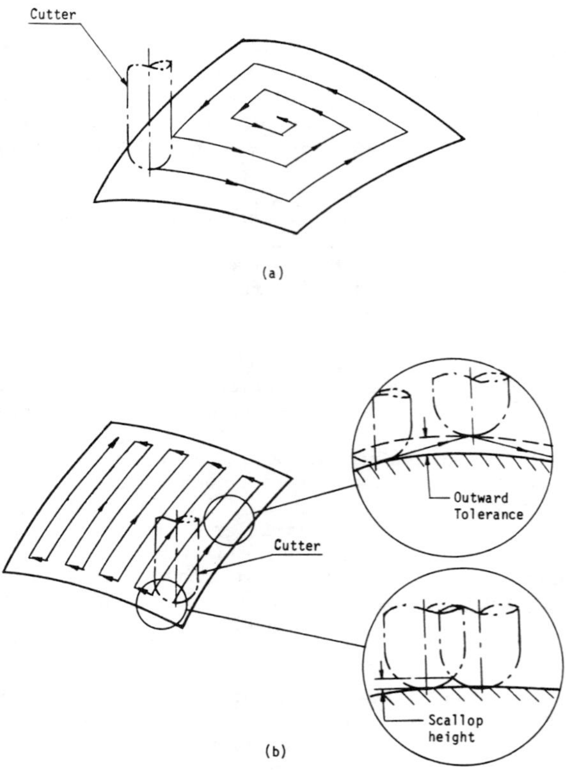

Figure 2. The approximation of a profile based on linear interpolation with a spiral (a) or zigzag (b) cutter path.

series of linear-interpolation-motion vectors. The cutter position data, called CLDATA (cutter-center-line data), can be transformed into NC codes consists either of the linear interpolation code, G01, and circular interpolation codes, G02 and G03, or, as in most cases, of the linear interpolation code, G01, only, depending on the structure of the CLDATA and the design of the postprocessor [1].

A three-dimensional surface is approximated by a zigzag or spiral type of cutter motion which consists of a large number of linear motion vectors (Fig. 2). In addition to the deviation of the cutter from the designed part profile due to linear interpolation, material between two parallel travels of the cutter is left uncut, and hence a corrugated surface results. The scallop height is determined by the selected feed, i.e., the distance between two parallel travels of the cutter. Therefore, both the deviation of the tool from the designed profile due to linear interpolation motion and the scallop height should be examined during verification of an NC program.

The cutter positions and path in an NC program are defined on the basis of an ideal coordinate system whose axes are mutually perpendicular. For

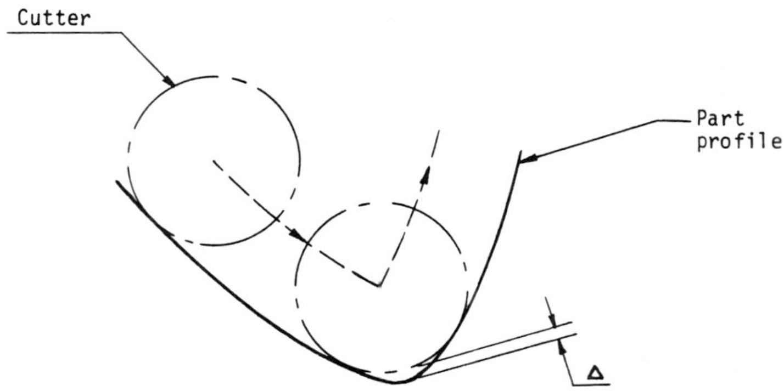

Figure 3. The cutter path generated by a computer-aided programming system or CAD/CAM system does not yield an correct profile when the cutter radius is larger than the radius of curvature of the part profile. The deviation Δ may be out of tolerance.

an ideal NC machine free from geometric and control errors, the machining error is the same as the deviation of the cutter position specified in the NC program from the designed profile when the deflections of the part, cutting tool and machine due to cutting force are neglected. Since a profile is usually approximated by linear interpolation, the machining error for profile milling operation is the maximum deviation of the starting and end positions of the motion vectors from the designed profile. Furthermore, undercut or gouging will results if the selected tool radius is greater than the radius of curvature of a concave part profile. In such case, the calculated cutter path by a computer-aided NC programming system or CAD/CAM system does not yield a correct profile (Fig. 3).

Real NC machines are far less accurate than what is commonly believed. Although NC machines can direct a tool to follow a complex profile without relying on human expertise and can attain remarkable repeatability, they still exhibit geometric errors and the positioning errors arising from their control systems. Since these errors are not considered during programming and, due to automatic nature of NC operation, cannot be compensated for during machining, NC machine tools can only attain IT10 class (the ISO system of tolerance) on the average [13], and are generally less accurate than conventional machine tools whose errors can be compensated for through human interference [13,14]. Machining of parts with tight tolerance requirement has still to be carried out on conventional machine tools if these errors cannot be counterbalanced on NC machines.

Machining errors arise also from the dynamic response of the servo control system during sequential, uninterrupted execution of NC statements on NC machines [1]. An overshoot results when the motion direction changes. Significant deflections of the machined part, cutting tool and machine tool may also occur when heavy depths of cut and high feed rates are programmed. These problems indicate that an effective NCPV system need also consider non-geometric factors, especially for high material removal rate.

Thus, the function of an NCPV system should not be limited to intepreting NC statements and displaying the designed part, fixtures and cutter path; factors that have significant effect on the accuracy of the machined part, such as machine tool geometric errors, the characteristics of the NC control system, the tool wear and the deflection of the tool-machine-part system, should also be considered.

A verified NC program must be free of errors of any type and immediately executable. It is worth noting that on the shop floor, an NC program is oftentimes rejected not because of incorrect definition of cutter geometry positions, but because of inappropriate process defined by the program. Furthermore, the machining specifications have significant effect on the surface finish, dimensions, and accuracy of a machined part. Consequently, the functions of an NCPV system should also include identification of errors in NC program syntax and in specified machining processes and specifications.

3. THE NCPV SYSTEMS FOR DIFFERENT NC PROGRAMMING SYSTEMS

Generally, there are three different approaches for generating an NC program for a designed part, namely, manual programming in machine-readable codes (NC codes), computer-assisted NC programming in higher-level languages (such as APT and COMPACT II), and NC program generation by means of CAD/CAM systems [1]. All of these are widely used in industry. An NCPV system can be designed for verification of NC programs generated from one or several NC programming systems, depending on the application environment.

Usually, the NC programs generated by different programming systems for a designed part have different contents although they serve the same purpose, i.e., correct definition of the same machining process and cutter path. The major differences among NC programs generated by different systems are found in the statements that define the cutter path or motion.

An NC programmer normally utilizes those NC codes that can define as long and as complex a cutter path as possible in order to simplify the programming and to shorten the program. NC codes that define circular interpolation, machining-cycle, and multiple-machining-cycle motions are common in manually written NC programs. NC programmers also make use of subprograms, codes defining the programming mode (e.g., programming in absolute or incremental coordinates) and the machine coordinate system, and codes defining calculations and decision operations (such as in parametric programming) to create a shorter program.

In computer-assisted NC programming systems, NC programs are translated and processed from the CLDATA, which consists of a series of discrete point data. The section of an NC program that defines the cutter path consists of much fewer types of codes than those used in manual NC programming. Normally they are G00 and G01 codes when NC programs are generated by CAD/CAM systems or G00, G01, G02 and G03 codes when computer-assisted NC programming systems (such as the APT system) are used. There is no calculation or decision statement in the resulting NC programs. All the cutter positions in an NC program are defined in one coordinate system and in one programming mode determined by the postprocessor design.

The codes and structure of a manually written NC program are usually more complex than those of programs generated by computer-assisted NC programming systems or CAD/CAM systems. Syntax errors are very common in manually written NC programs. Thus, syntax examination should be one of the basic functions of the NCPV systems for verifying manually written NC programs, in addition to verification of cutter positions and path. In contradistinction, syntax errors are much less likely in system-generated NC programs. For these systems, a customized postprocessor is utilized for each NC machine to convert the CLDATA into an NC program free of syntax error. Therefore, syntax examination is not necessary. Either the CLDATA or the resulting NC program can be used as input to the NCPV system. The CLDATA contains more information than provided by its corresponding NC program. In addition to the data defining cutter positions, it includes also data defining the geometric entities based on which the cutter path is defined. Such data can be used to create the geometric model of the designed part in the NCPV system.

Since a cutter path is verified through comparison of the part geometry with the cutter path, the geometric model of the designed part should be created. While a manually written NC program may provide partial information regarding the part to be machined (e.g., when the cutter radius compensation code is used to define the cutter path), an NC program generated by a computer-assisted programming system provides no information regarding the designed part. However, such information is completely contained in the CAD model when a CAD/CAM system is used, or is partially contained in the CLDATA and source program when a computer-assisted NC programming system, such as APT, is utilized.

In summary, the requirement on the design of an NCPV system varies with the NC programming system by which NC programs to be verified are generated. Since the verification should be carried out by the NCPV system itself, additional information regarding the designed part, fixtures and tools is required for manual and computer-assisted NC programming systems. Syntax examination should be a basic function of the NCPV system that verifies manually written NC programs.

4. THE APPROACH TO THE DESIGN OF NCPV SYSTEMS

The purpose of an NCPV system is to provide NC machines with NC programs which are free of errors and can be used to direct the machining processes for making the designed parts. Types of errors that should be examined include syntax errors and errors in cutter path and positions. Generally, factors that should be considered in verification of cutter path and positions include the geometric data of the designed part, fixtures and the defined cutter path, machine tool geometric errors, the characteristics of the NC control system, part material, and machining specifications including, for example, the feedrate, depth of cut, cutting speed and the selected cutting tools. The machining process defined by an NC program should also be considered as one of the aspects to be verified in an NC program since an incorrect machining process also results in rejection of the NC program.

4.1 Verification of NC Program Syntax

Verification of NC program syntax is required for manual programming of NC programs. Although computer-assisted NC programming systems and

CAD/CAM systems are widely used, a large proportion of NC programs are still written manually. These programs have to be verified before they are input into the NC machines and executed. Until now, most workers [3-13,15] have concentrated on cutter path verification only, with the exception of Rolstadas [2] who also discussed syntax verification.

With the increasing amount of software incorporated in CNC controllers, the geometric complexity of parts that can be manually programmed increases. Manual NC programming is very efficient for programming of multiple-hole drilling, peck-drilling and boring operation, rough and finishing turning operation, 2-D machining of simple profiles in the three coordinate planes, and simple die-surface machining [16]. Since the profiles of many engineering parts are composed of lines, circles, planes and cylinders only, their machining process can be programmed efficiently by NC programmers. However, syntax errors occur very often in manually programmed NC programs.

Manual NC programming should follow certain rules which depend on the design of NC controllers. Although the formats and codes for programming of modern CNC control systems of different makes have not been standardized, their programming principles are very similar. All modern CNC controllers adopt the word-address block format for specifying NC codes and statements. In the following, the FANUC 6MB and 6T CNC controllers are selected as examples to illustrate syntax verification.

A general rule of manual NC programming of modern CNC machines is that codes of the same type cannot be specified in one NC block with the exception of the G codes and that G codes of the same group cannot be specified in one NC block [1]. Thus, each code can be specified only once in an NC block (statement) with the exception of G codes of different groups. Other basic rules in manual NC programming are, for example,

- Once specified, most codes are effective until they are cancelled or replaced by the same codes of different values with the exception of some G codes (such as the G04 code specifying the dwelling time, which is effective only in the current block);
- If not specified, a code denoting absolute coordinates (e.g., the X, Y or Z code) has the same value as specified in the preceding block, whereas those denoting incremental coordinates (such as codes U, V, W, I, J and K, or the X, Y and Z codes in incremental programming mode) are zero if they do not appear in an NC block;
- Each NC block should end with a character denoting the end of a block; and so on.

There are also rules of programming cutter motion. For example, the FANUC 6MB CNC controller requires that the motion plane be specified (by a G17, G18 or G19 code) in the NC block that defines a circular interpolation motion or the start-up motion defined by a G41 or G42 code which specifies a tool motion with cutter radius compensation. For the FANUC 6T CNC lathe controller, the first cutter motion should be defined as perpendicular and parallel to the axis of the part in a multiple turning cycle (defined by a G71 code) and multiple facing cycle (defined by a G72 code) respectively. Failure to follow these programming rule results in an unexcutable program or an incorrect cutter path.

An NCPV system for verifying the syntax of a manually written NC program should include a number of routines each of which ascertains that one specific programming rule is followed in the NC program. The syntax verification can proceed as follows. An NC program is read into the NCPV system block by block. The system checks whether the first block contains only the O code which denotes the beginning of a program and the program number. If so, the system reads in the next block. Otherwise, an error message is issued together with the block sequence number, and then the system reads in the second block. Starting from the second block, the NCPV system examines whether the coordinate system, measurement system (inch or millimeter) and programming mode (incremental- or absolute-coordinate programming) are defined. The system also checks if there are codes other than G code or G codes of the same group appearing more than once in a block. For each block, the system uses various routines to examine whether the programming rules for the motion and operation specified in the current block are followed. For example, if the specified motion in an NC block is circular interpolation motion, the system will check whether a G17, G18 or G19 code and the center offset codes I, J and/or K (or the radius code R) are specified. Routines may also be included in the NCPV system to examine:

- if an NC block ends with an end-of-block character;
- if a reference-point-return motion is defined at the beginning of an NC program if such return motion is required for the machine;
- if the spindle has been started when a cutting motion is specified;
- if the tool has returned to the safe or required position when a tool change code is specified; etc.

Finally, the NCPV system should check whether the last code is M02 (end of program) or M30 (end of program and memory rewind) and whether the coolant and spindle are stopped before termination of the program.

NC controllers of older design require programming in block formats other than word-address one [1]. Therefore, special routines are needed to verify both the contents and the order of codes in a block.

NC programs generated by a computer-assisted NC programming system usually present no syntax error since the programming rules have been followed by the postprocessor during transformation of the CLDATA. Therefore, it is not necessary to verify the syntax.

4.2 Verification of Cutter Path

Verification of the cutter path defined in an NC program is concerned with small variations in dimensions. A part machining program is not acceptable even if the specified cutter path is out of tolerance by a very small amount, say 0.01 mm or 0.0004 in. Therefore, as pointed out previously, an NCPV system should not rely on NC programmers' visual judgement to locate, on the screen, an excessive deviation of tool position (which may very possibly be of a very small order, say 0.01 mm) at a certain position. The factors that should be considered in cutter path verification are well beyond those arising from geometry consideration only. An analysis of these factors and possible approaches to solve the problems induced by these factors are presented below.

4.2.1 <u>The representation and geometric verification of the cutter path defined by the NC program</u>. Verification of an NC program should be based on comparison of the cutter path with the designed part profile. Correct representation of a designed part and the cutter path is essential.

Graphic display helps NC programmers identify and locate the error(s) in an NC program, thus it is necessary for an NCPV system. Wire-frame modeling can be used for verification of 2-D and 2½-D parts and cutter paths. For a general three-dimensional part, solid modeling is appropriate for correct representation of the in-process workpiece and cutter path defined by the NC program [8,11,12]. Such representation and NC program verification are based on the interaction of the volume swept by the cutter in space during machining and that of the stock material. An obvious way to verify an NC program is to test if the formed workpiece is the same as the designed part, i.e.,

$$P = M - (M \cap S)$$

where P --- the designed part;
M --- the stock material; and
S --- the volume swept by the cutter in space during machining.

As a matter of fact, the cutter motion is determined based on linear interpolation according to the specified tolerance (Fig. 2). This means that certain deviations are allowable. Unless a profile is composed of lines or planes only, the machined profile can never be the same as the designed one. In most cases, such a test cannot yield a positive result. Therefore, tolerances should also be considered in verification of NC programs. A test criterion can be utilized whereby the envelope surface of the cutter motion should be located between the surfaces of outward and inward tolerances. Since it is difficult to create the tolerance surface by offsetting a general surface, Wang and Wang used instead the method of offseting the cutter path [12]. An outward and an inward offset cutter paths, S^+ and S^-, for a cutter path S are then created (Fig. 4a and 4b). Their algorithm for verifying the cutter path S is as follows:

> The volumes, T^- and T^+, swept by the cutter for the motion S^- and S^+ should and should not interfere with the designed profile, respectively.

However, such a verification criterion is not valid when the cutter path consists of a series of linear interpolation motions which are generated by a computer-assisted NC programming system for a profiled surface (Fig. 4c). This suggests that the amount of shift, δ, should not be the same as the tolerance Δ_{out} or Δ_{in}, but be corrected on the basis of the curvature of the profile. Thus, the verification algorithm becomes more complicated.

On the other hand, according to Wang and Wang's algorithm, the direction of the offset from S to S^- or S^+ can be selected as that of the normal to the profile at the middle point of a motion step (Fig. 4c). The offsets at the two end points of a motion step S_a are not in the directions of the normals to the profile at those points. Therefore, the offset surfaces formed by offsetting the consecutive motion steps do not constitute a continuous surface. This indicates that verification of cutter path can only be carried out step by step by comparing S^+ and S^- with the section of the designed profile corresponding to the motion step S_a. Generally, this method is approximate and has inherent errors at both ends of a motion step. Thus a cutter motion vector which is considered by the verification system as out of tolerance might still be acceptable. Difficulties will also be experienced when it is desirable to verify a cutter path on the basis of Boolean operations relating the

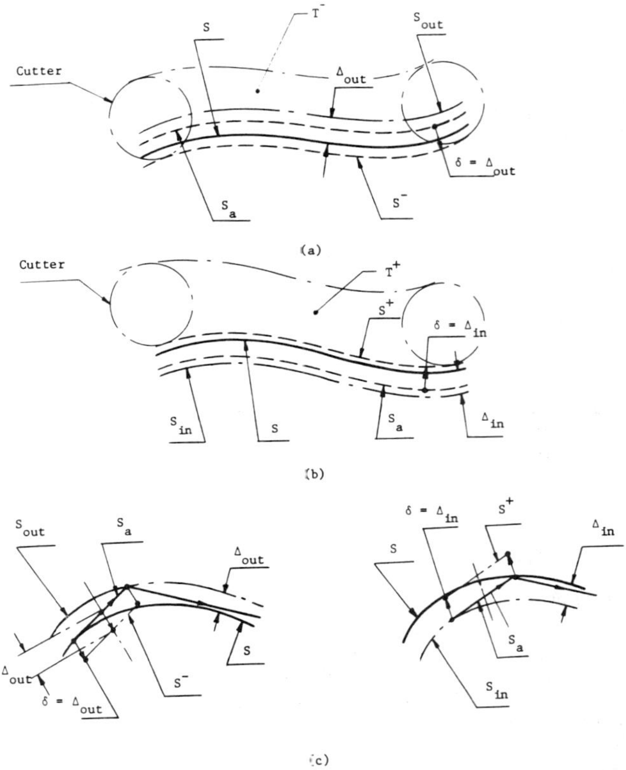

Figure 4. According to Wang and Wang's theory [12], a cutter path can be verified by utilizing the offset cutter paths S^+ and S^- which are the offsets from the actual cutter path S_a in the directions opposite to that of inward and outward tolerances, Δ_{in} and Δ_{out}, respectively. The volume swept by the tool moving along path S^- should interfere with the volume occupied by the designed profile S (a), whereas that swept by the tool moving along path S^+ should not (b). Such a verification criterion does not apply when the cutter path is composed of a series of linear interpolation motion instead of a curved one similar to the designed profile (c).

S_{out} -- the surface offset from the designed profile S by an outward tolerance Δ_{out};

S_{in} -- The surface offset from the designed profile S by an inward tolerance Δ_{in};

T^- and T^+ -- the volume swept by the tool moving along path S^- and S^+ respectively.

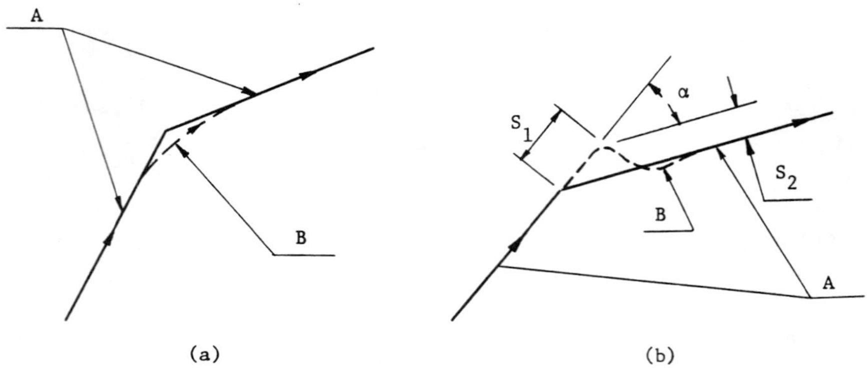

Figure 5. The deviation of the cutter path, B, from the programmed one, A, due to the dynamic response of the NC controller system with (a) or without (b) automatic acceleration and deceleration functions.

complete volume of the designed part to those swept by the consecutive cutter motion steps as a whole.

The current NCPV systems [11,12] verify a cutter path step by step. They compare a motion vector with the corresponding section of the designed profile. A manually programmed cutter path contains no or partial information regarding the designed profile. Thus, it is hardly possible for an NCPV system to identify the correspondence between a motion vector and the section of the profile machined in this motion step. Such correspondence must be provided by the NC programmer together with the designed profile in order for the NCPV system to verify the cutter path. For NC programs generated by computer-assisted NC programming systems, such as APT, the NCPV system can make use of the CLDATA or even the source APT program as input to determine the correspondence. A CAD/CAM system uses the CAD model of a designed part to generate the NC program. Theoretically, there should be no error in the definition of the cutter path provided that the CAD model, the specified machining specifications and process, and the algorithm for calculation of the cutter path are correct.

4.2.2 Consideration of factors other than part geometry in verification of cutter path. For high-accuracy parts, an NC cutter path programmed simply on the basis of part geometry might not yield a part to the specified dimensions and tolerances; factors that should be considered during its verification include dynamic response of the NC machine-controller system and the machine tool geometric errors, in addition to tool and part deflections. Their effect on the cutter path is discussed below.

Due to machining requirement, the NC statements in an NC program are uninterruptedly executed in sequence. For two NC statements defining, respectively, two consecutive linear interpolation motions, some deviation of the cutter path from the programmed one always exists in the vicinity of the end point of the first movement or the starting point of the second movement (Fig. 5). Such deviation is caused by the abrupt

Table 1 The deviation of cutter path in circular interpolation motion for the FANUC 6MB controller [1, 17] (see Fig. 6)

$$D = 1.5125 \bullet 10^{-7} \bullet (F^2/R) \quad \text{(mm)}$$

F (mm/min)	R (mm)		
	10	25	100
100	$0.02 \bullet 10^{-2}$	$0.01 \bullet 10^{-2}$	$0.002 \bullet 10^{-2}$
250	$0.09 \bullet 10^{-2}$	$0.04 \bullet 10^{-2}$	$0.01 \bullet 10^{-2}$
500	$0.38 \bullet 10^{-2}$	$0.15 \bullet 10^{-2}$	$0.04 \bullet 10^{-2}$
1000	$1.51 \bullet 10^{-2}$	$0.61 \bullet 10^{-2}$	$0.15 \bullet 10^{-2}$

change in direction of the cutter motion and is determined by the dynamic response of the NC control system, the angular change, α, of the motion vector, and the specified feedrate. For an NC controller with automatic acceleration and deceleration functions, the actual cutter path at a corner is a round-out of the designed cutter path (Fig. 5a) [1,17]. For an NC controller without automatic acceleration and deceleration functions, an overshoot results (Fig. 5b). The overshoot S_1 increases with the specified feedrate (e.g., S_1 can be 0.05 mm when the feedrate is 1000 mm/min [1]). The deviation S_2 due to overshoot S_1 is

$$S_2 = S_1 \bullet \sin \alpha$$

When machining a curved profile, the change in motion direction between two consecutive steps of the linear interpolation motion is small; thus the deviation S_2 is much less than the overshoot S_1 and is not significant. However, when two linear interpolation motion steps exhibit a large angular change in motion direction, such deviation should be considered in the definition of roughing and finishing cutter motion and in the selection of feedrate. Similarly, in circular interpolation motion, an error is generated in the radial direction due to the delay occuring between the input and output of the DC servo motors (Fig. 6) [17]. The radial errors for different feedrates and radii for the FANUC 6MB CNC controller system are listed in Table 1. The trend in modern

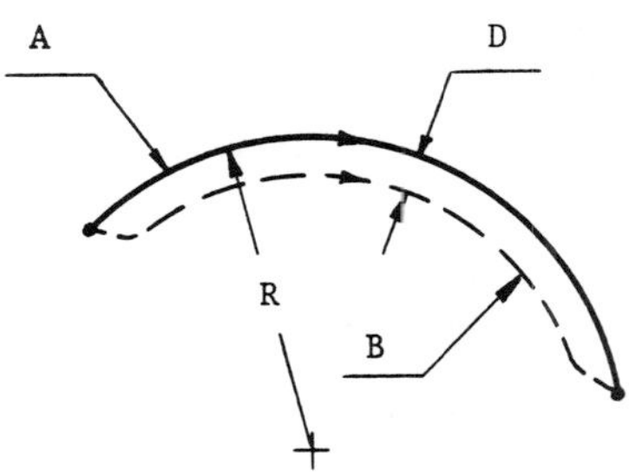

Figure 6. The deviation, D, of the cutter path, B, from the programmed cutter path, A, in circular interpolation motion due to delay of the DC motor response to the input.

machining technology is to use high material removal rates on CNC machines. Therefore, the deviation of the cutter position from the defined one due to dynamic response of the servo-control system should be considered in verification of NC programs.

The machining accuracy of an NC machine tool is more affected than that of conventional machine tools by such factors as geometric errors of the machine tools, weight deformation, clamping deformation, cutting force deformation, cutting tools, thermal effect, environment, etc. due to the automatic nature of its operation [13,14]. Although the effect of many of these factors can be reduced through proper design of machining process, tools, and clamping method, and through control of the environment, the geometric accuracy of an NC machine tool cannot be changed. High geometric accuracy of a machined part is especially difficult to attain when the part has to be machined in different spatial positions with respect to the machine table (usually this is realized through rotation of the rotary table or through repositioning the part on the machine table). The more motion axes a cutting operation includes, the more types of machine tool geometric errors effecting machining accuracy. For example, for a 4-axis (X, Y, Z, B) machining center with its tool axis (Z) in the horizontal direction, the tool position errors due to the geometric errors of the machine tool are [14]

$$\delta_x = \delta_x^{(1)} + \delta_x^{(2)} + \delta_x^{(3)} + \delta_x^{(5)} + \delta_x^{(6)}$$
$$\delta_y = \delta_y^{(1)} + \delta_y^{(2)} + \delta_y^{(3)} + \delta_y^{(4)} + \delta_y^{(6)}$$
$$\delta_z = \delta_z^{(1)} + \delta_z^{(2)} + \delta_z^{(3)} + \delta_z^{(5)} + \delta_z^{(6)}$$

where $\delta_x^{(1)}, \delta_y^{(1)}, \delta_z^{(1)}$ --- positioning errors along the X, Y and Z axes;

$\delta_x^{(2)}, \delta_y^{(2)}, \delta_z^{(2)}$ --- the position errors in X, Y and Z directions resulting from errors in perpendicularity of the X, Y and Z axes with respect to one another;

$\delta_x^{(3)}, \delta_y^{(3)}, \delta_z^{(3)}$ --- the position errors in X, Y and Z directions resulting from errors in parallelism of the rotary table top plane with respect to the X-Z plane (Note that the rotary table axis B is in the Y direction);

$\delta_y^{(4)}$ --- the position error due to the error in perpendicularity of the rotary table axis B with respect to the X-Z plane;

$\delta_x^{(5)}, \delta_z^{(5)}$ --- the position errors due to the error in angular positioning of the rotary table about the B axis;

$\delta_x^{(6)}, \delta_y^{(6)}, \delta_z^{(6)}$ --- the position errors resulting from the roll, yaw and pitch due to the coordinate motion in the X, Y and Z directions respectively

if the errors in spindle rotations are neglected. For a lathe, which has only X- and Z-axis motions, the tool position errors are

$$\delta_x = \delta_x^{(1)} + \delta_x^{(2)} + \delta_x^{(6)}$$

$$\delta_z = \delta_z^{(1)} + \delta_z^{(2)} + \delta_z^{(6)}$$

The calculation and measurement of these errors are discussed in reference [14]. The difficulty in the determination of cutter-position error is related to the calculation of type 1 and type 6 errors, $\delta^{(1)}$ and $\delta^{(6)}$. The positioning errors $\delta_x^{(1)}$, $\delta_y^{(1)}$, and $\delta_z^{(1)}$ are determined by the accuracy of servo-control systems and are also functions of the tool position with respect to the machine. They are not fixed but rather random variables. The means of these errors can usually be utilized in calculations. The errors $\delta_x^{(6)}$, $\delta_y^{(6)}$, and $\delta_z^{(6)}$ arise from the deviation in form of the slideways and vary with the carriage or slide position on the slideway, and cannot be cauculated on the basis of a few parameters or measurements. Thus, a practical approach to the estimation of these errors is to create a database or library that stores the error map or matrix for each NC machine. Such a database can be linked to the NCPV system for estimation of the tool position error resulting from the machine tool geometric errors.

Verification of a cutter path by taking machine tool errors and the dynamic response of the control system also into consideration results in longer processing times. As a matter of fact, these factors are of importance only to the sections of part profile with tight tolerance requirements. Thus, the processing time can be significantly reduced by limiting such calculations to those sections only.

4.3 Verification of Machining Process and Specifications

An NC program is a step-by-step description of the NC machining process in machine-readable language (NC codes). Programming of an applicable NC program includes definition of correct machining process and selection of proper tools, fixtures and machining specifications, in addition to the definition of a correct cutter path and positions. Failing to specify correctly any of these factors results in inferior quality of the machined part and in rejection of the NC program. This indicates that factors other than the cutter position and path are of equal importance for verification of NC programs although the majority of statements in an NC program are specified for the purpose of defining the cutter path and motion.

The machining process defined in an NC program should consist of machining operations arranged in a rational sequence. Each machining operation should be realizable after finishing the preceding operations and is necessary for carrying out the next and following operation(s). In order to reduce thermal effect and the effect of internal stress and the deflection due to cutting forces on machining accuracy, finishing operations should normally be carried out after all the major roughing operations have been finished. Fixtures and their locations should be properly selected in order to minimize the clamping deflection. Process planning and design also play a very important role in attaining the desired machining accuracy when a machining process involves utilization of several part surfaces as reference surfaces to define cutter positions. In such a case, the machining process should be carefully planned based on the datums, datum features and the specified orientation tolerances of the part. It should be pointed out that machining-process planning is a process consisting of a series of engineering selections, compromises and decisions. The rule of process planning has not been well defined since too many factors are involved. It seems that a process-verification routine in an NCPV system should be designed based on the existing knowledge and engineering expertise achieved in a company. An NCPV system can also take advantage of a CAPP system, if available, to realize the process verification.

The machining specifications that should be verified include the feedrate(s), feed(s), depth(s) of cut, cutting speed(s), coolant, method of cutting (e.g., up- or down-milling), the type(s) and dimensions of the cutting tool(s), etc. They all affect the surface finish and dimensional and form accuracy of the part. For a specific part material, these parameters can be selected according to a machining data handbook. The detection of interference of a cutting tool with stock material during rapid motion is very important for avoidance of collisions even though the cutter path specified in an NC program is geometrically correct.

The amount of tool wear is also a factor that effects the accuracy and surface finish of the machined part when a single tool is used for machining a large area.

5. CONCLUSIONS

The purpose of an NCPV system is to provide error-free NC programs which can be used as immediate input to NC machines to fabricate the designed parts. Therefore, the function of an NCPV system should not be limited to verification of cutter geometric positions and path only. Factors that affects executability of an NC program, the accuracy of the machined part, and the rationality of the machining process should also be examined by the NCPV system.

An NCPV system should be capable of verifying NC programs with minimum possible human judgement and intervention or, preferably, without relying upon them at all. It should perform three basic functions, namely, syntax, cutter path and machining process verification. Its design is determined by the NC programming system(s) used and the application environment.

For manual NC programming, syntax verification should be one of the basic functions of the NCPV system; but such function is usually not necessary for NC programs generated through computer-assisted NC programming systems and CAD/CAM systems since the NC program syntax has already been verified by the postprocessors.

Verification of a cutter path can be carried out based simply on the relationship between the geometric positions of the tool and the designed profile when the requirement on part accuracy is less strict and the programmed feedrate is low. Additional factors such as the dynamic response of the control system, machine tool geometric errors, and tool deflection should be taken into consideration for parts with tight tolerances and for machining at high material removal rate. CAD techniques and systems can be utilized to simulate and display the cutter path and the designed part as well. Solid modeling techniques is necessary for verifying 3-D and multiaxis profile machining, whereas the wire-frame model enhanced with surface representation can be used to verify 2-D and 2½-D cutter path.

The machining process and the sequence of operations specified in an NC program should also be verified as they are among the factors that cause rejection of an NC program. Automatic verification of process plans can be realized by linking the NCPV system with a CAPP system if available. Otherwise, it can be carried out through interaction between a human operator and the NCPV system. In such case, human judgement and intervention based on graphic display of the cutter motion and machining operations are necessary.

Errors in syntax, cutter positions and machining process should be detected automatically by an NCPV system which gives messages indicating the types of errors and the statements where the errors are located.

The trend in NC programming is to realize automatic generation of error-free NC programs directly from the CAD models of designed parts. Research in this area concentrates on automatic generation of the machining plan and cutter path based on the geometric features of a given part and the given stock material. An NC Program can be error free only when the factors that cause errors have been eliminated at the time it is generated. This suggests that the design of such a programming system inevitably involves application of the basic principles of NC program verification, thereby eliminating the need for postgeneration verification. The effectiveness of computer-aided NC programming systems can also be improved by applying these verification principles in the design of cutter-position calculation routines and postprocessors. Thus, the NCPV system should not be designed as a separate tool, but its functions should rather be incorporated into a total NC generation-verification system.

6. ACKNOWLEDGEMENTS

The author would like to thank Professor Michel A. Melkanoff for his review and comments.

REFERENCES

1. Chang, Chao-Hwa & Melkanoff, M. A. , NC Machine Programming and Software Design, to be published by Prentice-Hall, Inc.
2. Rolstadas, A., "Verification of Part Programs in a DNC System", The International Future of NC/CAM, Proc. 11th Annual Meeting and Technical Conf., ed. by M. A. DeVries, NC Society, Springlake, NJ, 1974, p. 166.
3. Schreiter, H., "Computer-Aided Preparation and Checking of Control Tapes for NCM", Advances in Computer-Aided Manufacture, D. McPherson ed., North-Holland Publishing So., Amstadam, 1977, p. 31.
4. Laffret, D., et al, "ELAN --- A Conversational Processor for Generation, Checking and Correction of NC tapes", Advances in Computer-Aided Manufacture, D. McPherson ed., North-Holland Pub. Co., Amstadam, 1977. p. 43.
5. Drewry, R. G., "NC Graphics for Programming Debugging and Tape

Verification", NC --- from the Users' Point of View, NC Society, Glenview, Illinois, 1977, p. 37.
6. McGoldrick, P. F., et al, "A Computer Technique for the Pictorial Representation of the Cutting Action of Three Axis Numerically Controlled Milling Machines", Proc. of CAD '80, The 4th Intl. Conf. on Computers in Design Engrg., IPC Science and Technology Press, Ltd., Guildford, Surrey, U.K., 1980, p. 650.
7. McGoldrick, P. F., et al, "NECTAR --- A Graphical Technique to Represent the Component Produced from the Cutter path of a Numerically Controlled Milling Machine", Annals of the CIRP, V. 33/1/1984, p. 225.
8. Wallis, A. F. and Woodwark, J. R., "Creating Large Solid Models for NC Tool Path Verification", Proc. of CAD '84, Int.. Conf. and Exhibition on Computers in Design Engrg., IPC Science & Technology Press, Ltd., Guildford, Surrey, U.K., 1984, pp. 455-9.
9. Ruberl, S. T., "Numerical Control Part Program Verification by Computer Simulation" (source unknown)
10. Fridshal, R., et al, "Numerical Control Part Program Verification System", Proc. of NC Society, 19th Annual Meeting and Tech. Conf., 1982, pp. 236-251.
11. Sungurtekin, V. A. and Voelcker, H. B., "Graphic Simulation and Automatic Verification of NC Machining Programs", Proc. of 1986 IEEE Intl. Conf. on Robotics and Automation, Vol. 1, p. 156.
12. Wang, W. P. and Wang, K. K., "Real-Time Verification of Multiaxis NC Programs with Raster Graphics", Proc. of 1986 IEEE Intl. Conf. on Robotics and Automation, Vol. 1, pp. 166-177.
13. Tlusty, J., "Testing and Evaluating the Accuracy of Numerically Controlled Machine Tools", 1981 Mfg. Engrg. Trans., 9th North Amer. Mfg. Research Conf., Proc., Soc. of Mfg. Engrs., 1981, pp. 121-136.
14. Chang, Chao-Hwa, "Automatic Monitoring NC Machine Tool Errors with Electronic Surface-Sensing Probe", Procceedings of Manufacturing International 1988, Vol.1, Symposium on Product and Process Design, ed. by G. Chryssolouris and R. Komanduri, ASME, New York, pp. 333-344.
15. Shpitalni, M. and Lemaistre, C., "The Problem of NC Program Verification --- Analysis and Approaches", Computer-Integrated Manufacturing, M. Martinez ed., ASME, New York, NY, 1983, pp. 82-88.
16. MAZAK CAM-CNC System --- MAZATROL CAM M-2 (catalogue), Yamazaki Machinery Works, Ltd., Japan, 1983.
17. FANUC System 6M-Model B -- Operator's Manual, FANUC Ltd., Japan.

AN AUTOMATIC POSITION DETECTOR FOR HANDHELD NDE PROBES

R. C. Savedra, P. H. Li, E. C. Johnson, and G. F. Hawkins
The Aerospace Corporation
Los Angeles, CA 90009

INTRODUCTION

The use of handheld NDE probes is prevalent in the aerospace community. This is especially true where data must be taken in the field, the part exhibits a complicated geometry, or when there is insufficient time to build an automated scanning apparatus. Data taken by hand is less reliable than that taken by automatic methods. There is no assurance that the entire part has been covered, since a hardcopy image of the part is not produced. The use of handheld NDE probes could be significantly facilitated by a versatile device designed to remotely sense the spatial position of the probe each time a data point is acquired. Such a device, which will be referred to as an Automatic Position Locator (APL), is currently under development and will be described in this paper.

THEORY OF OPERATION

The components which comprise the APL are depicted in Fig. 1. An ultrasonic transducer is mounted on the back side of the stipulated NDE probe. This transducer

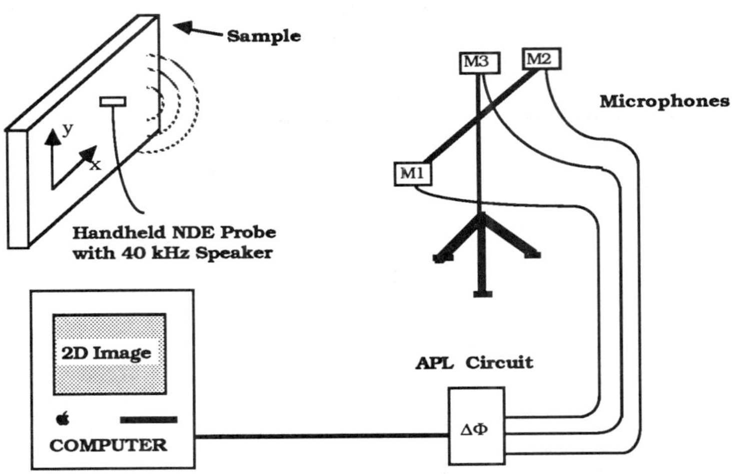

FIGURE 1. The components which comprise the automatic position locator.

serves to propagate 40 KHz sound waves in the direction opposite that of the surface being scanned. This acoustic signal is then detected by three microphones (labeled M1, M2 and M3) which are attached to a stand situated several meters away from the probe. The geometry associated with M1 and M2 is depicted in Fig. 2. The microphones are separated by a line of length 2L which runs parallel to the sample surface at a distance R. The origin is defined by the intersection of the sample surface and the perpendicular bisector of the line between M1 and M2 in the plane of this figure. The probe is located at x. The difference in phase between the signal detected by M1 and that detected by M2 is given by the expression:

$$\Delta\Phi_{12} = \frac{2\pi}{\lambda}\left\{\sqrt{R^2 + (L+x)^2} - \sqrt{R^2 + (L-x)^2}\right\} \tag{1}$$

where λ is the 40 KHz wavelength in air ($\approx$ 0.83 cm). If one arranges the device so that $x \ll L$, Eqn. 1 reduces to,

$$\Delta\Phi_{12} \approx \frac{4\pi x}{\lambda\sqrt{(R/L)^2 + 1}} \tag{2}$$

where $\Delta\Phi_{12}$ is directly proportional to x. It is important to note that the wavelength is such that $\Delta\Phi_{12}$ may include multiple cycles of the ultrasound. Similar expressions relate the phase difference between M1 and M3 to the y coordinate of the probe. The phases of the signal detected at each microphone can therefore be compared to determine the position of the probe relative to a fixed point on the surface. The heart of the APL is a circuit which performs this comparison and digitizes the result for input into a Macintosh II computer. This circuit will be discussed below. Software has been developed which permits one to display a grid overlaying a portion of the part being scanned. As one moves the probe, the computer will fill in each section of the grid with a color indicative of the amplitude of the associated NDE signal, resulting in a two dimensional image which can be inspected immediately and then stored for future use.

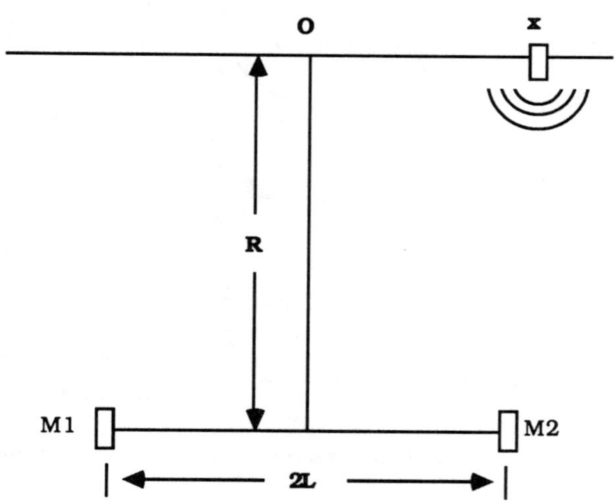

FIGURE 2. A top view of the geometry associated with M1 and M2.

CIRCUIT DESCRIPTION

A block diagram of the APL circuit is depicted in Fig. 3. The circuit contains a 40 KHz driver which outputs to the transducer on the handheld NDE probe and three receivers which take their input from M1, M2 and M3 (see Fig. 1). The received signals pass through a 60 dB amplification stage, a 40 KHz bandpass filter, and then a Phase Locked Loop (PLL) for final noise suppression. The synthesized 40 KHz square wave PLL outputs from M1 and M2 are then fed into the inputs of an up/down counter. These signals differ in phase because the positions of M1 and M2 are different with respect to the transducer. The counter outputs binary coded decimal (BCD) information which indicates the phase difference between the two signals. This difference may consist of more than one complete cycle. The BCD signal is then channeled into the D/A converter where each complete cycle adds a 10 mV step to the output. When the difference is not an integral number of complete cycles one of the BCD bits flickers resulting in a 40 KHz pulsed output as depicted in Fig. 4a. The pulse width is proportional to the phase difference (see Figs. 4a - 4c). Consequently, when this pulsed output is passed through a low pass filter, a DC signal which serves to interpolate inbetween the 10 mV steps is achieved. This signal is then sent to a scaling amplifier and digitized for input into a Macintosh II computer which calculates the x position of the probe. The y position of the probe is determined in an identical fashion through comparison of the signals detected by M1 and M3.

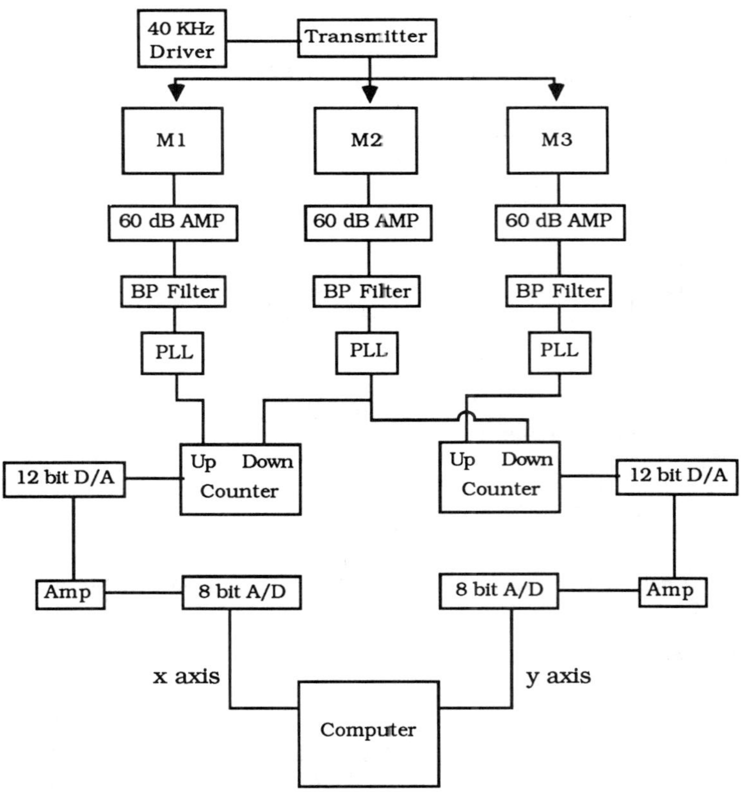

FIGURE 3. APL circuit block diagram.

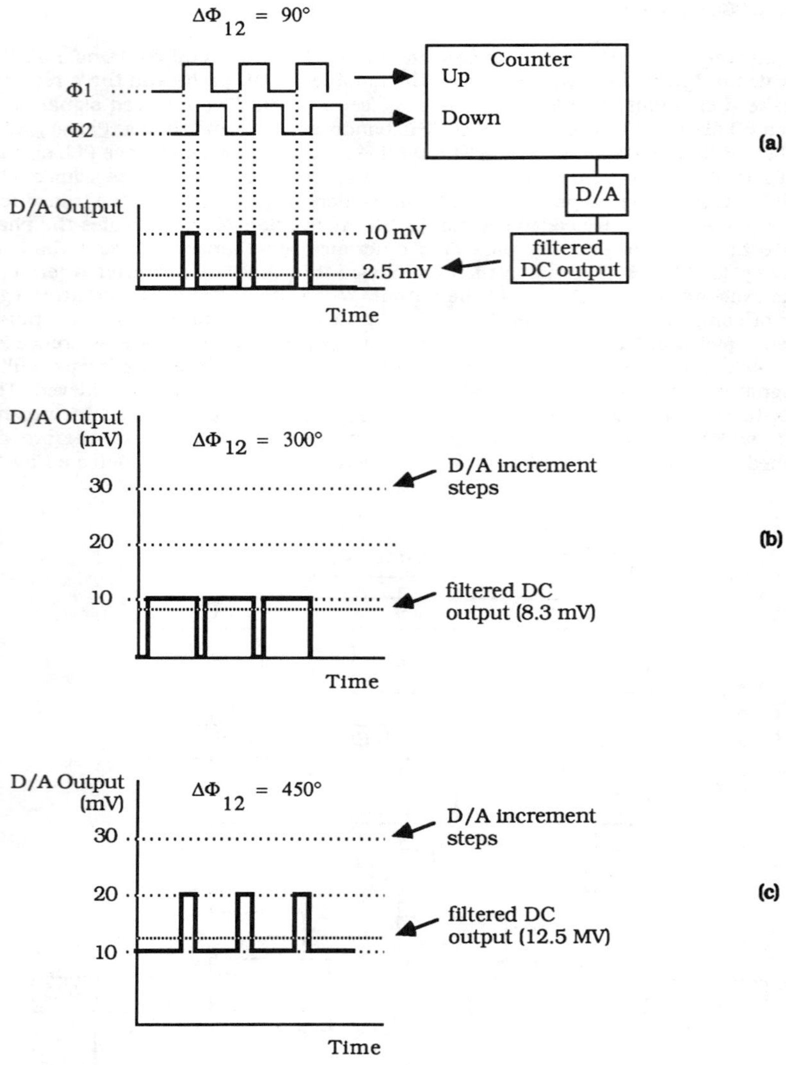

FIGURE 4. Changes which occur in the D/A output as $\Delta\Phi_{12}$ is varied from 90° to 450°.

SAMPLE MEASUREMENTS

The microphones and transmitter were arranged according to the geometry depicted in Fig. 2, with 2L = R = 30 cm. The APL output voltage was then measured as a function of x, the transmitter distance from the origin. The results are plotted as circles in Fig. 5. As predicted by Eqn. 2, the curve is linear for $x << L$. Deviations from linearity in this region are such that one would expect the APL to exhibit a spatial resolution of 1 mm or less. It might be noted on the basis of Eqn. 2, that as long as R and L are increased pro-

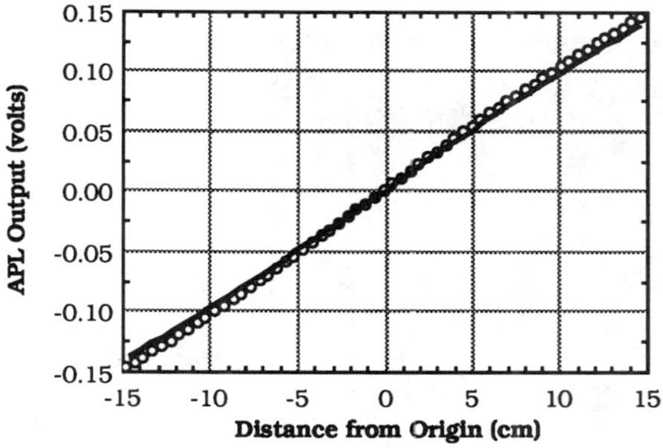

FIGURE 5. APL output as a function of x. The coordinate system is defined in Fig. 2. Experimental data are represented by open circles. The solid line is a theoretical curve.

portionately, this resolution should be maintained until R becomes too large for accurate detection of the 40 KHz signal. Successful tests were performed in the lab for values of R as large as 4.5 m. For comparison, a theoretical plot of $\Delta\Phi_{12}$ (given by Eqn. 1 multiplied by an appropriate amplification factor) is also included in this figure (solid line). Differences between the theoretical and experimental curves could possibly be traced to peculiarities in the directional response of the microphone/speakers used.

In order to test the utility of the APL for an actual scan, a test sample was prepared consisting of a 0.5 inch thick brass plate containing three blind holes of diameters 0.125, 0.250 and 0.375 inches respectively. In addition, a 0.375 inch end mill was used to cut a blind rectangular channel along the border of the plate. A 20 MHz ultrasonic transducer was then used in conjunction with the APL to scan the face of the plate opposite the holes and the rectangular channel. The results of this scan are depicted in Fig. 6.

DISCUSSION

One condition for optimum performance of the APL is that the medium (the air which serves to transmit the sound) between the transmitter and each of the three microphones should be uniform. Variations within the medium due to factors such as wind disturbance or thermal gradients could result in sound velocity changes leading to improper phase measurements. Tests performed in the lab indicate that the effects of thermal gradients are much more severe than those of wind currents. Such a thermal gradient might exist, for example, between the top and bottom of an enclosed room. In such a case, however, the microphones could be positioned appropriately to avoid the problem. The APL is also adversely affected by sources of stray 40 KHz sound. Such noise was found to accompany the jingling of a set of door keys on a ring. Lastly, it might be noted that the APL keeps track of the transmitter's location with respect to a reference position on the part being scanned. Any interruption of the sound path between the transmitter and the microphones (by an operator's hand for instance) could result in a reference loss. The APL must then be referenced again before the scan can continue.

The basic operating principle by which the APL detects positional changes is similar to

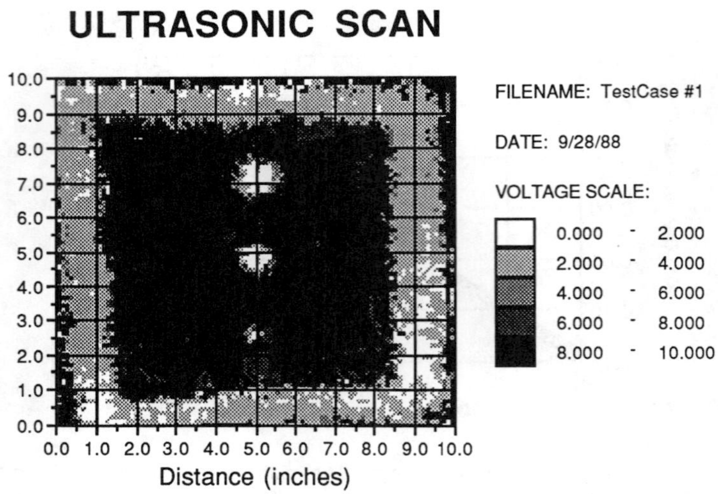

FIGURE 6. Computer Image resulting from an ultrasonic scan of the test sample described in the test.

that of a device called the Headmaster, which is currently marketed by The Personics Corporation.[1] The Headmaster is a device designed to interface with the mouse or joystick port of a computer in order to control the position of the cursor. The apparatus includes a headset which is worn by the computer operator and an ultrasonic transmitter which is set on top of the computer display. The transmitter propagates sound which is picked up by three microphones in the headset. The sound wave are compared for phase changes, and a circuit converts the phase changes into position data which is used by the computer to control the cursor.

Another device, called the Model GP-8-3D Sonic Digitizer and marketed by SAC[2] was developed for the same purpose as the APL . This device differs distinctively from the APL in that Time Of Flight (TOF) measurements of sonic pulses are used to determine the position of a transmitter on a handheld probe. The CW phase comparison technique employed in the APL as opposed to that of the TOF measurements offers several advantages. First, in order to produce the necessary TOF sonic pulses, the Model GP-8-3D Sonic Digitizer employs a stylus which emits a spark. Such a spark would preclude the use of this device for some applications such as the scanning of solid rocket motors. Second, the rate of data acquisition is limited by the TOF of the pulse whereas in principle, the APL could accumulate data at a 40 KHz rate. Thirdly, because the APL relies upon relative phase changes to compute positional information, it may be less sensitive to some thermal gradients. The Model GP-8-3D Sonic Digitizer utilizes four microphones to image in three dimensions. In like fashion, a fourth microphone could be added to the APL were a 3-dimensional capability required.

In conclusion, the APL is still in its development stage. On the basis of the data presented here, however, it appears quite likely that this device will prove to be quite useful for the acquisition of NDE data through the use of handheld probes.

ACKNOWLEDGEMENTS

The authors would like to thank Robert L. Corbin for the insight he contributed toward the development of the electronics used in this project.

REFERENCES

1. Personics Corporation, Concord, Massachusetts (Headmaster patent no. 4682159).

2. Science Accessories Corporation, Southport, Connecticut (GP-8-3D 3-Dimensional Sonic Digitizer patents no. 3626483 and no. 3821719).

ADVANCED ELECTRON PARAMAGNETIC RESONANCE STECTROSCOPY FOR NONDESTRUCTIVE TESTING AND EVALUATION OF ENGINEERING MATERIALS

G. Kordas
University of Illinois
Urbana, IL 61801

ABSTRACT

The fundamentals and applications of Electron Paramagnetic Resonance (EPR) spectroscopy are introduced to the scientists concerned with the non-destructive testing and evaluation of engineering materials. The principles behind Electron Nuclear Double Resonance (ENDOR), Electron Spin Echo (ESE), and Electron Paramagnetic Resonance Imaging (EPRI) techniques are also explained. The ESE and EPRI techniques have emerged in the last few years with the advent of very high speed digital electronics under computer control. Unlike the EPR and ENDOR instrumentation that has been in the market for nearly 20 years, the ESE and EPRI spectroscopies are not yet commercially available techniques and very few laboratories can perform these experiments, at present.

1. INTRODUCTION

The electron paramagnetic resonance (EPR) phenomenon was successfully observed in 1946. Since that time, the EPR spectroscopy has been used in the determination of the structure of paramagnetic centers (transition metal ions and defects) in solids. Structures of paramagnetic centers in crystalline, polycrystalline, and amorphous materials have been discovered and characterized, often with a precision and sensitivity unequalled by other spectroscopic techniques. EPR spectroscopy has added substantially to our understanding of the:

1) atomic structures,
2) formation and disappearance of defects, and
3) interactions of paramagnetic centers.

Recent developments in instrumentation and theory have made possible powerful extensions of the basic EPR spectroscopy, greatly enhancing its resolution and sensitivity to:

1) atomic arrangements,
2) bond angles,
3) structure of interfaces,
4) structure of pores in solids,
5) structure of composite materials, and
3) time-dependent phenomena such as motion of ions in solids.

These more powerful methods open up new opportunities for the precise determination of the structures of paramagnetic center in new advanced materials, where conventional EPR spectroscopy suffers from limited energy and resolution.

The present review article introduces the classical X-band (9.5 GHz) EPR and ENDOR spectroscopy and the new EPR techniques including ESE and EPRI spectroscopy to the materials scientists concerned with the non-destructive testing of advanced materials. Examples selected for description of these spectroscopies include materials in which these techniques have been successfully employed.

2. ELECTRON PARAMAGNETIC RESONANCE PHENOMENON

An electron is characterized by its mass, electric charge, and spin. The electron spin induces a magnetic moment μ. When an electron is placed into a magnetic field, Bext, an interaction between this magnetic dipole moment and an external magnetic field, Bext, occurs. This interaction splits the original degenerated state into two states. The energetic separation between these two states is proportional to the external magnetic field, Bext:

$$\Delta E = g\mu_\beta B_{ext}$$

where ΔE is the energy difference between the low and the upper state, g is the Lande factor, and μ_β is the Bohr's magneton.

An electron localized at the low energetic state can absorb electromagnetic energy, $h\upsilon$, and be excited to the upper state if the condition:

$$h\upsilon_{res} = g\mu_\beta B_{ext} = g\mu_\beta B_{res}$$

is fulfilled. These excitations can be realized by supplying microwave energy to the system. The electromagnetic energy, $h\upsilon$, is generated by a klystron and is conducted by microwave guides to the cavity. The sample is placed inside the cavity which is between the poles of a magnet. The magnetic field is homogeneous throughout the sample and increases linearly with the time of measurements. When the resonance condition is obtained, an absorption of the microwave energy occurs. The first derivative of the EPR absorption signal is usually recorded due to experimental reasons.

3. CONTINUOUS WAVE EPR SPECTROSCOPY

The EPR spectrum of a paramagnetic center or defect is described in terms of a spin Hamiltonian:

$$H = \sum_i bS_i \bullet \mathbf{g} \bullet \mathbf{B}(t) - \sum_k b_n I_k \bullet g_{nk} \bullet \mathbf{B}(t) + \sum_i \sum_k S_i \bullet \mathbf{A}_{ik} \bullet \mathbf{I}_k + H_{ss} + H_{S1} + H_{i1} + H_1$$

The first term describes the interaction of the unpaired electronic spins **{S}** with the dc and high frequency magnetic field **{B(t)}**, the second describes a similar interaction between nuclei with non-zero spin, $\{\mathbf{I}_k\}$, and the external field, $\mathbf{B}_{ext}$, and the third accounts for the electron-nuclear hyperfine interactions. Electron spins interactions $(\mathbf{H}_{ss})$, electron-lattice interactions **(Hil)**, and lattice energy $(\mathbf{H}_l)$ are assumed to be small or constant contributions to the spin hamiltonian which will be neglected in this elementary review.

The spin Hamiltonian is important because it facilitates the interpretation of spectroscopic data in terms of atomic structure. This can be illustrated by using the spectrum of a γ-ray irradiated silica glass [1].

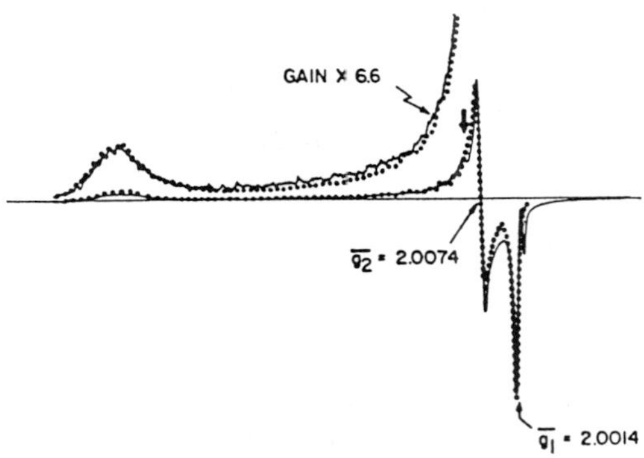

Figure 1. Experimental (__) and theoretical (.....) EPR spectrum of a γ-ray irradiated silica glass [1].

Figure 1 shows the experimental EPR spectrum together with a simulated spectrum. The simulation gives the g-matrix and the line width. The simulation technique involves a sophisticated algorithm that requires careful analysis and development. Such a spectrum is probably characterized by a g-matrix with g_x, g_y, g_z, and hence can be simulated under an appropriate theoretical model for powder EPR spectra and a correct choice of the three principal g-values. The dotted line in this figure shows the simulated spectrum for g-values of $g_1 = 2.0014$, $g_2 = 2.0074$, and a distribution centered at about $g_3 = 2.068$.

Knowing the g-values of the paramagnetic state, it is now possible to develop models for the paramagnetic centers. In this case, the model that successfully accounted for the observed g-matrix was a peroxy radical. Figure 2 shows the molecular orbitals of the superoxide ions. The peroxy radical ROO· is approximated theoretically as a perturbed superoxide anion radical, O_2^-, first described by Kanzig and Cohen [2] for the species defected in alkali halides.

Figure 2. Molecular orbitals of the superoxide ions.

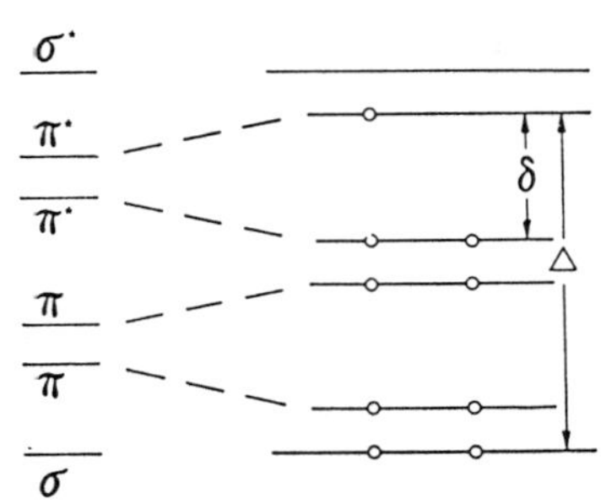

The g-matrix for the radical can be related by a quantum-mechanical model to its molecular orbital structure, resulting in predicated g-values, correct to the second order of perturbation, of:

$$g_{zz} = 2.0023 + 2\left(\frac{\lambda^2}{\lambda^2 + \delta^2}\right)^{1/2} \xi$$

$$g_{yy} = 2.0023\left(\frac{\delta^2}{\lambda^2 + \delta^2}\right)^{1/2} - \frac{\lambda}{\Delta}\left\{\left(\frac{\lambda^2}{(\lambda^2 + \delta^2)}\right)^{1/2} - \delta/(\lambda^2 + \delta^2)^{1/2} - 1\right\}$$

$$g_{xx} = 2.0023\left(\frac{\delta^2}{\lambda^2+\delta^2}\right)^{1/2} + \frac{\lambda}{\Delta}\left\{\left(\frac{\lambda^2}{\lambda^2 + \delta^2}\right)^{1/2} + \delta/(\lambda^2 + \delta^2)^{1/2} - 1\right\}$$

where, x is taken along the interatomic O-O line of centers, the term l is the oxygen spin-orbit coupling constant, ξ is a correction to angular momentum about the molecular z-axis, and δ and Δ are energies separating the antibonding π^* and the $\pi_y^* - \sigma^2$ orbitals respectively.

These energies are sensitive to the local environment of the radical [3]. For example, three superoxide ions detected in the pores of the sol-gel glasses exhibited three different δ and Δ parameters. A splitting, Δ, of about 0.2 eV generates g-values for the first center. For a moderate splitting of 0.6 eV, the g-values for the second defect can be obtained, while a strong splitting of 1 eV creates the g-values of the third defect.

This case shows that the x-band EPR spectroscopy is very sensitive to the local environment of the paramagnetic state and yields precise information concerning the structure and local environment of the defect.

The steps for the structural analysis include:

1) measurement of the spectrum,
2) simulation of the spectrum,
3) determination of g and A matrices and comparison with matrix values theoretically calculated for paramagnetic states.

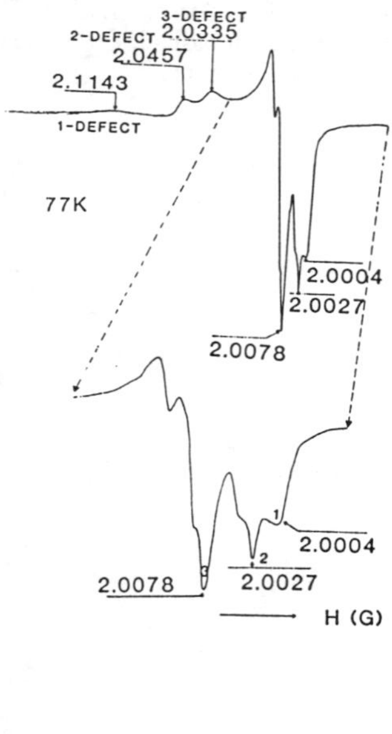

Figure 3. EPR Spectrum of irradiated sol-gel glasses.

4. ELECTRON NUCLEAR DOUBLE RESONANCE (ENDOR) SPECTROSCOPY.

An excellent introduction of ENDOR is given in the book of Kevan and Kispert [4]. ENDOR is often characterized as an NMR experiment with EPR detection, since the spectrum obtained is related to nuclear rather than electron magnetic dipole transitions.

Figure 4. Spin energy level diagram of the I=1/2 and S=1/2 system for $T_{1e}<T_{x1}<T_{1n}$, A>0, $g_n>0$, and $|A|/2>\upsilon_n$.

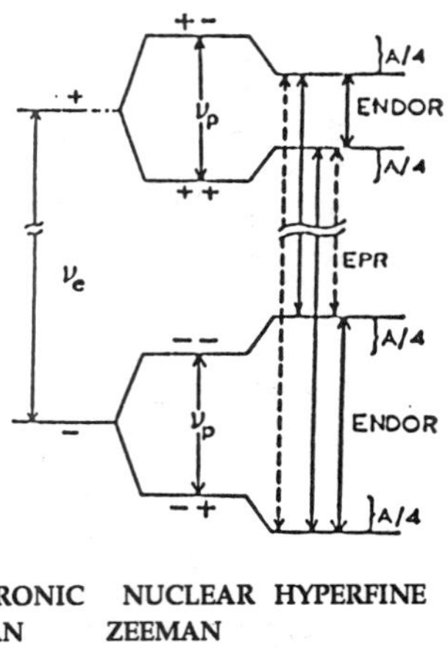

In case of an ENDOR experiment, both nuclear and electron spins are excited by subjecting the sample to both radio frequency and microwave radiation. This can be explained by using the simple spin energy level diagram of figure 4, which is the case for I = 1/2 and S = 1/2. First, microwave radiation is "pumped" to the electron spin system with a frequency appropriate to satisfy the transition from |-,+> to |+,+> and with a microwave power sufficient to saturate the transition. Second, the |+,-> to |+,+> nuclear dipole transition is excited by simultaneous radio frequency radiation while pumping the |-,+> to |+,+> transition. In this case, the rf power is high enough to nearly equalize the populations in the |+,-> and |+,+> states. At this stage, rapid relaxation from |+,-> to |-,+> via T_{1e} and |+,-> to |-,+> via T_{x1} will relieve the electron spin saturation and the EPR signal will grow in intensity.

As an example, we consider perylene radicals formed on the alumina surface exposed to 0.002 M perylene in benzene [5]. Figure 5 shows the experimental and theoretical ENDOR spectrum for the perylene radical. The theoretical calculation of the experimental spectrum allows the evaluation of the A-matrix from which the structure of the paramagnetic species can be deduced.

To conclude this discussion, ENDOR improves the resolution of inhomogeneously broadened lines in materials and provides information on the anisotropic as well as isotropic part of hyperfine interaction. This technique is very sensitive to very weak long range interaction (2-6 Å) allowing the paramagnetic state to act as sensor of their environment.

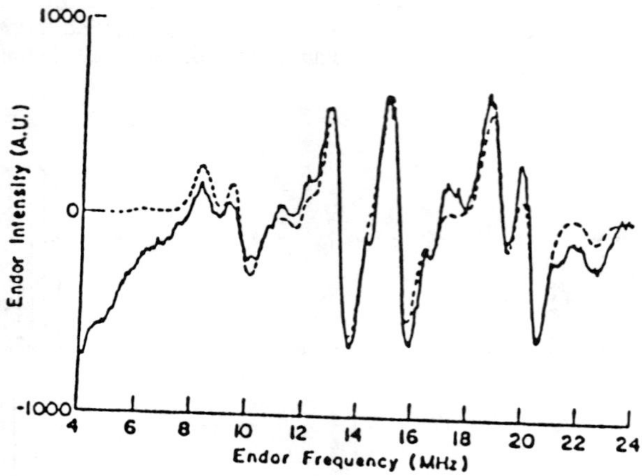

Figure 5. Experimental (-) and simulated (---) ENDOR spectra of perylene on silica-alumina perylene powder [5].

5. ELECTRON SPIN ECHO SPECTROSCOPY.

Unlike the continuous wave X-band EPR spectroscopy, ESE spectroscopy uses very short microwave pulses to excite the paramagnetic spin system and monitors the time evolution of the magnetization. The excitation of a spin system occurs by two resonant radio frequency pulses of definite durations separated by a time interval τ. An echo signal arises at a time 2τ after the first pulse. The ESE signal amplitude depends on the efficiency of spin relaxation. Figure 6 shows a typical Hahn or Two-pulse sequence diagram. The amplitude of the echo induced by the sequence is measured as a function of the time t between the first and second pulses.

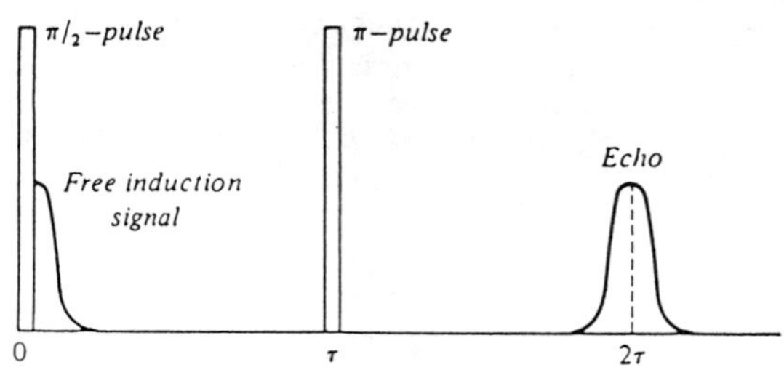

Figure 6. Two-pulse sequence echo experiment.

The variation of the echo amplitude V with the delay time t is given by the simple exponential function:

$$V_{echo}(2t) = V_o \exp[-(2t)/T_m]$$

where T_m is the phase memory time of the individual spin packets. This equation is valid only in the absence of any interactions between the unpaired electrons being observed and neighboring nuclei with non-zero magnetic moments. The exponential decay curve describing the echo amplitude as a function of t is called the electron spin echo envelope. ESE spectroscopy has many advantages over the more conventional cw EPR spectroscopy. For example, it reveals more distant interactions and therefore is sensitive to longer-range structure than either EPR or ENDOR. By observing the modulation due to anisotropic hyperfine interactions in the electron spin echo decay envelope (an effect known as ESEEM), one can get information on weakly coupled nuclei. Some features that make ESE particularly attractive for the nondestructive testing scientist are:

1) ESE improves resolution for inhomogeneously broadened lines by a factor $\approx\mu_e/\mu_n$,
2) The sensitivity of ESE is not dependent on the nuclear gyromagnetic ratio, and hence it is equally sensitive to all nuclei, and
3) ESE allows the direct measurement of the dynamical spin variables.

ESE can determine the structure and internal chemistry of the pores by detecting molecular oxygen (superoxide ions) trapped in the pores of materials. This is accomplished by measurements of the correlation time given by the equations:

$$\tau_r = \sigma/n(1-\sigma/\sigma_o)$$

where, σ is the number of molecules absorbed/cm2, σ_o is the number of molecules/cm2 at monolayer coverage, and n is the number of molecules competing for an adsorption site/cm2/sec. Thus by measuring τ_r thru T_1 and $(T_2)_r$ [7], one can develop an understanding of the interaction between adsorbate and surface of an isolated void. Measurements of T_1 and $(T_2)_r$ can be made over the range from 4.2 K to 273 K using Electron Spin Echo techniques. These data can be used to develop a picture of how tr varies with temperature which can lead to an understanding of the adsorbate-internal surface interactions.

More importantly, the ESE method can be used to determine the size and distribution of voids in cementeceous, ceramic, or sol-gel derived materials. This can be accomplished by measuring the decay envelop of Hahn echoes both with and without a constant field gradient, G. The echo amplitude is related to the diffusion coefficient [7] which in turn gives the time it takes a superoxide radical to diffuse across a void of a certain diameter. This technique is particularly important for voids in the range from 0.01 to 1 mm which otherwise are not accessible to other instrumentation techniques.

6. ELECTRON PARAMAGNETIC RESONANCE IMAGING TECHNIQUE (EPRI).

The pores might not be randomly distributed in composite, sol-gel, and cementitious materials. While ESE determines the size and distribution of pores, EPR imaging technique can also determine their location in these materials. Unlike the traditional EPR spectroscopy, the EPR imaging technique employes an external magnetic field superimposed by a field gradient. The magnetic field inside the sample, x, can be described with the equation:

$$B_{total}=B_{external}+B_x$$

The resonance in x can be observed if $B_{total} = B_{external} + B_x$ is equal to the resonance field, B_{res}. In order to obtain one-dimensional image, the field gradient sweeps along the x-axis generating an image thru the determination of the resonance intensity calculated at each location, x. This powerful technique which has not been exploited yet by the materials scientist provides information about the voids, whiskers, or fibers randomly distrusted in the materials. An EPR image of the paramagnetic states can reveal the chemistry in the interfaces, distribution of fibers, pores, and whiskers in the materials. EPR imaging spectroscopy can also determine the motion of ions in the liquids and solids. For example, figure 7 shows the EPR image of perchlorotrimethyl in decalin at 1, 24, 48, 72, and 96 hours [8]. Based on a diffusion model [8], the analysis of these EPR images revealed for D the value of 0.5010^{-6} cm^2/sec. Diffusion of oxygen has also been studied in polytetrafluoroethylene (PTFE). Figure 8 shows a 2-D tomograph of oxygen in PTFE. This tomogram allows the detection of the diffusion front of oxygen after heating of PTFE in air converting the fluoroalkyl radicals, R·, into peroxy radicals, $RO\cdot_2$ [8]. The effective diffusion coefficient, D_{O2}, is proportional to $[O_2]^0/[R\cdot]^0$ corresponding to the membrane case diffusion kinetics [6].

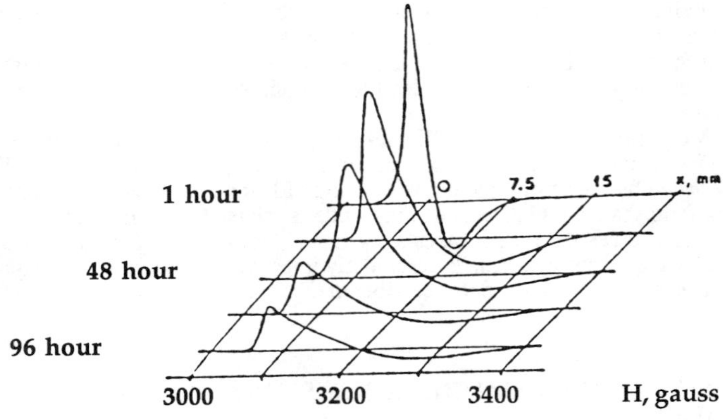

Figure 7. EPR image of perchlorotrimethyl radical in decalin as a function of the time.

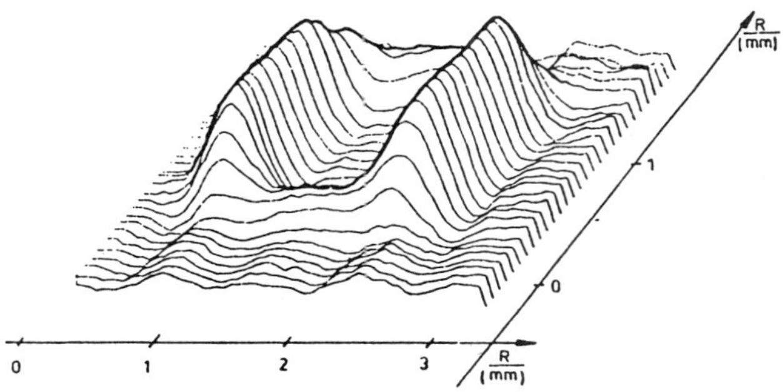

Figure 8. 2-D Tomograph of oxygen in PTFE [9].

In conclusion, the EPRI technique when applied to diffusion of ions in liquid or solids offers the potential of nondestructive evaluation of the kinetics of diffusion. This potentials needs further exploitation by the non-destructive testing scientists.

REFERENCES

1. E.J. Friebele, D.L. Griscom, M. Stapelbroek, and R.A. Weeks, Fundamental Defect Centers in Glass: The Peroxy Radical in Irradiated High-Purity Fused Silica,Phys. Rev. Lett., 42(1979)1346.
2. Kanzig and R.G. Cohen, Paramagnetic Resonance of Oxygen in Alkali Halides, Phys. Rev. Lett., 3(1984)481.
3. G. Kordas, L.C. Klein, and R.A. Weeks, Electron-Spin-Resonance-in-Resonance (ESR) Study of Sol-Gel Gllasses, , j. Non-Cryst. Solids, 71(1985)327.
4. L.Kevan and L. Kispert, ENDOR Spectroscopy, J. Wiley, New York, 1976.
5. R.B. Clarkson, R.L. Belford, K.S. Rothenberger, and H.C. Crookham, J. Catalysis, 106(1987)500.
6. L. Kevan and R.N. Schartz, Time Domain Electron Spin Resonance, J. Wiley, New York, 1979.
7. R.B. Clarkson and R.G. Koosen, Surface Sci., 74(1978)325.
8. E.V. Galtseva, O.Ye. Yakimchenko, and Ya.S. Lebedev, Diffusion of Free Radicals as Studied by Tomography, Chem. Physics Lett., 99(1983)301.
9. Akademie Der Wissenschaften, Zentrum fuer wissenschaftlichen Geraetebau, Geraete Information, 1199 Berlin, Rudower-Chaussee 6, DDR.

FAN AND VENTILATION EQUIPMENT TESTING

K. Helmink, L. L. Christianson, B. Hughes, G. L. Riskowski, and R. Korthals
University of Illinois
Urbana, IL 61801

INTRODUCTION

A fan test chamber was constructed at the University of Illinois to enable us to test fans used in livestock and horticulture applications. Uses will be to cooperate with manufacturers in product development and evaluation, and to conduct research on factors that affect fan performance. The capacity of the fans we are able to test range up to 48" diameter inlet and up to 36" diameter exhaust fans, and with airflow rates up to 30,000 cfm and pressures up to 20 in H_2O.

A series of factors were studied for their effects on fan performance (e.g., number of fan blades, radial position of motor with respect to fan axis (drive shaft), blade position on horizontal fan axis, orifice length, inlet screens, exhaust screens, angle of planar obstructions in front of fan, and slope of entrance to fan.

The objectives of this paper are to: 1) describe the fan test chamber and 2) present results on fan performance as affected by factors commonly encountered in agricultural applications.

REVIEW OF LITERATURE

Airflow patterns are difficult to monitor. Anemometers are used to measure air speeds, but these measurements can be deceivingly high when used to calculate air flow rates, because of air turbulence. Air straighteners can be used to limit the effect of turbulence, when anemometers are used to determine flow rates. Also, numerous air velocity measurements are needed to provide the integrated velocity times area needed to obtain volumetric airflow rates.

Because measuring volumetric airflow rates with anemometers is labor intensive, pressure measurements are often used to provide an indication of the potential for airflow. If an airflow rate curve as a function of static pressure drop across the fan is available, one pressure measurement can be used to estimate the airflow rate accurately. Usually these curves (airflow versus static pressure) are available from the manufacturers.

The AMCA (Air Movement and Control Association) Standard 210-85 is a widely accepted standard for developing the airflow versus static pressure curves in a test lab. The standards specifies that pressures are read to $\pm$ .005 in. H_2O or 1% of full scale of manometer. Greater accuracy would be preferred in these measurements, but liquid manometers can only be read to the stated accuracies. British fan testing procedures were reviewed (Mouslsley et. al. 1987), and their standards are similar to AMCA standards.

AMCA (1985) Standard 210-85 (which is also ASHRAE Standard 51-1985) define two basic types of fan test chambers to test the performance of fans. These are the pitot transverse or nozzle bank option (Fig. 1). We chose the nozzle bank chamber type. There are several differences between the two options. The minimum air flow velocity required to use the pitot transverse option is 2400 fpm; whereas the nozzle bank chamber requires a minimum velocity of 2800 fpm at the nozzle exit. The velocity in the nozzle bank chamber is much lower everywhere else in the chamber except at the nozzle exit, therefore making airflow leveling easier and minimizing the effect of air compressibility in the chamber.

Simplicity of measurement is the biggest advantage of the nozzle bank system. The pitot tube chamber requires measuring air velocities at 24 points in the pitot transverse and averaging them to calculate a volumetric airflow rate. The nozzle bank chamber only requires 2 pressure measurements to calculate the volumetric air rate. The combination of the variable pitch supply fan plus the nozzle bank chamber provides us with a simple, convenient way of varying the airflow rates.

Once the fans have been rated in the lab, there are still problems to overcome in the field operation and set-up of the tested fans. Cruse (1976) and ASHRAE (1988) state that system design engineers and users are often faced with the unpleasant fact that, when a fan is installed in an air moving system, that the performance of the combined fan and system, is significantly below that expected by the designer. This reduction in the performance of the fan after installation in many cases is due to the effects of the inlet or exhaust connections between the fan and system.

Bundy et. al. (1988) reported on the importance of maintaining a fan in clean, running condition once it is installed. Bundy tested fans of 2 and 6 years of field use against new fans to see how the performance of the fans deteriorated with time. Bundy found that shutters and their maintenance had a large effect on fan performance, reducing airflow by as much as 43%. Ten inch fans were tested in all cases.

Albright (1976) reported on the economic and physical variation in the performance of axial fans of similar size. For example, Albright compared several 36 inch fans with 1/2 hp motors, with a nominal ventilation rate of 1623 ft^3/sec at .402 in. H_2O, at a electrical power rate of 6 cents per kWh. The annual operating costs of the fans tested ranged from $2686 to $4098 per year. An annual savings of $1412 per year was possible if the proper fan was chosen. Albright concluded that choosing the proper fan for a given situation would produce significant economic savings.

DESCRIPTION OF FAN TEST CHAMBER

The University of Illinois fan test chamber was built according to the Air Movement and Control Association (AMCA, 1985) Standard 210-85 (Figure 15 in the AMCA Standard). The cross-sectional area of the chamber is 8' x 7' -10 1/2" with an overall length of 20' -1 1/2"

(Figure 2). This size chamber allows us to test fans up to 48" inlet fans and 36" exhaust fans. The variable pitch supply fan installed on our fan test chamber is capable of producing 30,000 cfm at 20 inches H_2O.

The equipment used during the fan tests includes two manometers, one for measuring the static pressure differential across the nozzle board, and the second for measuring the negative total pressure developed in the pitot tube in front of the test fan. A tachometer is used to measure the rotational speed of the test fan. The power, voltage, and current draw of the test fan are also recorded. We can also measure the torque required to turn the fan shaft at varying pressures. Figure 3 provides a more complete equipment listing.

The chamber is in the process of being AMCA certified. Airflow uniformity in the chamber has been thoroughly checked and screens have been patched in the chamber to get approximately level airflow. A repeatability test was performed on the chamber (Test VIII). The leakage rate of the chamber was found to be approximately 13 cfm with a corresponding pressure across the nozzle board of 5.3 in. H_2O. AMCA states that the chamber should be sufficiently tight so that measurements are not affected by more than one-half the allowable instrument error. The manometers are our limiting instrument with an error of $\pm$ .005 in. H_2O, which corresponds to a maximum allowable leakage of approximately 50 cfm. Therefore our chamber is well within AMCA standards.

TEST PROCEDURE

The fan performance was determined by measuring airflow rates at various static pressures. Fan static pressure points generally were 0, .05, .10, .15, and .20 in. H_2O. Sometimes more data were taken to ensure or verify accuracy. Additionally, fan instabilities caused some data to be taken near but not at .05 in. H_2O increments. Other parameters measured during a fan test were: delta P nozzle board, RPM of the test fan, and the volts, amperage, and watts required to operate the test motor. Once the data were collected, the data were converted to a standard airflow cfm, or an equivalent airflow with standard atmospheric air (AMCA, 1985). Test 1 compared the performance of an ACME 36 inch 6-bladed fan versus a Chicago Climate Systems 36 inch 4-bladed fan. Tests 2 thru 7 were conducted with the Chicago Climate Systems fan.

Background of fan environment tested

Test I: Test to determine performance effects of 4-blades vs. 6-blades per fan (Figure 4).

Test II: Test to determine how the radial position of motor affects fan air delivery (Figure 5).

Test III: Test to determine how the blade position on horizontal fan axial affects fan air delivery (Figure 6).

Test IV: Test to determine how the orifice length affects fan air delivery (Figure 7).

Test V: Test to determine how intake screens affect fan air delivery (Figure 8).

Test VI: Test to determine how exhaust screens affect fan air delivery (Figure 9).

Test VII: Test to determine how angled planar obstructions (plywood sheets) affect fan air delivery. Angled plywood sheets were set up in front of fan to provide resistance to airflow (Figure 10).

Test VIII: Test to determine how slope of entrance to fan affects air delivery (Figure 11).

DATA

Airflow as effected by the number of blades

Test	Description	Static Pressure (in. H_2O)	Fan Speed (rpm)	Airflow (cfm)
Ia.	6 blades	.02	529	11279
	ACME 36 in. fan	.05	528	11166
	(no guard)	.10	526	10448
		.15	525	9576
		.20	523	8430
b.	4 blades	.05	516	11572
	36 in. fan;	.10	514	10926
	Chicago Climate Systems	.20	512	7748
	(no guard)			

Airflow as affected by the radial position of motor

Test	Description	Static Pressure (in. H_2O)	Fan Speed (rpm)	Airflow (cfm)
IIa.	10.5 in. from fan axial to	.01	511	10792
	motor center	.05	505	10246
		.10	497	8482
		.15	487	6449
		.19	481	3840
b.	14.75 in. from fan axial to	.00	512	11237
	motor center	.05	503	10131
		.10	498	8837
	Notes:	.15	490	6694
	1) fan wobble noted			

Airflow as affected by blade position on horizontal fan axial. Factors changed include the blade to housing clearance and placement of fan in orifice.

IIIa. Forward position			
3/8 in. clearance fan tip	.00	513	10919
to housing: 13/32 in. bearing	.05	513	9936
to fan hub (minimum blade	.10	514	8988
clearance 3/16 in.)	.15	516	6678
	.18	516	2343
b. Middle position	.01	511	10944
3/8 in. clearance fan tip	.05	509	10058
to housing: 2 in. bearing	.10	509	8992
to fan hub	.15	507	7314
	.19	502	3585
c. Back position	.00	514	11074
5/16 clearance fan tip	.05	511	9959
to housing: 3 in. bearing	.10	510	9069
to fan hub	.15	507	7840
(fan mounted at end of shaft)	.20	499	4917

Airflow as affected by orifice (housing) length.

IVa. Orifice length reduced by 16%	.05	519	10627
original orifice length	.10	518	9732
4.75 in.	.15	516	8503
	.20	514	6656
b. Original orifice	.01	517	11179
plus standard guard	.05	517	10682
	.10	515	9263
	.15	514	7806
	.20	513	6048

Airflow as affected by mesh of screen on the intake side of the fan.

Va. 1/2 x 1 in. mesh screen plus	.05	513	9901
exhaust side standard guard	.10	512	8577
	.15	512	7336
	.20	510	5133
b. 1 x 2 in. mesh screens plus	.05	511	10593
exhaust side standard guard	.10	510	9493
	.15	510	8122
	.20	508	5705
c. Exhaust side standard guard	.01	517	11179
	.05	517	10682
	.10	515	9263
	.15	514	7806
	.20	513	6048

Airflow as affected by mesh of screen on exhaust side of fan.

VIa. 1/2 x 1 in. mesh screen plus standard guard	.00	517	11558
	.05	517	9735
	.10	516	9322
	.15	514	8030
	.20	513	5029
b. 1 x 2 in. mesh screen plus standard guard	.01	512	11310
	.05	512	10800
	.10	512	9673
	.15	513	8464
	.20	512	5069
c. standard guard	.01	517	11179
	.05	517	10682
	.10	515	9263
	.15	514	7806
	.20	513	6048
d. no guard	.05	516	11572
	.10	514	10926
	.20	512	7748

Airflow as affected by angle of planar obstruction on exhaust side of fan.

VIIa. 15° angle from horizontal plus exhaust side standard guard	.01	518	11271
	.05	517	10726
	.10	516	9506
	.15	516	7926
	.20	515	6199
b. 30° angle from horizontal plus exhaust side standard guard	.01	517	11142
	.05	517	10350
	.10	515	9230
	.15	515	7946
	.20	513	6027
c. 60° angle from horizontal plus exhaust side standard guard	.00	517	9756
	.05	515	8540
	.10	513	7664
	.16	512	5757
	.18	510	4614
	.20	511	4092

Airflow as affected by slope of entrance to fan.

VIIIa. Sloped entrance to fan 19° plus standard guard (test 1)	.01	525	10805
	.05	521	10520
	.10	518	9364
	.15	515	9582
	.20	503	2824
b. Sloped entrance to fan 19° plus standard guard (test 2)	.04	515	10546
	.05	516	10288
	.10	517	9575
	.15	518	7788
	.20	519	5473
c. Sloped entrance to fan 19° plus standard guard (test 3)	.02	519	10828
	.06	518	9883
	.11	515	8576
	.15	513	7324
	.18	512	5004
d. Straight entrance plus standard guard	.02	514	10532
	.05	514	10060
	.10	514	8989
	.15	515	7671
	.19	516	5039

RESULTS AND CONCLUSIONS

The primary objective of this section is to analyze how ventilation rates change over normal operating pressures for alternative fan system design configurations. The scope of tests performed were as follows:

Factors of fan construction
- a. Number of blades
- b. Radial motor position
- c. Placement of blades on horizontal fan axis
- d. Orifice length

Factors of fan installation
- a. Screens on intake side of fan
- b. Screens on exhaust side of fan
- c. Angled planar obstructions
- d. Slope of entrance to fan

The factors listed above were distinguished as important in the fan system design. These factors were tested to determine the degree of their effects on performance and to determine the optimal design for each of these factors. All tests were performed on 36 inch diameter fans.

Number of Blades

Based on the data taken, at fan pressures below 0.13, the four-bladed fan performed better and at pressures above 0.13 the six-bladed fan provided more stable airflow rates over a range of higher pressures. The average air flow rate increased 12% from the six-bladed fan to the four-bladed fan (Test 1, Figure 4).

Radial Motor Position

The fan performed best with the motor mounted closer to the fan axis rather than away for the positions tested. The two distances tested were 10 1/2 in. and 14 3/4 in. from the fan drive shaft to the center of the fan motor. A 2% average reduction in air flow occurred when moving the motor from the more central position to the outer edge of the fan housing (Test 2, Figure 5).

Blade Position

At pressures below 0.10 the blade position did not seem to have much effect on the air flow (cfm). When static pressures exceeded 0.10, the middle blade position was the best, the forward position was intermediate, and the back position was the poorest. Three blade positions tested were: 1) 13/32 in. from bearing to fan hub, 3/8 in. clearance fan tip to housing (minimum blade clearance was 3/16 in.); 2) 3 in. from bearing to fan hub, 5/16 in. clearance fan tip to housing; 3) 2 in. from bearing to fan hub, 3/8 clearance fan tip to housing. From the first to the second position, there was a 5% increase in air flow. From the second to the third there was a 2% reduction in air flow (Test 3, Figure 6).

Orifice Length

Throughout the pressure range the longer orifice averaged 1.7% greater air flow rates compared to the shorter orifice. Above 0.15 in. H_2O the air flow delivery of the shorter orifice declined more sharply than the longer orifice. The orifice was reduced 16% in length from 4.75 in. to 4.0 in. to obtain the shorter orifice (Test 4, Figure 7).

Intake Screens

On average the addition of 1/2 x 1 in. and 1 x 2 in. intake screens reduced the air flow by 8.9% and 3.3% respectively as compared to the standard guard case. The smaller the openings in the screen, the more the air flow was affected. The cases tested were a 1/2 x 1 in. screen, and a 1 x 2 in. screen with both cases having a standard exhaust guard installed (Test 5, Figure 8).

Exhaust Screens

The addition of any size screen to the exhaust side of the fan affected the air flow. Adding the standard guard (which was a wire mesh consisting of concentric circles spaced about an inch apart) decreased the air flow compared to no guard by 15%. Adding the 1 in. x 2 in. mesh guard with the standard guard decreased air flow 19%. Replacing the 1 in. x 2 in. by 1/2 in. x 1 in. screen decreased the air flow by 23%. For pressure ranges of 0.10 to 0.15, the magnitude of the difference in air flow was constant between the tests that were done. The cases tested were no guard, standard guard, 1/2 x 1 in. guard, and a 1 x 2 in. guard (Test 6, Figure 9).

Angle of Planar Obstructions On Exhaust Side of Fan

This test consisted of putting sheets of plywood with 15°, 30°, and 60° angles in front of the fan (this is referred to as the planar angle of obstruction). A planar angle of obstruction between 0° and 30° in front of the fan has little effect on the air flow. Between 15° and 30° there was only a 2% decrease in air flow. Obstructions that have angles greater then 30° have a dramatic effect on the air flow. A 20% decline resulted from the use of an obstruction between 30° and 60° (Test 7, Figure 10).

Slope of Entrance to Fan

The difference between the straight box and angled box for the range tested was not as great as the effects of the other parameters tested. Only a 5% decrease in air flow was measured from the straight box to the angled box. At lower pressures the angled box allowed more air to flow than the square box, but at higher pressures the straight box allowed a greater air flow than the angled box (Test 8, Figure 11).

REFERENCES

Albright, L. D. 1975. Air moving efficiencies of ventilating fans ASAE Paper NA75-304, St. Joseph, MI. 49085

AMCA 1985. Laboratory methods of testing fans for rating. Air Movement Control Association, Inc., 30 West University Drive, Arlington Heights, Il 60004.

ASHRAE 1988. Air handling equipment: fans American Society of Heating, Refrigeration, and Air Conditioning Engineers Equipment Handbook Chapter 3 1791 Tullie Circle, N.E., Atlanta, GA. 30329

Bundy, D. S., Mangold, D.W. 1988. The airflow performance of confinement ventilation fans measured in use. Proceedings of the Third International Livestock Environment Symposium. April 25-27, 1988

Cruse, E. A. 1976. The AMCA certified ratings program for fans ASAE Paper 76-3533, St. Joseph, MI. 49085

Moulsley, L.J., Randall, J.M., Hartshorn, R.L., Houghton, C.J., and Randle, D.G. Facilities for measuring fan performance Divisional Note DN 1408, AFRC Inst. Engng, Res, Silsoe, April 1987

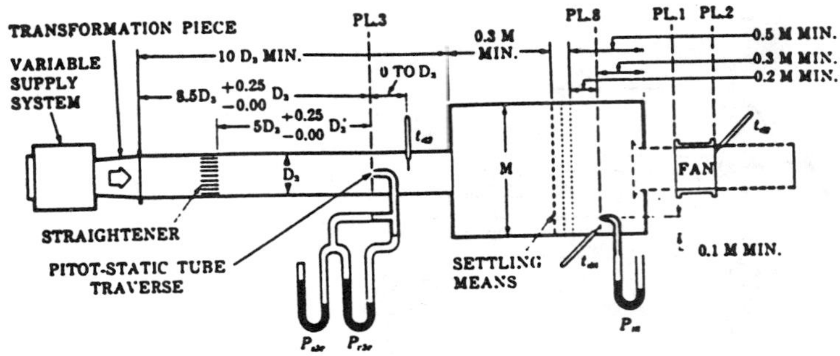

Inlet Chamber Setup—Pitot Traverse in Duct

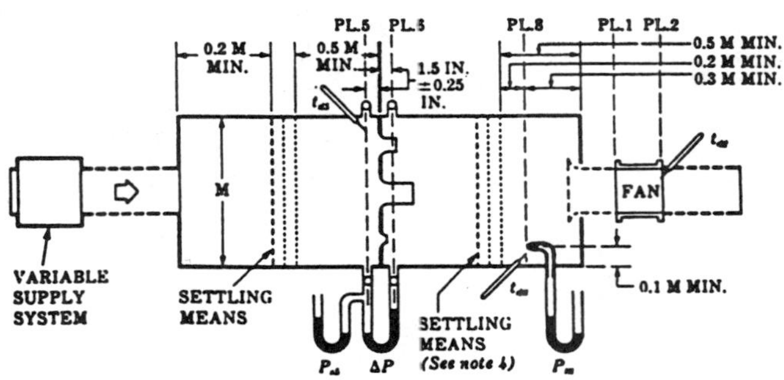

Inlet Chamber Setup—Multiple Nozzles in Chamber

ANSI/AMCA STANDARD 210-85 ANSI/ASHRAE STANDARD 51-1985

Figure 1. Inlet Chamber setup

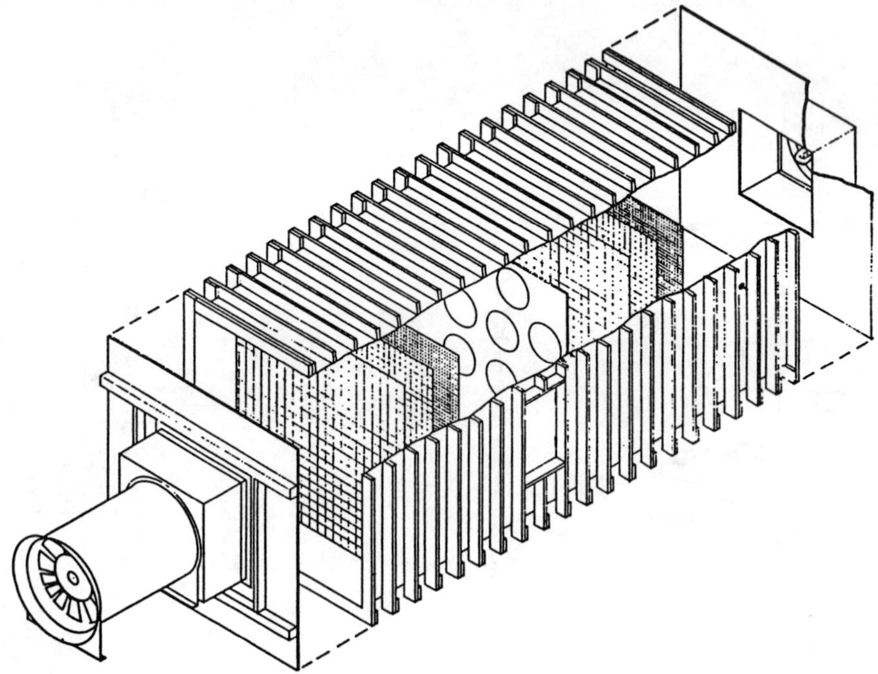

Figure 2. Fan test chamber

Figure 3. Specifications of fan test chamber instruments.

1. Nozzle bank - twelve 8-inch nozzles made by Helander Inc.

2. Static and delta pressure gauges - manufactured by Dwyer Instrument, Inc. with 1% accuracy at full scale of maximum observed or .005 in. H_2O with a range of 0-5 in. of water, model #-250.5-AF

3. Settling screens - the three types of screens used were 5/64 in. on 7/64 in. with 46% open area, 3/16 in. on 1/4 in. with 50% open area, and 5/32 in. on 3/16 in. with 63% open area

4. Temperature probes - room temperature and humidity were measured with a mechanically aspirated barometer, psychrometer, model #-J-3312-20, which is accurate to + or - .3 ^{o}F full scale and the relative humidity is accurate to + or - 1% full scale; all temperature probes were obtained from Cole-Palmer

5. Pitot probe - 5/6 diameter, 12 in. insert, made by Dwyer Instruments, Inc., model #-160-12 (figure 6)

6. Volt meter - volt-ohm-milliammeter Simpson 360 with a + or - 1 volt accuracy

7. Ampmeter - General Electric ampmeter with a + or - 5% accuracy, model #-942D

8. Wattmeter - 2-OSI Electronic Precision AC watt transducer, Model EW5-20B measures up to 4kw for a range of 0-300 volts and 0-20 amps; Model EW5-10B measures up to 1kw for a range of 0-150 volts and 0-10 amps: <u>Accuracy + or - 0.2% of reading</u>

9. Tachometer - a Ultra Sensors, Inc. tachometer was used with a + or - 1 count accuracy, model #-DT-2232A

10. Anemometer - TSI model 1650 Air Velocity/Temperature Meter; velocity range 0-6000 feet per minute; accuracy + or - 2% full scale (calibrated 4-4-86)

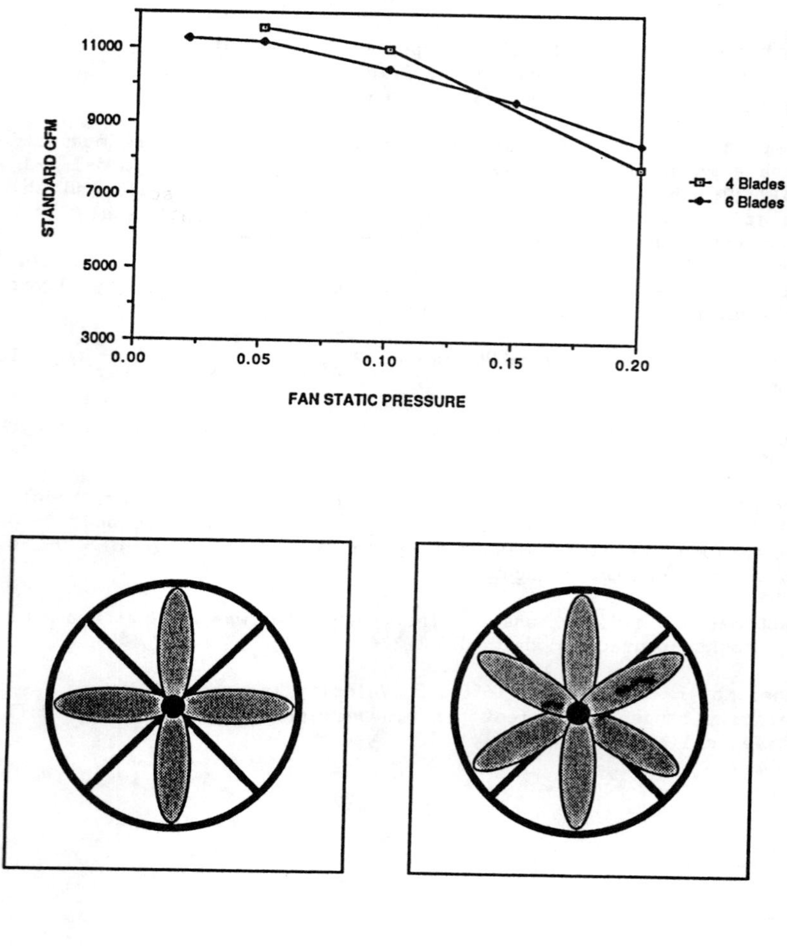

Figure 4. Number of blades

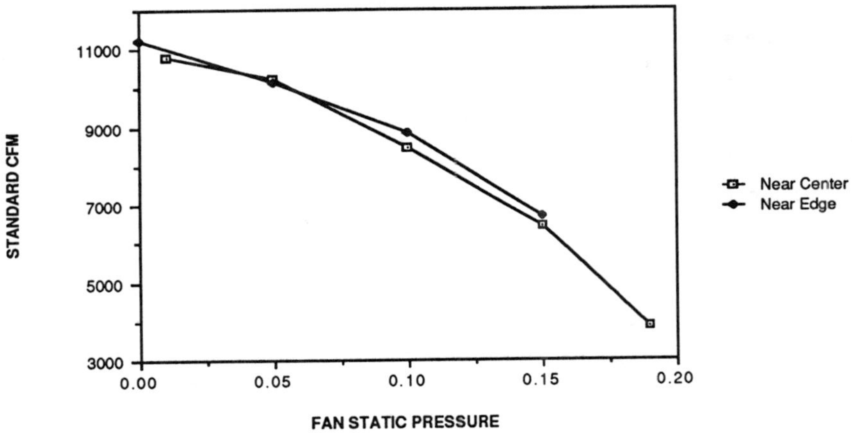

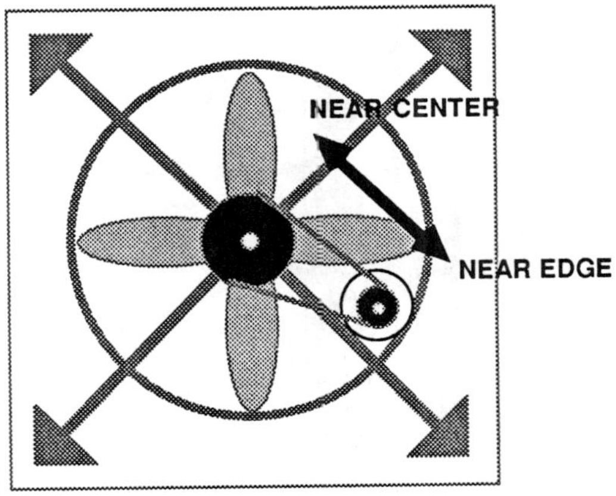

Figure 5. Radial position of motor

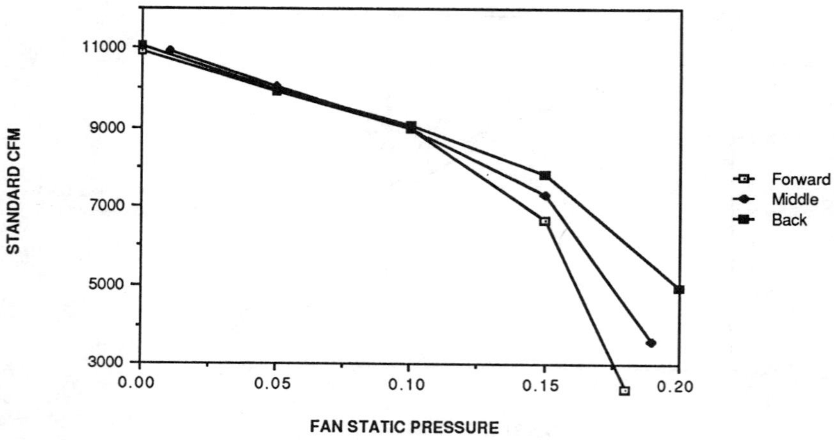

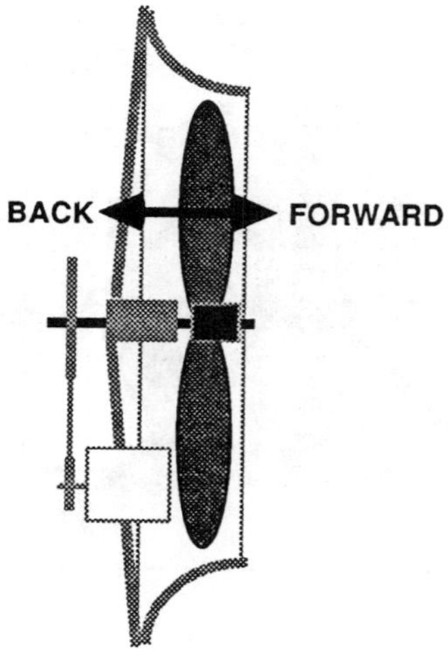

Figure 6. Blade position on horisontal fan axel

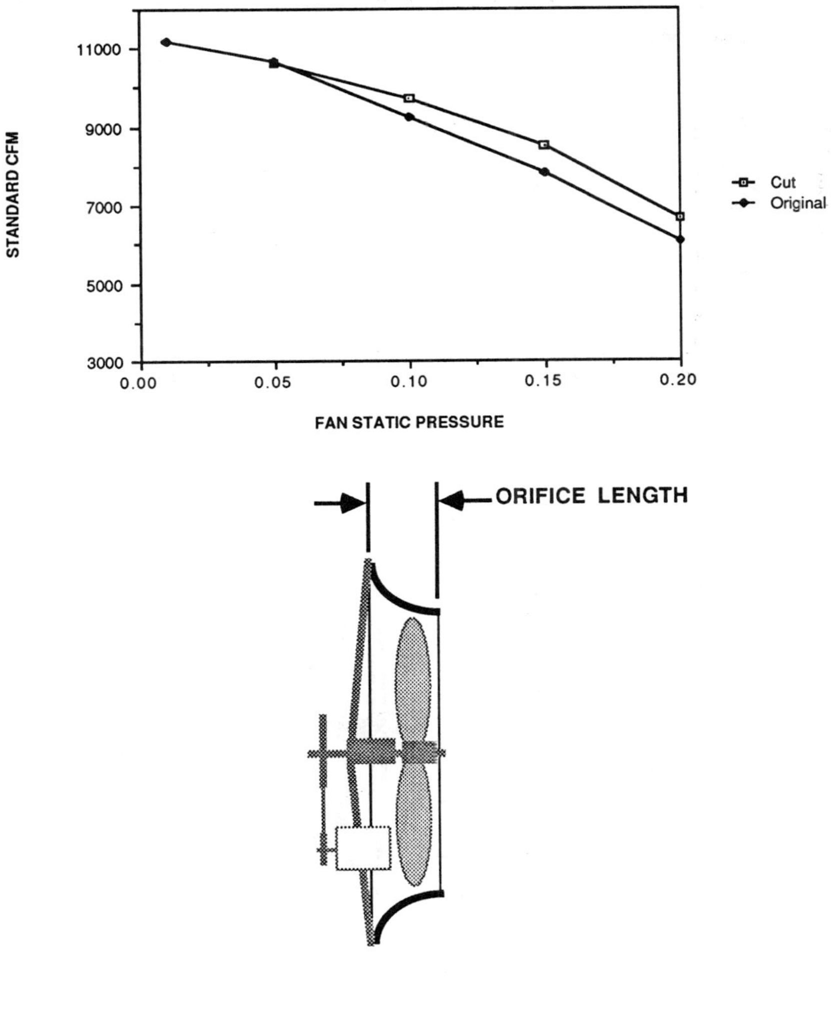

Figure 7. Orifice length

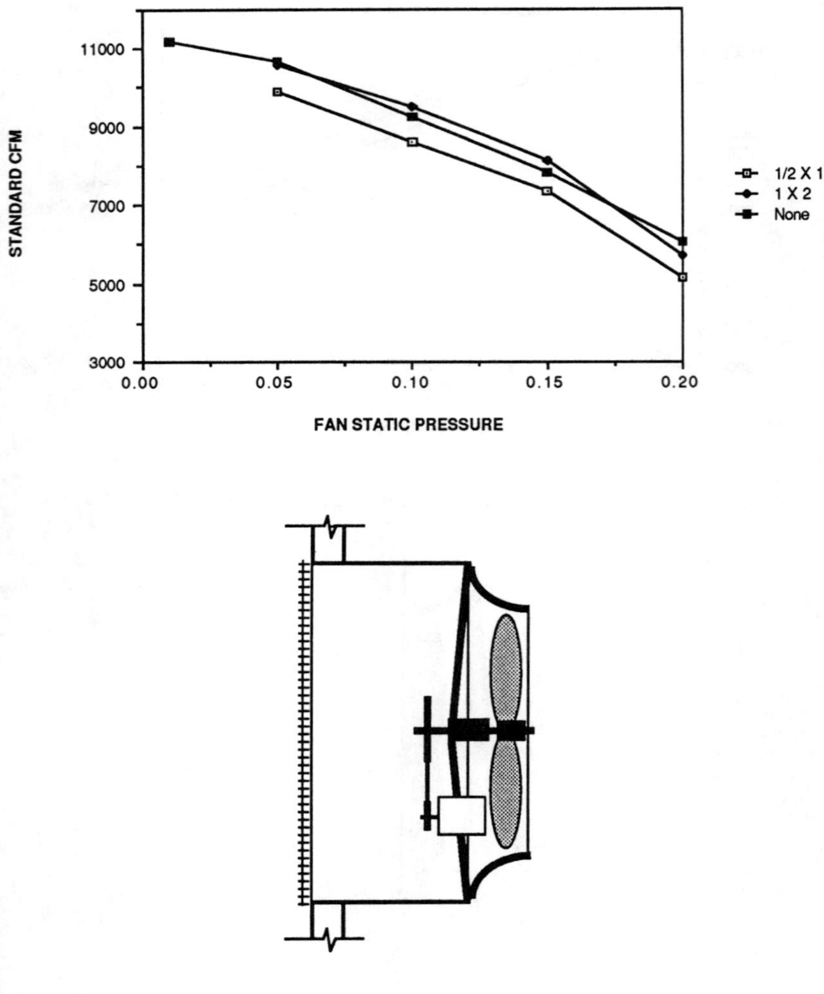

Figure 8. Mesh of screen on the intake side of fan

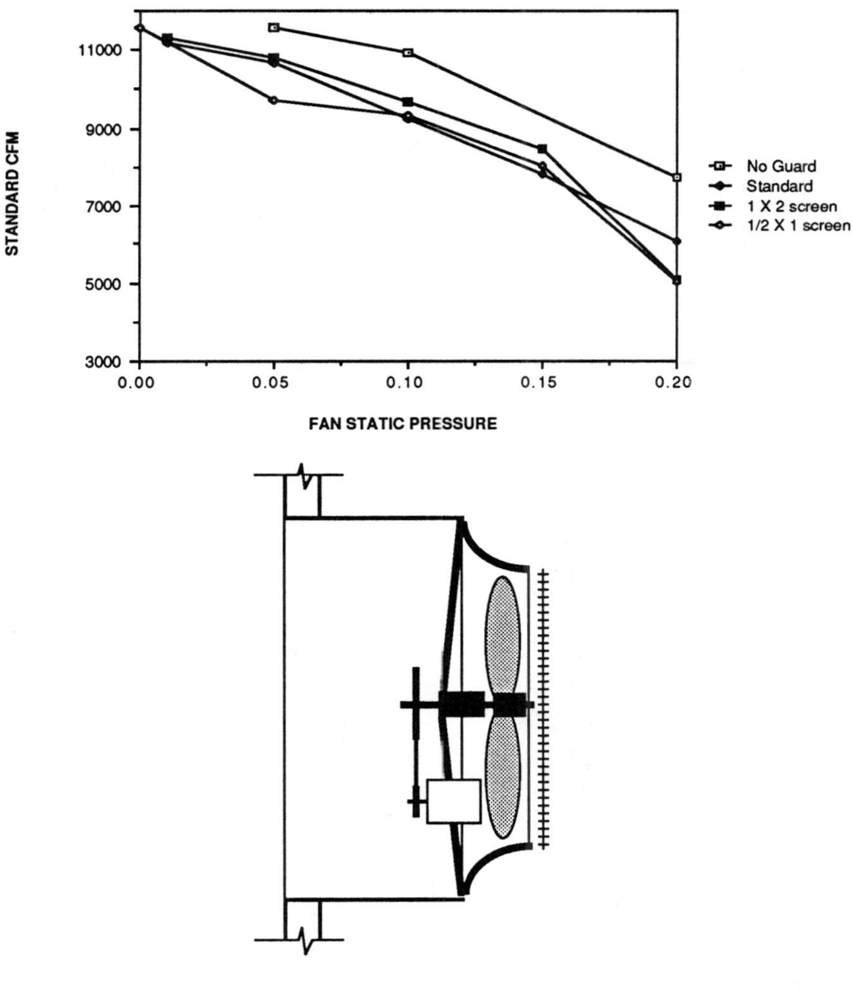

Figure 9. Mesh of screen on exhaust side of fan

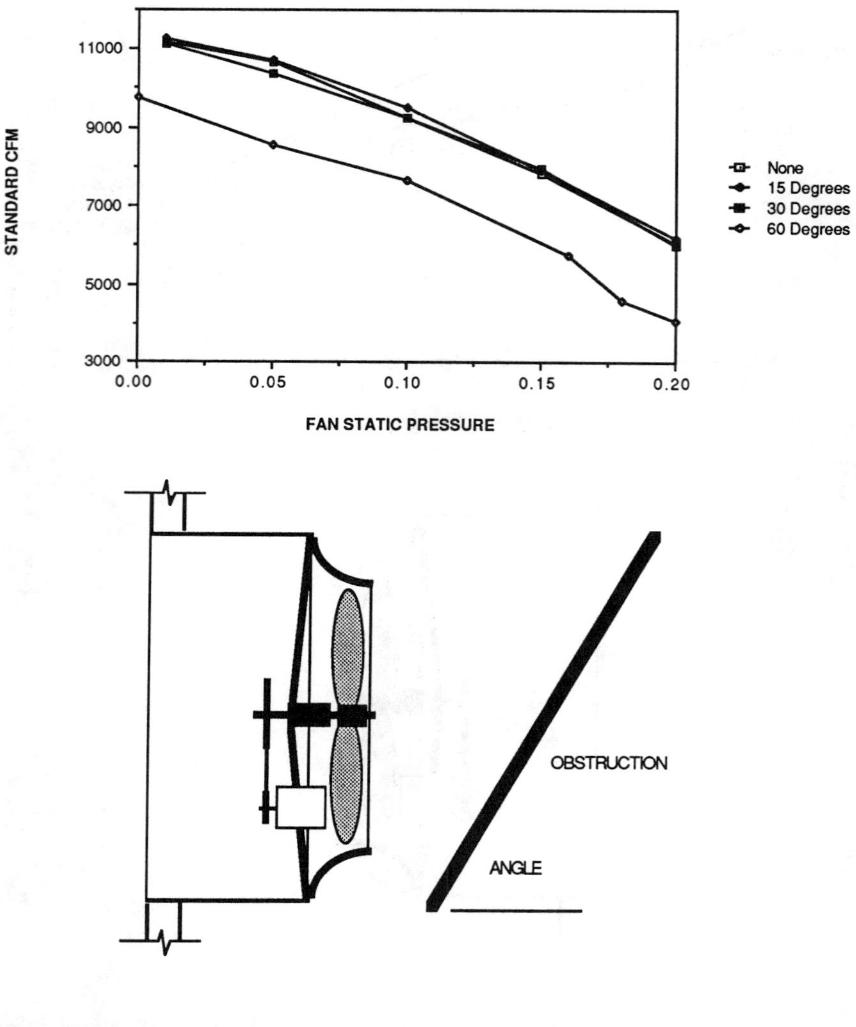

Figure 10. Angle of planer obstruction on exhaust side of fan

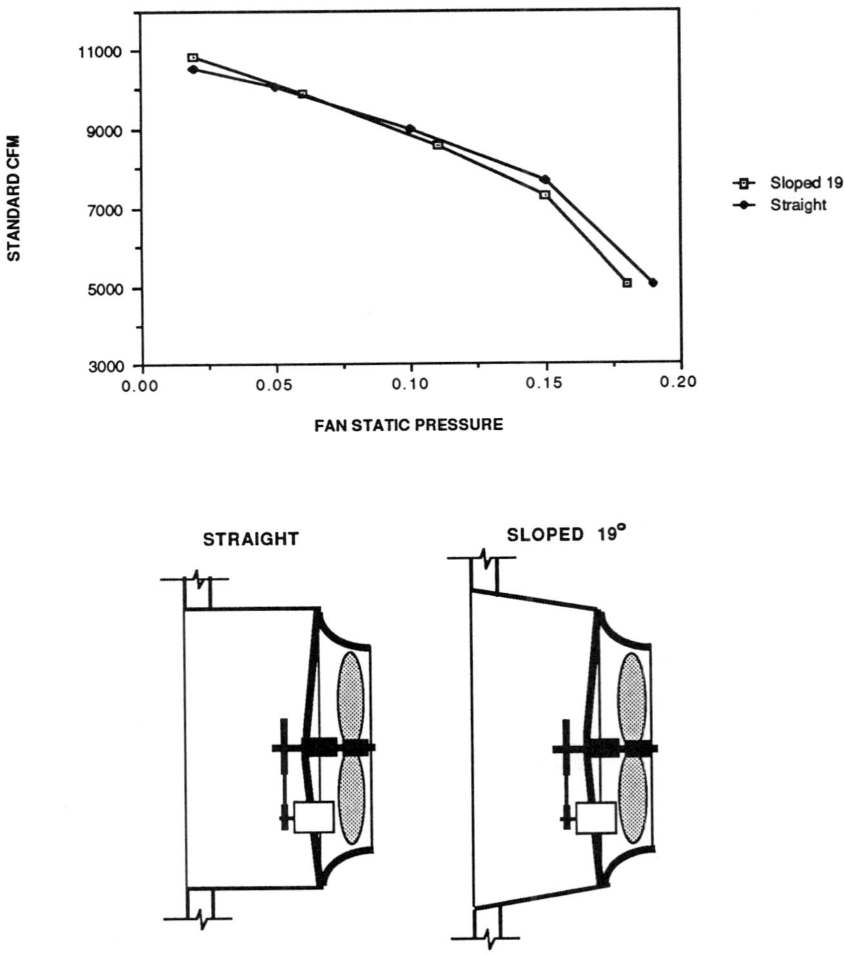

Figure 11. Slope of entrance to fan

NONDESTRUCTIVE EVALUATION OF BOND STRENGTH FOR EXPLOSIVELY WELDED JOINTS

R. C. Gupta
Mutah University
Jordan

ABSTRACT

The strength of an explosively welded joint depends on its interface characteristics which depend on the welding parameters. Mechanism of explosive welding process and interface wave generation are discussed. The interface wave characteristics are investigated and permissible range of parameters are established. Welding is possible if the welding parameters such as the obliquity angle β and the plate velocity V_p are within the permissible range. However, bond strength may be poor even if welding is found sound by NDT. Thus, bond strength data from several references are investigated and a semi-empirical condition is established which if satisfied would result in a reasonably good strength bond.

1. INTRODUCTION

The nondestructive testing of welds, in itself, does not provide evidence of weld strength. However, if NDT reveals the weld to be crack/defect free, evaluation of strength may be possible from consideration and analysis of the mechanical state and weld joint design; or 'if designed correctly it must be safe'.

But the above statement does not apply to explosively welded joints. Even if the weld (explosively welded) is defect/crack free, strength of the weld may be poor, because strength of the weld depends mainly on its interface-characteristics. Usually a wavy interface is generated and is preferred . However, a moderate size wave is expected to give better results. If the interface wave characteristics which depend on the welding parameters could be predicted, evaluation of weld strength could be possible. For this, an understanding of explosive welding process, interface wave generation, aspects relating to its interface characteristics and permissible range of parameters are essential. Aspects relating to metallurgy and strength of bond are also studied and investigated.

2. EXPLOSIVE WELDING PROCESSES

Explosive welding is done by progressive high velocity oblique impact of one(flyer) plate to another(parent) plate by impact created by detonating the explosive charge. Surface-to-surface welding is produced which is not usually possible with other welding processes.

Explosive welding has come to play an important role in many industries[1] such as the chemical, atomic energy, aircraft, space, cryogenic and power generating industries. The explosive bonding process is extensively used for flat plate cladding of dissimilar metals required for fabricating chemical equipments used in corrosive service conditions. Explosive welding is applied to large plate cladding[2], welding of tubes to tube-plates of heat exchangers[3], tube plugging[4], welding of pipes[5], production of duplex[6] and triplex[7] tubing and fibre reinforced materials[8] and for many other applications.

2.1 Mechanism of Explosive Welding

A schematic diagram of asymmetric explosive welding is shown in Fig.1.a where the flyer plate is kept inclined at a set-up angle α to the parent plate. Explosive is laid on the flyer plate while a buffer plate is kept in between to avoid damage to the plate. The explosive is ignited from one end with an electric detonator, from a distance. Figure 1.b indicates the configuration of the flyer plate during the process of welding. Figure 1.d represents kinematically equivalent system for Fig.1.c such that the impact point S is considered at rest(stagnation point). Pressure encountered[9] in explosive welding is very much higher than the yield strength of metal, and strain rates are also

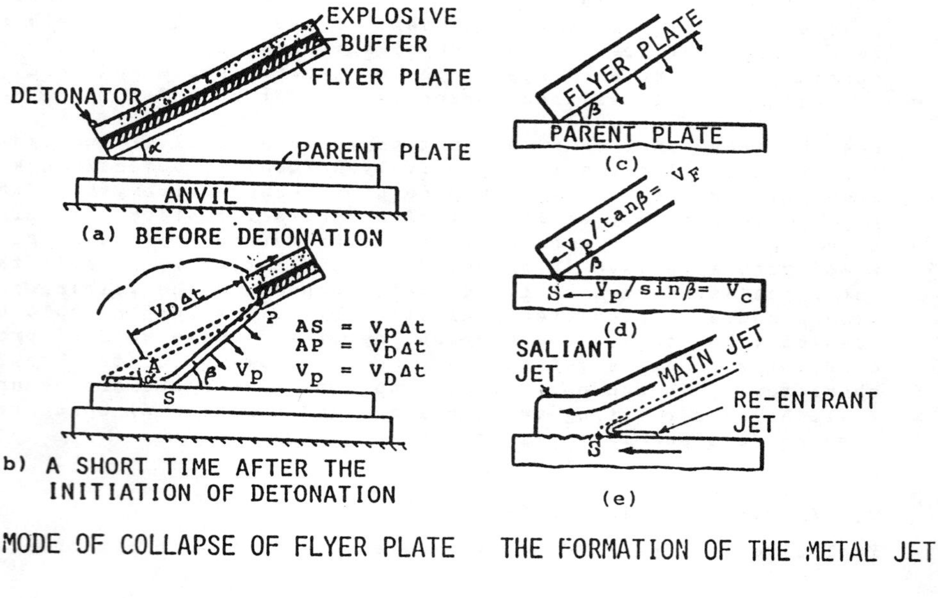

FIGURE.1 Explosive welding:collapse of plate and jet formation

extremely high[10], and under such conditions metals behave as fluids; thus a hydrodynamic model is usually considered. Hence the flyer plate is considered to impinge on the parent plate as a main fluid jet which divides into (i)salient jet and (ii) re-entrant jet, as shown in Fig.1.e. The re-entrant jet usually contains contaminated layers from the inside of the plates and thus cleans the surfaces to be welded. The virgin plate surfaces come in contact under high pressure resulting in surface-to-surface bonding.

The followings are the well known relations

$$V_p = \frac{0.612\ e/m}{2 + e/m} V_D \tag{1}$$

$$V_F = V_p/\tan\beta \tag{2}$$

$$\phi = \tan^{-1}(V_p/V_D) \tag{3}$$

$$\beta = \alpha + \phi \tag{4}$$

where V_p, V_D, V_F are the plate velocity, detonation velocity of explosive and the stream velocity; α , ϕ and β are the set-up angle, bend angle and the obliquity angle; e/m is the explosive loading(the ratio of the mass of explosive to that of the flyer+buffer plate).

Figure 2 shows the arrangements for the welding of tube to tube-plate of a heat exchanger. The parallel arrangement(Fig. 2.a) is suitable with low detonation velocity explosive, and the inclined arrangement(Fi.2.b) is preferred with high detonation velocity explosive.

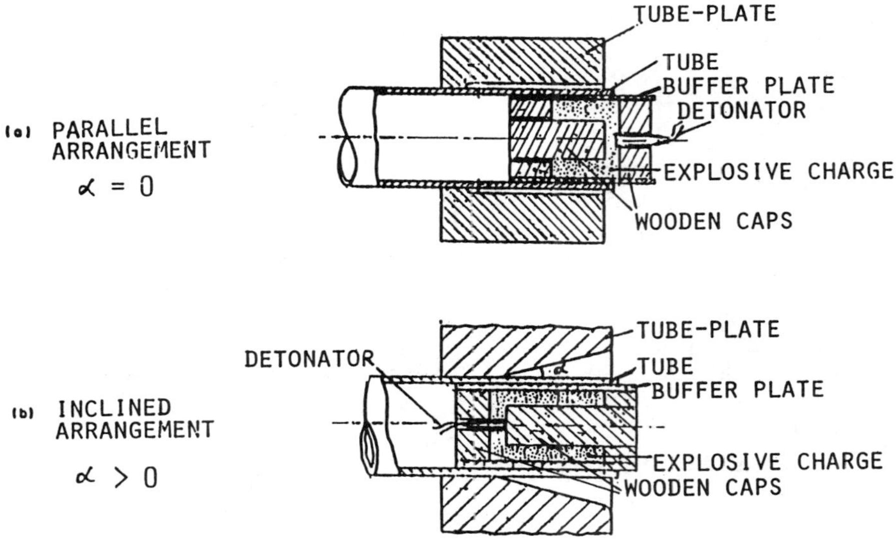

FIGURE.2 Tube to tube-plate welding of heat exchangers

3. WELD INTERFACE

3.1 Types of Weld Interface and Importance of Wavy Interface

Depending on the welding parameters(such as V_p and β) the nature of the weld interface[11] may be
(i) straight(direct bond)
(ii) wavy interface
(iii) uniform layer of solidified melt

Type(iii) interface is not considered good as it gives rise to formation of brittle intermetallic compounds thus lowering the strength. As compared to a straight bond the wavy interface is considered to be better because[12] (a)the waves increase the area of contacting surfaces several fold and thus expose a great deal of fresh surface on the plates which readily bond under pressure,(b)the waves provide very high strains in the thin surface layers and (c) the waves provide a mechanical interlocking. However,large size waves could cause stress concentration and reduce strength, thus a moderate size wave could give optimum bond strength.

3.2 Mechanism of Interface Wave Generation

It is intriguing that under such a violent condition of explosion,regular periodic waves are formed in explosive welding. Many theories have been proposed to explain the interface wave generation. Indentation model of Bahrani et.al.[13], stress wave mechanism of Godunov et.al.[14] and vortex shedding model of Kowalic and Hay[15] are a few of these. The author's swinging wake model[16] is similar to that of Kowalic and Hay and is based on hydrodynamic analogy, and explains the interface wave generation and the interface wave characteristics.

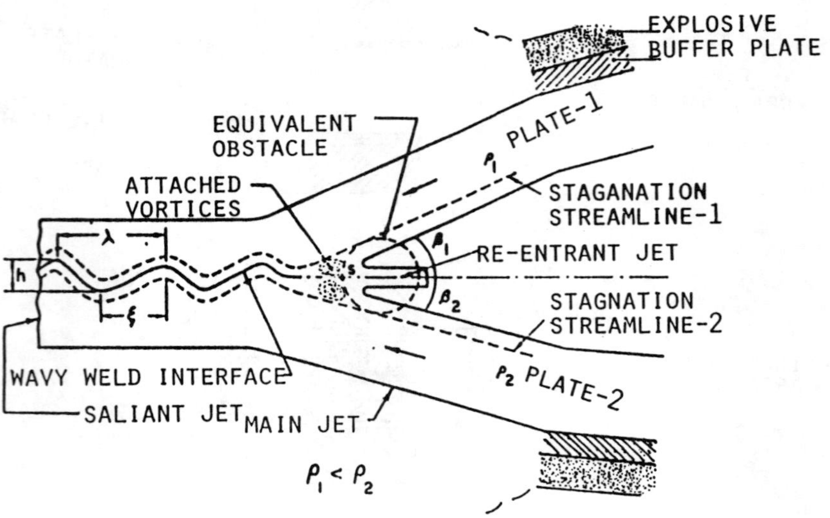

FIGURE.3 Hydrodynamic model for general arrangement of explosive welding: the swinging wake mechanism

Figure 3 shows a hydrodynamic model for a general arrangement of explosive welding. The re-entrant jet is important for cleaning the surfaces.The re-entrant jet is also important for wave generation as it introduces an equivalent obstacle in the flow of jets, behind which there may be attached vortices. The vortices usually shed and the wake behind the obstacle swings from side to side producing a wavy interface.

3.3 Interface Wave Characteristics

Interface wave profile is a sine-wave for welding of similar metals whereas it is a distorted-wave for dissimilar metal combination. Amplitude to wave-length ratio h/λ and the degree of distorsion(fig.3)defined as $\eta = (2\xi - \lambda)/\lambda$ are given by[16]

$$h/\lambda = 0.75 \frac{\rho_1/\rho_2}{(1 + \rho_1/\rho_2)(0.75 + 0.25\ \rho_1/\rho_2)} \qquad (5)$$

$$\eta = \frac{1 - \rho_1/\rho_2}{1 + \rho_1/\rho_2} \qquad (6)$$

The amplitude of the wave h is important but quite difficult to be predicted. It may be considered that the diameter of the obstacle d should be proportional to the re-entrant jet thickness t_r , thus taking an average value of the proportionality constant[17] as 3 ,

$$d = 3\ t_r \qquad (7)$$

The wave amplitude h around a circular obstacle of diameter d in fluid flow is given by

$$h = 1.25\ d \qquad (8)$$

4. EXPLOSIVE WELDING PARAMETERS AND WELDABILITY WINDOW

4.1 Explosive Welding Parameters

Various explosive welding parameters of importance are

(i) Explosive detonation velocity V_D
(ii) Explosive loading e/m
(iii) Plate velocity V_p
(iv) Obliquity angle β
(v) Collision velocity $V_c (=V_p/\beta)$
(vi) Material properties of the flyer plate such as density and strength of the metal
(vii) Set-up angle, stand-off distance etc.

Specific pressure p_s and modified specific pressure p_s^* are defined as

$$p_s = 2\sigma_c/\rho V_p^2 \qquad (9.a)$$

$$p_s^* = 2\sigma_o/\rho V_p^2 \qquad (9.b)$$

where σ_c and σ_o are the theoretical and the actual strength of the(flyer)plate material and ρ its density and V_p the plate velocity. Subsequently we could use p_s (or p_s^*) and β as the governing parameters as most of the welding parameters are embedded in these.

4.2 Permissible Range of Welding Parameters

Bahrani and Crossland[18] carried out explosive welding tests to study the phenomenon of wave formation and its effect on strength of bond. Experimental variation of interface wave amplitude h with obliquity angle β for different explosive loading are shown in Fig.4. The general nature of the curve is also represented in the figure. It is seen that below a critical angle β_L and above an angle β_E welding does not take place. Welding with straight interface occurs when $\beta_W < \beta < \beta_E$ and a wavy interface is produced if $\beta_L < \beta < \beta_W$.

Considering that there is a limited extent of fluidized zone[19] within the plate and that there is some energy loss[20] in collision, it has been derived[16] that the thickness of the re-entrant jet t_r is given by,

$$t_r = \frac{r_f}{2}\left[1 - \cos\beta \Big/ \sqrt{1 - 1.25\, p_s^{0.5}\beta\sin^2\beta}\right] t \qquad (10)$$

where r_f is the ratio of the fluidized thickness to the flyer plate thickness and is given by

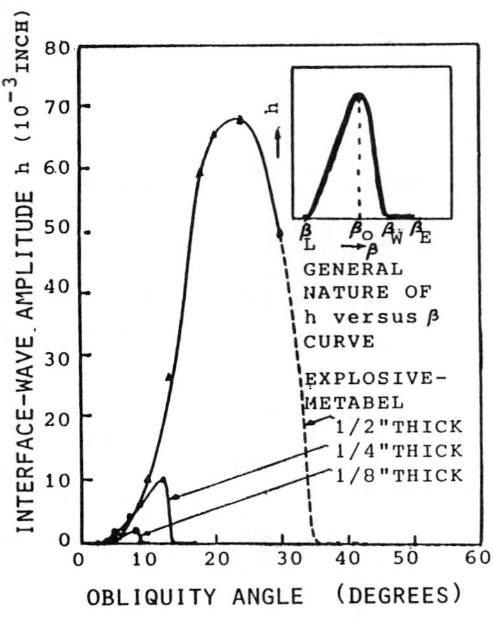

FIGURE.4 Variation of wave amplitude with obliquity angle for different explosive loading indicating the limiting obliquity angles

$$r_f = 1 \quad \text{for} \quad \beta < \beta_o \tag{10.a}$$

$$r_f = 4(1-\beta^2 p_s)^{0.5}/\pi\beta^2 \quad \text{for} \quad \beta_o < \beta < \beta_F \tag{10.b}$$

$$r_f = 0 \quad \text{for} \quad \beta > \beta_F \tag{10.c}$$

where β_o is the obliquity angle below which total thickness of the plate is fluidized and above which only a fraction of the plate thickness is fluidized; and β_F is the obliquity angle above which the plate does not show any fluidlike behaviour.

<u>Limiting obliquity angles</u> There are a few limiting angles such as β_L, β_W, β_o and β_E as mentioned before. Considering the existence and the thickness of the re-entrant jet and the hydrodynamic analogy for interface wave generation, expressions for these angles are obtained[16] as under

$$\beta_L = 6/p_s^{0.4} \tag{11.a}$$

$$\beta_o = 30/p_s^{0.5} \tag{11.b}$$

$$\beta_W = 32/p_s^{0.5} \tag{11.c}$$

$$\beta_E = 46/p_s^{0.5} \tag{11.d}$$

where the specific pressure $p_s = 2\sigma_c/\rho V_p^2$ in which σ_c is theoretical strength of the flyer plate metal, ρ its density and V_p is the plate velocity; β_L is the lower obliquity angle for jet-no-jet(weld-no-weld) boundary, β_o is the optimum obliquity angle for maximum wave height, β_W is the obliquity angle for wave-no-wave boundary and β_E is the upper limit for the obliquity angle for jet-no-jet(weld-no-weld) boundary.

However, since it is not possible to find theoretical strength σ_c for all metals and alloys in particular; it is desirable to find equivalent formulae for these limiting angles in terms of a modified specific pressure $p_s^* = 2\sigma_o/\rho V_p^2$ where σ_o is the actual strength of the flyer plate material. The modified formulae are found to be as follows

$$\beta_L = 3/p_s^{*0.4} \tag{12.a}$$

$$\beta_o = 8/p_s^{*0.5} \tag{12.b}$$

$$\beta_W = 14/p_s^{*0.5} \tag{12.c}$$

$$\beta_E = 20/p_s^{*0.5} \tag{12.d}$$

In M.K.S. units these formulae yield the limiting obliquity angles in degrees. All these formulae are found to be in reasonable agreement with the experimental results of various research workers for variety of metals and alloys at different welding conditions[13,16,18,21,23].

<u>Limiting values for plate velocities</u> Flyer plate velocity must exceed a certain value $V_{p\ min.}$ to ensure a minimum value of

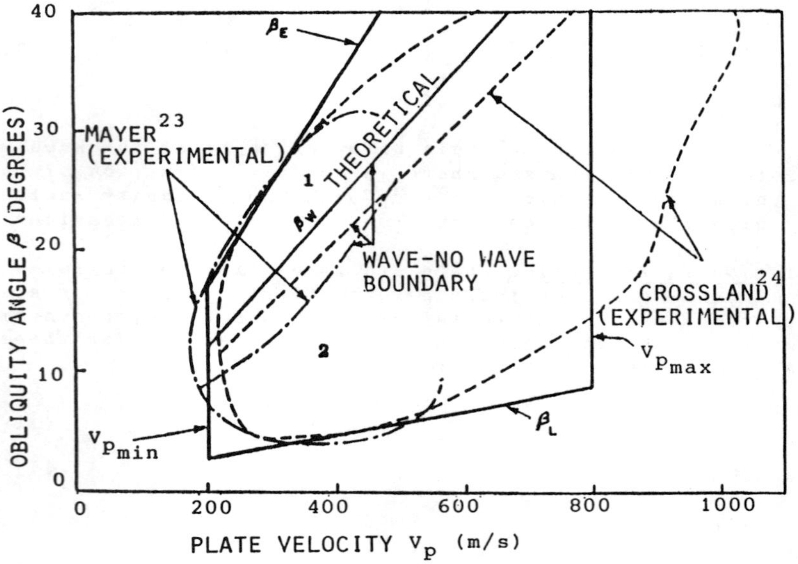

FIGURE.5 Weldability window (for copper)

contact pressure for dynamic yielding. Plate velocity should, however, be below a certain value $V_{p\,max.}$ to avoid melting and formation of intermetallic compounds. Although formulae for $V_{p\,min.}$ and $V_{p\,max.}$ are available in literature[22], it is reasonable and practical to take,

$$200\ m/s < V_p < 800\ m/s \qquad (13)$$

4.3 Weldability Window

The region of parameters within which welding is possible is called weldability window. Weldability window plot between β vs V_c, β vs V_p, e/m vs V_c etc. are reported in literature. Here, using the limiting values for β and V_p a theoretical weldability window plot between β and V_p is constructed in Fig.5. The window is even subdivided into two regions: (1) region of straight interface and (2) region of wavy interface. It is seen that the theoretical predictions are in reasonable agreement with the experimental results of Mayer[23] and Crossland[24]. Equation(11) is used in Fig.5 for copper. Equation(12) could also be used; for other materials and specially for alloys.

5. METALLURGY AND STRENGTH OF THE WELD

Explosive welding is essentially a pressure welding process, metallurgical problems are much less severe than with a fusion welding process as shown by Lucas et.al.[25] . However, in some cases, under incorrect welding conditions, formation of continuous molten interlayer can lead to brittle intermetallic compounds as in the case of titanium welded to steel[25]. In the area of the vortices on the interfacial waves also, there is some molten zone in which there is a mixture of the two metals being welded[26]. Away from the weld interface, metallurgical defects associated with shock loading[27] are found, such as slip and twinning. Normally, if the formation of harmful intermetallics or of a continuous cast inter-layer are avoided, then the strength of weld is equal to the strength of the weaker of the two materials being welded[18,28]. Fatigue strength[28] is normally enhanced due to shock loading, though fatigue failure usually originates at the weld interface. In general, the materials after welding become harder and their ductility is reduced. Subsequent heat treatment improves the properties[29,30].

The essential problem is that there is no completely adequate destructive test; ASTM side shear or tensile-shear or bend test can be carried out. The only nondestructive test currently available is ultrasonic testing, but though this can detect unbonded zones, it can not distinguish poorly bonded and good -strength welds. Explosive welding is a surface to surface welding. Thus a full scanning may be required and in some cases even this is not completely feasible.

6. EVALUATION OF THE BOND STRENGTH

There are many parameters in explosive welding which affect the bond strength. The dependence of the bond strength on the welding parameters are not known. It is expected that a wavy interface gives better results as the waves provide mechanical interlocking. However, too big a wave may lead to stress concentration and reduce the strength of the joint. The re-entrant jet is essential for bonding. However, too much of re-entrant jet could lead to larger wave size, thereby may reduce the strength. There is a minimum thickness of re-entrant jet which is essential for proper cleaning of plates for satisfactory weld. It is therefore considered that a moderate amount of re-entrant jet producing a moderate size wave could give best results for bond strength.

Even if NDT shows the weld to be crack/defect free, bond strength may be poor, because strength of the weld depends on its interface characteristics. Although it is possible to evaluate the interface characteristics accurately, which in turn possibly govern the bond strength, it is difficult to find an explicit relationship between the bond strength and the welding parameters.

However, the permissible range of parameters and the weldability window discussed before provides reasonably good guidelines for choosing the welding parameters to get a desired (straight or wavy) weld. An attempt is made to establish an additional condition, which if satisfied, gives optimum or near optimum bond strength; thus a good-strength bond could be assured without any destructive test.

6.1 Bond Strength and the Welding Parameters

It is useful to define following non-dimensional parameters (f_σ and f_w) one relating to the bond strength and the other relating to the welding parameters.

Bond strength ratio f_σ is defined as a ratio of the bond strength σ to the strength of the weaker metal(flyer/parent plate) σ_o, as

$$f_\sigma = \sigma / \sigma_o \tag{14}$$

usually $f_\sigma \leq 1$, but sometimes it may even exceed 1. However, f_σ upto 0.8 may be considered satisfactory.

Wave factor f_w is defined as the ratio of the wave amplitude h to the flyer plate thickness t, as

$$f_w = h/t \tag{15}$$

From equations (7) to (10) it is noted that the wave factor f_w (denoting the wave size as compared to the plate thickness) depends on two parameters: the obliquity angle β and the specific pressure p_s. The parameters p_s and p_s^* (defined earlier) are proportional to each other for a particular value of V_p as the theoretical strength σ_c and the actual strength σ_o of the material are themselves proportional to each other. Hence, it could be said that the wave factor f_w is a function of β and p_s^*. It can be shown that f_w is related to another non-dimensional parameter β/p_s^*, named as jet factor f_j (as the re-entrant jet thickness too could be shown to depend on it).

$$f_j = \beta/p_s^* \tag{16}$$

$$f_w = \text{function of } (\beta/p_s^*)$$

It is expected that f_σ depends on f_w and therefore on β/p_s^*. Therefore a plot of f_σ versus β/p_s^* could give some indication of the variation of strength with welding parameters.

strength data are collected from various references[16,18,31,32,33] and are plotted as bond strength ratio f_σ versus jet factor β/p_s^* in Figs. 6 and 7 .

It is observed that the variation of bond strength ratio with β/p_s^* shows a definite pattern. The strength is low for very low values of jet factor(in fact no bonding will take place at zero velocity or for zero obliquity angle i.e., $f_\sigma = 0$ at $\beta/p_s^* = 0$). Bond strength ratio f_σ increases with jet factor f_j and reaches a maximum value and then slowly decreases. For brass and copper(Fig.7) there appears to be a definite peak for a certain value of β/p_s^* whereas for steel(Fig.6) the curves do not show a sharp peak rather it is flattened.

From these figures(6 and 7) it may be concluded that there is some optimum value of jet factor(β/p_s^*) for maximum bond strength . Thus for optimum or near optimum bond strength, the condition is,

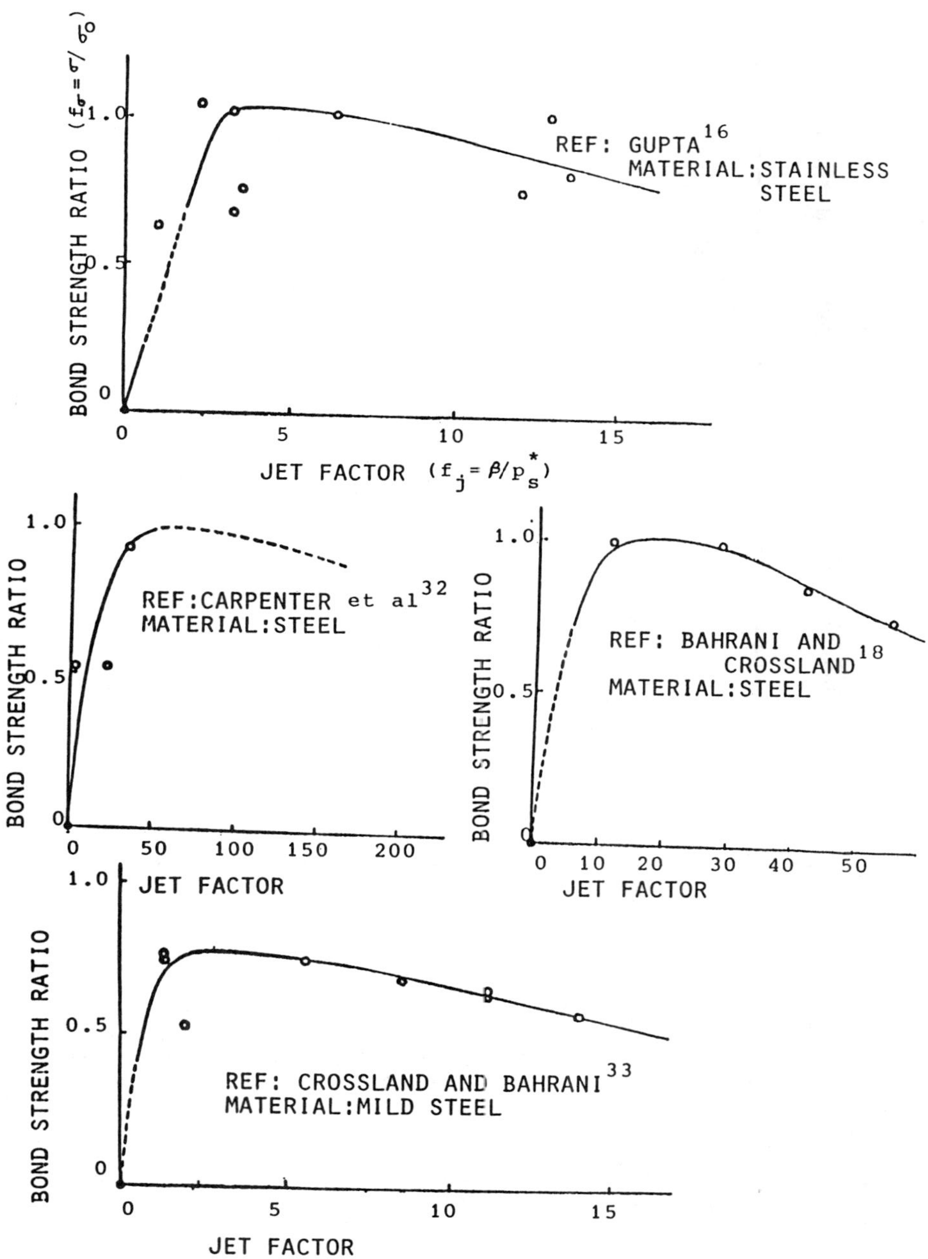

FIGURE.6 Variation of bond strength ratio versus jet factor for ferrous flyer plate materials

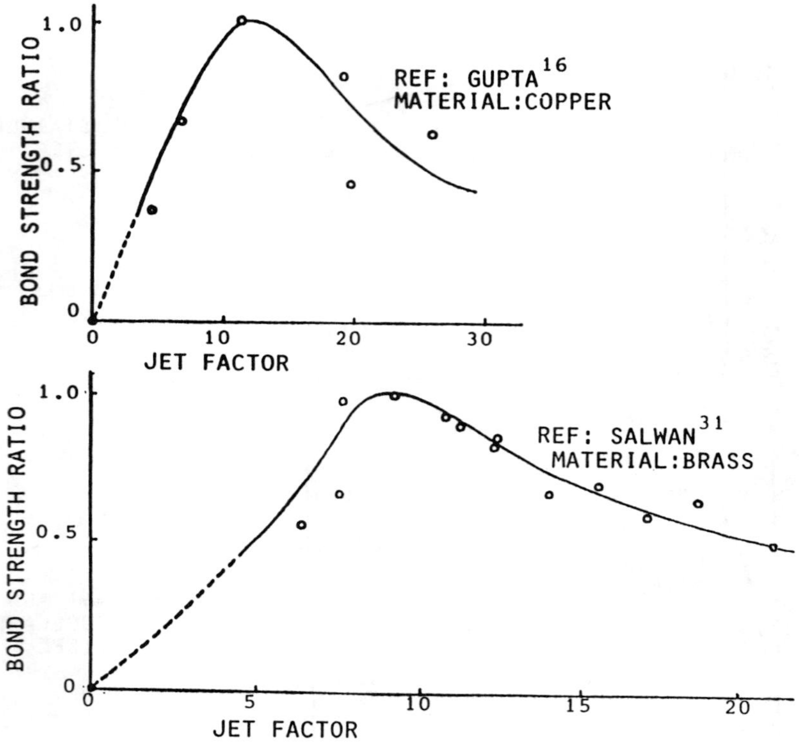

FIGURE.7 Variation of bond strength ratio versus jet factor for non-ferrous flyer plate materials

$$\beta/p_s^* \geqslant K \qquad (18)$$

where K is a constant and it is seen from these figures that for steels in most of the cases $K \approx 10$, for brass and copper too $K \approx 10$. It may be noted that the experimental data are from different sources and that the other conditions for these could be different, but all give $K \approx 10$; showing that the important parameter for bond strength is β/p_s^* .

Taking $K \approx 10$ and substituting $p_s^* = 2\,\sigma_o/\rho V_p^2$, the condition for optimum or near optimum bond strength is evaluated as

$$\frac{\beta \rho V_p^2}{2\sigma_o} \geqslant 10 \qquad (19)$$

where β is the obliquity angle(degrees), V_p is the flyer plate velocity(m/s), ρ is its density(Kg/m^3) and σ_o is its yield tensile strength(N/m^2).

6.2 Selection of Explosive Welding Parameters

It is mentioned before, that NDT can only reveal the unbonded zones and can not distinguish between poor bond and good strength bond. Even if the weld is found sound by NDT its strength may be poor. Weld strength depends upon the weld wave characteristics which depend on the re-entrant jet and which in turn depends on the welding parameters such as β and V_p . Hence, a careful selection of parameters is essential.

It is described earlier that there are limiting values for β and V_p . In fact β should lie between β_L and β_W for wavy interface, and that V_p should lie between 200 m/s to 800 m/s for proper welding. Also, for optimum or near optimum bond strength the parameters are related by: $\frac{1}{2}\beta\rho V_p^2/\sigma_o \geqslant 10$.

7. CONCLUSIONS

In order for explosive-welding to be successful the flyer plate velocity V_p should usually be kept within a limit such that 200 m/s $< V_p <$ 800 m/s. The obliquity angle β for welding to be done must lie within a certain limit such that $\beta_L < \beta < \beta_E$. Formulae for the angles β_L, β_W and β_E are given in Equations (11) and (12).

Strength of bond is found only by destructive tests,its soundness could,however, be judged by NDT. Explosive welded joints have a special feature that even if the NDT reveals the bond to be defect/crack free,its strength may be poor. A careful study suggests that a reasonably good bond strength could be obtained if the following condition is also satisfied,

$$\frac{\beta\rho V_p^2}{2\ \sigma_o} \geqslant 10$$

where σ_o is the actual strength(N/m^2), ρ is the density(Kg/m^3), V_p(m/s) is the velocity of the flyer plate and β is the obliquity angle(degrees).

Selection procedure for the welding parameters should be to select β ($\beta_L < \beta < \beta_W$) and V_p($200 < V_p < 800$ m/s) to satisfy the condition $\frac{1}{2}\beta\rho V_p^2/\sigma_o \geqslant 10$, to give a reasonably good-strength bond.

REFERENCES

1. Hay,D.R,Explosive Welding:Applications and Techniques,High Pressure Science and Technology,Plemn,New York,Vol.2,1979, pp.781-804.

2. Crossland,B.,Cave,J.A.,Banerjee,S.K.and Wylie,H.K.,Explosive Cladding of Large and Relatively Thick Flyer Plates, Proc.5th Int.Conf.on High Energy Rate Fabrication,Denver, 1975,pp.4.10.1-4.10.23.

3. Hardwick,R.,The Application of Explosive Welding to Indus-

trial Heat Exchangers,Proc.Int.Conf.-The use of High Energy Rate Methods for Forming,Welding and Compaction,University of Leeds,1973,paper 20.

4. Stanko,G.,Explosive Tube Plugging,Pressure Vessel and Piping Conf.,Aug.1980,San Francisco,ASME Special Publication, Explosive Welding,Forming,Plugging and Compaction.

5. Chadwick,M.D. and Evans,N.H.,Explosive Welding Techniques for Joining Steel Pipes,Pipes and Pipelines International, 12,1973,p.23.

6. Schetky,L.M.,The Explosive Fabrication of Duplex Tube,J. Inst.Metals,98,1970,p.364.

7. Wylie,H.K. and Crossland,B.,A Feasibility Study of Explosive Fabrication of Triplex Tubing,Proc.6th Int.Conf. on High Energy Rate Fabrication, Essen,1977.

8. Wylie,H.K.,Williams,J.D. and Crossland,B.,Explosive Fabrication of Fibre Reinforced Aluminium,Proc.3rd Int.Conf.of the Centr for High Energy Forming,1971,pp.2.2.1-2.2.26.

9. Birkhoff,G.,Mac Dougall,D.P.,Pugh,E.M. and Taylor,G.I., Explosive With Lined Cavities, J. of App.Phy.,Vol.19,1948, pp.563-583.

10. Robinson,R.L.,The Mechanism of Wave Formation in Impact Welding,Philosophical Magazine,Vol.31,1975,pp.587-597.

11. Walsh,J.M.,Shreffler,R.G. and Willing,F.J.,Limiting Conditions for Jet Formation in High Velocity Collisions,J.of App.Phys.,Vol.24,No.3.1953,pp.349-358.

12. Davenport,D.E. and Duvall,G.E.,Explosive Welding,ASTME High Rate Energy Forming Seminar,1961,paper SP-60-61.

13. Bahrani,A.S.,Black,T.J. and Crossland,B.,The Mechanism of Wave Formation in Explosive Welding,Proc.Royal Society of London,1967(A),pp.123-136.

14. Godunov,S.K.,Deribas,A.A.,Zabrodin,A.V. and Kozin,N.S. , Hydrodynamic Effects in Colliding Solids,J.of Computational Physics,1970,pp.517-539.

15. Kowalic,J.F. and Hay,D.R., A Flow Analogue for Explosive Bonding,Proc.3rd Int.Conf. of the Centre for High Energy Forming,1971,pp.1.1.1-1.1.21.

16. Gupta,R.C., Swinging Wake Mechanism for Interface Wave Generation and Permissible Range of Parameters for Explosive Welding,Ph.D. Thesis,I.I.T. Kanpur,1981.

17. Kowalic,J.F. and Hay,D.R.,A Mechanism of Explosive Bonding, Metallurgical Trans.,Vol.2,1971,pp.1953-1958.

18. Bahrani,A.S. and Crossland,B.,Further Experiments on Explosive Welding and Cladding with Particular Reference to the Strength of the Bond,Applied Mechanics Convention, England,Vol.80,1966,pp.1-16.

19. Gupta,R.C. and Kainth,G.S.,Permissible Range of Parameters for Interface Wave Formation,Proc.Pressure Vessel & Piping Technology,San Francisco,Aug.1980,ASME Special Publication, pp.99-119.

20. Gupta,R.C. and Kainth,G.S.,A Hydrodynamic Model of Explosive Welding Considering Kinetic Energy Loss,Proc. 9th AIMTDR Conf.,I.I.T. Kanpur,Dec.1980,pp.374-378.

21. Prasad,K.,Some Studies into Explosive Welding with Special Reference to Bond Strength,Master's Thesis,MNR Eng.College Allahabad,1983.

22. Wittman,R.,Proc.2nd Symp.on Use of Explosive Energy in Manufacturing Metallic Materials,Marianske Lanze,Oct.1973.

23. Mayer,M.D.,Impact Welding Using Magnetically Driven Flyer Plates, Proc.4th Int.Conf. of the Centre for High Energy Forming,Denver,July1973,pp.5.3.1-5.3.23.

24. Crossland,B.,The Present State of the Theory and Application of Explosive Welding,Inst,Phys.Conf.Ser.No.47,Chapter 4,1979,pp.394-409.

25. Lucas,W.,Williams,J.D. and Crossland,B.,Some Metallurgical Observations on Explosive Welding,Proc. of the 2nd Int. Conf. of the Centre for High Energy Forming,1969,p.8.1.

26. Williams,J.D.,Dhir,P. and Crossland,B.,The Nature of Vortices in Explosive Welds,Proc.of the 3rd Int.Conf. of the Centre for High Energy Forming,Denver,1971,pp.3.1.1-3.1.22.

27. Tamhankar,R.V. and Ramesan,J.,Metallography of Explosion Welds,Mater.Sci.Eng.,1974,13,3,p.245.

28. Crossland,B., Review of the Present State of the Art in Explosive Welding,Metal Technology,1976,pp.8-20.

29. Pruemmer,R.A.,Metallurgical Micro-radiographic and X-ray Residual Stress Investigation of Explosively Welded Aluminium - Steel,Proc. 3rd Int.Conf. of the Centre for High Energy Forming,Denver,July 1971.

30. Nilsson,T.,Hagwall,T. and Stahlberg,U.,Reduction of Residual Stresses in Explosive Welded Plates by Cold Forging, Scandinavian J. of Metallurgy,9,1980,pp.31-33.

31. Salwan,S.K.,Prasad,S. and Despande,A.S.,Explosive Welding Parameters,Indian Welding Journal,Vol.11,No.1,Jan.1979.

32. Carpenter,S.,Wittman,R.H. and Carlson,R.J.,The Relationship of Explosive Welding Parameters to Material Properties and Geometry Factors,Inst.Conf.,June1967.

33. Crossland,B. and Bahrani,A.S.,A Review of Explosive Welding Research Carried out in the Queen's Univ. of Belfast,Proc. 1st Int.Conf. of the Centre for High Energy Forming,Denver, 1967,

INDEX